Contents

Preface to the second edition

The layout of the first edition has been retained, with all chapters, references and problems having been revised. Worked examples and points to remember are included in each chapter for the first time.

The main improvement in the second edition is a description of the computer simulation of semiconductor devices. The models in the SPICE program are introduced, with their parameters being related to the physical principles underlying device operation. SPICE has then been used to simulate basic bipolar and MOS logic gates, while for MOS gates there is further simulation with reduced devices to bring out the benefits of scaling.

This approach has led to an expansion of Chapters 3 and 4 to explain the simulation of junction diodes and bipolar transistors. Chapter 6 includes simulation of junction and MOSFETs, nMOS and CMOS logic gates and scaled MOS circuits. In addition further memory and power MOSFETs have been introduced and gallium arsenide FETs are now covered.

Chapter 7 on integrated circuits is now much larger, in line with the advances in modern electronics, and includes an appreciation of the manufacturing processes of MOS as well as bipolar integrated circuits. The properties of these circuits are compared using especially written and tested SPICE programs to illustrate the published characteristics and complete program listings are provided in Appendix 3. In this way the writing of simulation programs is illustrated and a basis provided for developing them. Emphasis is laid on CMOS circuits, including a discussion of latchup, and the semi-custom design approach is described, particularly for gate arrays. Gallium arsenide integrated circuits are also considered, with advanced device designs being included.

Chapter 5 on optoelectronics has extra material on optical-fibre communications and a section on integrated optics, while Chapter 9 has a section on gallium arsenide MOSTs. The remaining chapters are essentially unchanged, except that some material has had to be deleted from Chapters 8, 10 and 11 to avoid excessive length.

Once again I would like to thank my colleagues, this time in the School of Computing and Information Technology and the School of Materials Science and

Physics at Thames Polytechnic. Critical comments from my students and also from members of other institutions have helped to clarify many points and I would particularly like to mention the assistance provided by staff from the Open University and from Napier College, Edinburgh.

Further thanks are due to Plessey Semiconductors Limited for permission to reproduce drawings from their computer-aided design manuals and again to Mrs Moss who typed the new manuscript.

Finally, I am greatly indebted to my wife for her forbearance during the preparation of the new edition.

J SEYMOUR
London 1987

Preface

This book provides an introduction to the physical principles underlying the operation of present-day electronic devices and components. It has its origin in the book *Physical Electronics* published by Pitman in 1972 which, although concentrating on semiconductor devices, devoted appreciable space to vacuum and gas-filled devices. When the time came for revision the development of integrated circuits and optoelectronics had been so extensive that it was clear that a simple revision would be insufficient. Furthermore it was decided to extend the scope of the book to include dielectric and magnetic materials and the devices and components based on them.

The title *Electronic Devices and Components* reflects these considerations and although some sections of the previous text are included, it is a new book. As before it is based on lectures given in recent years to students at Thames Polytechnic, London studying for degrees awarded by the Council for National Academic Awards. It is written to provide a complete course which can extend from the first to the final year of a degree in electrical and electronic engineering or applied physics. It is also suitable for students preparing for the examinations of the Council of Engineering Institutions and the Institute of Physics. Since each chapter is followed by class-tested tutorial examples with answers it can be used as a text for self-study by practising engineers and physicists.

The general approach is to formulate initially a simple physical theory for the operation of a device or component. Then, where appropriate, an equivalent circuit is evolved whose elements depend on such external factors as applied voltage and current, temperature and radiation. In this way the advantages and limitations of devices and components can be deduced and in many cases common applications are described.

In devices the main emphasis is on the solid-state with a whole chapter devoted to integrated circuits although some consideration is given to those thermionic and gas-filled devices which are still in use.

The essential physical theory is covered in Chapters 1 and 2 in which the behaviour of electrons in atoms and crystals is considered with emphasis on silicon, germanium and gallium arsenide as common semiconductor materials. Only a descriptive treatment of wave mechanics is included here, which is reinforced by a more analytical treatment in Appendix 1, while Appendix 2 contains the mathematics of energy level densities in a semiconductor. In Chapter 3 contacts between materials are discussed, leading to the diffused p-n junction as the basic semiconductor diode. Bipolar transistors with graded bases are considered in Chapter 4 and hybrid-π, charge control and Ebers–Moll models are introduced. The thyristor and its derivatives are also covered in this chapter. The next chapter considers the interaction between radiation and devices but it is possible to proceed directly to field effect transistors in Chapter 6 since Chapter 5 can be read at a later stage. In Chapter 5 optoelectronics, solid-state light devices, solar cells, light emitting diodes and liquid crystal displays are covered, together with gas and solid-state lasers and fibre optics.

Seven different types of field effect transistor are described in Chapter 6, together with charge transfer devices based on the MOST. Chapter 7 covers the principles of integrated circuit manufacture and considers the various logic families as well as operational amplifiers. Present-day vacuum and gas-filled devices are discussed in Chapter 8 in terms of photodevices, cathode ray tubes and thyratrons. In Chapter 9 the majority of the microwave devices are solid-state but the currently used thermionic ones are also included. Dielectric materials are introduced in Chapter 10 and all the common types of capacitor are considered, together with the quartz crystal. Finally, in Chapter 10 magnetic components are covered including inductor cores, computer cores and magnetic bubble memories.

I should like to thank all those who have made this book possible, including my colleagues in the School of Electrical and Electronic Engineering and the School of Materials Science and Physics at Thames Polytechnic and also my students. They have provided stimulating discussion and critical comment and the final result has been considerably improved by this feedback. My thanks are further due to Mrs Moss who typed the manuscript and to my wife who has again showed great patience with my preoccupation while it was being written.

J SEYMOUR
London 1981

1 Electrons in atoms

Introduction

The principles governing the behaviour of electric charge carriers underly the design of electronic devices in which the flow of carriers is controlled by electric and magnetic fields.

Studies in this field may be said to originate with J.J. Thomson's experiment of 1897, which is analysed in Chapter 8. He produced *cathode rays* by applying a voltage across a gas-filled tube (Fig. 1.1) and showed they could be deflected by an electric field and a magnetic field. They were found to be subject to the Newtonian laws of dynamics and he postulated that they consisted of minute negatively charged particles, later called *electrons*. He determined their charge/mass ratio, *e/m*, and showed that this was the same when different gases were used.

Further studies showed that particles of the same charge/mass ratio were emitted from metal wires heated to a high temperature, a phenomenon called *thermionic emission*, and also when certain metallic surfaces were irradiated with ultraviolet light, called *photoemission*. The results suggested that the electron is universally present in matter. Thomson's value of *e/m* was close to the presently accepted value of 1.759×10^{11} C/kg. This very high ratio suggests that the effect of gravitational force on the electron is negligible compared to the influence of an electric field.

In 1909 Millikan determined the charge on the electron, the accepted value of which is now 1.602×10^{-19} C, so that the mass can be calculated to be 9.108×10^{-31} kg. This may be compared with the mass of the hydrogen atom, which is the lightest atom known and yet is 1837 times as heavy as an electron.

Rutherford's atomic model

If electrons are present in each atom of matter, which is itself electrically neutral, their combined negative charge must be balanced by an equal amount of positive charge. In 1911 Rutherford proposed a model of the atom in which all the positive charge and most of the mass was concentrated in a nucleus. The electrons rotated

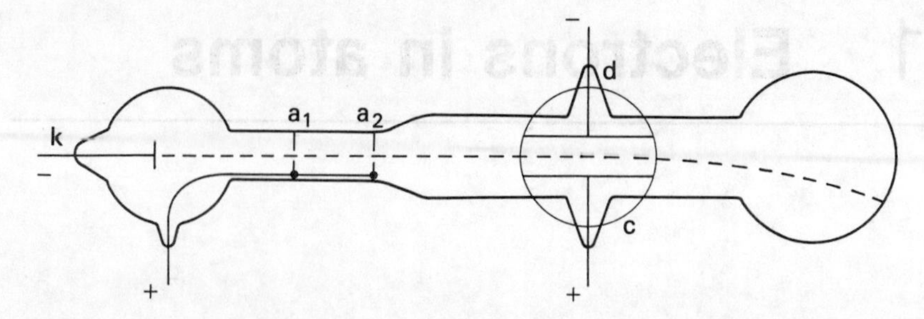

Figure 1.1 Section through J. J. Thompson's apparatus. The tube contains gas at a low pressure and when a high voltage is applied between the cathode k and anodes a_1 and a_2 a discharge occurs and electrons are produced. They pass through the holes in the anodes and travel down the tube, striking the bulb at the end and causing it to fluoresce. The resulting bright spot is deflected by applying a voltage between the plates d or by means of a magnetic field acting across the tube within the area of the circle.

round it due to the centripetal (centre-seeking) force exerted by the electrostatic attraction of the nucleus. He had determined by experiment that the diameter of a gold atom was about 10^{-10} m and the diameter of the nucleus about 10^{-14} m. Hence the atom consisted mainly of empty space, since its diameter was 10 000 times that of the nucleus.

As an electron moved in its orbit round the nucleus, its tangential direction of motion would change continuously so that it would experience acceleration. Theory shows that an accelerating electron radiates electromagnetic energy, a practical example being a radio transmitting aerial within which free electrons are accelerated to and fro, causing radiation from the aerial. Hence the electron in its orbit would radiate continuously, resulting in a continuous reduction of orbital velocity. In turn this would cause the electron to spiral into the nucleus due to electrostatic attraction, so that such an atom could not be stable. However, in practice many types of atom are stable, so that the model was not complete.

Planck's quantum theory of radiation

In 1913 Bohr resolved the inconsistency of Rutherford's atom by assuming that Planck's *quantum theory* of radiation could be applied to atomic structure. His postulates were confirmed by experimental evidence and justified by the later wave-mechanical theory. In 1901 Planck had shown that when a source was radiating electromagnetic waves its energy was quantized. This means that its magnitude was always a multiple of a unit known as a *quantum* and so could not change continuously. For electromagnetic radiation the quantum is called a *photon*, which may be regarded as a particle representing the interaction of radiation with matter. However, when the radiation is travelling through free space with the velocity of light it retains its wave properties. If the frequency of the

Table 1.1 Electromagnetic radiation
Velocity in free space, $c = 3 \times 10^8$ m/s

Type of radiation	Approximate wavelength	Frequency	Quantum energy
	m	Hz	eV
Radio	$10^4 - 10^{-3}$	$3 \times 10^4 - 3 \times 10^{11}$	$1.2 \times 10^{-10} - 1.2 \times 10^{-3}$
Thermal			
infra-red	$10^{-3} - 7 \times 10^{-7}$	$3 \times 10^{11} - 4 \times 10^{14}$	$1.2 \times 10^{-3} - 1.7$
visible light	$7 \times 10^{-7} - 4 \times 10^{-7}$	$4 \times 10^{14} - 7 \times 10^{14}$	$1.7 - 3$
ultra-violet	$4 \times 10^{-7} - 10^{-9}$	$7 \times 10^{14} - 3 \times 10^{17}$	$3 - 1.2 \times 10^3$
X-rays	$10^{-9} - 10^{-11}$	$3 \times 10^{17} - 3 \times 10^{19}$	$1.2 \times 10^3 - 1.2 \times 10^5$
γ-rays	$10^{-11} - 10^{-13}$	$3 \times 10^{19} - 3 \times 10^{21}$	$1.2 \times 10^5 - 1.2 \times 10^7$

radiation is f, each photon has energy hf joules, where h is the Planck constant. Thus h relates the dual wave and particle properties of the radiation and has the value 6.625×10^{-34} Js.

The frequencies of electromagnetic radiation cover a very wide range, as shown in Table 1.1, extending from long radio waves of frequency 30 kHz to γ-rays with an upper frequency of about 3×10^{21} Hz. The corresponding wavelengths λ, obtained from the relationship $\lambda = c/f$ (Appendix 1), range from 10^4 to 10^{-13} m. One quantum thus represents an extremely small amount of energy, indeed only about 2×10^{-12} J for γ-rays of the highest energy. A practical source of radiation may radiate many joules per second; i.e. its power will be measurable in watts. Hence it is not surprising that energy appears to be continuously variable on a macroscopic scale.

However, on an atomic scale the energy of an individual quantum is no longer negligible. An electron moving through a potential difference of 1 V acquires an energy of 1.602×10^{-19} J, which is called 1 *electronvolt* (1 eV). This is comparable with the energy of a photon, and the energies of electrons and photons are normally expressed in electronvolts, as in Table 1.1.

Worked example

A radio wave has a wavelength of 1 cm. Calculate its frequency and quantum energy.

Solution

In free space a radio wave has a velocity of 3×10^8 m/s. The frequency f is given by

$$f = \frac{c}{\lambda} = \frac{3 \times 10^8}{10^{-2}} = 3 \times 10^{10} \text{ Hz or } 30 \text{ GHz}$$

In joules the quantum energy is given by

$$W = hf = 6.63 \times 10^{-34} \times 3 \times 10^{10}$$
$$= 19.89 \times 10^{-24} \text{ J}$$

In electronvolts the quantum energy is given by

$$W = \frac{hf}{e} = \frac{19.89 \times 10^{-24}}{1.60 \times 10^{-19}} = 1.24 \times 10^{-4} \text{ eV}$$

Bohr's model of the hydrogen atom

Let us consider the application of Bohr's theory to the simplest atom, that of hydrogen, which has only one electron. The nucleus consists of a particle called a *proton* which carries a positive charge of the same magnitude as that of an electron, but whose mass is 1836 times that of the electron. Thus to a good approximation the proton may be considered fixed and unaffected by the rotation of the electron round it. Then the Coulomb electrostatic force of attraction between the charges is given by

$$F = \frac{e^2}{4\pi\epsilon_0 r^2} \tag{1.1}$$

where ϵ_0 is the permittivity of free space, which has a value of 8.854×10^{-12} F/m, and r is the distance between electron and proton (Fig. 1.2).

For the electron to move in a circular orbit of radius r with velocity u,

$$\frac{e^2}{4\pi\epsilon_0 r^2} = \frac{mu^2}{r} \tag{1.2}$$

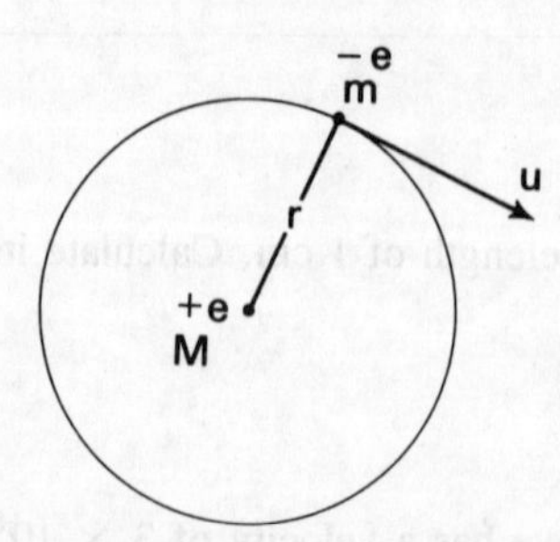

Figure 1.2 Diagram of hydrogen atom.

The energy associated with the atom will be the sum of the potential energy, V, of the two separated charges and the kinetic energy, K, of the electron rotating in its orbit. Let us assume for convenience that the potential energy is zero when the electron is an infinite distance from the nucleus. In effect this means a distance large compared with the atomic diameter of about 10^{-10} m. Then the work done in bringing the electron from infinity to distance r from the proton is

$$\int_{\infty}^{r} F\,\mathrm{d}r = \int_{\infty}^{r} \frac{e^2}{4\pi\epsilon_0 r^2}\,\mathrm{d}r = V$$

and

$$V = -\frac{e^2}{4\pi\epsilon_0 r} \qquad (1.3)$$

The kinetic energy is $\frac{1}{2}mu^2$, which is obtained from eq. (1.2), so that

$$K = \frac{e^2}{8\pi\epsilon_0 r} \qquad (1.4)$$

The total energy of the system is then $W = V + K$, i.e.

$$W = -\frac{e^2}{4\pi\epsilon_0 r} + \frac{e^2}{8\pi\epsilon_0 r}$$

$$= -\frac{e^2}{8\pi\epsilon_0 r} \qquad (1.5)$$

W is negative because zero energy is considered to be at infinity.

Up to this point Bohr's theory is the same as Rutherford's. But if the effect of radiation from the electron is included, W must decrease owing to conservation of energy, so that r decreases correspondingly and the electron spirals towards the nucleus. Hence Bohr first postulated that the electron could only move in orbits of certain radii corresponding to fixed amounts of energy, and while in one of these permitted orbits no radiation would occur. This idea is related to the quantum theory by considering the product of momentum and orbital circumference, which has the same dimensions as Planck's constant and gives

$$mu \times 2\pi r = nh$$

or

$$mur = n\frac{h}{2\pi} \qquad (1.6)$$

mur is called the *moment of momentum*, or *angular momentum*. n is 0, 1, 2, 3, etc., and is a *quantum number*.

Squaring and substituting in eq. (1.2) leads to

$$r = n^2 \frac{\epsilon_0 h^2}{\pi e^2 m} \tag{1.7}$$

and if this expression for r is substituted in eq. (1.5) we have

$$W_n = -\frac{1}{n^2}\frac{e^4 m}{8\epsilon_0^2 h^2} \tag{1.8}$$

for the energy of the nth orbit, which thus can be changed only in discrete steps, as illustrated in Fig. 1.3.

Inserting the values of the constants gives

$$W_n = \frac{13.58}{n^2}\ \text{eV} \tag{1.9}$$

and

$$r_n = n^2 \times 0.053\ \text{nm} \tag{1.10}$$

The corresponding values of n, W_n and r_n for hydrogen are given in Table 1.2. The minimum energy is −13.58 eV, which is the total energy of the stable or *ground* state of the atom. The higher permitted energy levels are unstable or *excited* states. The electron does not radiate while it is in any of the permitted levels so its energy remains constant. Figure 1.3 has symmetry about the energy axis and is said to represent a *potential well*. At the bottom of the well the atom is in the ground state with the electron tightly bound to the nucleus.

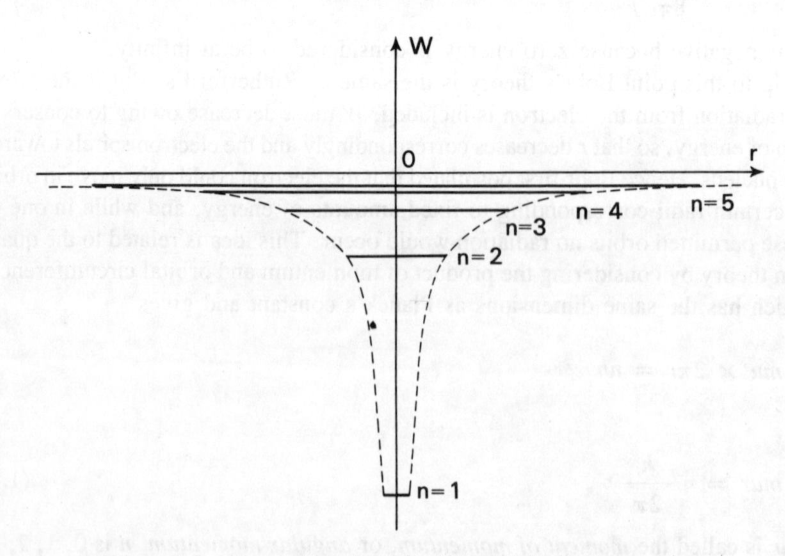

Figure 1.3 Energy levels and orbit radii in hydrogen atom.

Table 1.2 Energy levels and orbit radii for hydrogen atom

n	$-W_n$	r_n
	eV	nm
1	13.58	0.05
2	3.39	0.21
3	1.51	0.47
4	0.85	0.84
5	0.54	1.32
10	0.14	5.26

The electron makes a transition from the ground state to a higher level when it receives energy corresponding to the difference between the levels. Thus a transition from the ground state to level 2 requires 10.19 eV; from the ground state to level 3, 12.07 eV; and so on. If the electron were to receive 11 eV of energy, which does not correspond to a permitted level, it would not make any transition. An electron spends only a very short time in the excited state and then returns to the ground state, either directly or by way of intermediate levels.

Bohr's second postulate is that, when an electron jumps from an excited level to one of lower energy, a quantum of radiation is emitted (Fig. 1.4). The frequency of the radiation is given by

$$hf = W_2 - W_1 \tag{1.11}$$

where $W_2 - W_1$ is the difference in energy between the two levels. For instance, in the transition from level 4 to level 2 the energy difference is 2.54 eV, corresponding to a frequency of 6.13×10^{14} Hz and a wavelength of 489 nm. There are a number of possible transitions between the levels, some of which are shown in Fig. 1.5. When a sufficiently high electric field is set up in the gas the energy of the atoms will be raised and the transitions will occur, each causing emission of radiation at the corresponding frequency. The wavelength of the radiation may be measured by means of a *spectroscope*, which employs a prism to

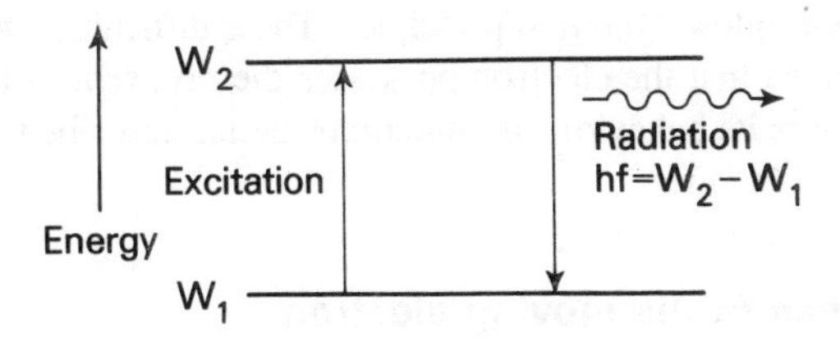

Figure 1.4 Excitation and radiation due to electron transitions between energy levels.

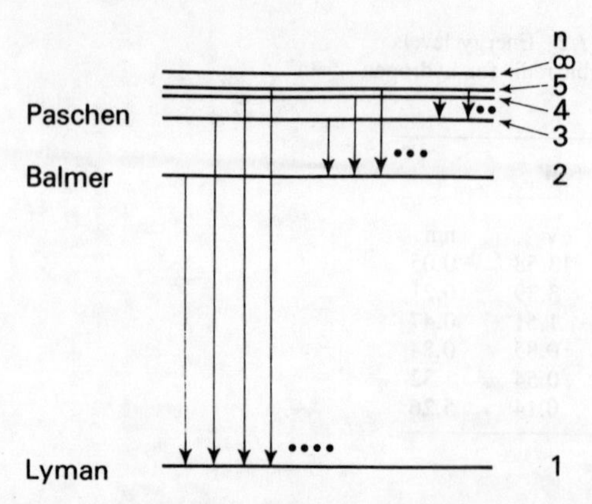

Figure 1.5 Some of the possible transitions between the energy levels of a hydrogen atom. Paschen, Balmer and Lyman were spectroscopists who observed spectral lines at the wavelengths corresponding to transitions ending at the appropriate level. The values of wavelength were later given by Bohr's theory.

separate the individual wavelengths into spectral lines, and it is found that these experimental values agree very closely with the theoretical values for hydrogen.

If the atom receives energy of 13.58 eV or more, the electron can become detached from the nucleus. This process is known as *ionization*, and 13.58 eV is the *ionization energy*, W_i, for hydrogen. Thus the larger the orbit radius, r, the smaller is the energy binding the electron to the nucleus, since the attractive force is reduced. As r is increased the energy difference between the levels becomes less and less until when the electron is freed from the nucleus its energy can change continuously.

Bohr's theory was extended by Sommerfeld in 1915, who showed that in general an orbit could be elliptical with the circular orbit as a special case having equal axes. The behaviour of the hydrogen atom, with only one electron, was then explained in considerable detail and attempts were made to extend the general principles to atoms with more than one electron. However, in these atoms repulsive forces occur between electrons, which means that a particular electron is affected by the repulsive forces of all the other electrons. This makes the calculation of the energy levels very difficult, and in addition Bohr's postulates are assumptions which do not follow from first principles. These difficulties were not resolved until it was realized that the electron possesses the properties of a wave as well as a particle, that is its behaviour is sometimes better described by wave motion than by particle dynamics.

Wave properties of the moving electron

The idea that a wave phenomenon, such as electromagnetic radiation, could also have particle properties was familiar from Planck's quantum theory, which had been confirmed by experiment. In 1924, de Broglie proposed that a moving particle such as an electron, and indeed all matter in motion, could possess wave properties and that the wavelength λ associated with a particle of momentum p ($= mu$) was given by

$$\lambda = \frac{h}{p} \tag{1.12}$$

as shown in Appendix 1. Thus the wave and particle properties of matter are linked by Planck's constant h, just as for the wave and particle properties of radiation, and in 1927 the wave nature of electrons was confirmed by experiments carried out by Davisson and Germer and also by G.P. Thomson, the son of J.J. Thomson.

It can easily be shown that Bohr's assumption of preferred orbits for an electron follows naturally from the concept of matter waves. Such a wave has an amplitude related to the probability of locating the electron at a point and is known as a *guiding wave*. For a non-preferred orbit there will not be a whole number of wavelengths around the orbit (Fig. 1.6(*a*)), which means that it will not be stable since the peaks will also travel round as the wave makes successive revolutions. However, for a preferred orbit Bohr's expression for momentum may be used and equated to eq. (1.12), i.e.

$$p = \frac{nh}{2\pi r} = \frac{h}{\lambda}$$

so that

$$n\lambda = 2\pi r \tag{1.13}$$

Thus there are a whole number of wavelengths and the orbit is stable (Fig. 1.6(*b*)), which gives a physical interpretation of Bohr's first postulate. The wave concept

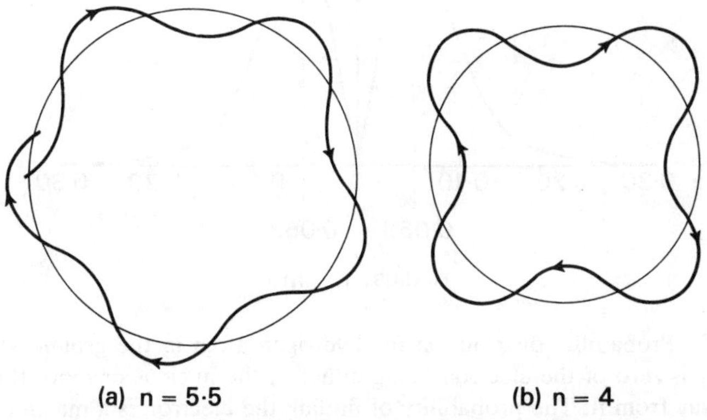

Figure 1.6 De Broglie wave round a circular orbit.

of electrons was expanded by Schrödinger in 1925 into *wave mechanics*, which leads automatically to preferred energy levels and has been used very successfully to obtain the structure of atoms with many electrons. Some of the basic ideas of wave mechanics are discussed in Appendix 1.

One result of wave mechanics is that the orbits are no longer precisely defined, and in a hydrogen atom, for example, the electron can traverse the whole of the space about the nucleus. However, it spends most of the time at a distance from the nucleus corresponding to a permitted Bohr radius; i.e. the probability of locating it is greatest at this distance (Fig. 1.7). The uncertainty in determining the properties of an electron was expressed by Heisenberg in 1927 in his *uncertainty principle*, which may be formulated as follows:

If the error in determining the momentum of a particle is Δp, then the error in determining its position is Δx, where

$$\Delta x \Delta p \geq \frac{h}{4\pi} \tag{1.14}$$

Putting $p = mu$,

$$\Delta x \Delta u \geq \frac{h}{4\pi m} \tag{1.15}$$

which shows that for a particle of very low mass, such as an electron, the errors involved can be large.

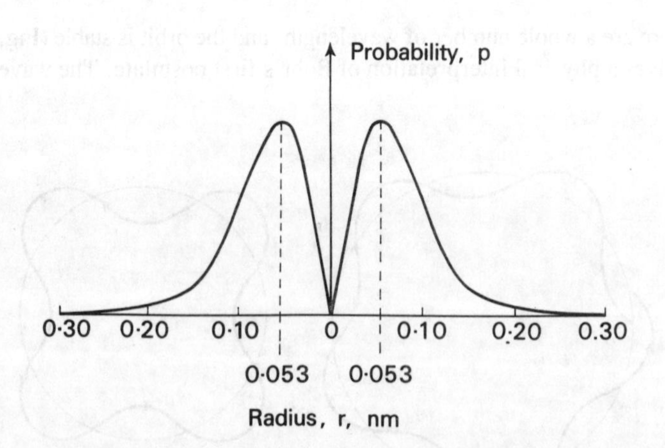

Figure 1.7 Probability distribution for hydrogen atom in the ground state. The probability is zero of the electron being either at the nucleus or more than about 0·3 nm away from it. The probability of finding the electron is a maximum at the first Bohr orbit of radius 0·053 nm.

Worked example

The error in determining the position of an electron is 2×10^{-10} m. Calculate the error in determining its speed.

Solution

The error in determining the speed of an electron with a position known within 2×10^{-10} m is

$$\Delta u \geq \frac{6.63 \times 10^{-34}}{4\pi \times 9.11 \times 10^{-31} \times 2 \times 10^{-10}}$$
$$\geq 2.9 \times 10^{5} \text{ m/s}$$

Thus, when the position of an electron is specified within a tolerance corresponding to the dimensions of an atom, the tolerance on its speed is very large and this uncertainty requires a statistical approach to events on an atomic scale. In fact, we can only find the probability that an electron will be in a certain place with a given speed, and this leads to the use of probability functions as described in the next chapter.

Atoms with many electrons

The total number of electrons in an atom is called the *atomic number* Z and is always equal to the number of protons, so that the atom is electrically neutral. In addition, the nucleus may contain neutrons, which have no electrical charge, but have almost the same mass as the proton, namely 1.67×10^{-27} kg. Thus for most purposes the total mass of the atom may be considered to be in its nucleus.

All the elements can be arranged in ascending order of atomic number, starting with hydrogen and adding one electron at a time. Although it might appear that all the electrons should occupy the orbit of lowest energy, the number of electrons in any one orbit is limited. When this limit is reached a new orbit of greater radius is started, the heaviest atoms having seven orbits. It is still convenient to retain the idea of an orbit, even though it has been replaced by a charge distribution having a maximum value at the preferred distance from the nucleus.

The term *orbital* is used in wave mechanics to denote a particular energy state from which the probability of finding an electron at a given point can be calculated.

Quantum numbers

Each possible electron orbital is uniquely defined by a set of four *quantum numbers*, which arise naturally from wave mechanics, and each number is related to a physical property of the electron. The first is the *principal quantum number*, n, which corresponds to the Bohr quantum number n (eq. (1.6)). It is related to the total energy of the atom and hence specifies the number of orbitals, or orbits, so that its allowed values are

$$n = 1, 2, 3, \ldots 7$$

The second is the *orbital angular momentum quantum number*, l, which specifies the angular momentum in units of $h/2\pi$, and wave mechanical theory shows that the angular momentum is given by

$$p = \sqrt{[l(l+1)]}\,\frac{h}{2\pi} \tag{1.16}$$

The allowed values of l are limited by the value of n being considered so that

$$l = 0, 1, 2, \ldots (n\text{–}1) \tag{1.17}$$

Angular momentum has direction as well as magnitude, i.e. it is a vector. This is illustrated in Fig. 1.8 for $l = 2$, with the z-axis taken as reference. The allowed directions of $\sqrt{[l(l+1)]}$ (whose magnitude is 2.45 in this case) are determined by taking integral steps along the z-axis. The component of angular momentum along this axis is then

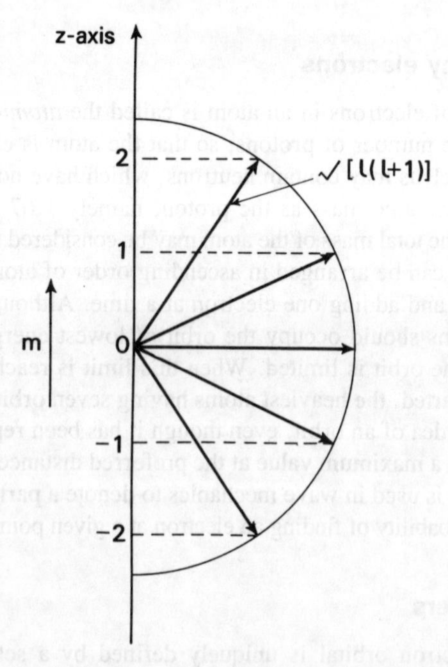

Figure 1.8 Preferred directions of the angular momentum vector for $l = 2$.

$$p_z = m \frac{h}{2\pi} \tag{1.18}$$

where m has the values –2, –1, 0, 1, 2 for $l = 2$. When $l = 0$, p_θ is also zero, which corresponds to a random orientation of the vector.

In fact, m is the *magnetic quantum number*, which specifies the magnetic moment produced along the z-axis by an electron moving round its orbit. Consider a circular orbit of radius r with an electron moving round it with velocity u (Fig. 1.9(*a*)). The time for one revolution is $2\pi r/u$ so that the moving electron is equivalent to a current

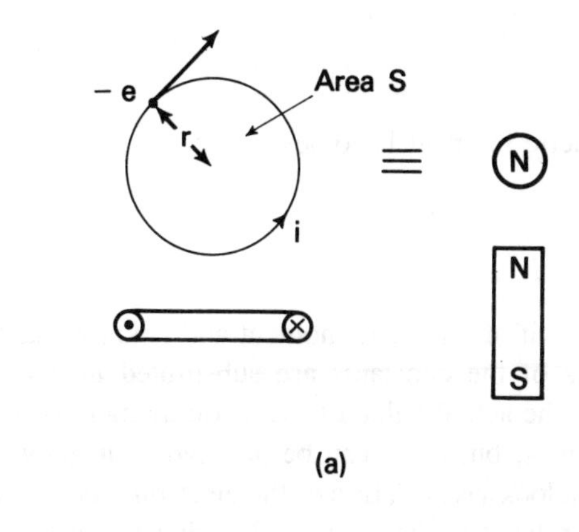

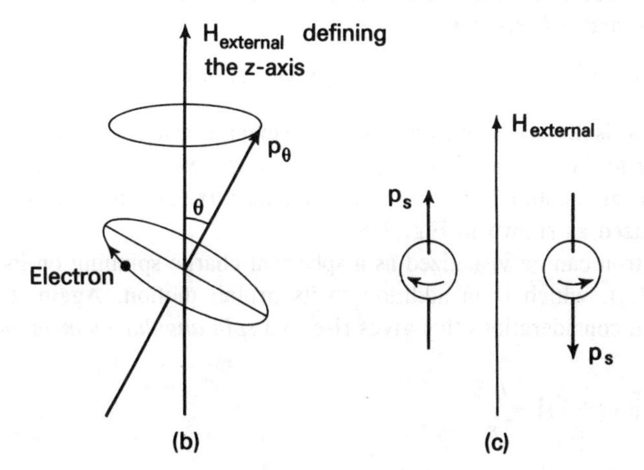

Figure 1.9 Orbital and spin angular momentum. (*a*) Magnetic dipole due to the rotation of an electron in an orbit; (*b*) precessional motion of the angular momentum vector p_θ about an external magnetic field; (*c*) angular momentum due to an electron spinning round its axis. The left-hand vector is parallel and the right-hand vector is antiparallel to the magnetic field vector.

$$i = \frac{eu}{2\pi r}$$

Now, a current i moving in a circle whose area is S is equivalent to a magnetic dipole of moment iS, so that the magnetic moment due to the electron is

$$iS = \frac{euS}{2\pi r} = \frac{eur}{2} \tag{1.19}$$

since $S = \pi r^2$. Using the Bohr condition for angular momentum (eq. (1.16)), $n = 1$ gives the smallest value, which is

$$mur = \frac{h}{2\pi} \tag{1.20}$$

Hence the magnetic moment becomes

$$\mu_B = \frac{eh}{4\pi m} \tag{1.21}$$

μ_B is the atomic unit of magnetic moment and is known as the *Bohr magneton.* When the values of the constants are substituted μ_B has a value of 9.27×10^{-24} A m^2, and the actual value of magnetic moment is then $m\mu_B$. m can only have integral values, but these can be positive or negative corresponding to a clockwise or anticlockwise rotation of the electron. As shown above the allowed values of m are related to the value of l in that it can have any integral value between $-l$ and $+l$, so that

$$m = -l, -(l-1) \ldots -1, 0, +1 \ldots +(l-1), +l \tag{1.22}$$

The z-axis is defined by the direction of an external magnetic field (Fig. 1.9(*b*)), and owing to the interaction of this field with the electronic magnetic moment the angular momentum vector rotates round the z-axis, or 'precesses'. The angle θ is quantized as shown in Fig. 1.8.

An electron can be visualized as a spherical charge spinning on its own axis (Fig. 1.9(*c*)), which is in addition to its orbital motion. Again, from wave mechanical considerations this gives rise to a *spin angular momentum* given by

$$s = \sqrt{[s(s+1)]}\,\frac{h}{2\pi} \tag{1.23}$$

where s is the *spin quantum number*, and the corresponding magnetic moment allows only two senses for p_s, parallel and antiparallel to the magnetic field. Thus the allowed values of s are $+\frac{1}{2}$ or $-\frac{1}{2}$, which differ by a whole quantum number. The total magnetic moment is then a combination of the values due to orbital motion and spin, the whole being expressed in Bohr magnetons.

Each electron is then specified by a set of the four quantum numbers and the number of electrons in an orbit is governed by Pauli's *exclusion principle*, introduced in 1925. This states that in any one atom no two electrons can have the same set of quantum numbers. Thus, when all the possible combinations of a particular set of numbers have been achieved with one value of n, a new value of n corresponding to a new orbit is required for any further electrons.

Electronic structure of atoms

The general ideas of atomic structure can now be illustrated by considering the elements of lowest atomic number and extending the results to those with higher values of Z. This assumes that the elements can be built up by starting with the simplest, hydrogen, and adding one electron at a time, which is only possible in theory.

Hydrogen, Z = 1

Hydrogen has only one electron and one orbit in the ground state. Hence $n = 1$, so that $l = 0$ and $m = 0$, and the electron may have either value of s (eqs (1.17) and (1.22)).

Helium, Z = 2

The next in order is helium, which is an inert gas, and the two electrons can be in the same orbit but with opposite spins. Their quantum numbers are $n = 1$, $l = 0$, $m = 0$, $s = +\frac{1}{2}$; and $n = 1$, $l = 0$, $m = 0$, $s = -\frac{1}{2}$. There are no other possible combinations of the numbers, so two electrons is the maximum number that can be contained in the orbit.

Lithium, Z = 3

Lithium is a metal, and the third electron has to go into a second orbit since the first one is full. The third electron is thus further from the nucleus than the two others and is less tightly bound to it, so that it can become detached giving rise to metallic properties. The two inner electrons have the same quantum numbers as the helium electrons and the third has $n = 2$, $l = 0$, $m = 0$, $s = \frac{1}{2}$.

With elements of increasing atomic number, further electrons can go into this orbit as shown in Table 1.3 for the first 18 elements. This shows that only two electrons, with opposite spin, can have $n = 2$, $l = 0$, but the second orbit is not filled until a further six electrons have been added.

These correspond to $n = 2$, $l = 1$, so that m can have the values −1, 0 and

Table 1.3 Electronic configurations of the first 18 elements

		1s	2s	2p	3s	3p
	n =	1	2		3	
	l =	0	0	1	0	1
Z	Element					
1	Hydrogen	1				
2	Helium	2				
3	Lithium	2	1			
4	Beryllium	2	2			
5	Boron	2	2	1		
6	Carbon	2	2	2		
7	Nitrogen	2	2	3		
8	Oxygen	2	2	4		
9	Fluorine	2	2	5		
10	Neon	2	2	6		
11	Sodium	2	2	6	1	
12	Magnesium	2	2	6	2	
13	Aluminium	2	2	6	2	1
14	Silicon	2	2	6	2	2
15	Phosphorus	2	2	6	2	3
16	Sulphur	2	2	6	2	4
17	Chlorine	2	2	6	2	5
18	Argon	2	2	6	2	6

+1, each of which corresponds to two electrons with opposite spins.* Thus the orbit can contain a maximum of eight electrons, each with a different set of quantum numbers, and a completely filled second orbit corresponds to the inert gas neon. If this process is continued it is found that each filled orbit contains a maximum of $2n^2$ electrons. Hence for n = 1, 2, 3, 4 . . . the maximum numbers of electrons in each orbit are 2, 8, 18, 32 . . . respectively. In addition when all the $l = 1$ states have been filled in a particular orbit the atom is exceptionally stable and these are the atoms of the inert gases neon, argon, krypton and xenon.

After neon the next element, sodium, requires three orbits and the elements of higher atomic number are obtained as further electrons are added. As the orbits of lower energy are filled, according to the exclusion principle, electrons are

* An individual orbit may be conveniently referred to by the system at the top of Table 1.3. Here the number is the value of n, and the symbol is related to the value of l, as follows:

l = 0, 1, 2, 3
s p d f

Thus the 3p orbit has $n = 3$, $l = 1$ and the 2s orbit has $n = 2$, $l = 0$, and in this context the symbol s has no connection with the spin quantum number. A superscript to each letter gives the number of electrons in that state. Thus the electronic configuration of silicon is given by $1s^2 2s^2 2p^6 3s^2 3p^2$, for example.

1s	1 H	2 He
	s 1	s 2

	IA	IIA		IIIA	IVA	VA	VIA	VIIA	← VIII		→	IB	IIB		IIIB	IVB	VB	VIB	VIIB	
2s	3 Li	4 Be												2p	5 B	6 C	7 N	8 O	9 F	10 Ne
3s	11 Na	12 Mg												3p	13 Al	14 Si	15 P	16 S	17 Cl	18 Ar
4s	19 K	20 Ca	3d	21 Sc	22 Ti	23 V	24 Cr	25 Mn	26 Fe	27 Co	28 Ni	29 Cu	30 Zn	4p	31 Ga	32 Ge	33 As	34 Se	35 Br	36 Kr
5s	37 Rb	38 Sr	4d	39 Y	40 Zr	41 Nb	42 Mo	43 Tc	44 Ru	45 Rh	46 Pd	47 Ag	48 Cd	5p	49 In	50 Sn	51 Sb	52 Te	53 I	54 Xe
6s	55 Cs	56 Ba	5d	57 La	72 Hf	73 Ta	74 W	75 Re	76 Os	77 Ir	78 Pt	79 Au	80 Hg	6p	81 Tl	82 Pb	83 Bi	84 Po	85 At	86 Rn
7s	87 Fr	88 Ra	6d	89 Ac	d s 2 2	d s 3 2 4 1 3 2	d s 5 1 5 1 4 2	d s 5 2 6 1 5 2	d s 6 2 7 1 6 2	d s 7 2 8 1 7 2	d s 8 2 10 0 9 1	d s 10 1	d s 10 2		s d 2 1	s p 2 2	s p 2 3	s p 2 4	s p 2 5	s p 2 6
	s 1	s 2		d s 1 2																

Figure 1.10 Periodic table of the first 89 elements. The outermost orbitals are given on the left of the rows where the number refers to the value of *n*, the principal quantum number. The letters refer to the value of *l*, the orbital angular momentum quantum number, where *s*, *p*, *d* and *f* correspond to *l* being 0, 1, 2 and 3, respectively. The maximum numbers of electrons for the *s*, *p*, *d* and *f* states are 2, 6, 10 and 14, respectively. The atomic numbers are given inside each box with the abbreviations.

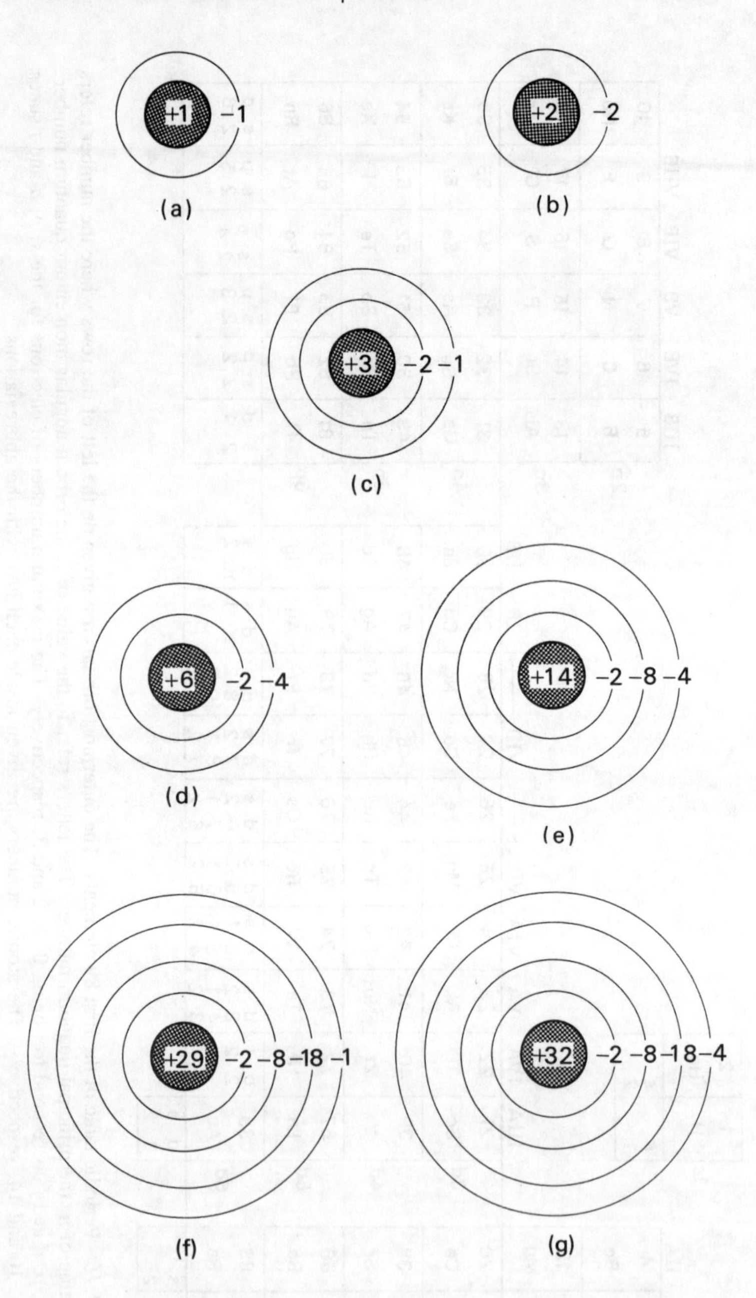

Figure 1.11 Electronic arrangement of some atoms. (*a*) Hydrogen, $Z=1$; (*b*) Helium, $Z=2$; (*c*) Lithium, $Z=3$; (*d*) Carbon, $Z=6$; (*e*) Silicon, $Z=14$; (*f*) Copper, $Z=29$; (g) Germanium, $Z=32$.

arranged in an outer orbit. The electrons in the outermost orbit are known as the *valence* electrons and determine the physical and chemical properties of the element. An arrangement of elements in terms of their valence electrons forms the basis of the *periodic table* (Fig. 1.10) and this is a striking configuration of Mendeleeff's similar classification of the elements by purely chemical properties. Here elements in Groups I to VII have from one to seven valence electrons respectively, with the inert gases having eight outer electrons and forming a group on their own. Thus lithium and sodium are in Group I, boron and aluminium in Group III and nitrogen and phosphorus in Group V. Further orbital arrangements are shown in Fig. 1.11. Both lithium and copper have one valence electron, which is typical of elements with strong metallic properties found in Group I. Carbon, silicon and germanium all have four valence electrons and are therefore in Group IV.

Points to remember

* Electrons in atoms are best described in terms of wave motion.
* The structure of an atom is explained through orbitals of discrete energy, each containing a limited number of electrons.
* The periodic table summarises the atomic structure of the elements by means of quantum numbers.

Problems

1.1 Calculate the quantum energy in electronvolts of (i) radio waves of frequency 100 MHz, (ii) infra-red radiation of wavelength 10 μm, (iii) blue light of wavelength 488 nm, (iv) X-rays of wavelength 1 pm.
[4.14×10^{-7} eV; 0.124 eV; 2.25 eV; 1.24×10^{6} eV]

1.2 Obtain the uncertainty in the velocity of an electron confined within a volume of (i) 10^{-6} m^3, (ii) 10^{-18} m^3, (iii) 10^{-30} m^3.
[5.8×10^{-3} m/s; 58 m/s; 5.8×10^{5} m/s]

1.3 Show that the expression for Heisenberg's uncertainty principle

$$\Delta x \Delta p \geq \frac{h}{4\pi}$$

can take the form

$$\Delta W \Delta t \geq \frac{h}{4\pi}$$

for the tolerances in energy and time respectively.

1.4 The time taken for an electron to drop from an excited state to the ground state is about 10^{-8} s. If the energy difference is 1 eV find the uncertainty in (i) the energy of the emitted radiation, (ii) the frequency of the radiation, (iii) the wavelength of the radiation.
[3.3×10^{-8} eV; 8 MHz; -4.1×10^{-14} m]

1.5 Determine the de Broglie wavelength of (i) 10 eV electrons, (ii) 1 MeV protons, (iii) neutrons at room temperature, having an energy of 0.025 eV, (iv) an aircraft of mass 10^5 kg flying at a speed of 600 m/s.
[3.88×10^{-10} m; 2.87×10^{-14} m; 1.81×10^{-10} m; 1.1×10^{-41} m]

1.6 A parallel beam of electrons of energy 3 keV passes through a thin metal foil. If the second diffraction maximum ring is found at an angle of 9.0° with the incident beam, what is the lattice spacing of the crystal planes from which reflection occurred?
[1.39×10^{-10} m]

2 Electrons in crystals

The combination of atoms

When large numbers of atoms are combined, as in a crystal, they are held in their relative positions by a balance between various types of interatomic force. In general the attractive forces are most important at large distances between atoms, and the repulsive forces between the nuclei predominate at close spacings. Thus there is an equilibrium distance between the nuclei corresponding to the actual spacing in the crystal, which is maintained to give a regular 3-dimensional array. This is illustrated in Fig. 2.1, where the total energy is the sum of the attraction and repulsion energies. The corresponding forces are given by the slope of the energy curve, dW/dr, at a point. Thus at the equilibrium spacing r_0 the total force is zero since $dW/dr = 0$, and at this point the energy W_0 is known as the *binding energy*, since it corresponds to an attraction between the atoms. The energy has to be greatly increased to move atoms closer than the spacing r_0 against nuclear repulsion, and rather less increased to separate them against the binding energy.

There are two very important ways in which the binding energy is provided to form a stable structure. These are the *covalent bond* and the *metallic bond* which are discussed below.

The covalent bond and the tetrahedral arrangement

As an example of this bond let us consider the simplest, which is the formation of molecular hydrogen. The hydrogen atom can attain the stable structure of the inert gas helium if it can acquire one more electron. It achieves this by sharing its electron with another hydrogen atom, so that both electrons are associated with one of the nuclei for part of the time (Fig. 2.2(*a*)). Thus each atom has a share in a structure like helium and most of the time the electrons are between the nuclei, screening them from each other and reducing their electrostatic repulsion. This screening therefore constitutes an effective attractive force between the atoms, which is the covalent bond.

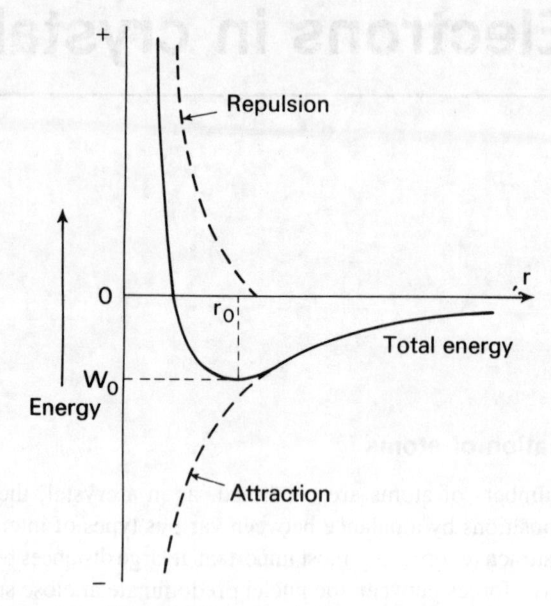

Figure 2.1 Interatomic energies as a function of spacing.

The covalent bond also occurs in crystals of carbon (diamond) and the elemental semiconductors germanium and silicon, where an atom of each element has four valence electrons. Each of these electrons follows an elliptical orbit and these orbits are symmetrically orientated so that they point to the corners of a regular *tetrahedron* (Fig. 2.2(*b*)). In a crystal there is another atom at each corner of the tetrahedron, so that each atom has a share in eight electrons and thus achieves the inert-gas type of structure for much of the time. This is illustrated in two dimensions in Fig. 2.2(*c*) for germanium and silicon, the nuclei plus completed inner electron orbits being drawn at atomic cores with a net charge of $+4e$, balanced by the valence electrons. r_0 refers to the distance between nearest neighbours along the arms of the tetrahedron and is 0.245 nm for germanium and 0.235 nm for silicon.

The tetrahedral arrangement of atoms also occurs in compound semiconductors such as gallium arsenide. Here the central atom is gallium with three valence electrons and the other four atoms are arsenic, each with five valence electrons. Thus, neighbouring atoms join together to share on average eight electrons and the material is known as a III-V compound. Another important III-V compound with a similar atomic arrangement is gallium phosphide. r_0 is 0.245 nm for GaAs and 0.236 nm for GaP, which indicates very similar interatomic spacings to germanium and silicon respectively.

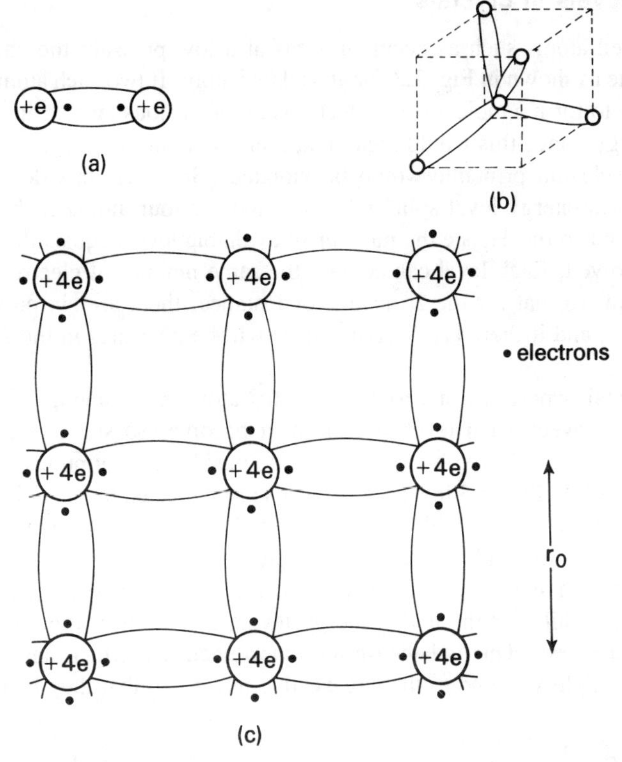

Figure 2.2 Examples of covalent bonding. (*a*) Molecular hydrogen; (*b*) tetrahedral arrangement of atoms; (*c*) two-dimensional diagram of germanium or silicon crystal.

The metallic bond

Consider now the elements in the first group of the periodic table, such as copper, with one valence electron. The valence electrons of neighbouring copper atoms can be shared in a similar manner to the electrons of a covalent bond, but since it is not possible for each copper atom to have seven close neighbours an inert-gas structure cannot be achieved. However, electron sharing does occur with the electrons not restricted to particular atomic cores but wandering freely through the crystal. Nearly every atom contributes one valence electron to the 'cloud' of charge which forms bonds between the atoms by screening nuclei from each other. The metallic bond is more flexible than is the covalent bond, and this leads to the ductile properties of a metal, and the presence of the freely moving cloud of electrons ensures good electrical and thermal conductivity. This type of bond also occurs with the elements in Groups II and III having metallic properties, aluminium being an important Group III metal in electronic devices.

Energy levels in crystals

For isolated atoms such as occur in a gas at a low pressure the energy levels are discrete as shown in Fig. 2.3 for an isolated atom. If two such atoms approach each other to form a molecule the electrons in the various levels cannot have the same energy, since this would mean that their quantum numbers were identical and the exclusion principle would be violated. However, this does not occur because each energy level splits into two, and for four atoms each level splits into four and so on. Hence the number of available levels equals the number of atoms involved. Each level can contain the same number of electrons as in the single atom, so that for the 2p level, for instance, there are six possible states (Table 1.3), and if there are N atoms there will be $6N$ states in the 2p group of energies.

In a crystal, where N is approximately 10^{29} atoms per cubic metre the energy difference between each level and the next becomes so small (approximately 10^{-8} eV) that the allowed energies may be considered continuous, forming an *allowed band*. In Fig. 2.3 the allowed bands are indicated by cross-hatching and are separated by regions with no allowed levels, which are known as *forbidden bands* (*see also* Appendix 1). It may be noted that the levels of highest energy, separated by small energy differences in a single atom, have overlapping bands in a crystal, while for the lower energy levels the width of the corresponding band is much less. This is because the inner electrons, which have the lowest energy, are tightly bound to their respective nuclei, so that they behave almost

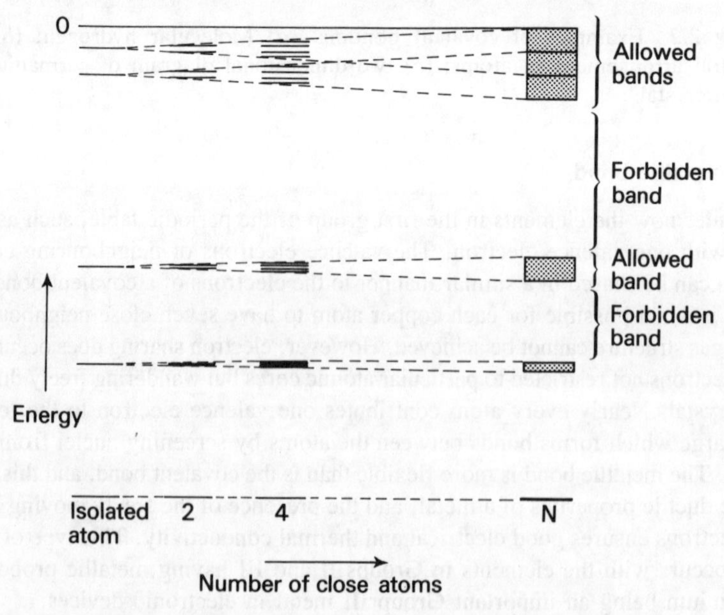

Figure. 2.3 Interaction of atoms leading to energy bands.

as though they were in a single atom. The potential energies of adjacent nuclei (Fig. 2.4(*a*)) combine to form a periodically varying potential through the crystal, which is shown in one dimension superimposed on the energy bands for a metal in Fig. 2.4(*b*). Here all the lower allowed bands or *core levels* are filled with the inner electrons, but the upper band is only partly filled with the higher energy levels unoccupied. The filled part is known as the *valence band* and extends above the periodic potential through the whole crystal. Thus electrons in this band are shared by all the atoms and form the metallic bond. The empty levels above the valence band are known as the *conduction band*, and if an electric field is applied across the crystal electrons from the upper levels in the valence band can easily gain enough energy to move into the vacant levels just above. Here they can move under the influence of the field and so give rise to the high electrical conductivity of a metal. For copper virtually all the valence electrons take part in conduction and since each atom has one valence electron the number density or concentration of free electrons is about $8 \times 10^{28}/\mathrm{m}^3$. This value may be obtained from the number of atoms in the atomic weight, *Avogadro's number*,

$$N_A = 6.023 \times 10^{26} \text{ atoms/kmol}$$

together with the kilogram atomic weight and density of copper given in Table 2.1.

Table 2.1

Element	Atomic weight	Density (kg/m^3)	Atoms per cubic metre
		$\times 10^3$	$\times 10^{28}$
Copper	63.6	8.96	8.5
Germanium	72.6	5.3	4.4
Silicon	28.1	2.4	5.0

Worked example

For copper determine the number of atoms in a cubic metre of the material. The atomic weight is 63.5 and the density is 8.96×10^3 kg/m^3.

Solution

Since there are 6.023×10^{26} atoms in the molecular weight in kilograms of any element the number of atoms in a cubic metre of copper is given by

$$\frac{6.023 \times 10^{26} \times 8.96 \times 10^3}{63.5} = 8.50 \times 10^{28} \text{ atoms/m}^3$$

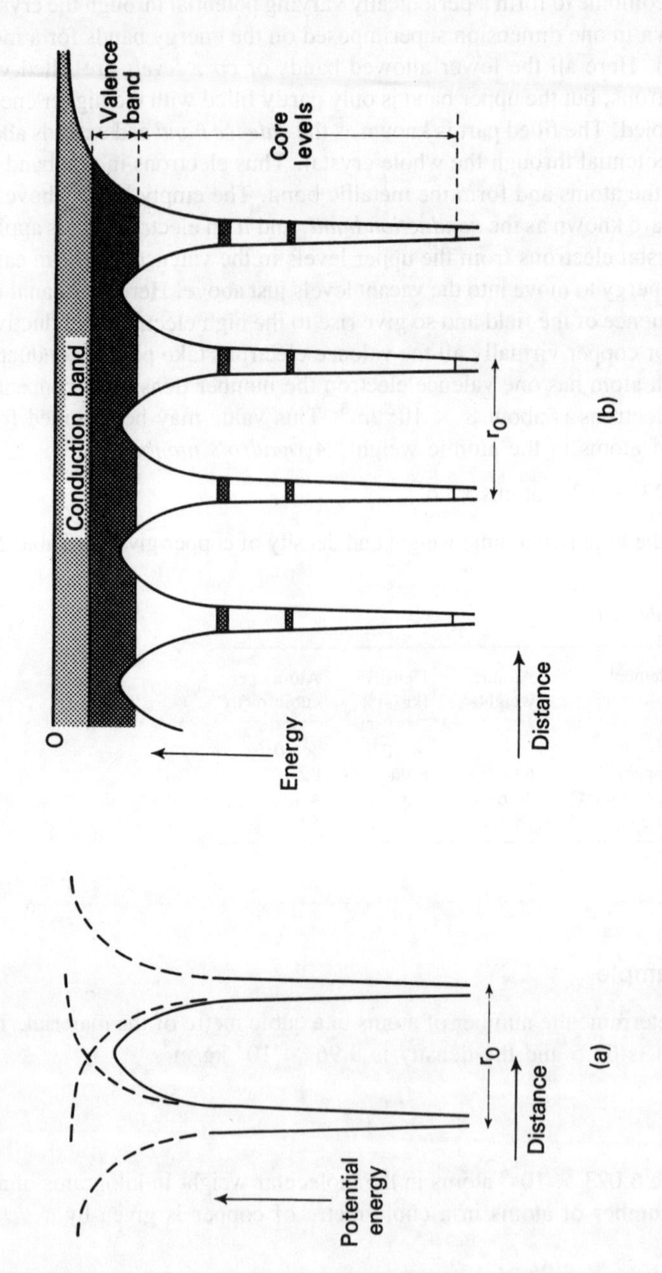

Figure 2.4 One-dimensional variation of energy in a crystal

Electrical Conduction in solids

Solids may be classified according to their electrical resistivity, which is very high for insulators and very low for good conductors such as metals. A semiconductor has a resistivity between these two extremes as shown in Table 2.2, and this wide variation in resistivity may be explained in terms of the energy band structure of the materials.

Table 2.2

Material	Resistivity (Ω m)
Metals	10^{-8}
(e.g. aluminium, copper, silver)	
Semiconductors	10^{-4} - 10^{7}
Insulators	10^{12} - 10^{20}
(e.g. glass, mica, polystyrene)	

The upper allowed bands of a metal, a semiconductor and an insulator are compared diagrammatically in Fig. 2.5, which shows that in both a semiconductor and an insulator the valence and conduction bands are separated by a forbidden band or *energy gap*, W_g. This is much larger for an insulator than for a semiconductor, typical room temperature values being 5.47 eV for diamond, which is an insulator, and 0.66 eV for germanium and 1.12 eV for silicon, which are semiconductors (Ref. [1]). Even for a semiconductor the energy required to move an electron from the top of the valence band is too great to be acquired from a normal electric field: in fact, electrons are moved into the conduction band by excitation due to temperature or the absorption of radiation. At room temperature for germanium only about 1 in every 10^9 of the valence electrons is free, giving a density of about $2.5 \times 10^{19}/m^3$. In silicon the energy gap is wider and the density of free electrons is only about $1.5 \times 10^{16}/m^3$. These densities are far less than for copper and account for the lower conductivity of semiconductors, since conductivity is proportional to the number of current carriers (eq. (2.81)). For insulators, the carrier densities are even less than for semiconductors owing to their much larger energy gap, so that the conductivity of insulators is practically negligible.

Intrinsic semiconductors

In any solid the atoms possess energy due to the temperature of the material, and this energy causes each atom to vibrate about its mean position in the crystal. Near and above room temperature this thermal energy is proportional to the temperature. Some of the energy is shared with the valence electrons, and in a semiconductor a few are ejected from the covalent bonds and become free to take

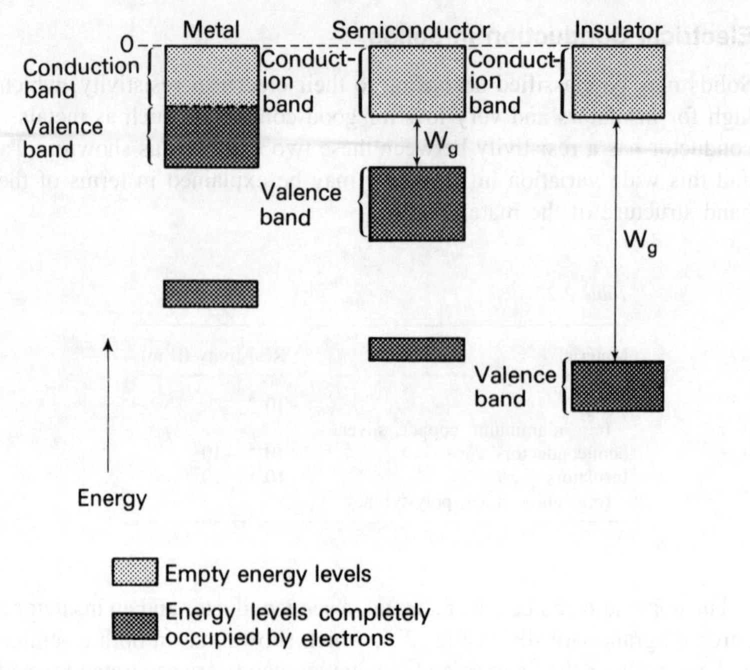

Figure 2.5 Energy bands in a metal, a semiconductor and an insulator.

part in conduction. This corresponds to an electron acquiring energy W_g and moving into the conduction band (Fig. 2.6(*a*)), and the number becoming free electrons increases rapidly with temperature. When an electron is released the charge of $+4e$ on each atomic core is only compensated by three electrons, so that a *hole* is left in the bond with an effective charge of $+e$. If the free electron moves away from the hole the positive charge in the bond can attract a neighbouring valence electron (Fig. 2.6(*b*)), so that it appears that the hole has moved. This process can continue through the crystal, so that the motion of valence electrons in one direction under the influence of an electric field can be considered as the movement of holes in the opposite direction (Fig. 2.6(*c*)). When a hole reaches the negative end of the crystal it is filled by an electron from the external circuit.

The free electrons also move through the crystal under the influence of the electric field, so that conduction takes place simultaneously in both the conduction band and the valence band, which is a phenomenon peculiar to semiconductors. The hole may be conveniently considered as a current carrier with charge $+e$ and a similar mass to the electron, and conduction is said to occur through the agency of *electron-hole pairs*. Two electrons flow in the external circuit for each electron-hole pair, and the process within the semiconductor is known as *intrinsic conduction* since it is a property of the pure material.

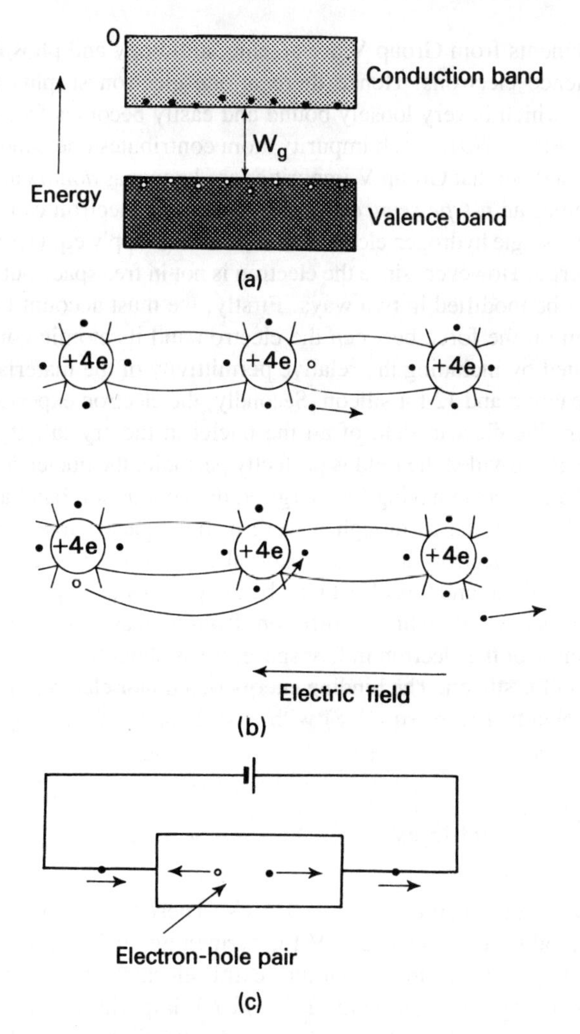

Figure 2.6 Intrinsic semiconductor. (*a*) Energy diagram; (*b*) electron-hole pair; (*c*) current flow.

Impurity or extrinsic semiconductors

The conductivity of a pure semiconductor is too low for most purposes and it is increased by the addition of impurities, which are chosen from elements in Group III or V of the periodic table having atoms of nearly the same size as germanium or silicon atoms. Thus an impurity atom can easily take the place of a germanium or silicon atom during the growth of a crystal from the liquid state and without distorting the crystal structure. Typically 1 atom in 10^7 is replaced by an impurity atom. The material is then called an *extrinsic semiconductor* and the addition of impurities is known as *doping*.

Suitable elements from Group V are arsenic, antimony and phosphorus, each with five valence electrons. Hence there is one electron surplus to the band requirements, which is very loosely bound and easily becomes free even at low temperatures (Fig. 2.7(*a*)). Each impurity atom contributes one conduction electron to the crystal, so that Group V impurities are known as *donors* and the doped material becomes an *n-type semiconductor*. This extra electron can be regarded as similar to the single hydrogen electron, so that we can apply eq. (1.9) to calculate its binding energy. However, since the electron is not in free space but in a crystal, eq. (1.9) must be modified in two ways. Firstly, we must account for the effect of the medium on the force between the electron and its atomic core. This can be approximated by including the relative permittivity of the material, ϵ_r, which is 16 for germanium and 12 for silicon. Secondly, the electron experiences a force due to the periodic electric field of all the nuclei in the crystal. It is shown in Appendix 1 that, provided the field is perfectly periodic, the nuclei do not impede the motion of an electron having its energy in the conduction band and so it can move through the crystal as though it were in free space. However, the force experienced by an electron due to an external electric field is modified by the internal periodic field, and this effect may be included by saying that the electron has an *effective mass*, m^*, which is different from its mass outside the material.

If m is the mass of the electron in free space, m^* is about $0.25m$ for germanium and about $0.8m$ for silicon. The binding energy of a donor electron in germanium may then be obtained from eq. (1.8) with $n = 1$, $m = m^*$ and ϵ_0 replaced by $\epsilon_0\epsilon_r$, which gives

$$\frac{0.25 \times 13.6}{16^2} = 0.013 \text{ eV} \tag{2.1}$$

A similar calculation for silicon gives 0.076 eV. These figures may be compared with practical values of about 0.01 eV for germanium and 0.05 eV for silicon. Such low binding energies mean that the 'extra' electron can be detached very easily. On the energy diagram (Fig. 2.7(*b*)) each impurity atom introduces an allowed energy level in the forbidden band and just below the conduction band. Thus the effective energy gap for donor electrons is very small, about 0.01 eV for germanium and 0.05 eV for silicon, so that at room temperature practically all the donors have lost their electrons. These are free to move in the conduction band, and since there is one for each donor atom the conductivity can be increased by an amount controlled by the donor concentration. The ionized donor atoms remain as fixed positive charges in the crystal lattice.

The corresponding elements from Group III are indium, gallium and boron, each with only three valence electrons. This causes a deficiency of one electron for each impurity atom, giving a hole in one bond (Fig. 2.7(*c*)). This hole can accept an electron from the valence band, so that impurity atoms from Group III are known as *acceptors* and the material becomes a *p-type semiconductor*.

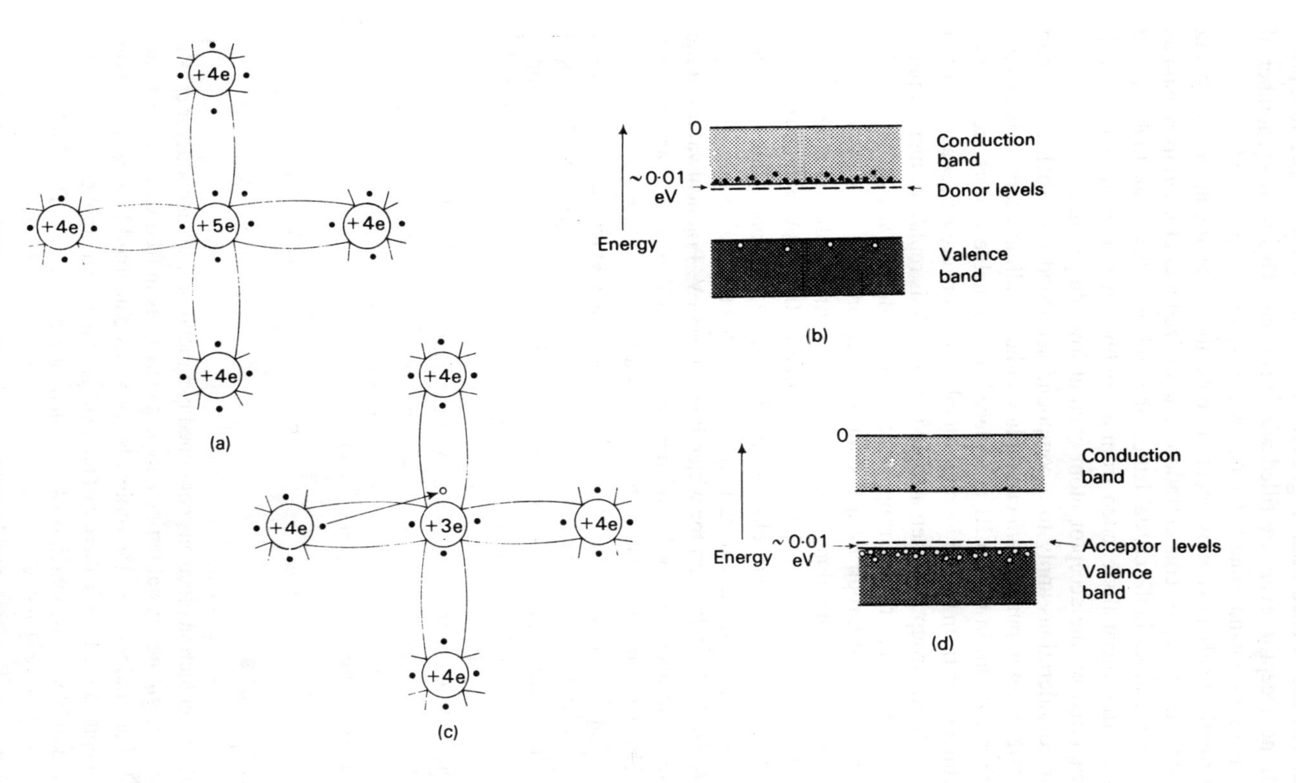

Figure 2.7 Extrinsic semiconductors. (*a*) Atomic arrangement, n-type; (*b*) energy diagram, n-type; (*c*) atomic arrangement, p-type; (*d*) energy diagram, p-type.

Each impurity atom introduces an allowed energy level in the forbidden band and just above the valence band (Fig. 2.7(*d*)), so that at room temperature practically all the acceptor levels are filled with electrons. This leaves a number of holes in the valence band equal to the number of acceptor atoms, and conduction occurs through the effective movement of holes, the value of conductivity being controlled by the acceptor concentration. The ionized acceptor atoms remain as fixed negative charges in the crystal lattice, the ionization being due to the charge $-e$ of an electron thermally excited from a neighbouring covalent bond, which will be captured by the acceptor atom even at low temperatures.

Similar considerations apply to a compound semiconductor such as gallium arsenide, where an impurity atom can replace either a gallium atom or an arsenic atom. When gallium in Group III is replaced then suitable donors are silicon, germanium and tin from Group IV and suitable acceptors are zinc, cadmium and magnesium from Group II. When arsenic in Group V is replaced, then sulphur, selenium and tellurium from Group VI are suitable donors and carbon, silicon and germanium from Group IV are suitable acceptors.

The binding energy of a donor electron may be estimated from eq. (2.1) with m^* about $0.07m$ and ϵ_r at 11. This gives a value of 0.008 eV, which indicates that the donor levels are very close to the bottom of the conduction band due to the very low electron mass in gallium arsenide. Practical values are all at about 0.006 eV, except for tellurium for which it is at 0.03 eV. Practical values of the gap between the acceptor levels and the top of the valence band are between 0.026 eV for carbon and 0.04 eV for germanium.

Undoped gallium arsenide has a very high resistance which can be increased further by creating energy levels near the centre of the energy gap. Any electrons and holes are trapped in these levels, which are so deep that the trapped charges cannot be released by thermal energy at room temperature, and the material becomes *semi-insulating*. A common method of achieving this is to damage the material at an atomic level by bombarding it with hydrogen ions (protons), using an ion implantation system described later in Chapter 7. Semi-insulating (SI) gallium arsenide is used in integrated circuits and microwave devices, as described in Chapters 7 and 9.

The electron gas

The cloud of conduction electrons contained in a metal or a semiconductor crystal can move through the crystal lattice almost as easily as in free space and so may be regarded as analogous to the molecules of a gas contained in a vessel. Many of the concepts of the kinetic theory of an ideal gas have been applied successfully to the cloud of electrons, which is often called an *electron gas*. The more useful concepts are discussed below.

The most important measurable property of a gas is its pressure, which was formerly expressed as the height of a column of mercury exerting the same pressure.* At standard temperature and pressure, namely 0 °C and 101.3 kN/m^2, the volume occupied by the kilogram molecular weight, or the kilomole, is 22.4 m^3. Whatever the gas being considered, this volume must contain a definite number of molecules, which is *Avogadro's number,* N_A. Hence the concentration, or number density, of any gas at standard temperature and pressure is

$$n = \frac{6.023 \times 10^{26}}{22.4} = 2.689 \times 10^{25} \text{ molecules/m}^3 \quad (2.2)$$

which is known as *Loschmidt's number.*

Boltzmann's constant, k

Under any conditions of pressure and temperature the ideal gas equation is applicable:

$$\frac{PV}{T} = constant \quad (2.3)$$

Thus at any pressure P and absolute temperature T the concentration is given by

$$n = 2.689 \times 10^{25} \times \frac{P}{1.013 \times 10^5} \times \frac{273}{T}$$

$$= 7.25 \times 10^{22} \frac{P}{T} \text{ molecules/m}^3 \quad (2.4)$$

This may be written in the form

$$P = nkT \quad (2.5)$$

where k is $1/7.25 \times 10^{22}$ or 1.38×10^{-23} J/K, which is *Boltzmann's constant.*

It may be shown (Ref. [1]) that the pressure of a gas is due to the transfer of momentum to the walls of the containing vessel and is given by

$$P = \tfrac{1}{3} nm\bar{u}^2 \quad (2.6)$$

where $\bar{u}$ is the average speed and m is the mass of each molecule. Hence, from eq. (2.5),

$$\tfrac{1}{3} nm\bar{u}^2 = nkT \quad (2.7)$$

* The average atmospheric pressure is taken to be 760 mmHg which corresponds to 101.3 kN/m^2. Another unit of pressure is the bar, equal to 100 kN/m^2.

or

$$\tfrac{1}{2}\, m\bar{u}^2 = \tfrac{3}{2}\, kT \tag{2.8}$$

The left-hand side of this equation is the average kinetic energy of the gas molecules, which is directly related to the temperature of the gas through Boltzmann's constant. Thus the kinetic energy of a gas molecule, or any particle to which the kinetic theory can be applied, is given at room temperature (20 °C or 293 K) by

$$\tfrac{3}{2} \times 1.38 \times 10^{23} \times 293 = 6.07 \times 10^{21}\ \text{J}$$

$$= \frac{6.07 \times 10^{-21}}{1.6 \times 10^{-19}} = 0.038\ \text{eV} \tag{2.9}$$

(The term kT on its own, which also occurs frequently, thus has a value of 0.025 eV.) In fact, as a result of the temperature the molecules are in constant motion, in all directions, with a range of velocities above and below the mean value.

Mean free path, $\bar{l}$

Due to this random motion collisions occur between the molecules as well as with the walls of the vessel. Hence there is a mean distance a molecule can travel before colliding with another, the *mean free path*, $\bar{l}$. The molecules are considered as hard spheres of diameter d and it may be assumed that for a short time one molecule is moving with speed $\bar{u}$ while all the others are at rest. This molecule travels along the axis of an imaginary cylinder whose diameter is $2d$, and if any other molecule has its centre on or within this cylinder a collision occurs (Fig. 2.8(*a*)). In one

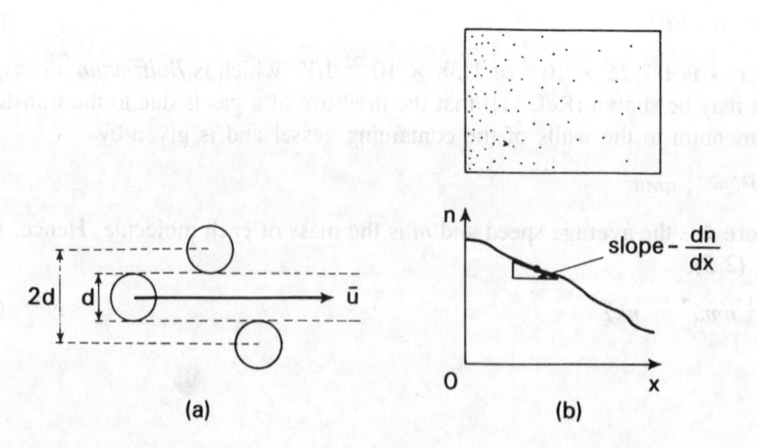

Figure 2.8 The ideal gas. (*a*) Derivation of mean free path; (*b*) diffusion and concentration gradient.

second a volume $\pi d^2 \bar{u}$ is swept out, and this volume contains $\pi d^2 \bar{u} n$ molecules, so that this will also be the number of collisions per second. Thus the average time between collisions t_c known as the *relaxation time*, is given by

$$t_c = \frac{\bar{l}}{\pi d^2 \bar{u} n} \tag{2.10}$$

The distance covered by a molecule between two collisions is called its *free path* and its average value over a large number of collisions is its mean free path, $\bar{l}$. Since both t_c and $\bar{u}$ are average quantities, we can write

$$\bar{l} = t_c \bar{u} \tag{2.11}$$

so that

$$\bar{l} = \frac{1}{\pi d^2 n} \tag{2.12}$$

If allowance is made for the movement of the molecules within the cylinder it may be shown that

$$\bar{l} = \frac{1}{\sqrt{2}\pi d^2 n} \tag{2.13}$$

Thus the mean free path is inversely proportional to the number density of the molecules and hence also inversely proportional to the pressure.

Diffusion

Since the molecules are in continuous random motion, there may be at any instant a high concentration of molecules in one part of the vessel and a low concentration in another part. A transport process then occurs which restores equilibrium conditions. This is known as *diffusion*, in which movement of molecules takes place away from the region of high concentration. A familiar example is that of a balloon which is blown up to an internal pressure above atmospheric. If the neck of the balloon is opened air rushes out until the pressure is equal both inside and outside. A *concentration gradient* is set up under these conditions with a slope from high to low concentration (Fig. 2.8(*b*)). At any distance x the slope of the concentration/distance curve is $-\mathrm{d}n/\mathrm{d}x$, and the rate at which molecules diffuse down the gradient at this point is found to be proportional to the slope. The constant of proportionality is called the *diffusion coefficient, D*, so that the law of diffusion is

$$\frac{1}{S}\frac{\mathrm{d}N}{\mathrm{d}t} = -D\frac{\mathrm{d}n}{\mathrm{d}x} \tag{2.14}$$

where S is the cross-sectional area of the boundary at distance x over which diffusion is taking place, N is the *total* number of particles diffusing over the boundary and dN/dt is the diffusion rate. We shall be concerned with numerical values of D when the idea of diffusion is applied to semiconductors, the units of D being m^2/s.

The Boltzmann factor

The velocities of the gas molecules cover a wide range which conforms to a distribution law evolved by Maxwell and Boltzmann. We are concerned here with one result of the Maxwell–Boltzmann distribution which is of great practical importance in the operation of electronic devices.

If we choose any energy W the required result is that the number density of molecules with energy W is given by

$$n_W = n \exp\left(\frac{-W}{kT}\right) \tag{2.15}$$

where n is the number density of molecules. Since kT has the dimensions of energy, the exponential term is a pure number, known as the *Boltzmann factor*. Hence eq. (2.15) becomes

$$\exp\left(-\frac{W}{kT}\right) = \frac{n_W}{n} = p_M(W) \tag{2.16}$$

Thus the Boltzmann factor gives the fraction of molecules with energy W. This is also expressed as $p_M(W)$, the statistical probability that a molecule will have an energy W. For example, if $n = 10^{20}/m^3$, $n_W = 10^{12}/m^3$, then n_W/n is 10^{-8} and $p(W)$ is 1 in 10^8.

The statistics of electrons and holes

The statistics of the molecules in an ideal gas, as exemplified by the Boltzmann factor, cannot be applied directly to the conduction electrons in a metal or a semiconductor. This is because the energy of gas molecules is not restricted to quantum values, but the energy of conduction electrons in a solid is governed by the number of available levels, which can only be filled according to the exclusion principle. The relevant statistics were first evolved by Fermi and Dirac and lead to a function in which the exponential form is retained, the *Fermi–Dirac function*, so that for solids

$$p_F(W) = \frac{1}{1+\exp\left(\dfrac{W-W_F}{kT}\right)} \tag{2.17}$$

Here $p_F(W)$ is the probability that a level of energy W will be occupied by an electron, and W_F is the Fermi energy or *Fermi level*, which is a purely mathematical parameter. It may not correspond to an allowed level, but it provides a reference with which other energies can be compared. If we put $W = W_F$ in eq. (2.17) the exponential term becomes unity and $p_F(W) = \frac{1}{2}$. Now, a probability of 1 implies that the event concerned is a certainty, while a probability of 0 implies that it can never happen. Hence a probability of $\frac{1}{2}$ can be interpreted in this case as meaning that an electron is equally likely to have an energy above the Fermi level as below it.

Let us now consider the effect of temperature on the Fermi function. At the absolute zero temperature, for all energies less than W_F, $p_F(W) = 1$, so that all these levels are full. As T is increased above zero the function changes as shown in Fig. 2.9. Thus the probability of a level above W_F being occupied increases with temperature, and at the same time the probability of a level below W_F being occupied decreases, as electrons are transferred from lower to higher levels. For equal energies above and below W_F the probabilities of the level being filled and empty respectively are also equal (*see* Problem 2.1).

Intrinsic semiconductors

In order to apply the Fermi function to an intrinsic semiconductor we have to decide where the Fermi level should be on the energy diagram. At $T = 0$ K the valence band is full of electrons and the conduction band is empty, so that we should expect the Fermi level to lie somewhere in the energy gap. As the temperature is increased electrons cross the gap to fill some of the levels in the conduction band, leaving an equal number of empty levels in the valence band. Hence the probabilities of a level being filled in the conduction band and empty in the valence band must be equal, and since the Fermi function changes symmetrically about W_F, the Fermi level should occur in the middle of the energy gap (Fig. 2.10). A detailed calculation (Appendix 2) shows that this is true provided that the effective masses of a hole and an electron are equal. It should be

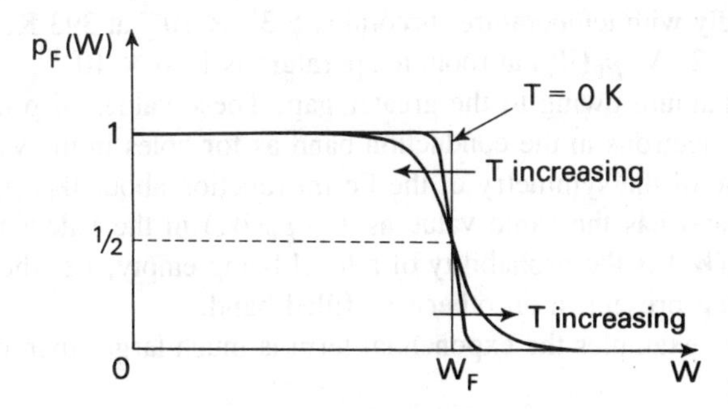

Figure 2.9 The Fermi-Dirac function.

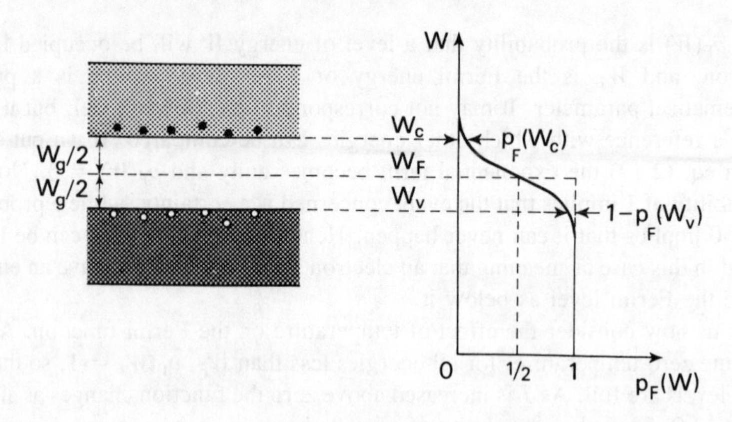

Figure 2.10 Energy diagram and probability function for an intrinsic semiconductor.

noted that in Fig. 2.10 energy is shown increasing upwards from an arbitrary zero. The actual energy level taken as zero is not required since W_F is now the reference level and energy differences from W_F are considered.

Let us now calculate the probability of an electron being excited into a state at the bottom of the conduction band at room temperature, 293 K. For germanium the energy gap, W_g, is 0.66 eV, so that $W_c - W_F = \frac{1}{2} W_g = 0.33$ eV, and $kT = 0.025$ eV, so that

$$p_F(W_c) = \frac{1}{1+\exp 13.2}$$

$$= \frac{1}{1+5.40 \times 10^5} = 1.85 \times 10^{-6} \qquad (2.18)$$

neglecting the 1 in the denominator. This is a very small probability but it does increase rapidly with temperature, becoming 5.32×10^{-5} at 393 K. For silicon, with $W_g = 1.12$ eV, $p_F(W_c)$ at room temperature is 1.86×10^{-10}, much smaller than for germanium owing to the greater gap. These values of probability are the same for electrons in the conduction band as for holes in the valence band, since, because of the symmetry of the Fermi function about W_F, $p_F(W_c)$ in the conduction band has the same value as $1 - p_F(W_v)$ in the valence band (Fig. 2.10). $1 - p_F(W_v)$ is the probability of a level being empty, i.e. the probability of a hole being present in an otherwise filled band.

In the above examples the exponential term is much larger than unity so that we can write

$$p_{\mathrm{F}}(W) \approx \frac{1}{\exp\left(\frac{W - W_{\mathrm{F}}}{kT}\right)} = \exp\left(-\frac{W - W_{\mathrm{F}}}{kT}\right) \tag{2.19}$$

This is in a similar form to the Boltzmann factor (eq. (2.16)), so that the Fermi–Dirac distribution has reduced to the Maxwell–Boltzmann distribution. This will be true provided that $W - W_{\mathrm{F}} > 3kT$, which occurs in most intrinsic semiconductors and lightly doped extrinsic semiconductors, so that Maxwell–Boltzmann statistics are generally applicable to semiconductors, up to a limiting value of T, the temperature. Here there are relatively few electrons available to fill a large number of empty states, so that the exclusion principle is comparatively unimportant. It may no longer hold at high temperatures or in extrinsic semiconductors with a large density of impurities.

The actual number density of electrons, n, will be limited by the density, N_{c}, of the energy levels in the conduction band, since each level can accommodate only one electron. Thus, as shown in Appendix 2,

$$n = N_{\mathrm{c}} p_{\mathrm{F}}(W_{\mathrm{c}}) \tag{2.20}$$

so that

$$n = N_{\mathrm{c}} \exp\left(-\frac{W_{\mathrm{c}} - W_{\mathrm{F}}}{kT}\right) \tag{2.21}$$

where W_{c} is the energy at the bottom of the conduction band. For an intrinsic semiconductor in which $N_{\mathrm{c}} = N_{\mathrm{v}}$, $m_{\mathrm{e}} = m_{\mathrm{h}}$ and the Fermi level is at the centre of the energy gap, we have

$$n = N_{\mathrm{c}} \exp\left(-\frac{W_{\mathrm{g}}}{2kT}\right) \tag{2.22}$$

Similarly the number density of holes, p, is governed by the density, N_{v}, of the energy levels in the valence band, since only one hole can appear in each level. The probability of a hole occurring is

$$1 - p_{\mathrm{F}}(W) = 1 - \frac{1}{1 + \exp\left(\frac{W - W_{\mathrm{F}}}{kT}\right)}$$

$$= \frac{1}{\exp\left(\frac{W_{\mathrm{F}} - W}{kT}\right) + 1}$$

$$\approx \exp\left(-\frac{W_F - W}{kT}\right) \tag{2.23}$$

Hence

$$p = N_v \exp\left(-\frac{W_F - W_v}{kT}\right) \tag{2.24}$$

where W_v is the energy at the top of the valence band. Thus, for an intrinsic semiconductor,

$$p = N_v \exp\left(-\frac{W_g}{2kT}\right) \tag{2.25}$$

For an intrinsic semiconductor as above $n = p = n_i$, the number density of electron-hole pairs, and an important result is obtained from the product of n and p, since

$$np = n_i^2 = N_c N_v \exp\left(-\frac{W_g}{kT}\right) \tag{2.26}$$

Thus

$$n_i = (N_c N_v)^{1/2} \exp\left(-\frac{W_g}{2kT}\right) \tag{2.27}$$

and n_i depends only on the type of semiconductor and the temperature. It is shown in Appendix 2 that, where G is a constant,

$$(N_c N_v)^{1/2} = GT^{3/2} \tag{2.28}$$

$$= 4.83 \times 10^{21} T^{3/2} \tag{2.29}$$

This leads to values at room temperature of N_c and N_v near 10^{25} carriers/m^3. The expression for n_i then becomes

$$n_i = GT^{3/2} \exp\left(-\frac{W_g}{2kT}\right) \text{ electron-hole pairs/m}^3. \tag{2.30}$$

At room temperature accepted values of n_i in electron-hole pairs/m^3 are

germanium	2.5×10^{19}
silicon	1.5×10^{16}
gallium arsenide	1.8×10^{12}

The very large differences in n_i are due to small differences between the materials in the size of the energy gap. Thus, for an insulator, where W_g is much larger than for silicon, n_i is negligible.

The temperature dependence of n_i can be obtained from eq. (2.30) by taking logarithms of both sides:

$$\ln n_i = (\ln G + 3/2 \ln T) - \frac{W_g}{2k}\frac{1}{T} \tag{2.31}$$

Above about 200 K the bracketed term varies much more slowly with temperature than the $1/T$ term, so that $\ln n_i$ increases almost linearly as $1/T$ decreases.

n- and p-type semiconductors

The position of the Fermi level for an extrinsic semiconductor must depend on the density of the impurity atoms, since the greater this density the greater is the probability of electrons appearing in the conduction band, for n-type material. For an extrinsic semiconductor with an impurity density up to about $10^{23}/\text{m}^3$, i.e. about 1 atom in 10^6 replaced by an impurity atom, there are few enough carriers for Maxwell–Boltzmann statistics to be applied.

Thus for an n-type semiconductor we can use eq. (2.21) to give the number density, n_n of conduction electrons:

$$n_n = N_c \exp\left(-\frac{W_c - W_F}{kT}\right) \tag{2.32}$$

Hence

$$W_c - W_F = kT \ln \frac{N_c}{n_n} \tag{2.33}$$

so that the position of the Fermi level is fixed relative to the bottom of the conduction band (Fig. 2.11(*a*)). Since the binding energy of the donor electrons is so low, near room temperature, practically all the donor atoms are ionized, so that $n_n \approx N_d$, the number density of the donor atoms. Thus as N_d is increased, $W_c - W_F$ is reduced and the Fermi level moves closer to the *bottom* of the conduction band.

Similarly for a p-type semiconductor the number density of holes, p_p, is given by

$$p_p = N_v \exp\left(-\frac{W_F - W_v}{kT}\right) \tag{2.34}$$

and

$$W_F - W_v = kT \ln \frac{N_v}{p_p} \tag{2.35}$$

p_p is approximately equal to N_a, the number density of acceptor atoms, and as

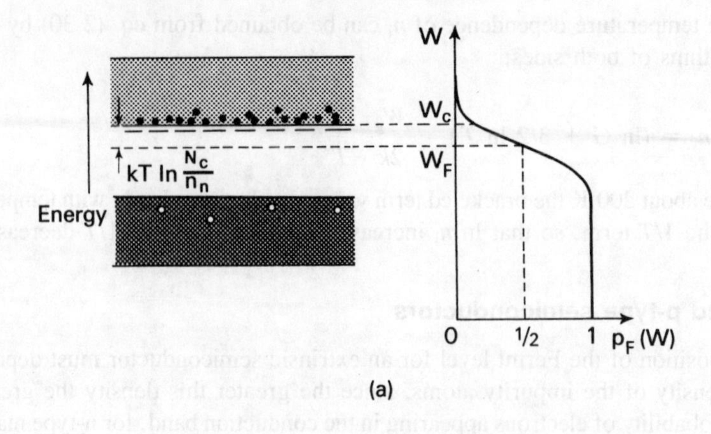

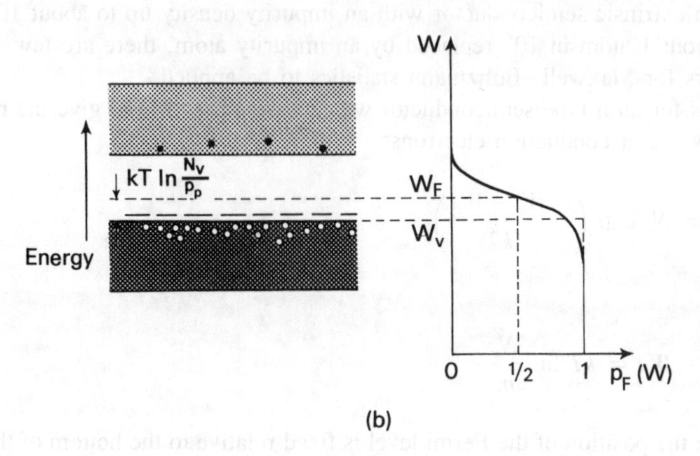

Figure 2.11 Energy diagrams and probability functions. (*a*) n-type semiconductor; (*b*) p-type semiconductor.

N_a is increased the Fermi level moves closer to the *top* of the valence band (Fig. 2.11(*b*)).

During the manufacture of semiconductor devices, n-type material may have to be changed to p-type by the addition of acceptor impurities, and the reverse may also be required. Hence both donor and acceptor impurities can be present at the same time, and if $N_d > N_a$ the material is n-type, while if $N_a > N_d$ it is p-type. In order to preserve electrical neutrality there must be equal densities of positive and negative charges in the material, so that

$$p + N_d = n + N_a \tag{2.36}$$

This assumes that all the impurity atoms are ionized, so that there is a density

N_a of fixed negative charges in the lattice. Also eq. (2.26) still holds, since it is independent of the impurity densities, so that

$$np = n_i^2 \tag{2.37}$$

Equation (2.37) is also known as the law of mass-action. Hence eqs (2.36) and (2.37) must be solved simultaneously to find n and p. Putting $n - p = N_d - N_a = 2x$ and substituting from eq. (2.37),

$$n - \frac{n_i^2}{n} = 2x$$

so that

$$n^2 - 2nx - n_i^2 = 0 \tag{2.38}$$

Hence

$$n = x + \sqrt{(x^2 + n_i^2)} \tag{2.39}$$

and similarly,

$$p = -x + \sqrt{(x^2 + n_i^2)} \tag{2.40}$$

where x is positive for n-type and negative for p-type material. When doping an intrinsic semiconductor, it is normal practice to make either N_a or N_d much larger than n_i at room temperature. Thus, applying the binomial theorem to eqs (2.39) and (2.40),

$$n = x + \sqrt{x^2}\left[1 + \frac{1}{2}\left(\frac{n_i}{x}\right)^2\right] \tag{2.41}$$

and

$$p = -x + \sqrt{x^2}\left[1 + \frac{1}{2}\left(\frac{n_i}{x}\right)^2\right] \tag{2.42}$$

since $(n_i/x)^2 \ll 1$. Hence for n-type material $N_d \gg N_a$ and $x = N_d/2$, so that

$$n_n = N_d + \frac{n_i^2}{N_d} \approx N_d \tag{2.43}$$

and

$$p_n = \frac{n_i^2}{N_d} \tag{2.44}$$

Similarly, for p-type material, $N_a \gg N_d$ and $x = -N_a/2$, so that

$$n_p = \frac{n_i^2}{N_a} \tag{2.45}$$

and

$$p_p = N_a + \frac{n_i^2}{N_a} \approx N_a \tag{2.46}$$

taking $\sqrt{x^2}$ as always positive.

Majority and minority carriers

In n-type silicon with a donor density of $10^{22}/\text{m}^3$ and an electron-hole density of about $10^{16}/\text{m}^3$ at room temperature, the electron density will be $10^{22}/\text{m}^3$ and the hole density about $10^{10}/\text{m}^3$ from eqs (2.43) and (2.44). Thus there are 10^{12} electrons for each hole, so that in an n-type semiconductor the electrons are called *majority* carriers and the holes, *minority* carriers. Similarly in a p-type semiconductor with a comparable impurity density the holes are the majority carriers and the electrons the minority carriers.

Degenerate semiconductors and metals

If the impurity density of an n-type semiconductor is increased, the Fermi level moves closer to the conduction band. When a density of about $5 \times 10^{23}/\text{m}^3$ is reached, N_c/n_n is about 20 and $\ln (N_c/n_n) = 3.0$, so that $W_c - W_F = 3kT$. Now, the approximate limit at which the Fermi function reduces to the Boltzmann factor is $W_c - W_F \geq 3kT$, so that, for doping levels above about $5 \times 10^{23}/\text{m}^3$, $W_c - W_F < 3kT$ at room temperature and Maxwell–Boltzmann statistics are no longer valid. The semiconductor is then said to be *degenerate*, while in a semiconductor with a lower doping level the free electrons obey the classical Maxwell–Boltzmann statistics and the material is said to be *non-degenerate*. Similarly a p-type semiconductor becomes degenerate when $W_F = W_v < 3kT$, that is when $N_a \geq 5 \times 10^{23}/\text{m}^3$.

With a further increase in impurity density above $5 \times 10^{23}/\text{m}^3$ the Fermi level continues to move until at a density of about $10^{26}/\text{m}^3$ it is in fact *within* the conduction band in an n-type semiconductor and *within* the valence band in a p-type semiconductor (Figs. 2.12(*a*) and (*b*)). The semiconductor now has so many majority carriers that it has nearly degenerated into a metal, and some of the levels available for conduction are completely filled, both at room temperature and below. The energy diagram for a metal, given in Fig. 2.12(*c*), shows the Fermi level at the top of the valence band. Virtually all the free electrons with energies near the top of the valence band can acquire sufficient energy from an applied field to appear in the adjacent levels in the conduction band. The distribution of energies obeys Fermi–Dirac statistics since only those electrons with energies close to

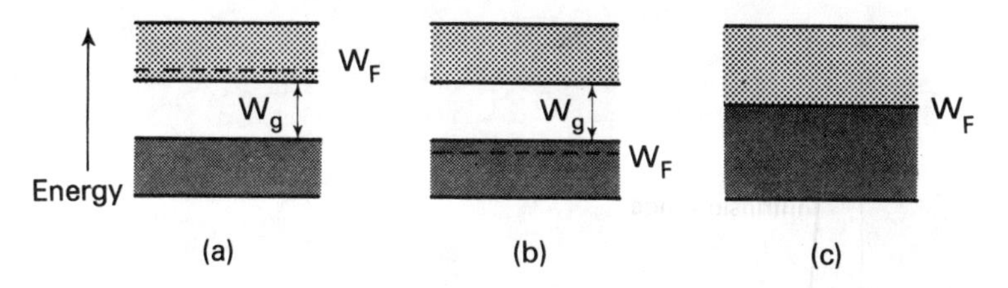

Figure 2.12 Energy diagrams. (*a*) Degenerate n-type semiconductor; (*b*) degenerate p-type semiconductor; (*c*) metal.

the Fermi level will take part in conduction. For metals the value of W_F lies between 3 and 14 eV above the zero level at the bottom of the valence band, which corresponds to an average velocity near 10^6 m/s. The value of W_F for copper is 7 eV, which is much larger than the thermal energy at room temperature of 0.038 eV. Thus temperature has a negligible effect on both the number and energy of the current carriers in a metal. This contrasts with non-degenerate semiconductors, whose current carriers have energies of at least $3kT$ from the Fermi energy and hence obey Maxwell–Boltzmann statistics.

Effect of temperature on n- and p-type semiconductors

In an n-type semiconductor at room temperature $n_i \ll N_d$, but n_i increases rapidly with temperature (eq. (2.30)) until $n_i = N_d$ at the *transition temperature*. Then, from eqs (2.43) and (2.44), $n_n = 2N_d$ and $p_n = N_d$. Similarly for a p-type semiconductor at the transition temperature, $p_p = 2N_a$ and $n_p = N_a$. Above this temperature the semiconductor reverts to the intrinsic type, so that ln n or ln p becomes proportional to $1/T$ (Fig. 2.13), while below it the carrier density is practically constant down to a temperature which depends on the density of impurities. As the temperature is further reduced, fewer and fewer of the impurity atoms are ionized. The transition temperature increases with N_d or N_a and can extend up to about 500 K, while for a degenerate semiconductor it is even higher, so that over a useful operating range the carrier density of extrinsic semiconductors is constant. During the operation of semiconductor devices, the internal temperature must be kept well below the transition temperature, since if it is exceeded the sudden increase of current carriers can lead to further heating and rapid failure of the device.

Lattice scattering and mobility

For a single crystal with a perfectly periodic arrangement of atoms the effect of the periodic field of the crystal lattice has been included by giving the electron an effective mass (Appendix 1). A free electron can move through such a crystal as though it were in free space, so that the atoms will not impede its motion and

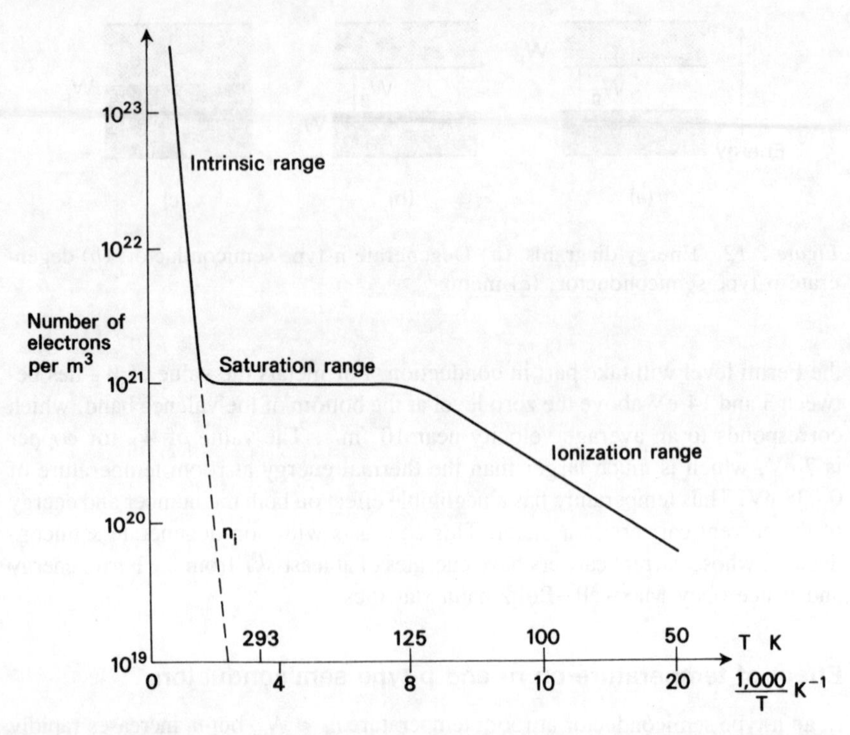

Figure 2.13 Electron density as a function of temperature for silicon (after Ref. [2]). $N_d = 10^{21}/\text{m}^3$.

the crystal will have zero electrical resistance. However, in practice no crystal is perfect and there are two main ways in which the perfectly periodic field may be disturbed in semiconductors such as silicon and germanium. Firstly, there may be sites scattered at random through the lattice from which an atom is missing. Such a site may also be occupied by an impurity atom which will have a charge associated with it. In either case there is a local disturbance of the periodicity of the total electric field within the crystal. Secondly, at temperatures above absolute zero, the atoms are not stationary but are vibrating about their mean positions with an energy proportional to the temperature. These thermal vibrations result in the propagation of elastic waves through the crystal lattice at the speed of sound appropriate to the material. Where an elastic wave has frequency f its energy is quantized in units of hf joules with the quantum being known as a *phonon*. The phonons can be considered as particles which are analogous to the photons of electromagnetic waves. The number of phonons flowing through the crystal will rise as the amplitude of the atomic vibrations is increased by a rise in temperature. An electron may be considered to collide with the phonons leading to a process known as *lattice scattering*. The resulting thermal motion

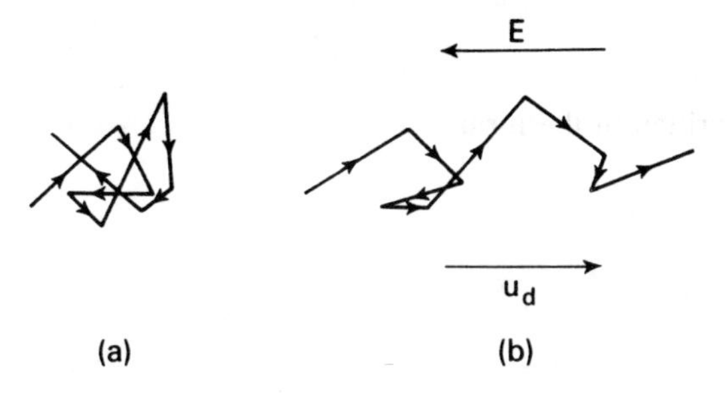

Figure 2.14 Thermal motion of an electron. (*a*) Electric field zero; (*b*) electric field applied.

of the electrons is completely random in direction, and an electron is supposed to move in a straight line at constant acceleration between collisions (Fig. 2.14(*a*)). In the absence of an applied field the velocity of the electrons moving across any given plane at any instant is zero, owing to the random nature of the motion.

In a semiconductor the density of free electrons corresponds to the density of molecules in a gas at low pressure. Hence the electrons will have an average energy of $(3/2)kT$ due to their thermal motion, and their average speed may be obtained from eq. (2.8), which gives

$$\bar{u} = \sqrt{\frac{3kT}{m}} \tag{2.47}$$

At room temperature, 293 K, and assuming that $m^* = m$, this leads to an average speed of about 10^5 m/s. The path of an electron may pass many atoms before scattering occurs and, as in a gas, the average distance travelled between collisions is known as the *mean free path,* $\bar{l}$, which is about 10^{-7} m. Hence the *relaxation time,* t_c, between collisions is $\bar{l}/\bar{u}$, about 10^{-12} s. For metals, although Maxwell–Boltzmann statistics are not applicable, a relaxation time can be found which is about 10^{-14} s; this is only 1% of the value for semiconductors owing to the far higher density of free electrons (compare eq. (2.10)) and the greater packing density of metals.

If a constant electric field E is applied to the crystal the electron will experience a force eE moving it towards the positive terminal. This results in a *drift velocity,* u_d being superimposed on the thermal motion (Fig. 2.14(*b*)). For small values of E the energy acquired by the electron from the field is dissipated by collision with the atoms. The amplitude of the atomic vibrations is thereby increased and the temperature of the crystal is raised, causing 'joule heating' of the material.

The acceleration given to the electron by the field is eE/m, which is constant. Since drift velocity is much less than the thermal velocity, as shown below, the average time between collisions is still t_c, so that the average drift velocity is given by

$$u_d = \frac{eEt_c}{m} \tag{2.48}$$

This can be written in the form

$$u_d = \mu E \tag{2.49}$$

where

$$\mu = \frac{et_c}{m} \tag{2.50}$$

μ is known as the *mobility* and represents the drift velocity per unit electric field, so that it is measured in metres per second divided by volts per metre. For small electric fields the value of μ is independent of the value of E.

Inserting the appropriate values of t_c in eq. (2.50) yields mobilities of 1.8×10^{-1} m²/V s for semiconductors and 1.8×10^{-3} m²/V s for metals. These figures are about right as may be seen from the experimental values in Table 2.3, a more complete list being given at the end of the book. μ_p for holes is less than μ_n for electrons since the process of charge transfer for valence-band electrons is slower than for conduction-band electrons. A typical value of E in the operation of semiconductors is 1000 V/m, so in silicon at room temperature the drift velocity of electrons is 150 m/s and of holes 60 m/s, only about 0.1% of the thermal velocity. Hence application of the field has a negligible effect on the thermal velocity.

When E exceeds about 2×10^5 V/m (Fig. 2.15) the motion of an electron becomes much less random and more in the direction of the field. Since it moves through an effective potential $E\bar{l}$ between collisions it gains kinetic energy $Ee\bar{l}$ from the field. This energy imparts a drift velocity approaching the thermal velocity and is lost at each collision. Then the maximum velocity is u, given by

$$\tfrac{1}{2} mu^2 = Ee\bar{l} \tag{2.51}$$

and

$$u_d = \tfrac{1}{2} u = \frac{1}{2}\left(\frac{2Ee\bar{l}}{m}\right)^{1/2} = \left(\frac{e\bar{l}}{2m}\right)^{1/2} E^{1/2} \tag{2.52}$$

Thus in this region the drift velocity increases with the square root of the field,

Table 2.3 Mobilities at 20 °C

			m²/Vs
Aluminium			1.2×10^{-3}
Copper			3.2×10^{-3}
Silicon:	electrons	μ_n	1.5×10^{-1}
	holes	μ_p	4.8×10^{-2}

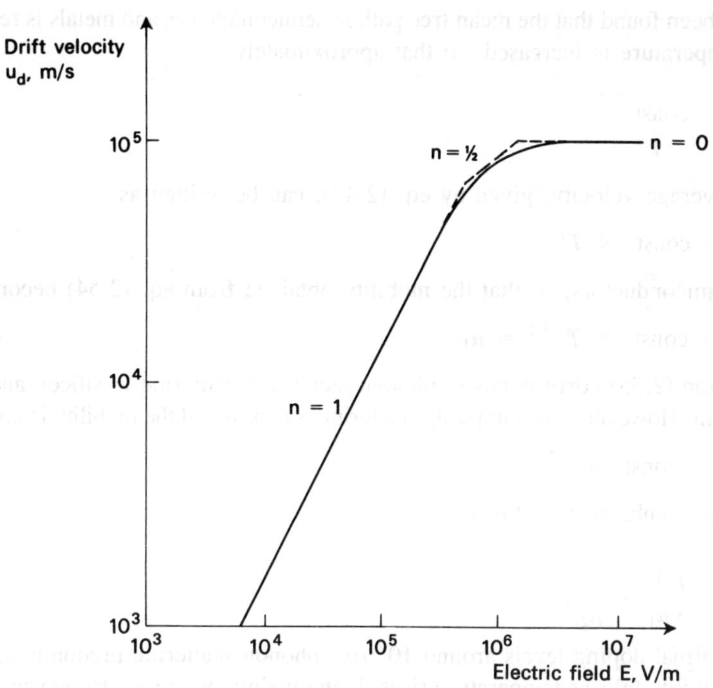

Figure 2.15 Electron drift velocity and electric field, $u_d = \text{const.} \times E^n$, for electrons in silicon.

as shown by the $n = \frac{1}{2}$ asymptote in Fig. 2.15 and the mobility is dependent on the field, since

$$\mu = \frac{u_d}{E} = \left(\frac{e\bar{l}}{2mE}\right)^{1/2} \tag{2.53}$$

As E increases μ decreases until at fields above about 2×10^6 V/m the drift velocity reaches a limiting value of about 10^5 m/s at room temperature for silicon. At these very high fields the electron gives up so much energy to the crystal that new modes of atomic vibration are set up, so that an increase in field is compensated by a fall in mean free path. Eventually breakdown of the material occurs between 10^7 and 10^8 V/m.

Mobility and temperature

The mobility in the low-field region is a function of temperature which can be deduced by putting $t_c = \bar{l}/\bar{u}$ in eq. (2.50), which gives

$$\mu = \frac{e\bar{l}}{2m\bar{u}} \tag{2.54}$$

It has been found that the mean free path in semiconductors and metals is reduced as temperature is increased, so that approximately

$$\bar{l} = \frac{\text{const.}}{T}$$

The average velocity, given by eq. (2.47), can be written as

$$\bar{u} = \text{const.} \times T^{1/2}$$

for semiconductors, so that the mobility obtained from eq. (2.54) becomes

$$\mu = \text{const.} \times T^{-3/2} = \mu_1 \tag{2.55}$$

Equation (2.55) corresponds to phonon (acoustic) scattering in silicon and germanium. However, when impurity scattering is considered the mobility is given by

$$\mu_2 = \text{const.} \times T^{3/2} \tag{2.56}$$

and the combined mobility is

$$\mu = \left(\frac{1}{\mu_1} + \frac{1}{\mu_2}\right)^{-1} \tag{2.57}$$

For normal doping levels around $10^{22}/\text{m}^3$ phonon scattering predominates and the mobility falls as temperature rises, being mainly due to μ_1. However, as the doping level is increased impurity scattering becomes more important and for degenerate semiconductors mobility is nearly independent of temperature. This is because μ_2 rises with temperature and so compensates the fall due to μ_1. In gallium arsenide the phonon scattering is such that the combined mobility (Ref. [3]) is given by

$$\mu = \text{const } T^{1/2} \tag{2.58}$$

For metals only electrons with energies close to the Fermi energy take part in conduction, so that their energy and average velocity are mainly determined by the value of W_F. Hence $\bar{u}$ in eq. (2.54) is virtually independent of temperature so that for metals

$$\mu = \frac{\text{const.}}{T} \tag{2.59}$$

which is also confirmed by experiment.

Mobility and Ohm's law

Consider a block of metal, of length L and cross-sectional area S, which contains n free electrons per cubic metre (Fig. 2.16). If a voltage V is applied across it a field $E = V/L$ will be set up, causing the electrons to move with a drift velocity

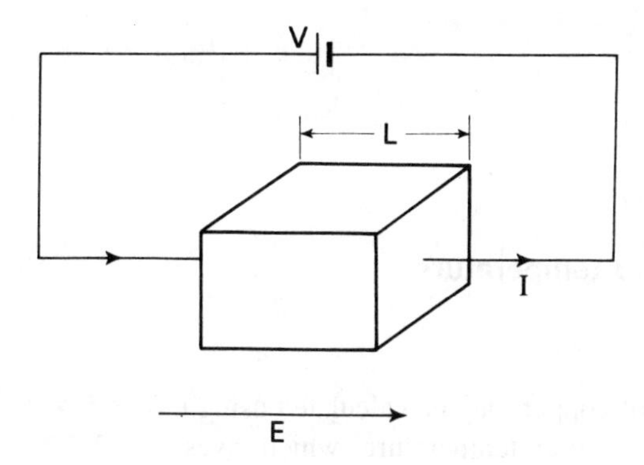

Figure 2.16 Current through a metal block.

u_d metres per second, and in one second they will sweep out a volume u_dS cubic metres. Hence the total charge passing through a given plane in one second will be neu_dS coulombs, which is also the current I, i.e.

$$I = neu_dS \tag{2.60}$$

The current density, $J = I/S$ amperes per square metre, is therefore given by

$$J = neu_d = ne\mu E \tag{2.61}$$

since $u_d = \mu E$. This may be shown to be a form of Ohm's law, corresponding to the familiar expression

$$I = \frac{V}{R} \tag{2.62}$$

where R is the resistance of the block in ohms, given by

$$R = \rho \frac{L}{S} \tag{2.63}$$

ρ being the *resistivity* of the material in ohm-metres, or the resistance of a metre cube. Hence

$$J = \frac{V}{RS} = \frac{V}{\rho L} = \frac{E}{\rho} \tag{2.64}$$

This may be written in the form

$$J = \sigma E \tag{2.65}$$

where σ is the conductivity of the material in siemens per metre. Hence from eq. (2.61),

$$\sigma = ne\mu \tag{2.66}$$

and

$$\rho = \frac{1}{\sigma} = \frac{1}{ne\mu} \tag{2.67}$$

Resistivity and temperature

Metals

The resistivity of copper may be calculated using $n = 8.5 \times 10^{28}/\text{m}^3$, $\mu = 3.2 \times 10^{-3}\ \text{m}^2/\text{V s}$ at room temperature, which gives $\rho = 2.3 \times 10^{-8}\ \Omega\ \text{m}$, using eq. (2.67). The mobility of the free electrons in a metal varies inversely with temperature according to eq. (2.59), so that the resistivity as a function of temperature becomes

$$\rho = \text{const.} \times T \tag{2.68}$$

In practice the resistivity of metals increases nearly linearly with temperature above about 100 K (Fig. 2.17).

Intrinsic semiconductors

Both the electrons and the holes contribute to the total current (Fig. 2.6(*c*)), and although they flow in opposite directions their contributions add since their charges are of opposite sign. The electron current density is

$$J_n = n_i e \mu_n E \tag{2.69}$$

and the hole current density is

$$J_p = n_i e \mu_p E \tag{2.70}$$

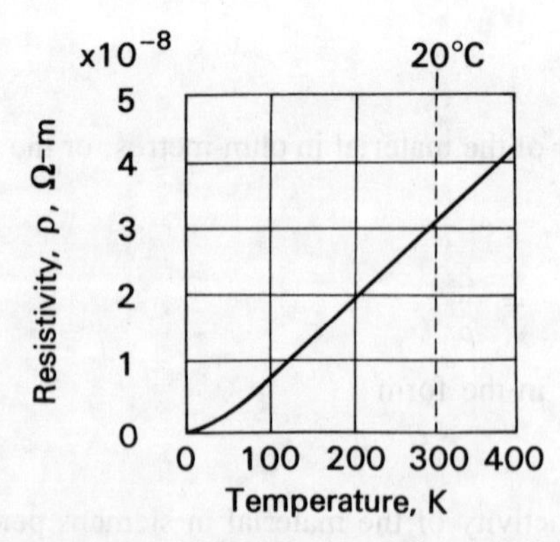

Figure 2.17 Temperature dependence of the resistivity of a metal.

since the densities of electrons and holes are both equal to the density of electron-hole pairs, n_i. Thus the total current density is

$$J_n + J_p = J = n_i e\mu_n E + n_i e\mu_p E$$

or

$$J = E n_i e(\mu_n + \mu_p) \tag{2.71}$$

and the intrinsic conductivity is

$$\sigma_i = \frac{J}{E} = n_i e(\mu_n + \mu_p) \tag{2.72}$$

Worked example

Calculate the conductivity and resistivity of an intrinsic silicon specimen at room temperature.

Solution

From page 40 n_i for silicon is 1.5×10^{16} electron-hole pairs per m^3 and from page 48 the electron mobility is $0.15\ m^2/Vs$ and the hole mobility is $0.05\ m^2/Vs$ for silicon. Then

$$\sigma_i = 1.5 \times 10^{16} \times 1.6 \times 10^{-19}(0.15 + 0.05)$$
$$= 4.8 \times 10^{-4} \tag{2.73}$$

and

$$\rho_i = 1/\sigma_i = 2083\ \Omega\ m \tag{2.74}$$

In eq. (2.72) both n_i and the mobilities vary with temperature, according to eqs (2.30) and (2.55), (2.56) or (2.58), respectively. Since σ_i depends on the product of n and μ, if μ depends on $T^{-3/2}$ the conductivity becomes

$$\sigma_i = \text{const.} \times \exp\left(-\frac{W_g}{2kT}\right) \tag{2.75}$$

and if μ depends on $T^{1/2}$ the conductivity becomes

$$\sigma_i = \text{const.} \times T^2 \exp\left(-\frac{W_g}{2kT}\right) \tag{2.76}$$

in which the exponential term varies much more rapidly with temperature than does the T^2 term. Hence in either case we may write

$$\rho_i = \text{const.} \times \exp\left(\frac{W_g}{2kT}\right) \tag{2.77}$$

so that the resistivity falls rapidly as the temperature rises. Taking logarithms in eq. (2.77),

$$\ln \rho_i = \ln \text{const.} + \left(\frac{W_g}{2k} \times \frac{1}{T}\right) \tag{2.78}$$

and above about 200 K, $\ln \rho_i$ varies almost linearly with $1/T$. The slope of the line is proportional to W_g and measurement of the resistance as a function of temperature is often used to determine the energy gap of an intrinsic semiconductor.

Equation (2.77) may be written in the form

$$R = R_0 \exp\left(\frac{b}{T}\right) \tag{2.79}$$

where R_0 and b are constants. This is the characteristic of a negative-temperature-coefficient device called a *thermistor* (Fig. 2.18), formed normally from a mixture of materials having the properties of an intrinsic semiconductor. The thermistor is widely used as a temperature-sensitive element in the measurement and control of temperature.

Extrinsic semiconductors

In an extrinsic semiconductor the current is again carried by electrons and holes, so that

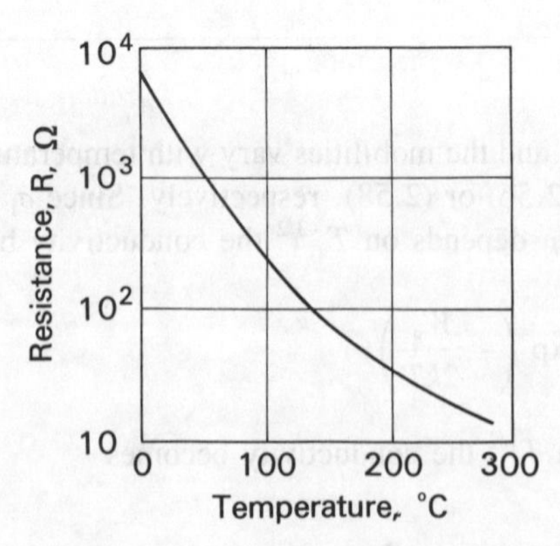

Figure 2.18 Temperature dependence of the resistance of an intrinsic semiconductor (thermistor).

$$J = ne\mu_n E + pe\mu_p E \tag{2.80}$$

and

$$\sigma = e(n\mu_n + p\mu_p) \tag{2.81}$$

where n and p are no longer equal. The appropriate expressions for n and p are given in eqs (2.45) and (2.46), so that for an n-type semiconductor eq. (2.81) becomes

$$\sigma_n = e\left(N_d\mu_n + \frac{n_i^2}{N_d}\mu_p\right) \tag{2.82}$$

$$\approx eN_d\mu_n$$

at room temperature; and for a p-type semiconductor

$$\sigma_p = e\left(\frac{n_i^2}{N_a}\mu_n + N_a\mu_p\right) \tag{2.83}$$

$$\approx eN_a\mu_n$$

at room temperature. These expressions are true down to the temperature above which nearly all the impurity atoms are ionized (Fig. 2.13). The corresponding expressions for resistivity are

$$\rho_n = \frac{1}{eN_d\mu_n} \tag{2.84}$$

for n-type semiconductors, and

$$\rho_p = \frac{1}{eN_a\mu_p} \tag{2.85}$$

for p-type semiconductors. Since for moderate doping μ decreases as temperature rises according to eq. (2.57) or (2.58), the resistivity increases with temperature until the transition temperature is reached. At this point the semiconductor reverts to the intrinsic type and the resistivity falls again (Fig. 2.19), so that the transition temperature is related to the temperature of maximum resistivity. Figure 2.19 also shows the transition temperature increasing with the density of impurities and hence of current carriers.

The Hall effect and measurement of semiconductor properties

In 1879 Hall showed that a current flowing in a conductor would be deflected by a magnetic field, in a similar way to the deflection of a beam of electrons moving in the vacuum in a cathode-ray tube. The effect is very small in metals, but may be easily observed in semiconductors and forms the basis of measuring

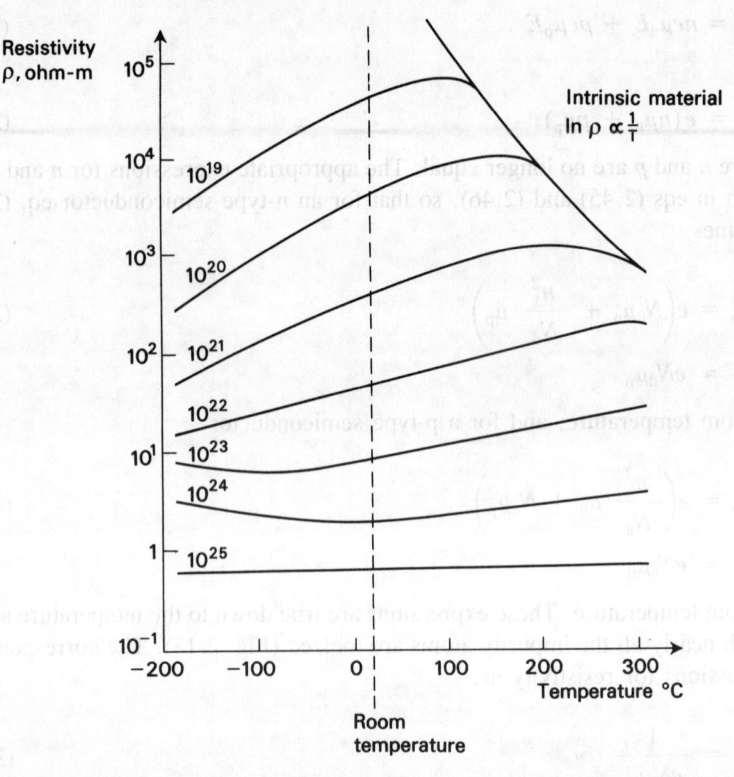

Figure 2.19 Temperature dependence of the resistivity of n-type silicon. Donor density is in atoms/m^3.

the current carrier density and mobility. In addition the nature of the carriers, whether electrons or holes, may be established.

Consider a conducting plate whose length l is at least twice its breadth b to ensure uniform current flow across the plate (Fig. 2.20). It has thickness t and carries a steady current I_x, and a uniform magnetic field B_y normal to its plane will deflect the current carriers according to the left-hand rule. This causes either type of carrier to crowd against the same side of the strip, (Fig. 2.21(*a*) and (*b*), setting up an electric field E_H which will oppose further deflection of the carriers. E_H is called the *Hall field* and its establishment is the *Hall effect*, while as a result of the crowding the resistance of the conductor increases in the magnetic field.

Where the charge on each carrier is q, which will be $-e$ for an electron and $+e$ for a hole, we have at equilibrium

$$qE_H = B_y q u_x \tag{2.86}$$

or

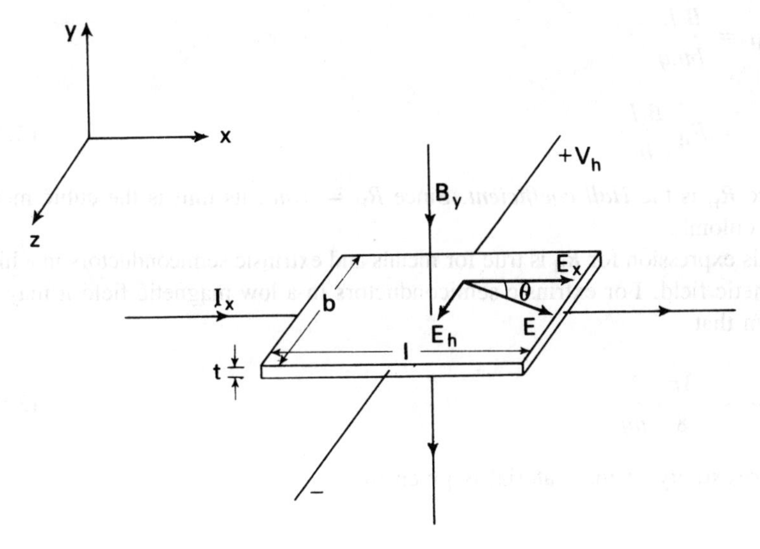

Figure 2.20 Hall plate and positive axes.

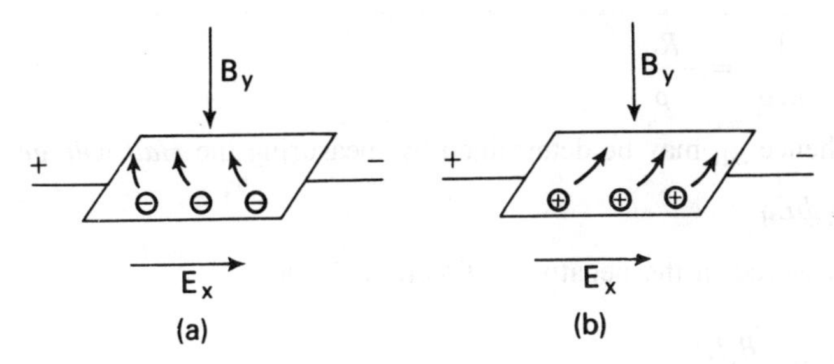

Figure 2.21 The Hall effect (*a*) deflection of electrons; (*b*) deflection of holes.

$$E_H = B_y u_x \tag{2.87}$$

The current density is given by

$$\frac{I_x}{bt} = nqu_x$$

where there are n current carriers per cubic metre, so that

$$u_x = \frac{I_x}{btnq} \tag{2.88}$$

Hence

$$E_H = \frac{B_y I_x}{btnq}$$

$$= R_H \frac{B_y I_x}{bt} \qquad (2.89)$$

where R_H is the *Hall coefficient*. Since $R_H = 1/nq$, its unit is the cubic metre per coulomb.

This expression for R_H is true for metals and extrinsic semiconductors in a high magnetic field. For extrinsic semiconductors in a low magnetic field it may be shown that

$$R_H = \frac{3\pi}{8} \frac{1}{nq} \qquad (2.90)$$

The resistivity of the material is given by

$$\rho = \frac{1}{ne\mu} \qquad (2.91)$$

and the *Hall mobility* is

$$\mu_H = \frac{1}{ne\rho} = \frac{R_H}{\rho} \qquad (2.92)$$

R_H and hence μ_H may be determined by measuring the *Hall voltage*, given by

$$V_H = bE_H$$

when measured in the negative z-direction. Hence

$$V_H = R_H \frac{B_y I_x}{t} \qquad (2.93)$$

and is directly proportional to the product of B_y and I_x. This property is also useful as the basis of a multiplying device, although the main application is the use of a Hall plate (or probe) to measure magnetic flux density.

The sign of V_H depends on the sign of R_H, which is negative for electrons and positive for holes. Hence the predominating current carrier may be identified and its density and mobility determined, provided that the density of the opposite carrier is negligible, which occurs in most n- or p-type specimens. The above expressions are true for the ratio $l/b \geq 2$.

The total electric field within the material is E, the vector sum of E_H and E_x (Fig. 2.20). The angle θ between E and E_x is called the *Hall angle*:

$$\tan\theta = \frac{E_H}{E_x} = \frac{R_H B_y (I_x/bt)}{(I_x/bt)\rho}$$

$$= \frac{R_H B_y}{\rho} = \mu_H B_y \tag{2.94}$$

Thus the angle of deflection increases with the mobility of the charge carriers and the strength of the magnetic field. Since R_H is inversely proportional to n, it will be small for metals and large for semiconductors, whose current carriers thus experience a large deflection. They also produce a Hall voltage up to about 0.5 V as compared with about 1 μV for metals.

Recombination and lifetime

We have seen that in a semiconductor at constant temperature there are fixed densities of electrons and holes, given by the expression $np = n_i^2$. This is in fact due to a dynamic equilibrium between the generation of carriers due to excitation and the *recombination* of holes and electrons, which may be expressed in terms of the equation.

$$\frac{dn_i}{dt} = g - r \tag{2.95}$$

Here g is the rate of generation and r the rate of recombination of electron-hole pairs, and at equilibrium $g = r$, so that $dn_i/dt = 0$, or n_i is constant.

Recombination through direct transition of a conduction electron to the valence band is unlikely in germanium or silicon, as explained in Chapter 5. Not only must the electron lose energy W_g but also its momentum must be dissipated, which requires the simultaneous creation of many phonons. Instead, the electron returns to the valence band through one or more intermediate energy levels (Fig. 2.22). These are due to sites in the crystal lattice where there is a local discontinuity in the periodic potential caused by missing atoms, impurity atoms or other crystal defects. In an n-type semiconductor such a defect can lead to a localized level below the bottom of the conduction band which is normally empty and acts as an *electron trap*, with a high probability of capturing a free electron. A similar defect in a p-type semiconductor can lead to a localized level above the top of the valence band which is normally filled by an electron. When a free hole approaches the defect it will be filled by this electron, which thus returns to the valence band, so that the defect acts as a *hole trap* with a high probability of capturing a free hole.

Where the defect leads to a localized level nearer the middle of the energy gap it acts as a *recombination centre*, which has a high probability of capturing both an electron and a hole. If the centre is occupied by an electron it will attract a hole and recombination will occur, with the electron returning to the valence band. An electron-hole pair is removed and the centre is empty and ready to receive another electron. before entering a recombination centre an electron or a hole may enter or leave several traps, energy being released as by radiation on entering and being absorbed from the surroundings on leaving the trap. The average

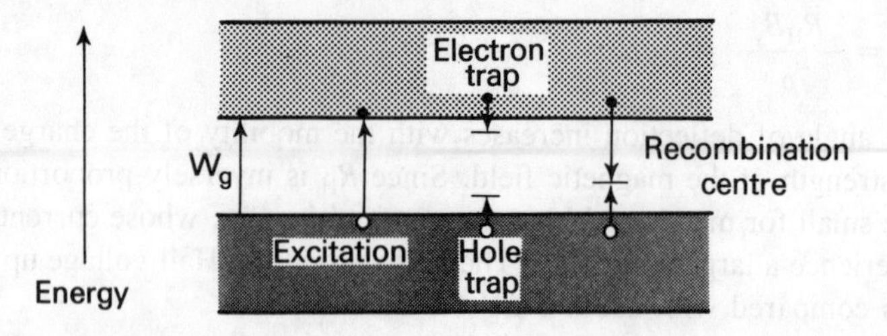

Figure 2.22 Excitation, trapping and recombination in a semiconductor.

time an electron or hole can exist in the free state is known as the *lifetime*, τ, and may be different for the two types of carrier. Thus electron and hole traps have a relatively long capture lifetime and a recombination centre has a short capture lifetime.

Excess current carriers in semiconductors

The equilibrium density of current carriers may be increased, at constant temperature, in two main ways, resulting in either a uniform or a non-uniform distribution of carriers in the crystal. These are in addition to the carriers already present and so are known as *excess* carriers.

A uniform distribution may be obtained by illuminating the crystal with light of a suitable wavelength. In this case the excess carriers are due to excitation of electrons from the valence band to the conduction band, so that energy W_g is required. If the light has a frequency f, excitation will occur when $hf \geq W_g$ (Fig. 2.22), so that each photon of energy hf will generate one electron-hole pair and the rate of generation will be directly proportional to the number of photons being absorbed per second or the intensity of the light. A new dynamic equilibrium is set up with the total generation and recombination rates being equal, the number of electron-hole pairs generated optically being added directly to the number generated thermally. The density of the excess conduction electrons Δn per cubic metre will be equal to the density of the excess holes, Δp, and will cause the conductivity to increase to a value $\sigma + \Delta\sigma$, where

$$\Delta\sigma = e\ \Delta n(\mu_n + \mu_p) \qquad (2.96)$$

This increase in σ is known as *conductivity modulation*, and the excess conductivity $\Delta\sigma$ is proportional to the excess density of holes and electrons, which in this case depends on the light intensity. This is the principle of a *photoconductive cell*, the current through it being proportional to the intensity of light falling on it (*see* Chapter 5).

If the illumination is removed, the dynamic equilibrium is disturbed, and electron-hole pairs recombine until the equilibrium due to thermal generation is restored. The rate of recombination, $-\mathrm{d}\,\Delta n/\mathrm{d}t$, is proportional to the excess carrier density Δn and is given by

$$\frac{d\,\Delta n}{dt} = -\frac{1}{\tau}\,\Delta n \qquad (2.97)$$

which defines the lifetime of the excess carriers. At time t after the illumination is removed, the excess density, $\Delta n(t)$, is obtained by solving eq. (2.96), which gives

$$\Delta n(t) = \Delta n(0)\exp\left(-\frac{t}{\tau}\right) \qquad (2.98)$$

Here $\Delta n(0)$ is the excess density just before the illumination is removed, so that τ is the time taken for the excess density to fall to 37% of this initial value (Fig. 2.23). It may be shown that τ is also the average time any electron stays in the conduction band, whether excited there thermally, optically or owing to donor impurities. Similar considerations also apply to the lifetimes of holes in the valence band.

In impurity semiconductors, illumination produces a large change in the density of the minority carriers, i.e. holes in n-type material and electrons in p-type, while hardly affecting the much larger numbers of majority carriers. Thus τ_n is the lifetime of electrons in p-type material and τ_p the lifetime of holes in n-type material. Practical values of τ are a few hundred microseconds and depend on the probability of recombination. This is increased by imperfections in the crystal structure such as grain boundaries. The surface of the material also has many recombination sites and has to be chemically cleaned and kept extremely dry to ensure that as little recombination occurs as possible.

Measurement of lifetime

If a filament of extrinsic semiconductor is irradiated by a pulsed light source the lifetime of the excess carriers can be obtained by observing the growth and decay of current as the source is switched on and off. The principle of the method is

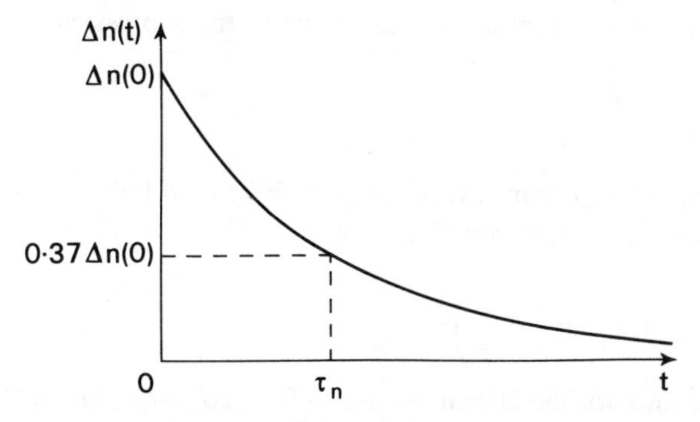

Figure 2.23 Time decay of excess electron concentration due to recombination.

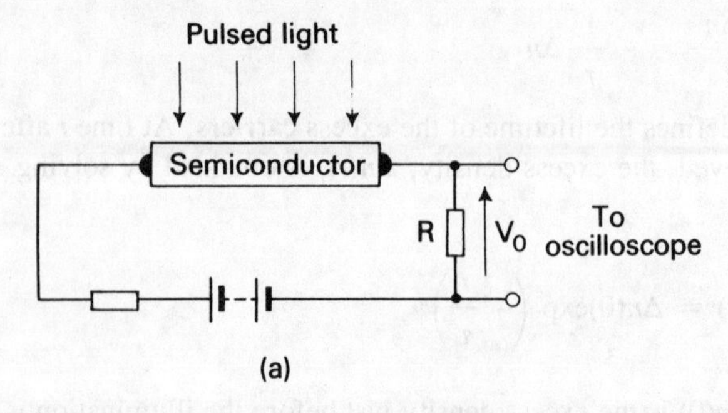

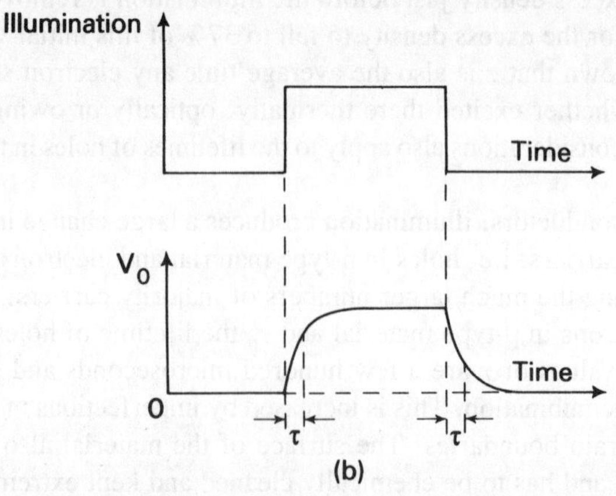

Figure 2.24 **Principle of lifetime measurement.**

illustrated in Fig. 2.24(*a*), where the light source may be pulsed by means of a fast-acting shutter. A current is passed through ohmic contacts along the filament, and this current is increased by the excess carriers to give a voltage waveform across the resistance R as shown at (*b*). Since the excess drift current ΔI at any instant is proportional to the excess carrier density, when the light source is switched off the current decays according to the expression

$$\Delta I = \Delta I_{\mathrm{m}} \exp\left(-\frac{t}{\tau}\right) \tag{2.99}$$

where ΔI_{m} is the maximum excess current. When the light source is switched on the current rises exponentially, so that

$$\Delta I = \Delta I_{\mathrm{m}} \left[1 - \exp\left(-\frac{t}{\tau}\right)\right] \tag{2.100}$$

τ is then the time for the current to rise to 0.63 ΔI_{m} or to fall to 0.37 ΔI_{m} and can be measured from the waveform by means of an oscilloscope.

The measured lifetime is due, not only to recombination in the bulk of the filament, but also to recombination at the surface. Since the surface contains many more imperfections than the bulk its recombination rate is higher and it is normally the controlling factor. If it is assumed that the rates of recombination are additive, then

$$\frac{1}{\tau_{\text{measured}}} = \frac{1}{\tau_{\text{bulk}}} + \frac{1}{\tau_{\text{surface}}} \tag{2.101}$$

where $\tau_{\text{bulk}} \gg \tau_{\text{surface}}$. The lifetime at the surface is increased by etching, which has the effect of making it smoother and hence removing recombination sites. Reference [4] has further details of the measurement of lifetime and other semiconductor parameters.

Diffusion of minority carriers

If excess carriers are injected at one surface of a crystal, for instance at the contact between a semiconductor and a metal, or between two semiconductors, the carrier density near the contact will be higher than elsewhere in the crystal. This non-uniform distribution is similar to that occurring in a gas and will cause diffusion of carriers away from the high-density region. Since they are charged they repel one another and a field is set up within the conductor also forcing the carriers away from the injecting contact. The number passing through unit area in one second is given by eq. (2.14), so that for excess electrons this becomes

$$- D_{\text{n}} \frac{\text{d}}{\text{d}x} \Delta n \text{ electrons/m}^2 \text{ s}$$

where D_{n} is the *electron diffusion coefficient*, and for excess holes the corresponding number is

$$- D_{\text{p}} \frac{\text{d}}{\text{d}x} \Delta_p \text{ holes/m}^2 \text{ s}$$

where D_{p} is the *hole diffusion coefficient*. The motion of the charged particles constitutes a *diffusion current*, which for electrons with charge $-e$ has a density of

$$J_{\text{n}} = eD_{\text{n}} \frac{\text{d}}{\text{d}x} \Delta n \quad \text{A/m}^2 \tag{2.102}$$

and for holes with charge $+e$ has a density of

$$J_{\text{p}} = -eD_{\text{p}} \frac{\text{d}}{\text{d}x} \Delta p \quad \text{A/m}^2 \tag{2.103}$$

It should be noted that no external field is required to cause a diffusion current to flow since there is an internal field set up due to the individual charges. This has the same effect on the current carriers as the pressure difference in a gas between regions of high and low density has on the gas molecules. For a gas in the steady state, eq. (2.5) gives

$$P = nkT \tag{2.104}$$

so that when a density gradient is set up the corresponding pressure gradient providing a force to expand the gas is

$$\frac{\mathrm{d}P}{\mathrm{d}x} = kT\frac{\mathrm{d}n}{\mathrm{d}x} \tag{2.105}$$

For a semiconductor with a non-uniform excess of electrons, at a certain distance from the injecting contact the excess density will be Δn per cubic metre. If at the same point the internal electric field is E the force on one electron is eE and the force on Δn electrons is

$$\Delta neE = \frac{\Delta neu}{\mu_n} = \frac{J_n}{\mu_n} \tag{2.106}$$

Here u is the corresponding velocity and Δneu is the current density J_n (eq. 2.102), so that we can write

$$\Delta neE = \frac{eD_n}{\mu_n}\frac{\mathrm{d}}{\mathrm{d}x}\Delta n \tag{2.107}$$

This force corresponds to the pressure gradient in a gas, so that, comparing eqs (2.105) and (2.107), we have

$$kT = \frac{eD_n}{\mu_n}$$

or

$$\frac{D_n}{\mu_n} = \frac{kT}{e} \tag{2.108}$$

Similarly for a semiconductor with a non-uniform excess of holes,

$$\frac{D_p}{\mu_p} = \frac{kT}{e} \tag{2.109}$$

These are expressions of *Einstein's law* (1905) for the diffusion of charged particles. At room temperature, $kT/e = 0.025$ V, so that, using $\mu_n = 0.15\ \mathrm{m^2/Vs}$ and $\mu_p = 0.05\ \mathrm{m^2/Vs}$, we obtain $D_n = 3.75 \times 10^{-3}\ \mathrm{m^2/s}$ and $D_p = 1.25 \times 10^{-3}\ \mathrm{m^2/s}$.

Diffusion length

Consider a long bar of p-type semiconductor with an excess concentration $\Delta n(0)$ of electrons maintained at one end. As the electrons diffuse away from the end they will recombine with the holes, but since they are continuously replaced the concentration gradient is maintained. Hence in the steady state there is a continuous flow of electrons through the bar, the rates of change due to diffusion and recombination being equal at any point. They may be obtained for an element of thickness δx and distance x from the injecting contact (Fig. 2.25(a)), with a cross-sectional area S and volume $S\delta x$. The rate at which electrons enter the left-hand side of the element is then

$$- D_{\mathrm{n}} \frac{\partial}{\partial x} \Delta n S \tag{2.110}$$

and the rate at which they leave the right-hand side depends also on the rate of change of density gradient across the element and becomes

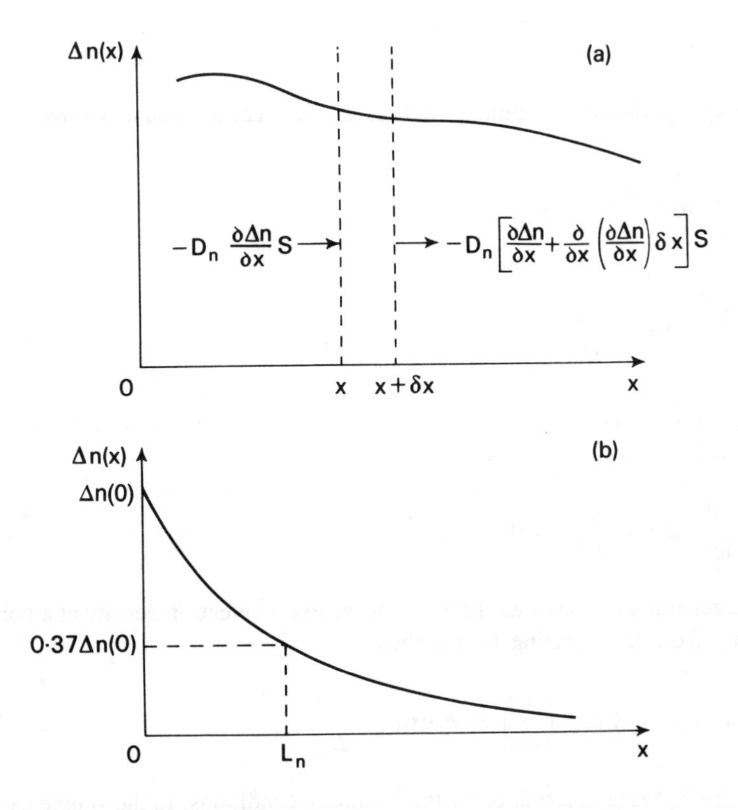

Figure 2.25 **Diffusion and recombination. (*a*) Excess electron flow through an element; (*b*) fall of excess electron concentration.**

$$- D_{\mathrm{n}} \left[\frac{\partial}{\partial x} \Delta n + \frac{\partial}{\partial x} \left(\frac{\partial}{\partial x} \Delta n \right) \delta x \right] S$$

Hence the net rate of change of the number of electrons is the difference between the rates of entering and leaving it, or the rate at which electrons are *gained* within the element is

$$D_{\mathrm{n}} \frac{\partial^2}{\partial x^2} \Delta n \, \delta x \, S \tag{2.111}$$

The rate of change of electron density due to recombination is given by eq. (2.97), so that

$$\frac{\partial}{\partial t} \Delta n = - \frac{\Delta n}{\tau_{\mathrm{n}}}$$

The rate at which electrons are *lost* within the element is then

$$\frac{\Delta n}{\tau_{\mathrm{n}}} S \, \delta x \tag{2.112}$$

Under equilibrium conditions (2.111) and (2.112) are equal, so that

$$D_{\mathrm{n}} \frac{\partial^2}{\partial x^2} \Delta n = \frac{\Delta n}{\tau_{\mathrm{n}}} \tag{2.113}$$

or

$$\frac{\mathrm{d}^2}{\mathrm{d}x^2} \Delta n - \frac{\Delta n}{D_{\mathrm{n}} \tau_{\mathrm{n}}} = 0 \tag{2.114}$$

We can write $D_{\mathrm{n}} \tau_{\mathrm{n}} = L_{\mathrm{n}}^2$, where L_{n} is the *diffusion length of electrons* in p-type material, which gives

$$\frac{\partial^2}{\partial x^2} \Delta n - \frac{\Delta n}{L_{\mathrm{n}}^2} = 0 \tag{2.115}$$

The general solution of eq. (2.115) for the excess electron density at a point distant x from the injecting face is then

$$\Delta n(x) = A_{\mathrm{n}} \exp \left(\frac{x}{L_{\mathrm{n}}} \right) + B_{\mathrm{n}} \exp \left(- \frac{x}{L_{\mathrm{n}}} \right) \tag{2.116}$$

A_{n} and B_{n} being determined by the boundary conditions. In the simple case considered at $x = 0$, $\Delta n(x) = \Delta n(0)$, and at $x = \infty$, $\Delta n(x) = 0$. Hence $A_{\mathrm{n}} = 0$, $B_{\mathrm{n}} = \Delta n(0)$, so that

$$\Delta n(x) = \Delta n(0)\exp\left(-\frac{x}{L_n}\right) \tag{2.117}$$

Thus $\Delta n(x)$ decreases exponentially with distance (Fig. 2.25(*b*)), and L_n is the distance from the injecting face at which the density of excess electrons has fallen to 37% of its initial value. Similar considerations apply to the injection of excess holes into an n-type semiconductor, so that their density as a function of x is

$$\Delta p(x) = A_p \exp\left(\frac{x}{L_p}\right) + B_p \exp\left(-\frac{x}{L_p}\right) \tag{2.118}$$

where L_p is the *diffusion length of holes*, and in the simple case as considered above,

$$\Delta p(x) = \Delta_p(0) \exp\left(-\frac{x}{L_p}\right) \tag{2.119}$$

If we suppose that $\tau_n \approx \tau_p = 10^{-5}$ s, then

$$L_n = \sqrt{(3.8 \times 10^{-3} \times 10^{-5})} = 0.19 \text{ mm}$$

and

$$L_p = \sqrt{(1.25 \times 10^{-3} \times 10^{-5})} = 0.11 \text{ mm}$$

and using these values in eqs (2.117) and (2.119) respectively, the corresponding carrier densities at any point in the bar can be obtained. Thus the diffusion length is a parameter which determines the length of any region in a device through which excess carriers must pass; it is discussed further for the p-n junction in Chapter 3.

It may be noted that, when excess carriers are introduced uniformly into a material, the balance of charge is preserved since carriers of both signs are produced. It might appear that the injection of carriers of one sign would upset this balance. However, in a practical device this does not occur since an electric field is set up which attracts carriers of opposite sign from the surroundings, so that the space charge of the excess carriers is rapidly neutralized. Thus injection of excess holes is always accompanied by a flow of electrons, and vice versa. This is again discussed further in connection with the p-n junction.

Points to remember

* The many overlapping discrete energy levels of electrons in crystals lead to the formation of energy bands.
* Valence electrons move through a crystal close to the atoms in the valence band of energy and the concept of a hole is used to describe their motion.

* Conduction electrons move further away from the atoms in the conduction band, which is higher than the valence band.
* In semiconductors acceptor impurities are used to increase the number of holes, and donor impurities to increase the number of electrons.
* Important semiconductors are silicon, gallium arsenide and germanium.
* The number of electrons and holes in a metal or a semiconductor is determined by the Fermi–Dirac statistics and the position of the Fermi energy level.
* Resistivity is defined by the mobility of electrons and holes as well as by their numbers, with temperature being an important factor.
* Examples of electronic devices based on single semiconductors are the thermistor, the strain gauge and the Hall probe.
* When excess charge carriers are introduced into a semiconductor their motion is due to diffusion processes and their number is affected by the recombination of electrons and holes.

References

[1] Park, D. *Introduction to Quantum Theory*, (McGraw-Hill), 1974.
[2] Smith, R.A. *Semiconductors*, (Cambridge University Press), 1979.
[3] Ehrenreich, H. 'Band structure and electron transport in GaAs', *Phys Rev.*, 1960, vol. 120, p. 1951.
[4] Gise, P. and Blanchard, R. *Modern Semiconductor Fabrication Technology*, (Prentice Hall), 1986.

Problems

2.1 If ΔW is a change in energy measured from the Fermi level show that

$$p_F(W_F + \Delta W) = 1 - p_F(W_F - \Delta W)$$

2.2 A pure semiconductor has an energy gap of 1.0 eV. For temperatures of 0 K and 293 K respectively, calculate the probability of an electron occupying a state near the bottom of the conduction band. State, with reasons, whether the probabilities at each of these two temperatures will be increased if the semiconductor receives radiation of wavelength (i) 1.0 or (ii) 2.0 μm.
[0, 2.55×10^{-9}; (i) Both probabilities increased since $\lambda < hc/W_g$; (ii) Neither probability increased since $\lambda > hc/W_g$]

2.3 Explain the significance of eq. (2.26) in both intrinsic and extrinsic semiconductors. If $N_c = N_v = 10^{25}$ levels/m^3 and $W_g = 1.0$ eV, calculate the density of electron-hole pairs in an intrinsic specimen at 150 °C.
[1.10×10^{19}/m^3]

2.4 Sketch a graph showing how the number density of current carriers varies with temperature in an n-type semiconductor containing about 10^{22} donor atoms per m^3.

Describe briefly the main processes responsible for this variation over a temperature range from below 50 K to above 500 K.

For an intrinsic semiconductor with energy gap 1.1 eV, estimate the temperature at which the number of current carriers becomes the same as in the n-type material above. Discuss the practical significance of this result. You may assume that the density of electron-hole pairs per m^3 is given by

$$n_i = 3 \times 10^{26} \exp\left(-\frac{eW_g}{2kT}\right)$$

[619 K]

2.5 At room temperature the conductivity of a crystal of pure silicon is 5×10^{-4} S/m. If the electron mobility is 0.14 m^2/V s and the hole mobility is 0.05 m^2/V s, determine the density of electron-hole pairs in the crystal.

If doping with donor atoms to give an impurity density of $10^{22}/m^3$ is carried out, calculate the new conductivity and the fraction of this conductivity due to holes at room temperature. Assume that all the donor atoms are ionized and that the electron and hole-mobilities are unchanged.

[$1.64 \times 10^{16}/m^3$; 224 S/m; 9.65×10^{-13}]

2.6 Describe the variation of drift velocity u_d with electric field E in a semiconductor and derive a relationship between u_d and E for low values of E.

An intrinsic semiconductor specimen at room temperature has resistance 8 MΩ which falls to 15 Ω when p-type impurities are introduced. Estimate the density of both holes and electrons in the p-type material, given that $n_i = 5 \times 10^{16}$ electron-hole pairs per m^3 and that the relaxation time t_c is 1.5×10^{-12} s for electrons and 5.5×10^{-13} s for holes.

[1.0×10^{23} holes/m^3; 2.5×10^{10} electrons/m^3]

2.7 Explain briefly why the temperature coefficient of the resistivity of a doped semiconductor is negative at low and high temperatures but positive at intermediate temperatures.

The resistance at room temperature of an intrinsic semiconductor specimen is 5.5 MΩ but after doping with acceptor atoms the resistance falls to 125 Ω. If there are 10^{22} acceptor atoms/m^3 and the electron mobility is 2.8 times the hole mobility, estimate the density of electron-hole pairs in the intrinsic specimen.

[$6 \times 10^{16}/m^3$]

2.8 A Hall probe having thickness 0.2 mm contains 4×10^{22} current carriers per m^3. Determine the probe current if the output is to be 0.02 V per tesla when the magnetic field is normal to the probe.

[25.6 mA]

2.9 A flat probe of width 4 mm and cross-section 0.2 mm^2 consists of n-type material containing 3×10^{22} electrons/m^3. If the probe is operated with a current of 50 mA and the Hall voltage is 100 mV determine the magnetic flux density.

[0.48 T]

2.10 A semiconductor resistor carries a current which is increased by ΔI due to illumination, with the p.d. across the resistor remaining constant. A uniform excess of current carriers Δn per m^3 is introduced by the illumination, and the rate of recombination of electrons and holes is proportional to Δn at any instant. Obtain a relationship between ΔI and Δn and derive an expression for ΔI at a time t after the light source has been switched off.

When the light source is switched on again after a long time, the initial rate of rise of current is 7 mA/s and the steady-state value of ΔI is 5 μA. Determine the minority carrier lifetime.

[0.7 ms]

2.11 Discuss briefly the physical processes which occur when excess charge carriers are introduced into a semiconductor at one surface of the crystal.

An electron current of 1 mA is injected into a p-type semiconductor of cross section 0.1 mm^2 and it may be assumed that the excess carrier concentration decays exponentially with distance measured from the injection contact. The diffusion coefficient is $3.4 \times 10^{-3}\ \text{m}^2/\text{s}$ and 7 mm from the contact the excess electron concentration is 50% of that at the contact. Determine the value of the excess electron concentration at the contact.

[$1.86 \times 10^{23}/\text{m}^3$]

3 Contacts between materials and p-n junctions

Contact between two materials

When two crystals of different material first make contact, there will be a flow of electrons from one to the other. This is because the electrons meeting the junction from one side will generally have more energy than those meeting it from the other side.

Consider two metals, A and B, with different Fermi levels W_{FA} and W_{FB} (Fig. 3.1(*a*)). Before they come into contact the energies at the surfaces of the metals will be equal and may be taken as zero. It is convenient to consider the energy difference between the Fermi level and the surface, which is known as the *work function*, ϕ. With metals this is also the depth of the conduction band. If $\phi_A < \phi_B$, filled states in the valence band of A will be at the same energy as empty states in the conduction band of B, so that, when the metals come into contact, electrons can flow from A to B. This results in the surface of A becoming positively charged, owing to the ionized atoms, and the surface of B becoming negatively charged (Fig. 3.1(*b*)). Flow proceeds until the two Fermi levels coincide (Fig. 3.1(*c*)), which implies that an electron has an equal probability of moving from A to B or from B to A. There is then a difference between the surface energy levels of $\phi_B - \phi_A$ and also a potential difference V_{AB} due to the surface charges (Fig. 3.1(*d*)).

If the ends of A and B not in contact are brought face to face without touching, work would have to be done in transferring an electron from A to B owing to the energy difference or *energy barrier* $\phi_B - \phi_A$, measured in electronvolts (Fig. 3.1(*e*)). This energy barrier must also equal V_{AB} electronvolts, or in terms of potential difference,

$$V_{AB} = \phi_B - \phi_A \text{ volts} \tag{3.1}$$

V_{AB} is known as the *contact potential* between A and B and is given by the numerical difference between the two work functions.

If the circuit is completed by joining the free ends of A and B, the contact potentials at the two junctions will be equal and opposite so that no current will flow

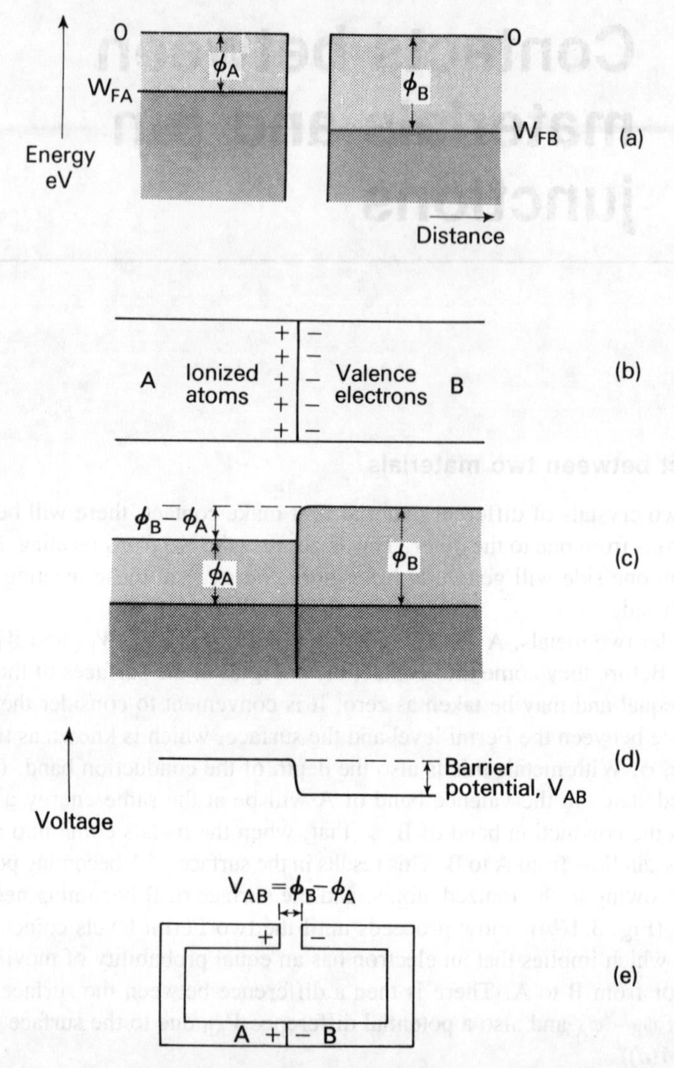

Figure 3.1 Contact between two metals. (*a*) Energy diagram before contact; (*b*) surface charges after contact; (*c*) energy diagram after contact; (*d*) barrier potential; (*e*) contact potential.

after the initial connection. However, if a battery is then inserted between A and B, there will be a flow of electrons in the two conduction bands, which have become continuous. The Fermi levels will still coincide at the junction of A and B and the contact potential will be unaffected by the applied battery voltage. This is because the potential difference occurs in a very short distance on each side of the contact corresponding to the interatomic spacing, about 10^{-10} m, and the contact has a low resistance.

Other contacts between a metal and a semiconductor or between two semiconductors can be considered by applying the general principle of the alignment of Fermi levels. The next section on metal-to-semiconductor contacts may be omitted at a first reading, and a jump made directly to the p-n junction on page 85. This is because the p-n junction appears as a diode in the majority of semiconductor devices.

Metal-to-semiconductor contacts

Consider a metal with work function ϕ_m and an extrinsic semiconductor with work function ϕ_s. There are four possibilities, depending on whether the semiconductor is n- or p-type and whether ϕ_m is less or greater than ϕ_s. In general, after the initial contact has been made, electrons flow from the material with the *smaller* work function.

n-type semiconductor, $\phi_m < \phi_s$

Initially electrons flow from the metal to the semiconductor until the Fermi levels coincide and, as before, surface charges appear on each side of the junction (Fig. 3.2). Since the energy levels in the two conduction bands overlap, electrons can flow easily in either direction when $\phi_s - \phi_m$ is small, as often occurs. This type of contact between a metal and a semiconductor is called an *ohmic contact* and the contact potential $\phi_s - \phi_m$ is again unaffected by an applied voltage. The height χ of the conduction band of the semiconductor is also known as the *electron affinity*.

n-type semiconductor, $\phi_m > \phi_s$

In this case the initial flow of electrons is from the semiconductor to the metal. When the Fermi levels coincide, the metal has acquired a negative surface charge, but the positive charge on the semiconductor is due to uncompensated donor atoms

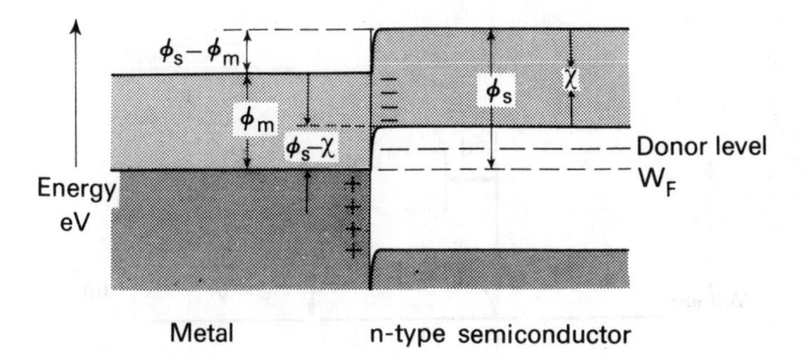

Figure 3.2 Energy diagram for contact between a metal and an n-type semiconductor, $\phi_m < \phi_s$.

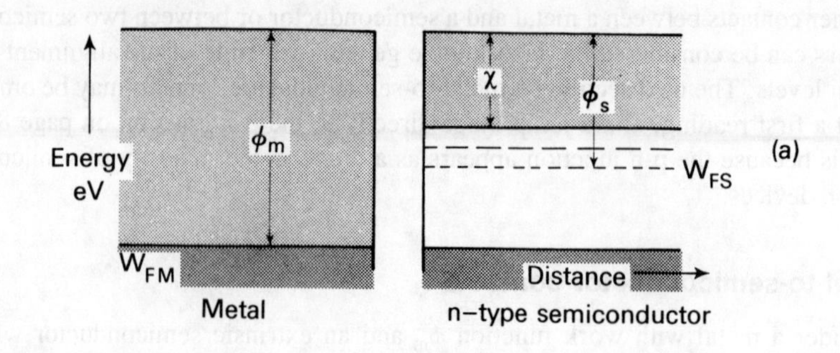
Energy
eV
ϕ_m
χ
ϕ_s
(a)
W_{FS}
W_{FM}
Distance
Metal
n–type semiconductor

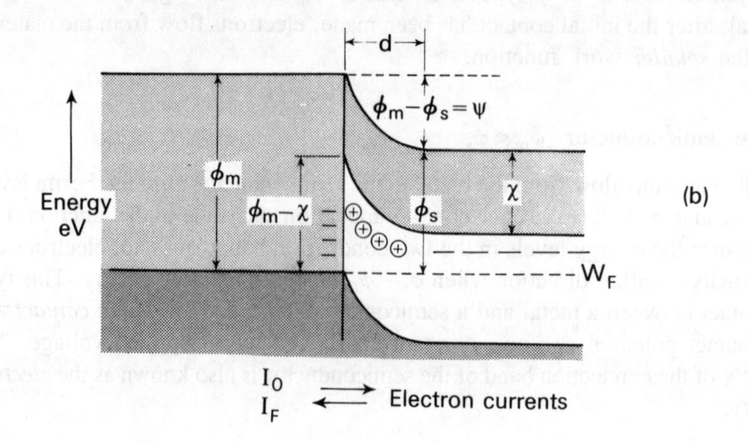
d
$\phi_m - \phi_s = \psi$
Energy
eV
ϕ_m
$\phi_m - \chi$
ϕ_s
χ
(b)
W_F
I_O
I_F
Electron currents

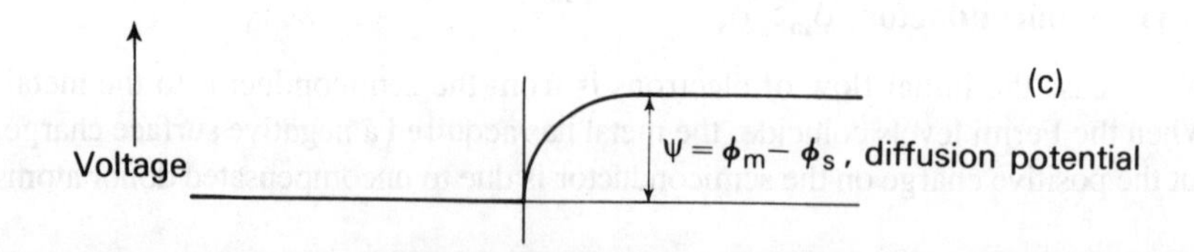
(c)
Voltage
$\psi = \phi_m - \phi_s$, diffusion potential

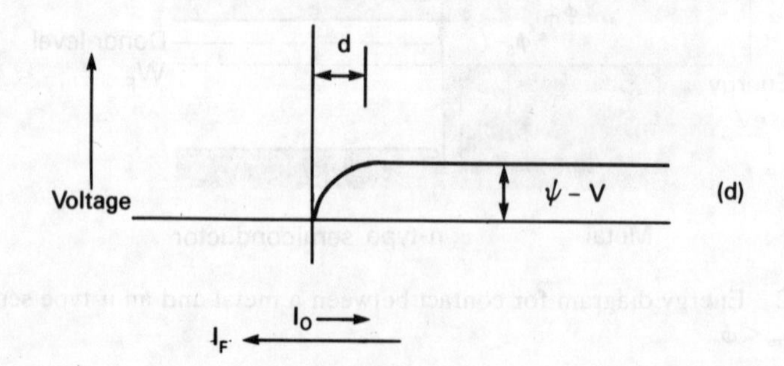
d
Voltage
$\psi - V$
(d)
I_O
I_F

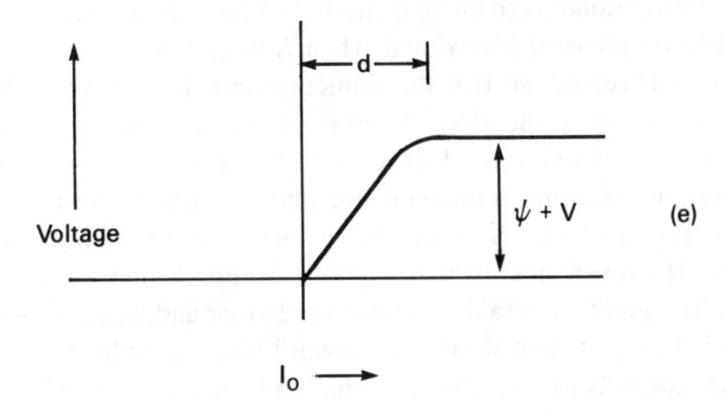

Figure 3.3 Contact between a metal and an n-type semiconductor, $\phi_m > \phi_s$. (*a*) Energy diagram before contact; (*b*) energy diagram after contact; (*c*) diffusion potential for zero bias; (*d*) potential diagram for forward bias; (*e*) potential diagram for reverse bias.

and so is distributed for a distance *d* from the junction (Fig. 3.3(*b*)). After contact, electrons can diffuse from semiconductor to metal if they have sufficient thermal energy to overcome the contact potential barrier $\phi_m - \phi_s$. This would result in an increase in the height of the barrier, but there is also diffusion of electrons from the metal over the somewhat larger barrier $\phi_m - \chi$. At equilibrium these two currents are equal and opposite, so that the contact potential remains constant and $\phi_m - \phi_s$ is called the *diffusion potential*, ψ (Fig. 3.3(*c*)). The region of width *d* is called the *depletion layer*, since there are virtually no free charges within it, the uncompensated atoms being fixed and the diffusing electrons passing through quickly. The resistance of the depletion layer is therefore much greater than that of the bulk of the metal or semiconductor.

The rate of diffusion of electrons across the deplation layer depends on the height of the potential barrier across it. An external bias voltage V applied between metal and semiconductor will appear almost entirely across the relatively high resistance of the depletion layer, so that this voltage adds algebraically to the diffusion potential ψ and current flow depends on both the magnitude and polarity of V. In general, the current has two components, I_F from semiconductor to metal, and I_0 in the reverse direction. The total current I is the difference between them, i.e. $I = I_F - I_0$, and the effect of the bias may be found by considering the voltage in the junction region.

For zero bias, $V = 0$, there is no net current since $I_F = I_0$ and $I = 0$. If the semiconductor is biased negatively with respect to the metal the energy of all the electrons in the semiconductor is raised and so the potential barrier is reduced to $\psi - V$ (Fig. 3.3(*d*)). Thus more electrons can diffuse from the semiconductor and I_F becomes greater than I_0, since the height of the barrier $\phi_m - \chi$ is unchanged. Fewer donors are uncompensated and so d is also reduced. The junction is said to be *forward biased* and when $I_F \gg I_0$, $I = I_F$.

If the bias is reversed, so that the semiconductor is positive with respect to the metal, the energy of the electrons in the semiconductor is lowered and the potential barrier is raised to $\psi + V$ (Fig. 3.3(*e*)). This greatly reduces the diffusion of electrons from the semiconductor and I_F tends to zero. Electrons are drawn away from the junction, so that more donor atoms are uncompensated and d increases. However, the barrer $\phi_m - \chi$ is unaffected so that I_0 is still unchanged. The junction is said to be *reverse biased* and when $I_F \rightarrow 0$, $I = -I_0$. Since there is heavy current flow with forward bias and light flow with reverse bias, the junction acts as a rectifier, so that with $\phi_m > \phi_s$ a *rectifying contact* is formed. Practical applications of this effect occur in the so-called 'hot carrier', or *Schottky-barrier diode*.

These are used in integrated circuits (Chapter 7), where aluminium is in contact with n-type silicon, and in microwave diodes (Chapter 9) where gold or platinum is in contact with silicon or gallium arsenide.

n^+-type semiconductor, $\phi_m > \phi_s$

A special case arises when the n-region is so heavily doped that it becomes degenerate, with the Fermi level within the conduction band as for a metal. The metal-n^+ contact is similar to the contact between two metals and is therefore ohmic even when $\phi_m > \phi_s$. It has great practical importance as the contact used for making connection to integrated circuits, using aluminium with $\phi_m \simeq$ 4.25 eV and n^+-type silicon with $\phi_s \simeq 3.1$ V.

p-type semiconductor, $\phi_m > \phi_s$

Again the initial flow of the electrons is from the semiconductor to the metal. This results in a surface charge of electrons on the metal and a surface charge of holes on the semiconductor (Fig. 3.4). This occurs because holes are the majority carriers in p-type material and so are free to collect at a surface. This is no depletion layer and external bias does not affect the small contact potential, $\phi_m - \phi_s$. There is a free flow of holes (due to electrons moving within the valence band) in either direction so that the contact is ohmic.

In integrated circuits the contact between aluminium and the p-regions is ohmic, since $\phi_s \simeq 4.2$ eV for p-type silicon.

p-type semiconductor, $\phi_m < \phi_s$

Here the initial flow of electrons is from metal to semiconductor, so that a positive charge is formed on the metal. The electrons are captured by acceptor atoms near the junction and a depletion layer of width d is formed (Fig. 3.5(*b*)). The action of the contact is similar to that of the metal-to-n-type contact with $\phi_m > \phi_s$, but

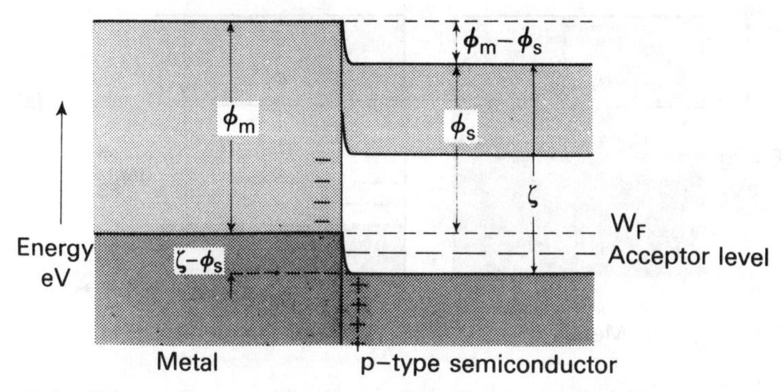

Figure. 3.4 Energy diagram for contact between a metal and a p-type semiconductor, $\phi_m > \phi_s$.

with holes as current carriers. With zero external bias, equal and opposite currents I_F and I_0 are set up, where I_F is due to holes from the semiconductor which possess sufficient thermal energy to cross the barrer $\phi_s - \phi_m$. I_0 is due to holes from the metal crossing the larger barrier $\zeta - \phi_m$, where ζ is the depth of the top of the valence band in the semiconductor.

Thus the contact is rectifying and forward bias occurs with the semiconductor positive, which reduces the potential barrier to $\psi - V$, where $\psi = \phi_s - \phi_m$ (Fig. 3.5(*d*)). This allows I_F to increase, while I_0 remains constant since the somewhat larger barrier $\zeta - \phi_m$ is unaffected by the bias. Reverse bias occurs with the semiconductor negative, which increases the barrier height to $\psi + V$ so that $I_F \to 0$, but I_0 is unaffected (Fig. 3.5(*e*)). It should be noted that, for contacts with both n- and p-type materials, I_0 increases with temperature since it is due to thermally generated carriers. Practical applications of this rectifying contact are the cuprous-oxide and selenium rectifiers. Both cuprous oxide and selenium are p-type semiconductors, and a rectifier is formed by the contact of copper with cuprous oxide, while an alloy of tin, cadmium and bismuth form a metallic contact with selenium.

Current/voltage characteristic of a rectifying contact

The relationship between the current and voltage of a rectifying contact may be obtained by considering the probability that a current carrier will have sufficient energy to cross a barrier of height W. Assuming that Maxwell–Boltzmann statistics are applicable, the number of carriers crossing the barrier per second from the semiconductor is $N(t)$, which is proportional to exp $(-W/kT)$. For a rectifying contact with zero bias $W = e\psi$ and the current flowing in one direction is $I_F = eN(t)$, which is balanced by I_0 flowing in the opposite direction. Hence

$$I_F = \text{const.} \times \exp\left(-\frac{e\psi}{kT}\right) = I_0 \tag{3.2}$$

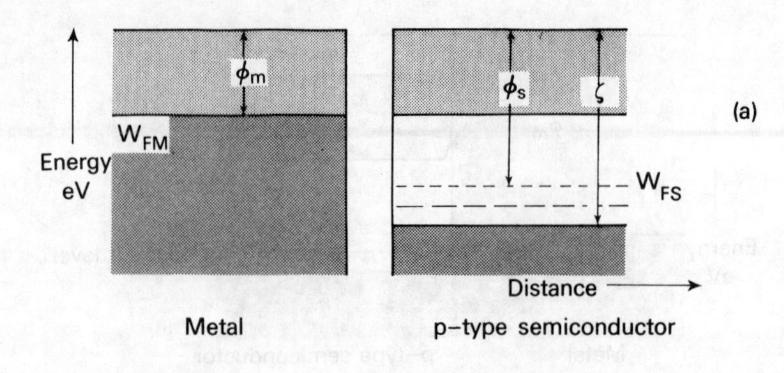

ϕ_m
W_{FM}
Energy
eV
ϕ_s
ζ
(a)
W_{FS}
Distance
Metal
p–type semiconductor

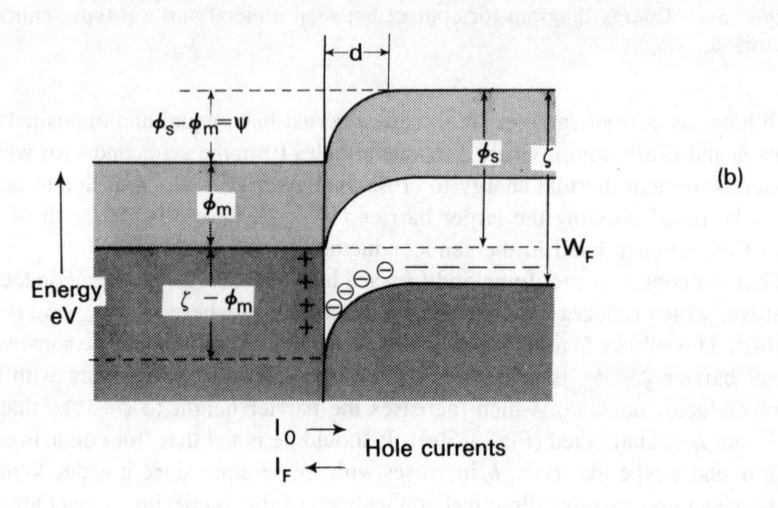

d
$\phi_s - \phi_m = \psi$
ϕ_s
ζ
(b)
ϕ_m
W_F
Energy
eV
$\zeta - \phi_m$
I_0
Hole currents
I_F

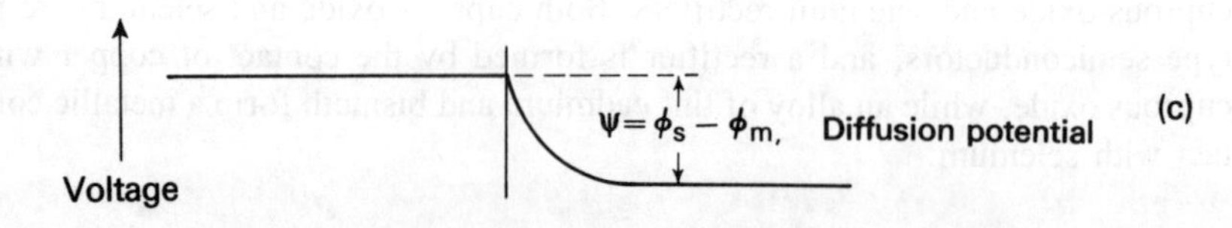

(c)
$\psi = \phi_s - \phi_m,$
Diffusion potential
Voltage

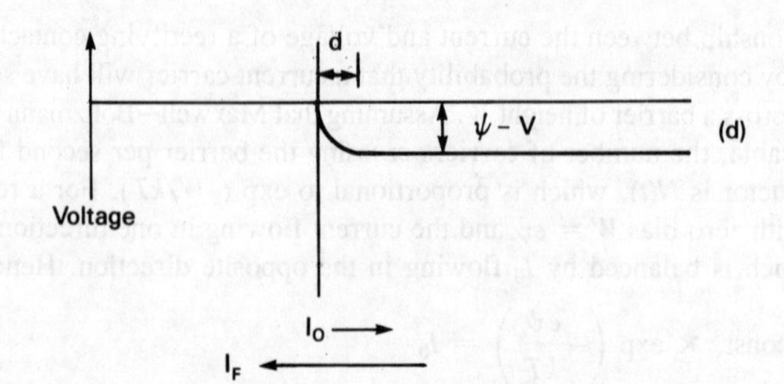

d
$\psi - V$
(d)
Voltage
I_0
I_F

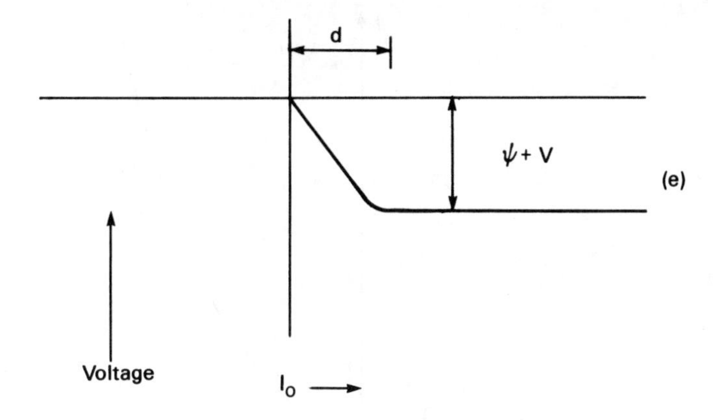

Figure 3.5 Contact between a metal and p-type semiconductor, $\phi_m < \phi_s$. (*a*) Energy diagram before contact; (*b*) energy diagram after contact; (*c*) diffusion potential for zero bias; (*d*) potential diagram for forward bias; (*e*) potential diagram for reverse bias.

With forward bias, V is positive and $W = e(\psi - V)$, so that

$$I_F = \text{const.} \times \exp\left[-\frac{e(\psi - V)}{kT}\right] \tag{3.3}$$

$$= I_0 \exp\frac{eV}{kT} \tag{3.4}$$

The total current is $I = I_F - I_0$, so that

$$I = I_0\left[\exp\left(\frac{eV}{kT}\right) - 1\right] \tag{3.5}$$

for the characteristic of a contact with negligible IR drops on either side. This is illustrated in Fig. 3.6. With reverse bias V is negative and the exponential term becomes much less than unity, so that $I = -I_0$. For the metal-to-n-type-semiconductor contact I_0 is an electron current, while for the metal-to-p-type-semiconductor contact it is due to holes. It will be shown in a later section that for a junction between p- and n-type semiconductors I_0 is due to both electrons and holes.

The reverse current I_0 will be independent of the voltage across an ideal diode for voltages above a small value. At room temperature kT/e is about 25 mV so that if V is −75 V, exp (eV/kT) is 0.05. Hence in eq. (3.5) the exponential term is much less than unity and $I = -I_0$ for values of reverse bias greater than about 75 mV.

Similarly, for forward bias when V is 75 mV, exp (eV/kT) is 20, which may be taken as much greater than unity. Hence, for values of forward bias greater

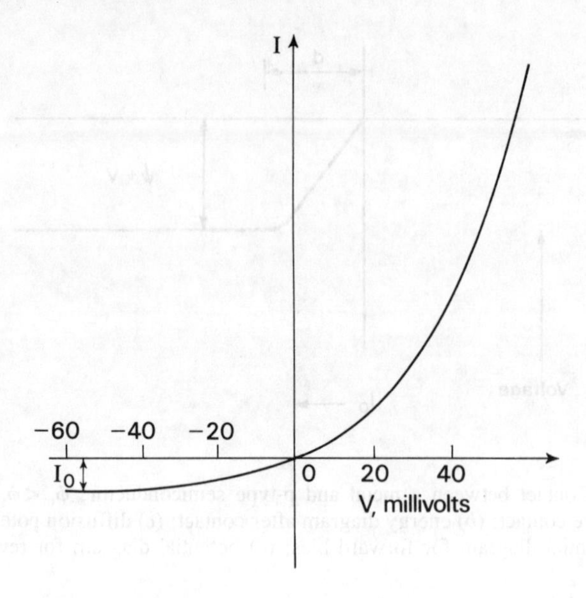

Figure 3.6 Current/voltage characteristic of a rectifying contact. The current scale is determined by the value of I_0. The curve corresponds to $T = 293$ K, giving $e/kT = 40$ V^{-1}.

than about 75 mV, $I = I_0 \exp(eV/kT)$ and current rises exponentially with voltage.

Thermoelectric effects

When electrons flow from one material to another, energy is also transported in the form of heat, the *Peltier effect* (Fig. 3.7(*a*)). It is found that the quantity of heat transferred is proportional to the quantity of electricity flowing. The constant of proportionality is the differential *Peltier coefficient*, $\alpha_{P\,ab}$, given by

$$\alpha_{P\,ab} = \frac{W}{Q} = \frac{P}{I} \text{ volts} \tag{3.6}$$

where W is the energy in joules tranferred to or from the junction between two materials, a and b, by a charge of Q coulombs. $\alpha_{P\,ab}$ is often more conveniently expressed in terms of the power P (watts) transferred by a current I (amperes).

If the two materials are joined at two points held at different temperatures, an open-circuit potential difference ΔV is produced as a result of a temperature difference ΔT between the junctions, the *Seebeck effect* (Fig. 3.7(*b*)). This leads to the differential *Seebeck coefficient*, $\alpha_{S\,ab}$, given by

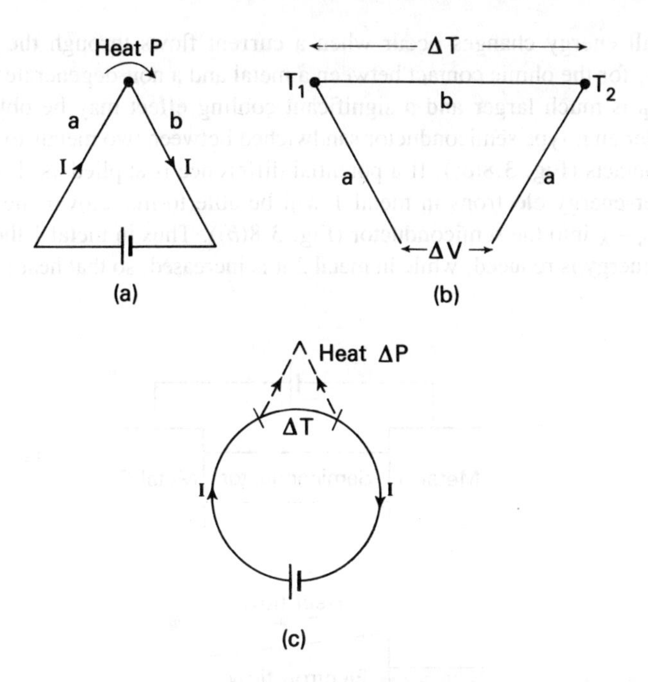

Figure 3.7 Thermoelectric effects. (*a*) Peltier effect; (*b*) Seebeck effect; (*c*) Thomson effect.

$$\alpha_{S\ ab} = \lim_{\Delta T \to 0} \frac{\Delta V}{\Delta T} \text{ volts per degree Celsius} \tag{3.7}$$

The e.m.f. generated when $\Delta T = 1$ °C is sometimes called the *thermo-electric power*. The two coefficients are related by Kelvin's law:

$$\alpha_{S\ ab} = \frac{\alpha_{P\ ab}}{T} \tag{3.8}$$

where T is the absolute temperature of the cold junction.

Finally, where there is a temperature difference ΔT over part of a *single* conductor the passage of current I leads to thermal power ΔP being generated (Fig. 3.7(*c*)). This is the *Thomson effect*, related to the Peltier and Seebeck effects, but of small practical importance and not considered in this book.

The junction of two metals to form a thermocouple has been used for a long time as a method of measuring temperature, with copper-constantan or iron-constantan couples having values of α_S up to about 50 μV/°C. Correspondingly low values of α_P occur, so that little energy is transferred when a current is passed through the junction, with a consequently small cooling effect. This is because the conduction electrons all have energies close to the Fermi level, and very small energy changes occur when a current flows through the junction. However, for the ohmic contact between a metal and a non-degenerate semiconductor α_P is much larger and a significant cooling effect may be obtained.

Consider an n-type semiconductor sandwiched between two metals to form two ohmic contacts (Fig. 3.8(*a*)). If a potential difference is applied as shown, only the higher-energy electrons in metal 1 will be able to move over the potential barrier $\phi_S - \chi$ into the semiconductor (Fig. 3.8(*b*)). Thus in metal 1 the average electron energy is reduced, while in metal 2 it is increased, so that heat is transfer-

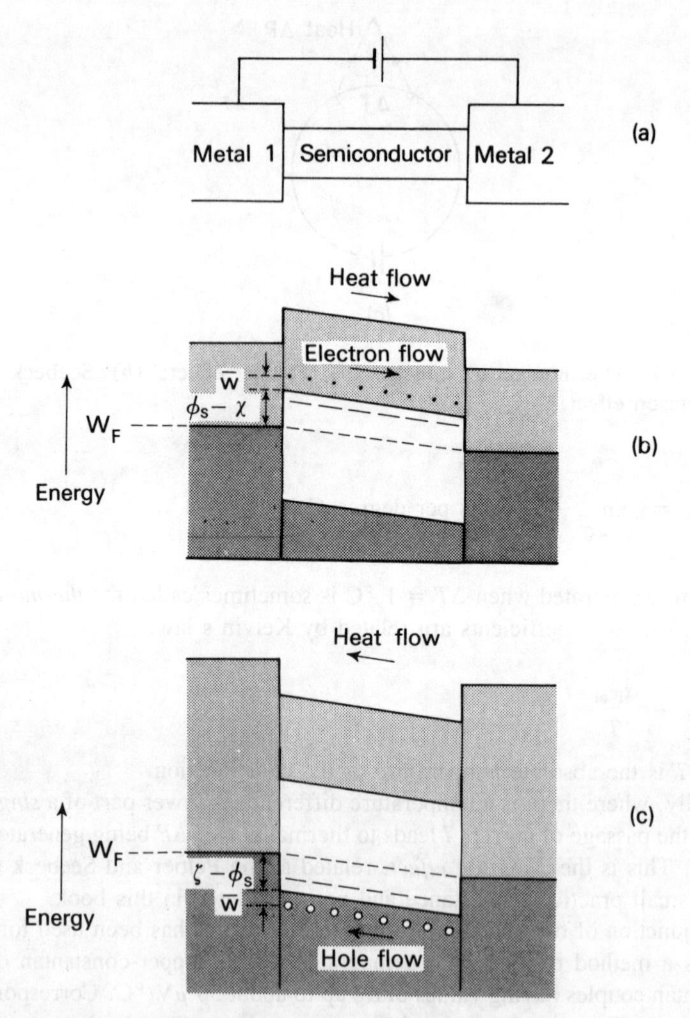

Figure 3.8 Thermoelectric cooling. (*a*) Metal-semiconductor-metal ohmic contacts and applied voltage; (*b*) energy diagram using n-type semiconductor in (*a*); (*c*) energy diagram using p-type semiconductor in (*a*).

red from metal 1 to metal 2. If a p-type semiconductor is substituted and the same voltage applied (Fig. 3.8(*c*)), a hole current will flow due to movement of electrons in the valence bands under the potential barrier $\zeta - \phi_s$. Thus low-energy electrons are removed from metal 1, increasing its average energy and reducing the average energy of metal 2, so that in this case heat is transferred from metal 2 to metal 1. The Peltier coefficients may be obtained from the energy diagrams, since the electrons crossing from a metal to an n-type semiconductor possess potential energy $(\phi_s - \chi)$ and mean kinetic energy $\bar{w}$, which is proportional to temperature (eq. 2.8). Thus the energy transported per unit charge is

$$\alpha_{P\ mn} = -\frac{\bar{w} + (\phi_s - \chi)}{e} \tag{3.9}$$

the minus sign indicating removal of energy from the metal. Similarly, for a metal-to-p-type-semiconductor contact,

$$\alpha_{P\ mp} = +\frac{\bar{w} + (\zeta - \phi_s)}{e} \tag{3.10}$$

the plus sign indicating energy transfer to the metal. Due to the temperature dependence of the quantities in eqs (3.9) and (3.10) α_P rises with temperature.

A thermoelectric cooling device is obtained by arranging n- and p-type materials in couples (Fig. 3.9). The passage of current due to the indicated applied voltage will cause all the top metal surfaces to be cooled and the lower ones to be heated, while reversal of the current will cause reversal of the direction of heat flow. Thus if one side of the device is fixed to a suitable heat sink maintained at room temperature, refrigeration of an article attached to the other side will occur. A p-n bismuth-telluride couple has a Seebeck coefficient of about 400 μV/°C, and for a well heat-insulated device with 16 couples, for example, a current of 10 A

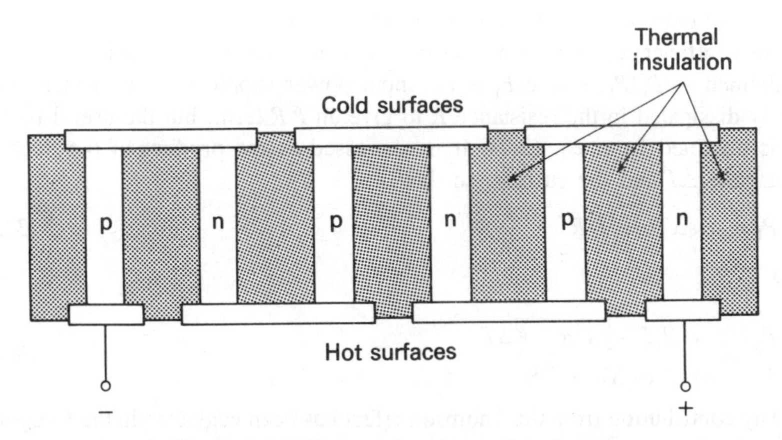

Figure 3.9 Thermoelectric cooling device.

will cause a heat flow of about 3 W, maintaining a temperature difference of about 30 °C between the two surfaces. From eq. (3.6) the higher the current passed through the device the greater will be the rate of heat flow, but a limit is set by the heat dissipating due to the electrical resistance of the device and by the heat flowing in from the surroundings. It may be shown that the Joule heat produced in the resistance flows equally to the hot and cold surfaces, so that for a cooling unit of resistance R with the cold surface at temperature T_c, the equation governing the thermal condition of the load is

$$P_c = \alpha_S I T_c - \tfrac{1}{2} I^2 R - K\Delta T \qquad (3.11)$$

P_c	$\alpha_S I T_c$	$\frac{1}{2} I^2 R$	$K\Delta T$
Net heat absorbed at cold junction	Peltier heat transferred from cold junction	Joule heat flowing to cold junction	Heat conducted from surroundings and hot junction

K is the thermal conductance of the device which is reduced by efficient thermal insulation, and ΔT is the temperature difference between the surfaces. A high value of α_S is desirable to give as large a drop in temperature as possible for a given current; α_S is used in the above equation since it is less dependent on temperature than α_P. Also the thermal conductance K and the electrical resistance R should be as small as possible in order to minimize the negative terms in eq. (3.11).

The suitability of a material for use as a thermoelectric device depends on the above considerations and may be deduced from a figure of merit, Z given by

$$Z = \frac{\alpha_S^2}{RK} \text{ kelvin}^{-1} \qquad (3.12)$$

At room temperature, for metal junctions Z is about 0.1×10^{-3} K^{-1}, while for bismuth-telluride it is about 2×10^{-3} K^{-1}, which indicates that semiconductors are better than metals for thermoelectric applications.

The coefficient of performance or the efficiency of a thermoelectric refrigerator is defined as P_c/P_I, where P_I is the input power supplied to the device. P_I is partly dissipated in the resistance R to give an I^2R term, but the useful part of P_I is obtained from eq. (3.7). It is expressed as the product of the Seebeck voltage $\alpha_S \Delta T$ and the current, so that

$$P_I = \alpha_S \Delta T I + I^2 R \qquad (3.13)$$

and

$$\frac{P_c}{P_I} = \frac{\alpha_S T_c I - \frac{1}{2} I^2 R - K\Delta T}{\alpha_S \Delta T I + I^2 R} \qquad (3.14)$$

Any contribution from the Thomson effect has been neglected in the foregoing analysis. Typical orders of magnitude for a single p-n couple are 10^{-3} Ω for R

and 10^{-2} W/K for K. For n couples the relevant quantities are $n\alpha_S$, nR and nK so the cooling power becomes nP_c. Further details on thermoelectric materials and devices are given in Ref. [1].

The p-n junction

The contact between p- and n-type semiconductors is particularly important since it forms the basis of electronic devices such as semiconductor diodes and transistors. It is a rectifying contact and is formed in a single crystal whose impurity atoms are changed from donors to acceptors at the junction.

A p-n junction is formed by allowing a Group III material to diffuse into an n-type region at high temperature. Similarly an n-p junction would be formed by diffusion of a Group V material into a p-type region. The type of impurity atom then changes at a well-defined cross-section of the composite crystal called the *metallurgical junction* (Fig. 3.10(*a*)). This diffusion process has superseded the earlier process in which a small piece of p-type impurity, for instance, was alloyed with an n-type semiconductor at high temperature.

The manufacturing process is described in Chapter 7 and results in a gradual transition from one type of impurity to the other across the junction. Where the impurity density is much higher on one side of the junction than the other the transition occurs over a very short distance to form a steeply graded or *abrupt* junction, which is more easily analysed than the general graded junction.

At the junction electrons and holes have recombined so that in the junction region, the number of ionized acceptors in the p-region equals the number of ionized donors in the n-region. Thus in the charge density diagram of Fig. 3.10(*b*) the areas enclosed on each side of the junction are also equal, N_d being the donor density in the n-region. This situation may be described approximately by the rectangular charge density diagram of Fig. 3.11(*b*), where N_a represents the mean acceptor density of Fig. 3.10(*b*). d is the width of the depletion layer in which there are virtually no free charges. N_a is then considerably greater than N_d and the depletion layer extends further into the n-region than the p-region. As a result of the change from the fixed negative to the fixed positive charges shown in Fig. 3.11(*a*), a potential difference occurs across the junction, the diffusion potential ψ volts (Fig. 3.11(*c*)).

Where ϕ_p and ϕ_n are the work functions of the p- and n-regions respectively, $\psi = \phi_p - \phi_n$ giving a fall in energy from the p- to the n-region of ψ electron-volts, arising as a result of the alignment of the Fermi levels. There are then equal probabilities of electrons and holes diffusing across the junction with zero applied bias.

In fact there are now two electron currents and two hole currents. The hole current consists of (*a*) minority holes from the n-region drifting into the p-region under the influence of the electric field E at the junction, I_{0p} and (*b*) majority holes from the p-region diffusing down the concentration gradient, I_{Fp}. Similarly the electron current consists of minority carriers drifting from the p-region I_{0n}

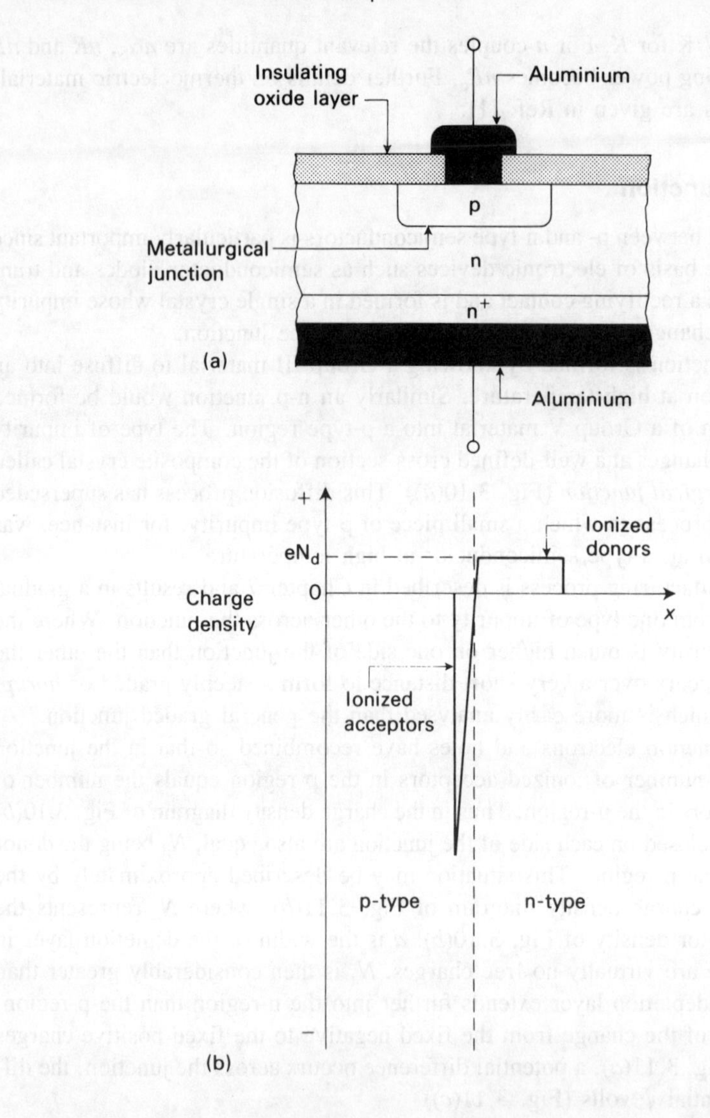

Figure 3.10 (*a*) p-n diode construction; (*b*) charge density in a steeply-graded p-n junction.

and majority carriers diffusing from the n-region I_{Fn}. The corresponding current densities are

$$J_p = ep\mu_p E - eD_p \frac{dp}{dx} \tag{3.15}$$

and

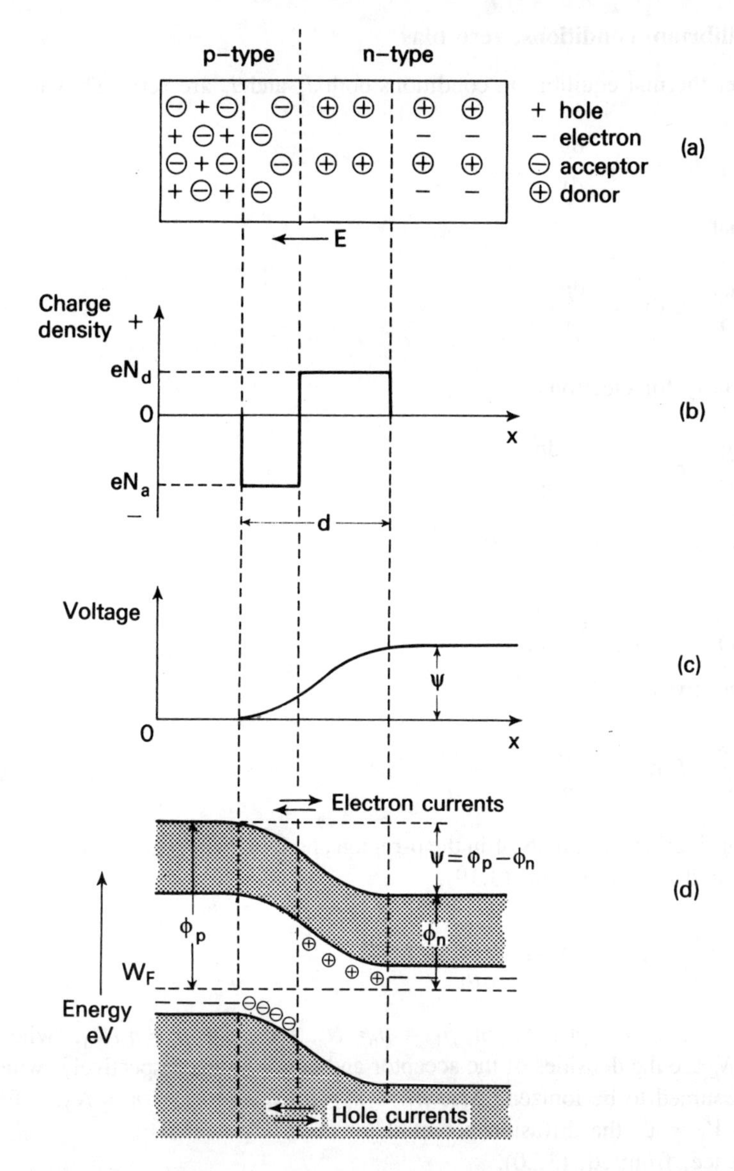

Figure 3.11 The abrupt p-n junction. (*a*) Impurities and current carriers; (*b*) charge density; (*c*) junction potential; (*d*) energy diagram.

$$J_n = en\mu_n E + eD_n \frac{dn}{dx} \tag{3.16}$$

for holes and electrons respectively, from eqs (2.80), (2.102) and (2.103).

Equilibrium conditions, zero bias

Under thermal equilibrium conditions both J_p and J_n are zero. Thus for holes,

$$ep\mu_p E = eD_p \frac{dp}{dx}$$

so that

$$\frac{\mu_p}{D_p} E\, dx = \frac{dp}{p} \tag{3.17}$$

Similarly for electrons,

$$\frac{\mu_n}{D_n} E\, dx = -\frac{dn}{n} \tag{3.18}$$

Substituting from eqs (2.107) and (2.108),

$$\frac{\mu_p}{D_p} = \frac{\mu_n}{D_n} = \frac{e}{kT}$$

which gives

$$\frac{e}{kT} E\, dx = \frac{dp}{p} = -\frac{dn}{n} \tag{3.19}$$

Considering two points, 1 in the p-region and 2 in the n-region outside the junction, and integrating eq. (3.19),

$$-\frac{e}{kT}(V_2 - V_1) = \ln \frac{p_2}{p_1} = \ln \frac{n_1}{n_2} \tag{3.20}$$

Now, for an abrupt junction, $p_1 = p_p \approx N_a$, and $p_2 = p_n \approx n_i^2/N_d$, where N_a and N_d are the densities of the acceptor and donor atoms respectively, which are all assumed to be ionized. Also $n_1 = n_p \approx n_i^2/N_a$ and $n_2 = n_n \approx N_d$. Finally $V_2 - V_1 = \psi$, the diffusion potential, or *built-in* potential.

Hence, from eq. (3.20),

$$\exp\left(-\frac{e\psi}{kT}\right) = \frac{p_n}{p_p} = \frac{n_p}{n_n} = \frac{n_i^2}{N_a N_d} \tag{3.21}$$

and

$$\psi = \frac{kT}{e} \ln \frac{N_a N_d}{n_i^2} \tag{3.22}$$

ψ may thus be calculated from the doping densities and the energy gap which occurs in the expression for n_i (eq. (2.30)).

Worked example

A silicon p-n junction has impurity concentrations of 3×10^{22} acceptor atoms/m^3 on the p-side and 3×10^{23} donor atoms/m^3 on the n-side. Determine the diffusion potential at room temperature.

Solution

From page 40 n_i for silicon is 1.50×10^{16} electron-hole pairs per m^3 and at room temperature $kT/e = 0.025$ V. Then

$$\psi = 0.025 \ln \frac{3 \times 10^{22} \times 3 \times 10^{23}}{(1.50 \times 10^{16})^2}$$

$$= 0.025 \ln \frac{9 \times 10^{45}}{2.25 \times 10^{32}}$$

$$= 0.025 \ln 4.00 \times 10^{13}$$

$$= 0.025 \times 31.32$$

$$= 0.78 \text{ V}$$

Similar calculations give $\psi = 0.41$ V for germanium and 1.14 V for gallium arsenide.

The relationship between the electron densities on each side of the junction and also the corresponding relationship between the hole densities may be obtained from eq. (3.21):

$$\frac{n_p}{n_n} = \frac{p_n}{p_p} = \exp\left(-\frac{e\psi}{kT}\right)$$

Thus

$$p_n = p_p \exp\left(-\frac{e\psi}{kT}\right) \tag{3.23}$$

and

$$n_p = n_n \exp\left(-\frac{e\psi}{kT}\right) \tag{3.24}$$

These equations specify the equilibrium conditions in which drift and diffusion currents balance in the depletion layer. If low-resistance ohmic contacts are made to the p- and n-regions (Fig. 3.12(*a*)) there will be changes in potential at these contacts exactly compensating the junction potential, so that the potential difference between the contacts is zero (Fig. 3.12(*b*)). Thus when they are joined no net current flows.

The total diffusion current $I_F = I_{Fn} + I_{Fp}$ and the total drift current $I_0 = I_{0n} + I_{0p}$. Then, as in the previous diodes, the total current $I = I_F - I_0$ and at zero bias $I_F = I_0$ so that $I = 0$.

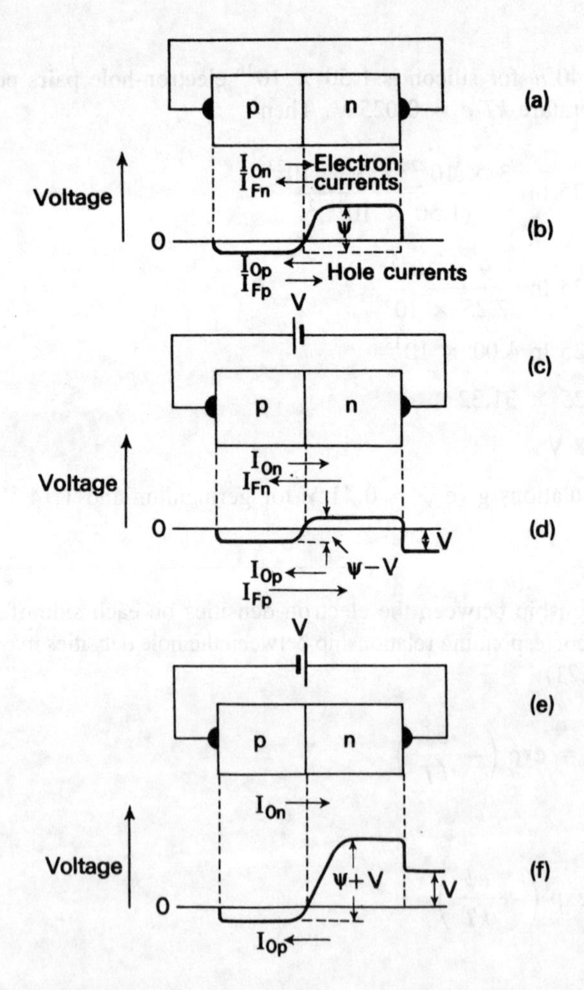

Figure 3.12 Operation of the p-n junction. (*a*) Zero bias; (*b*) voltage distribution for zero bias; (*c*) forward bias; (*d*) voltage distribution for forward bias; (*e*) reverse bias; (*f*) voltage distribution for reverse bias.

Forward bias

Suppose the p-region is now made positive with respect to the n-region (Figs. 3.12(*c*) and (*d*)). The height of the junction potential will be reduced, so that more electrons can diffuse across the junction from the n-region and more holes from the p-region, increasing the total diffusion current I_F. The current I_0 will remain unchanged, since this is due to the drift of holes from the n-region and electrons from the p-region across the junction under the influence of the field E. This field is sufficient to extract all the minority carriers from each region, even when the junction potential is reduced.

The total diffusion current $I_F = I_{Fn} + I_{Fp}$ and the total drift current $I_0 = I_{0n} + I_{0p}$. Then, as in the previous diodes, the total current $I = I_F + I_0$ and at zero bias $I_F = I_0$ so that $I = 0$. If the current I_F is large there will be a voltage drop in the bulk of the semiconductor, away from the junction region. This will be mainly in the n-region when the p-region is the more heavily doped. This voltage drop may often be ignored in comparison with the drop across the depletion layer, and it may then be assumed that the whole of the applied voltage V reduces the junction potential to $\psi - V$. Consider first the effect of lowering the junction potential on the hole component of I_F. The effect on the electron component can then be deduced by comparison.

The equilibrium hole density in the n-region is p_n, which is increased to a new value p_e due to the injection of excess holes from the p-region (Fig. 3.13(*a*)). It is assumed in the following analysis that the density of holes injected into the n-region is small compared to that of the majority carriers, which are electrons. This is known as *low-level injection* in which the charge neutrality in the n-region is undisturbed and the holes move only by diffusion to the ohmic contact. Recombination with the electrons also occurs, but since the holes are replaced continuously a constant excess density $p_e - p_n = \Delta p(0)$ is maintained at the boundary between the depletion layer and the n-region. It is reasonable to suppose that p_e depends on the junction potential in a similar manner to p_n, so that, replacing ψ with $\psi - V$ in eq. (3.23), gives

$$p_e = p_p \exp\left[-\frac{e(\psi - V)}{kT}\right] \tag{3.25}$$

Hence

$$p_e = p_n \exp\frac{eV}{kT} \tag{3.26}$$

again using eq. (3.23). Then

$$\Delta p(0) = p_e - p_n = p_n\left[\exp\left(\frac{eV}{kT}\right) - 1\right] \tag{3.27}$$

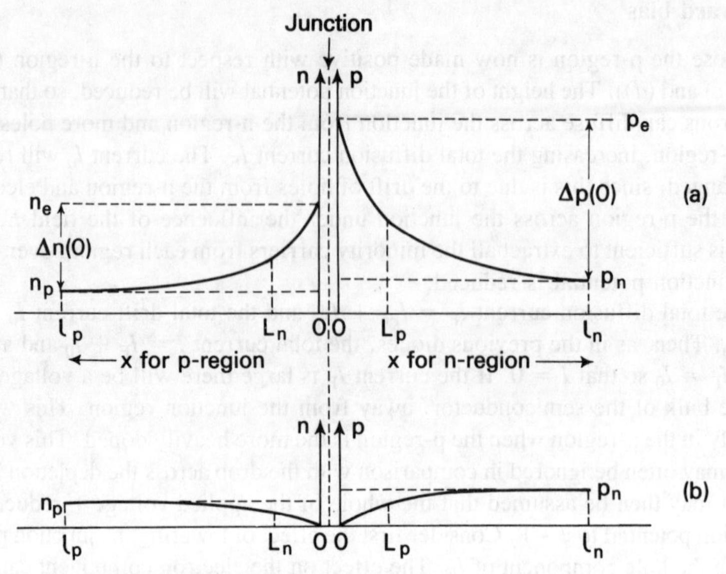

Figure 3.13 Excess carrier distributions for the p-n junction. (*a*) Forward bias; (*b*) reverse bias.

which does not depend on p_p but on p_n the minority carrier density in the n-region.

The current density carried by the injected holes at the edge of the depletion layer is then

$$J_p(0) = -eD_p \left(\frac{\partial \Delta p}{\partial x} \right)_{x=0} \tag{3.28}$$

Recombination occurs in the n-region, which has a length $l_n > L_p$, the diffusion length for holes. Thus an exponential decay of excess hole density will occur with increasing values of x as given by eq. (2.119) and

$$\Delta p = \Delta p(0) \exp\left(-\frac{x}{L_p}\right)$$

so that

$$\frac{\partial \Delta p}{\partial x} = \frac{-1}{L_p} \Delta p(0) \exp\left(-\frac{x}{L_p}\right) \tag{3.29}$$

and

$$\left(\frac{\partial \Delta p}{\partial x}\right)_{x=0} = -\frac{\Delta p(0)}{L_p} \tag{3.30}$$

Hence

$$J_p(0) = \frac{eD_p}{L_p}\Delta p(0) \tag{3.31}$$

and, substituting from eq. (3.27),

$$J_p(0) = \frac{eD_p\, p_n}{L_p}\left[\exp\left(\frac{eV}{kT}\right) - 1\right] \tag{3.32}$$

which is of the same form as eq. (3.5).

The electrons in the n-region which recombine with the injected holes are replaced by electrons from the external circuit, entering through the ohmic contact. At the junction the current is carried mainly by holes, but as x increases into the n-region the current is increasingly carried by electrons. At the ohmic contact all the current is carried by electrons and the hole density has fallen to the equilibrium value, p_n, which is one way of defining an ohmic contact. Thus the *total* current is constant throughout the n-region, at the value given by eq. (3.32), but the proportion carried by holes or electrons changes as x is increased. Under low-current conditions the space charge carried by the injected holes is always neutralized by an equal number of excess electrons, which is small compared with the equilibrium density n_n.

The effect of lowering the junction potential on the flow of electrons from the n- to the p-region may be deduced in a similar way. An excess electron density, n_e, is set up at the boundary between the depletion layer and the p-region, Fig. 3.13(*a*). n_e is obtained from eq. (3.24) by replacing ψ with $\psi - V$, so that

$$n_e = n_n \exp\left[-\frac{e(\psi - V)}{kT}\right] \tag{3.33}$$

$$= n_p \exp\left(\frac{eV}{kT}\right) \tag{3.34}$$

and

$$\Delta n(0) = n_e - n_p = n_p\left[\exp\left(\frac{eV}{kT}\right) - 1\right] \tag{3.35}$$

Recombination occurs in the p-region, so that

$$\Delta n = \Delta n(0) \exp\left(-\frac{x}{L_n}\right) \tag{3.36}$$

with $x = 0$ taken at the depletion layer boundary and x increasing positively to the left of the boundary. These equations lead to the expression

$$J_n(0) = \frac{eD_n n_p}{L_n}\left[\exp\left(\frac{eV}{kT}\right) - 1\right] \tag{3.37}$$

for the electron current density in the p-region. This current remains constant, but an increasing proportion of it is carried by holes as x increases into the p-region, until at the ohmic contact it is entirely a hole current and the electron density has fallen to the equilibrium volume n_p.

The processes of hole and electron injection occur simultaneously and independently so that the total current density, J, is the sum of the two components. Putting $I = JS$, where S is the junction area, the junction current is given by eqs (3.32) and (3.37) and

$$I = eS\left(\frac{D_p p_n}{L_p} + \frac{D_n n_p}{L_n}\right)\left[\exp\left(\frac{eV}{kT}\right) - 1\right] \tag{3.38}$$

where for forward bias, V is positive. This is the basic equation for the I/V characteristic of a p-n junction diode, (the Shockley equation Ref. [2]) and by comparison with eq. (3.5) it may be seen that

$$eS\left(\frac{D_p p_n}{L_p} + \frac{D_n n_p}{L_n}\right) = I_0 \tag{3.39}$$

This is the current that will flow when reverse bias is applied, i.e. with V negative.

In practice two changes may be made in the structure of the device. Firstly, in order to keep the forward voltage drop low at high currents the diode is shortened so that the length of one or both regions is less than the diffusion length (Fig. 3.14(*a*)). This also reduces the transit time through the device. The excess carrier distributions then become approximately linear, since they correspond to the initial part of an exponential decay. The slopes of the distribution are determined by the lengths l_n and l_p respectively, so that

$$\frac{\partial \Delta p}{\partial x} = -\frac{\Delta p(0)}{l_n} \tag{3.40}$$

and

$$\frac{\partial \Delta n}{\partial x} = -\frac{\Delta n(0)}{l_p} \tag{3.41}$$

giving

$$I = eS\left(\frac{D_p p_n}{l_n} + \frac{D_n n_p}{l_p}\right)\left[\exp\left(\frac{eV}{kT}\right) - 1\right] \tag{3.42}$$

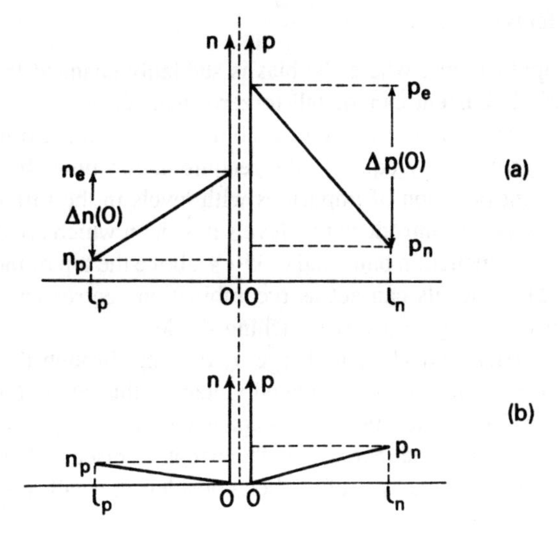

Figure 3.14 Excess carrier distributions for a short p-n junction. (*a*) Forward bias; (*b*) reverse bias.

This equation is of the same form as eq. (3.38), but I_0 has increased since $l_n < L_p$ and $l_p < L_n$. Thus a smaller value of V is required to produce the same forward current I at the expense of an increased leakage current.

Secondly, a diode is usually constructed by starting with a material of relatively high resistivity and then doping with impurity to convert part of it to a lower-resistivity region of the opposite type. If the starting material is n-type this gives a p^+-n structure, with resistivities of say 0.001 Ω m for the p-region and 0.1 Ω m for the n-region. The excess carrier distributions would be similar to Figs 3.13 or 3.14 but with p_e much further increased over n_e. If the starting material is p-type an n^+-p structure results, with the lower resistivity in the n-region and n_e much larger than p_e in Figs 3.13 or 3.14. The nature of the current carriers may be deduced from eqs (3.38) or (3.42) where

$$p_n = \frac{n_i^2}{n_n} \quad \text{and} \quad n_p = \frac{n_i^2}{p_p} \tag{3.43}$$

Thus in a p^+-n diode $p_p \gg n_n$ so n_p is negligible and the current is carried mainly by holes. In an n^+-p diode $n_n \gg p_p$ so p_n is negligible and the current is carried mainly by electrons.

Such diodes may be used alone or as part of a transistor (Chapter 4); the emitter-base diode, which approximates to an abrupt junction, would then be p^+-n for a p-n-p transistor and n^+-p for an n-p-n transistor.

Transient effects

In switching applications, where the bias is suddenly changed from forward to reverse, the diode current cannot fall to zero instantaneously. This is because the excess carriers are removed by recombination and so the time required for switching is approximately equal to the recombination time. It is substantially reduced by the introduction of impurities with levels in the forbidden gap. For instance, gold is used to introduce two levels in silicon which are 0.54 eV below the bottom of the conduction band and 0.35 eV above the top of the valence band respectively. These levels can act as recombination centres and so reduce the recombination time to give a fast switching diode.

The excess carriers introduce a charge q_d flowing through the diode to support the current I. A change in the applied voltage V thus alters I and q_d together and this effect can be represented by a capacitance $C_d = q_d/V$. C_d is known as the diffusion capacitance and increases with current, as discussed under transistors in Chapter 4. It is particularly useful when considering small, transient changes in voltage.

Quasi-Fermi levels*

When injection of minority carriers occurs the product of the electron and hole concentrations is no longer n_i^2, since the charge of the excess minority carriers is compensated by that of an equal number of excess majority carriers drawn into the region. For low-level injection the relative increase in the density of the majority carriers is very small, but the density of the minority carriers is greatly increased. On the energy diagram this is shown by using different Fermi levels for electrons and holes, which are illustrated in Fig. 3.15 for a forward-biased p-n junction. They are called *quasi-Fermi levels*, since they apply when the equilibrium conditions have been disturbed, and express the increased probability of finding minority carriers in otherwise empty levels.

In the p-region remote from the junction the Fermi level corresponds only to holes, since all the injected electrons have recombined. Towards the junction the hole Fermi level remains constant, but since the probability of an electron occupying a given level increases with W_F (eq. (2.21)) and the electron density also increases exponentially with distance (Fig. 3.13)(*b*), the electron Fermi level W_{F_n} rises linearly. In the depletion layer the quasi-Fermi levels remain constant since it is assumed that no recombination is occurring here and the excess electron and hole densities are determined by the applied voltage. In the n-region the probability of a hole occupying a given level falls as W_F increases (eq. (2.24)) and the hole density also decreases exponentially away from the junction. Thus the hole Fermi level W_{F_p} rises linearly, finally reaching the equilibrium level for electrons only when all the injected holes have recombined.

* The term 'imref' for a quasi-Fermi level has been coined from the word Fermi spelt backwards.

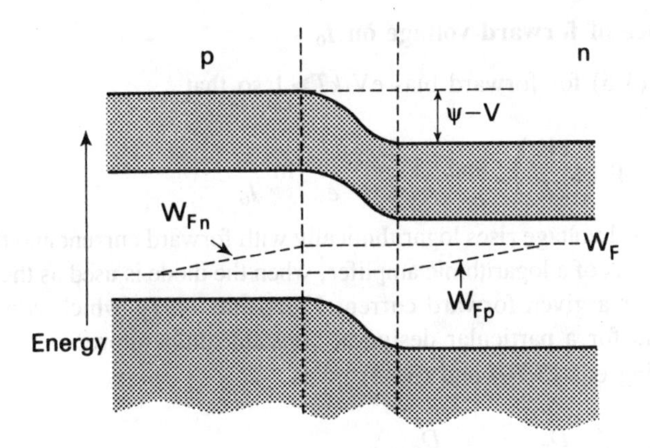

Figure 3.15 Quasi-Fermi levels.

Reverse bias

If the polarity of the bias is reversed, so that the p-region is negative with respect to the n-region, the junction potential is increased (Fig. 3.12(*e*) and (*f*)). This will prevent the diffusion of holes into the n-region and electrons into the p-region, but will increase the field E across the junction. Thus the extraction of minority carriers from each region, which constitutes the current I_0, will continue and diffusion of holes from the n-region and electrons from the p-region will occur to maintain this current across the junction. E is large enough to extract all the minority carriers near the junction even at low reverse bias voltages, so that the densities of minority carriers are zero at the edges of the depletion layer (Figs 3.13(*b*) and 3.14(*b*)). Since diffusion is occurring away from the junction an exponential distribution is set up on each side, defined by the same diffusion lengths L_p and L_n as for forward bias, and the corresponding linear distribution for a short diode. Hence on average all the minority carriers which are generated thermally within distances L_p and L_n from the edge of the depletion layer are extracted by the field across the junction.

The reverse current I_0 will be independent of the voltage across a perfect diode for reverse voltages greater than 75 mV, as shown from eq. (3.5), and below the breakdown value, discussed later. The value of I_0 at room temperature is a function of the material, length and area of the diode. For small diodes, with maximum current 100 mA say, I_0 would be a few μA for germanium, in the pA-region for silicon and much less for gallium arsenide. However, I_0 will rise rapidly with temperature, which is significant for germanium diodes but much less so for silicon and gallium arsenide diodes since the room temperature value is so much lower. The effect of I_0 on forward bias voltage is, however, significant for all three types of diode (Fig. 3.16).

Dependence of forward voltage on I_0

From eq. (3.5) for forward bias $eV/kT \gg 1$ so that

$$I = I_0 \exp\left(\frac{eV}{kT}\right) \quad \text{and} \quad V = \frac{kT}{e} \ln \frac{I}{I_0} \tag{3.44}$$

Thus, forward voltage rises logarithmically with forward current and this property forms the basis of a logarithmic amplifer, when the diode is used as the logarithmic element. For a given forward current V depends on I_0, which is a function of the material for a particular design at constant temperature.

Combining eqs (3.39) and (3.43) gives

$$I_0 = eSn_i^2\left(\frac{D_p}{L_pN_d} + \frac{D_n}{L_nN_a}\right)$$

and substituting for n_i^2 from eq. (2.30)

$$I_0 = eS\left(\frac{D_p}{L_pN_d} + \frac{D_n}{L_nN_a}\right) N_cN_v \exp\left(\frac{-W_g}{kT}\right)$$

This can be written in the form

$$I_0 = \text{const.} \exp\left(\frac{-eV_g}{kT}\right) \tag{3.45}$$

where V_g is the energy gap in the volts and substituting eq. (3.45) in eq. (3.44) gives

$$I = \text{const.} \exp\left[\frac{e(V - V_g)}{kT}\right] \tag{3.46}$$

For a germanium diode with $I = 10$ mA and $I_0 = 3$ μA, $V = 0.20$ volt from eq. (3.44). Corresponding values of V for silicon and gallium arsenide can be obtained using eq. (3.46), assuming equal constants in each case obtained by adjusting the area S.

For equal forward currents $V - V_g$ must be constant so taking V_g as 0.66 V for germanium, 1.12 V for silicon and 1.43 V for gallium arsenide, and assuming other factors are the same in each diode, leads to $V = 0.66$ volt for silicon and 0.97 volt for gallium arsenide (see Fig. 3.16).

The effect of temperature on diode characteristics

The temperature dependent terms in eqs (3.38) and (3.42) are p_n and n_p. Both are proportional to n_i^2 which thus controls the temperature dependence of I_0.

Squaring both sides of eq. (2.30), differentiating and rewriting in terms of n_i^2 leads to

$$\frac{dn_i^2}{dT} = \frac{W_g}{kT^2} n_i^2 + \frac{3}{T} n_i^2$$

or

$$\frac{dn_i^2}{n_i^2} = \left(\frac{W_g}{kT} + 3\right) \frac{dT}{T} = \frac{dI_0}{I_0} \tag{3.47}$$

Putting $W_g = 0.66$ eV for germanium, 1.12 eV for silicon and 1.43 eV for gallium arsenide and taking $kT = 0.025$ eV at room temperature.

$$\frac{dI_0}{I_0} = 29 \frac{dT}{T} \text{ for germanium, } 48 \frac{dT}{T} \text{ for silicon and}$$

$$60 \frac{dT}{T} \text{ for gallium arsenide} \tag{3.48}$$

As a guide to the rate of change from room temperature, 293 K, we can consider the case where $dI_0 = I_0$ or the current has doubled itself. The corresponding change in temperature, dT, then becomes 10 °C for germanium, 6.1 °C for silicon and 4.9 °C for gallium arsenide.

The rise in I_0 will result in a decrease in the value of V to keep the forward current constant, so from eq. (3.44)

$$\frac{dV}{dT} = \frac{k}{e} \ln \frac{I}{I_0} - \frac{kT}{e} \frac{1}{I_0} \frac{dI_0}{dT} \tag{3.49}$$

$$= \frac{V}{T} - 29 \frac{k}{e} \text{ for germanium}$$

$$= \frac{V}{T} - 48 \frac{k}{e} \text{ for silicon}$$

$$= \frac{V}{T} - 60 \frac{k}{e} \text{ for gallium arsenide} \tag{3.50}$$

Substituting the values of V derived above for $I = 10$ mA into eq. (3.50) gives the rate of change of V to keep I constant at room temperature to be between −1.8 and −1.9 mV/°C rise in temperature for similar diodes in all three materials.

Thus the effect of temperature is to cause the reverse current to rise and forward current will also rise in applications where the forward bias may be considered constant. Both of these effects are also important in the operation of transistors above room temperature.

Diode equivalent circuits

Current-voltage characteristics for the three types of diode are shown in Fig. 3.16, based on eq. (3.5) with $I_0 = 3\ \mu A$ for germanium and assuming negligible resistance in the p- and n-regions away from the junction. In all three characteristics I_0 is negligible on the current scale shown, in contrast to Fig. 3.6. The characteristics of an *ideal* diode, having zero resistance for forward bias and infinite resistance for reverse bias, are shown in Fig. 3.17(*a*). The symbol may be used to represent the ideal diode in this context but, in general, represents any diode.

The three I/V characteristics may be replaced by the linear approximation of Fig. 3.17(*b*). Here V_T is obtained from the asymptotes shown in Fig. 3.16, being typically 0.2 V, 0.7 V and 1.0 V for germanium, silicon and gallium arsenide respectively, r_f is dV/dI, the reciprocal of the slope of the line, which will depend on the current scale used but is typically in the region of a few ohms. This leads to the equivalent circuit of Fig. 3.17(*b*) for all three types of diode. The diode voltage is then given by

$$V = V_T + Ir_f \tag{3.51}$$

using the values of V_T and r_f for the diode being considered. In many applications r_f may be negligible compared with external series resistance, which gives the characteristic and equivalent circuit of Fig. 3.17(*c*). In all these equivalent circuits I_0 is assumed zero, giving an infinite reverse resistance which would otherwise appear across the terminals.

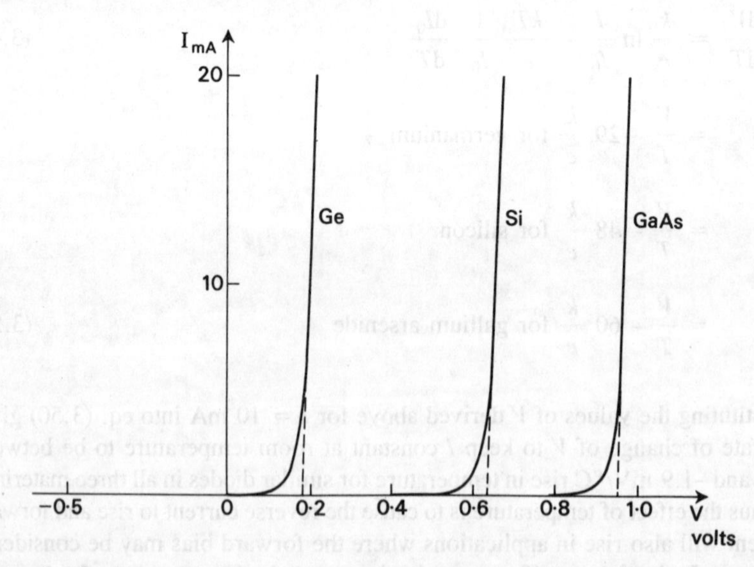

Figure 3.16 **Current/voltage characteristics of germanium, silicon and gallium arsenide p-n junctions.**

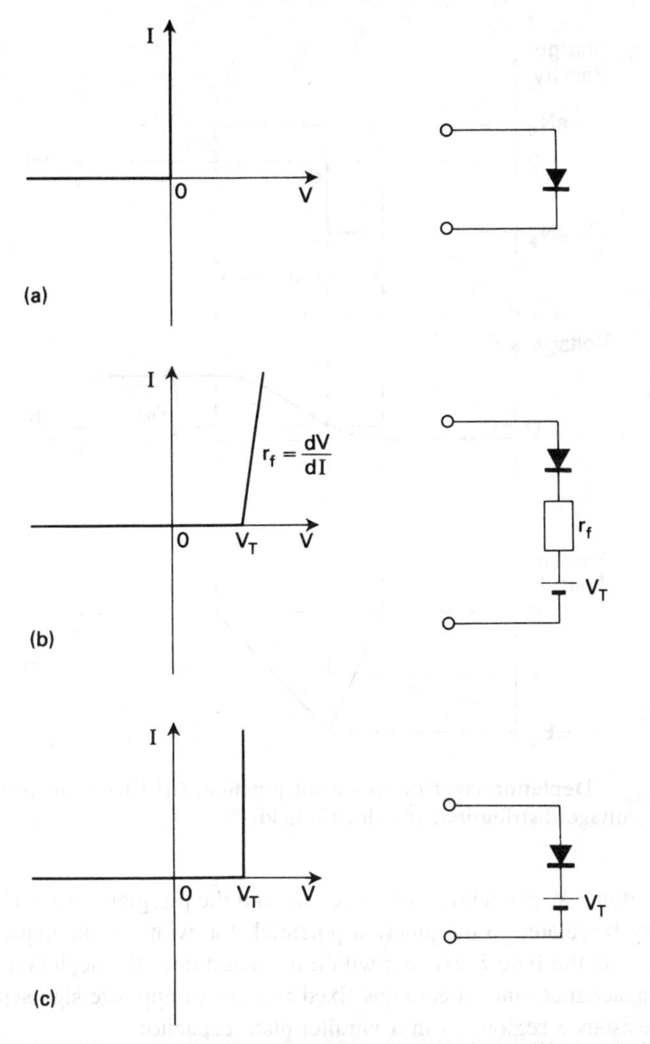

Figure 3.17 Diode equivalent circuits using (*a*) ideal diode; (*b*) threshold voltage V_T and resistance r_f; (*c*) threshold voltage only. Reverse resistance is assumed infinite.

Properties of the depletion layer

Abrupt junction

The charge densities on each side of the depletion layer are shown in Fig. 3.18(*a*), where $x = 0$ at the junction between the p- and n-regions. In order to simplify the analysis it is assumed that there is a constant density of ionized donor atoms N_d extending a distance w_n into the n-region and a constant density of ionized

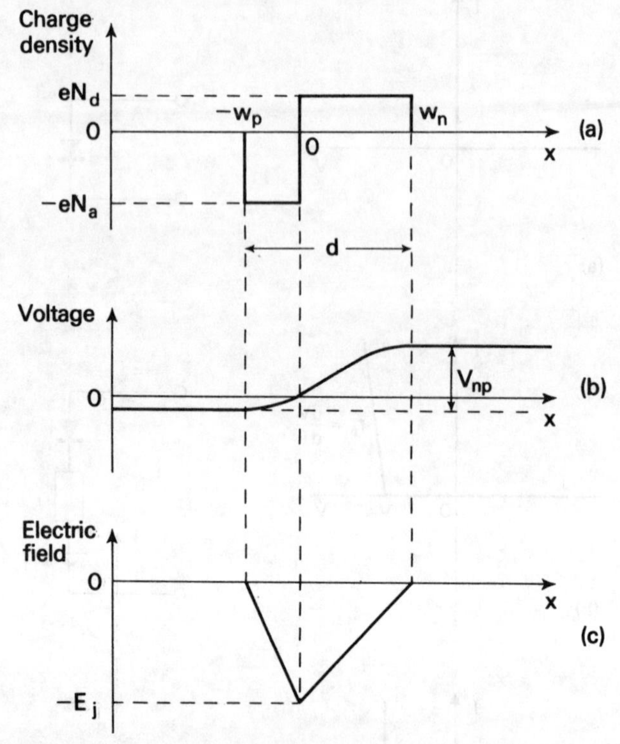

Figure 3.18 **Depletion layer of an abrupt junction. (*a*) Charge density distribution; (*b*) voltage distribution; (*c*) electric field.**

acceptor atoms N_a extending a distance $-w_p$ into the p-region. These charge densities may be related to the junction potential, the width of the depletion layer $w_n + w_p$, and the field E existing within it. In addition, the depletion layer will have a capacitance since it contains fixed charges of opposite sign separated by a high-resistance region, as in a parallel-plate capacitor.

Consider a surface within the depletion layer normal to the x-axis, where the electric flux density is D coulombs per square metre. The electric field due to this surface is

$$E = \frac{D}{\epsilon} \tag{3.52}$$

where $\epsilon = \epsilon_r \epsilon_0$ is the permittivity and ϵ_r the relative permittivity. Hence

$$\frac{dE}{dx} = \frac{d}{dx}\frac{D}{\epsilon} \tag{3.53}$$

and in the p-region of the depletion layer,

$$\frac{dE}{dx} = -\frac{eN_a}{\epsilon} \quad (-w_p<x<0) \tag{3.54}$$

while in the n-region of the depletion layer,

$$\frac{dE}{dx} = \frac{eN_d}{\epsilon} \quad (0<x<w_n) \tag{3.55}$$

which are forms of Poisson's equation for a region containing space charge. Integrating eqs (3.54) and (3.55),

$$E_p = -\frac{eN_a\,x}{\epsilon} + C_1 \text{ in the p-region} \tag{3.56}$$

and

$$E_n = \frac{eN_d\,x}{\epsilon} + C_2 \text{ in the n-region} \tag{3.57}$$

where C_1 and C_2 are constants.

The electric field is continuous and exists only within the boundaries of the depletion layer. Hence $E_p = 0$ when $x = -w_p$ and $E_n = 0$ when $x = w_n$, so that

$$C_1 = -\frac{eN_a w_p}{\epsilon} \tag{3.58}$$

and

$$C_2 = -\frac{eN_d w_n}{\epsilon} \tag{3.59}$$

Then

$$E_p = -\frac{eN_a}{\epsilon}(x+w_p) \tag{3.60}$$

$$E_n = \frac{eN_d}{\epsilon}(x-w_n) \tag{3.61}$$

Since x is negative in the p-region and positive in the n-region, the field in each region increases as x approaches zero (Fig. 3.18(*c*)) until, when $x = 0$,

$$E_p = E_n = E_j = \frac{eN_a w_p}{\epsilon_e\epsilon_0} = -\frac{eN_d w_n}{\epsilon_r\epsilon_0} \tag{3.62}$$

E_j is thus the maximum field existing in the depletion layer, and is a controlling factor in reverse breakdown. Also from eq. (3.62),

$$N_a w_p = N_d w_n \tag{3.63}$$

which express the fact that the total negative charge on one side of the junction equals the total positive charge on the other side, and also that the depletion layer penetrates a shorter distance into the more heavily doped region.

Junction potential

The junction potential as a function of the width of the depletion layer may be found by integrating eqs (3.60) and (3.61), since $V = -\int E\,dx$. Thus

$$V_p = \frac{eN_a}{2\epsilon}x^2 + \frac{eN_a}{\epsilon}w_p x + C_3 \tag{3.64}$$

$$V_n = \frac{eN_d}{2\epsilon}x^2 - \frac{eN_d}{\epsilon}w_n x + C_4 \tag{3.65}$$

and the variations of V_p and V_n with x are shown in Fig. 3.18(*b*).

Since the voltage is continuous across the junction it must be zero at $x = 0$, so that $C_3 = C_4 = 0$. Also the voltage will be constant for values of $x \le -w_p$ and for $x \ge w_n$. Putting $x = -w_p$ in eq. (3.64) and $x = w_n$ in eq. (3.65),

$$V_p(-w_p) = -\frac{eN_a}{2\epsilon}w_p^2 \tag{3.66}$$

$$V_n(w_n) = \frac{eN_d}{2\epsilon}w_n^2 \tag{3.67}$$

The junction potential is then $V_{np} = \psi - V$, where V is the external bias voltage, so subtracting eq. (3.66) from eq. (3.67),

$$V_{np} = \psi - V = \frac{e}{2\epsilon_r\epsilon_0}(N_a w_p^2 + N_d w_n^2) \tag{3.68}$$

Width of depletion layer

Since $N_a w_p = N_d w_n$ eq. (3.68) yields

$$\psi - V = \frac{e}{2\epsilon}N_a^2 w_p^2\left(\frac{1}{N_a} + \frac{1}{N_d}\right)$$

$$= \frac{e}{2\epsilon}\frac{N_a^2 w_p^2}{N_j} \tag{3.69}$$

and

$$\psi - V = \frac{e}{2\epsilon}\frac{N_d^2 w_n^2}{N_j} \tag{3.70}$$

where

$$\frac{1}{N_j} = \frac{1}{N_a} + \frac{1}{N_d} \tag{3.71}$$

Hence

$$w_p = \frac{1}{N_a}\left[\frac{2\epsilon N_j(\psi - V)}{e}\right]^{1/2} \tag{3.72}$$

$$w_n = \frac{1}{N_d}\left[\frac{2\epsilon N_j(\psi - V)}{e}\right]^{1/2} \tag{3.73}$$

and the width of the depletion layer is

$$d = w_n + w_p = \left[\frac{2eN_j(\psi - V)}{eN_j}\right]^{1/2} \tag{3.74}$$

Equation (3.74) shows that d decreases as the doping density is increased. In a p^+-n diode, where $N_a \gg N_d$, $N_j \approx N_d$ and in a n^+-p diode with $N_d \gg N_a$, $N_j \approx N_a$, so that in each case w_d is controlled by the impurity density of the high-resistivity side. w_d also depends on the applied bias and increases with the reverse voltage (V negative). This effect is particularly important in transistors (Chapter 4).

Electric field at the junction

The field at the junction is obtained by substituting the expression for w_p or w_n into eq. (3.62), which gives

$$E_j = \left[\frac{2eN_j(\psi - V)}{\epsilon_r\epsilon_0}\right]^{1/2} \tag{3.75}$$

E_j increases with the reverse voltage and also with the doping density. Thus, the voltage applied to achieve a given field for breakdown of the junction will be reduced as N_j is increased.

Depletion layer capacitance

The depletion layer capacitance per unit area of junction, C_j is given by the ratio of the change of the charge per unit area to the change of applied junction potential. Thus

$$C_j = \frac{dQ}{dV_{np}} \tag{3.76}$$

where $Q = eN_a w_p$ on the p-side. Hence

$$C_j = eN_a \frac{dw_p}{dV_{np}} \tag{3.77}$$

and differentiating eq. (3.72) with respect to $(\psi - V)$,

$$C_j = \left[\frac{e\epsilon_r\epsilon_0 N_j}{2}\right]^{1/2} (\psi - V)^{-1/2} \text{ farads/m}^2 \tag{3.78}$$

It may be noted that eq. (3.78) is also obtained by assuming that the depletion layer behaves as a parallel plate capacitor with $C_j = \epsilon_r\epsilon_0/d$. Thus C_j increases with the doping density and for a rectifier diode has a value in the order of 10 pF/mm^2. Also $C_j \propto V_R^{-1/2}$ for the abrupt junction (Fig. 3.19), where V_R is the reverse voltage, and so the junction may be used as a voltage-dependent capacitor for remote tuning purposes or as a parametric amplifier (Chapter 9). However, if the diode is to be used as a rectifier at very high frequencies, C_j degrades its performance by bypassing the junction, so that a diode with a small contact area between a metal and an n-type semiconductor is preferred.

Graded junctions

In a graded junction there is a gradual transition across the junction from one type of impurity to the other, resulting in complementary error function and Gaussian distributions as described in Chapter 7. For ease of analysis an exponential approximation may be made for these distributions near the metallurgical junction and a linear distribution or *grade* across the junction itself, as illustrated in Fig. 3.20.

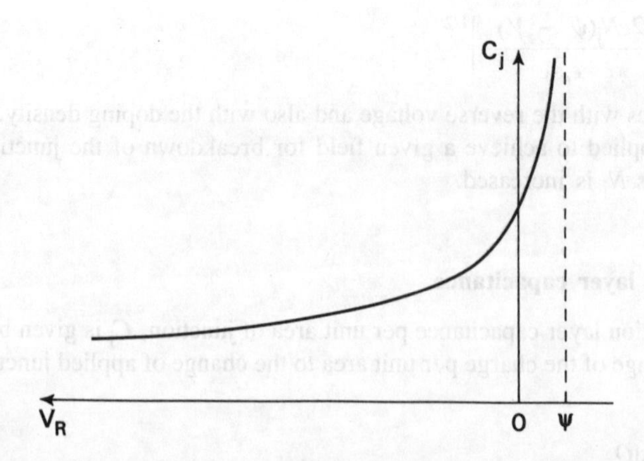

Figure 3.19 **Depletion layer capacitance and reverse voltage.**

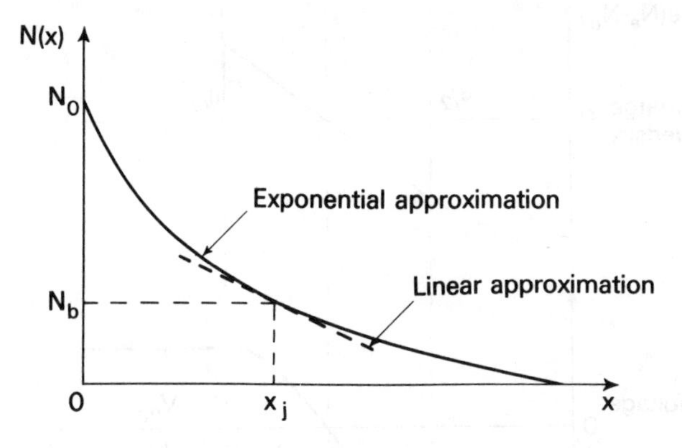

Figure 3.20 Impurity distribution in a graded junction

The impurity density is N_0 at the surface and falls approximately exponentially with penetration. At the depth x_j the impurity density of the diffused material equals the impurity density of the base material, N_b. Thus, if p-type impurities have been diffused into an n-type semiconductor at high temperature, the material changes from p- to n-type at $x = x_j$, which defines the junction. An approximate expression for the distribution is

$$N(x) = N_0 \exp\left(-\frac{ax}{N_b}\right) \tag{3.79}$$

where a is the *grade constant* in units of atoms/m^4. At $x = x_j$,

$$N(x_j) = N_b = N_0 \exp\left(-\frac{ax_j}{N_b}\right) \tag{3.80}$$

and the slope of the distribution is

$$\left.\frac{dN(x)}{dx}\right|_{x=x_j} = -a\frac{N_0}{N_b}\exp\left(-\frac{ax_j}{N_b}\right) \tag{3.81}$$

If we assume that the distribution at $x = x_j$ varies approximately linearly with distance, then

$$N(x) = -ax \tag{3.82}$$

and the slope is $-a$. Putting eq. (3.81) equal to $-a$ and taking logarithms leads to

$$a = \frac{N_b}{x_j}\ln\frac{N_0}{N_b} \tag{3.83}$$

for the grade constant.

The corresponding charge density in the region of the junction is illustrated in Fig. 3.21(a) and is also represented by a linear function

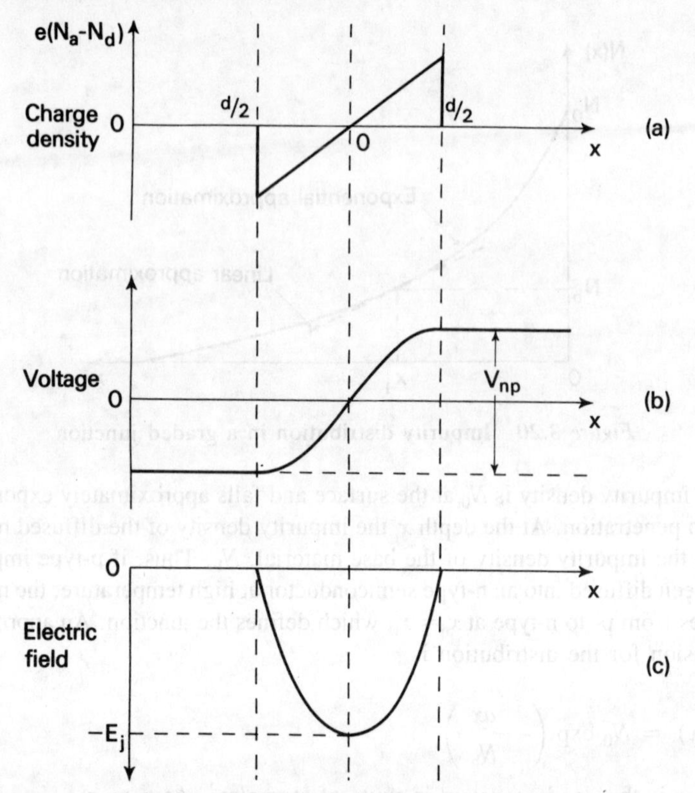

Figure 3.21 Depletion layer of a graded junction. (*a*) Charge density distribution; (*b*) voltage distribution; (*c*) electric field.

$$Q(x) = eax \tag{3.84}$$

The number of negative charges due to ionized acceptors in the p-region is equal to the number of positive charges due to ionized donors in the n-region. Then, taking x_j as the origin, the depletion layer extends an equal distance $d/2$ into each region owing to the linear distribution. When Gauss's theorem is applied,

$$\frac{dE}{dx} = \frac{eax}{\epsilon} \tag{3.85}$$

and

$$E = \frac{eax^2}{2\epsilon} + C_1 \tag{3.86}$$

When $x = \pm d/2$, $E = 0$, since $Q(x) = 0$, so that

$$C_1 = -\frac{ead^2}{8\epsilon} \tag{3.87}$$

and

$$E = \frac{eax^2}{2\epsilon} - \frac{ead^2}{8\epsilon} \tag{3.88}$$

which shows that the field within the depletion layer is a parabolic function of distance (Fig. 3.21(*c*)). It is a maximum at $x = 0$, where

$$E_j = -\frac{ead^2}{8\epsilon_r\epsilon_0} \tag{3.89}$$

The junction voltage is obtained by integrating eq. (3.88)

$$V = -\frac{eax^3}{6\epsilon} + C_1x + C_2 \tag{3.90}$$

which shows that it is a cubic function of distance (Fig. 3.21(*b*)). The voltage across the graded junction is

$$V_{np} = V_n - V_p = \psi - V \tag{3.91}$$

where V_n is the potential at $x = d/2$, and V_p is the potential at $x = -d/2$. If we assume that all the applied voltage is developed across the depletion layer so that $V_{np} = (\psi - V)$, evaluation of eq. (3.90) leads to

$$d = \left(\frac{12\epsilon_r\epsilon_0(\psi - V)}{ea}\right)^{1/2} \tag{3.92}$$

Thus, as the grade constant *a* is increased, the maximum junction field is increased and the depletion layer width is decreased.

Finally, the capacitance per unit area of the depletion layer is obtained from

$$C_j = \frac{\epsilon}{d} \tag{3.93}$$

the expression for the equivalent parallel-plate capacitance. Substituting for *d* from eq. (3.92) leads to

$$C_j = \left(\frac{\epsilon_r^2\epsilon_0^2 ea}{12}\right)^{1/2} (\psi - V)^{-1/3} \tag{3.94}$$

which indicates that C_j is proportional to $V_R^{-1/3}$ for a reverse biased graded junction. This is confirmed by experiment for low reverse voltages, but as the voltage is increased C_j becomes proportional to $V_R^{-1/2}$ as for the abrupt junction.

An expression for ψ may be obtained at thermal equilibrium from the zero bias condition for a p-n junction, since eq. (3.20) may be generally applied. In the graded junction the impurity density varies according to eq. (3.84), so that

$$N_a - N_d = -ax \tag{3.95}$$

and the impurity changes from an effective acceptor to an effective donor type as x is increased from a negative value. Then, assuming that all the impurity atoms are ionized, the hole density at the edge of the depletion layer on the p-side is

$$p_1 = \frac{ad}{2} \tag{3.96}$$

The hole density at the corresponding point on the n-side is related to the effective donor density by eq. (2.37), so that

$$p_2 = \frac{2n_i^2}{ad} \tag{3.97}$$

Then rearranging eq. (3.20) and putting $V_2 - V_1 = \psi$,

$$\psi = \frac{kT}{e} \ln \frac{p_1}{p_2} = \frac{kT}{e} \ln \frac{a^2 d^2}{n_i^2} \tag{3.98}$$

which shows that ψ increases with the grade constant a.

Reverse breakdown mechanisms

Avalanche breakdown

Until now no restriction has been placed on the magnitude of the reverse voltage across the juntion but in practice I_0 remains constant only until a voltage V_B is reached, the *reverse breakdown* voltage. Here I_0 increases very rapidly due to *avalanche multiplication* of the reverse current. This occurs because the field across the junction has become so large, owing to the applied bias, that an electron or hole can acquire sufficient energy to ionize a lattice atom on collision and form a new electron-hole pair for each carrier, the energy required being at least W_g eV. This process is repeated by the new carriers and is illustrated in Fig. 3.22. Current I_0 is then multiplied by a factor M given by

$$M = \frac{1}{1 - (V_R/V_B)^n} \tag{3.99}$$

so that the reverse current is

$$I_R = \frac{I_0}{1 - (V_R/V_B)^n} \tag{3.100}$$

n is a constant whose value depends on the material and is typically 6 for a silicon junction, as shown in Fig. 3.23. Equation (3.51) suggests that $I_R = \infty$ when V_R

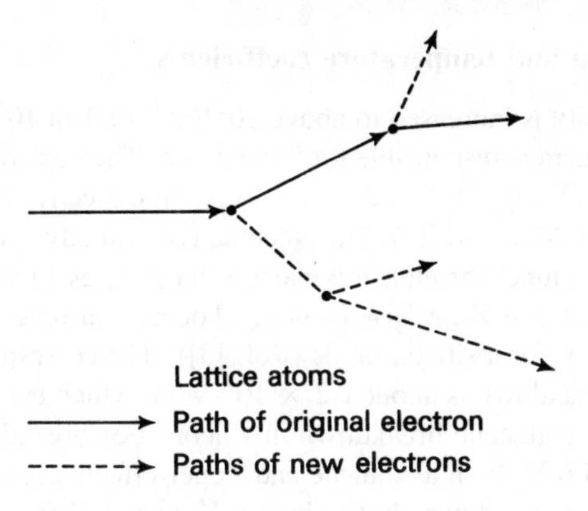

Figure 3.22 Avalanche mechanism.

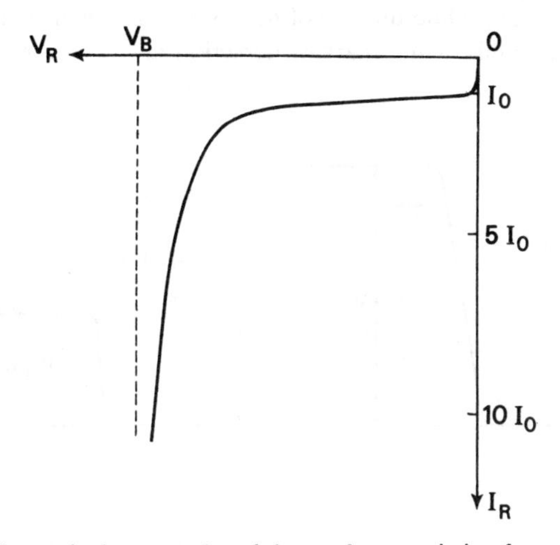

Figure 3.23 Theoretical reverse breakdown characteristic of a p-n junction, plot of eq. (3.100) with $n = 6$.

$= V_B$, but in fact I_R is limited by the resistance external to the junction. Putting $E_j = E_B$ and $\psi - V = V_B$ in eq. (3.75) leads to $V_B = \epsilon_r \epsilon_0 E_B^2 / 2eN_j$ so that for constant E_B, V_B is inversely proportional to the impurity density, and thus increases with the resistivity of the side with fewer impurity atoms. In practice V_B may be as high as 500 V in germanium and 2 kV in silicon diodes. Thus when breakdown does occur there will be a power dissipation within the diode of $I_R V_B$, which may be sufficient to cause destruction of the diode by overheating if I_R is not restricted. However, if the dissipation is kept within safe limits the breakdown curve is reversible, and when V_R is less than V_B, $I_R = I_0$ once more.

Zener breakdown and temperature coefficients

If the doping density is increased to above $10^{24}/m^3$, or 1 in 10^5 atoms replaced, a different mechanism is responsible for breakdown. The depletion layer becomes very narrow (eqs (3.74), (3.92)) and so a very high electric field exists across the junction (Fig. 3.24(*a*) and (*b*)). The potential rises rapidly over a short distance and electrons can tunnel through this narrow barrier, as in the metal-to-metal contact. This is known as *Zener breakdown* and occurs for reverse voltages below about 4 V_g or 4.5 V for a silicon diode (Ref. [3]). The corresponding minimum field for Zener breakdown is about 1.2×10^8 V/m, which compares with about 5×10^7 V/m for avalanche breakdown in silicon. For breakdown voltages between about 4 and 6 V_g both avalanche and Zener effects occur simultaneously, while in diodes breaking down above about 6 V_g or 6.7 V for silicon the avalanche effect predominates (Fig. 3.24(*c*)). It should be noted that avalanche breakdown requires a sufficient mean free path for carriers to acquire the energy needed for ionization, while this is not necessary for Zener breakdown, which therefore occurs mainly within narrow depletion layers. As a consequence the

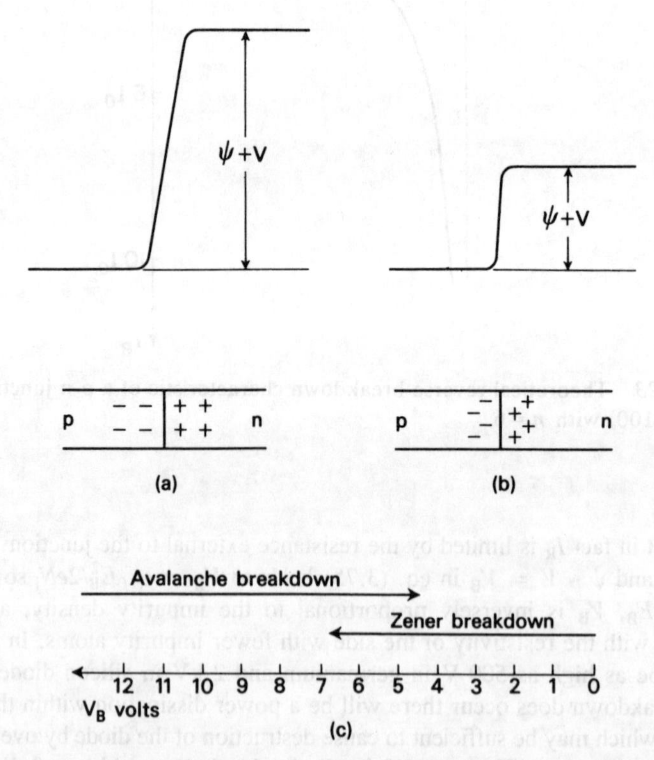

Figure 3.24 **Diode breakdown. Potential barrier and charge distribution for (*a*) avalanche; (*b*) Zener breakdown; (*c*) breakdown voltage ranges in silicon.**

temperature coefficient of V_B is positive for avalanche breakdown, since the increased amplitudes of the atomic vibrations reduce the mean free path, so that the breakdown field increases with temperature. By contrast, the temperature coefficient of V_B is *negative* for Zener breakdown, since the energy band gap W_g decreases as temperature is increased in germanium, silicon and gallium arsenide. Thus the breakdown field falls as temperature is increased when Zener breakdown is occurring. For a diode breaking down at a reverse voltage of about 5.2 V both mechanisms are equally important and the temperature coefficient is zero.

Breakdown diodes

The temperature coefficient is in the order of 0.01% per kelvin and in view of this stability and the constancy of V_B for large changes in reverse current, a diode working at its breakdown voltage may be used as a reference voltage source. Such a *breakdown diode* is designed to withstand a fairly large reverse current without damage and is commonly called a *Zener diode*, even though avalanche multiplication is occurring.

The current-voltage characteristic of a breakdown diode is shown in Fig. 3.25(*a*) and the diode operates between current I_{Zk} and I_{max}. The reverse current I_R increases rapidly with V_R above V_B, as predicted by eq. (3.100), but instead of a vertical rise in I_R the characteristic shows a finite slope resistance $r_r = dV_R/dI_R$. A maximum dissipation P_{max}, corresponding to the product of current I_{max} and the corresponding diode voltage V_Z, is specified above which the low voltage reverse current I_0 increases rapidly as the junction temperature rises. This leads to a negative slope resistance and thermal instability, so that destructive breakdown occurs above a turnover voltage V_U.

The diode may be used as a reference voltage source or voltage regulator in the circuit of Fig. 3.25(*b*), supplying a load R_L. Here R_B is chosen to prevent the dissipation exceeding P_{max} over the expected range of supply voltage V_S. Changes in V_S cause corresponding changes in diode current I_Z across the diode and load remains almost constant, since the diode current and voltage are linked through a nearly vertical line. The diode may be replaced by the equivalent circuit of Fig. 3.25(*c*) where r_r represents the slope resistance about the operating voltage V_Z. r_r has a typical value of about 20 Ω and decreases as I_Z is increased. Then

$$V_Z = V_B + I_Z r_r \tag{3.101}$$

and V_Z is almost equal to V_B since the $I_Z r_r$ term is relatively small.

$$I_Z = \frac{V_S - V_Z}{R} - \frac{V_Z}{R_L} \tag{3.102}$$

and R_B is chosen so that $I_Z < I_{max}$ for all values of V_S. The diode dissipation is given by

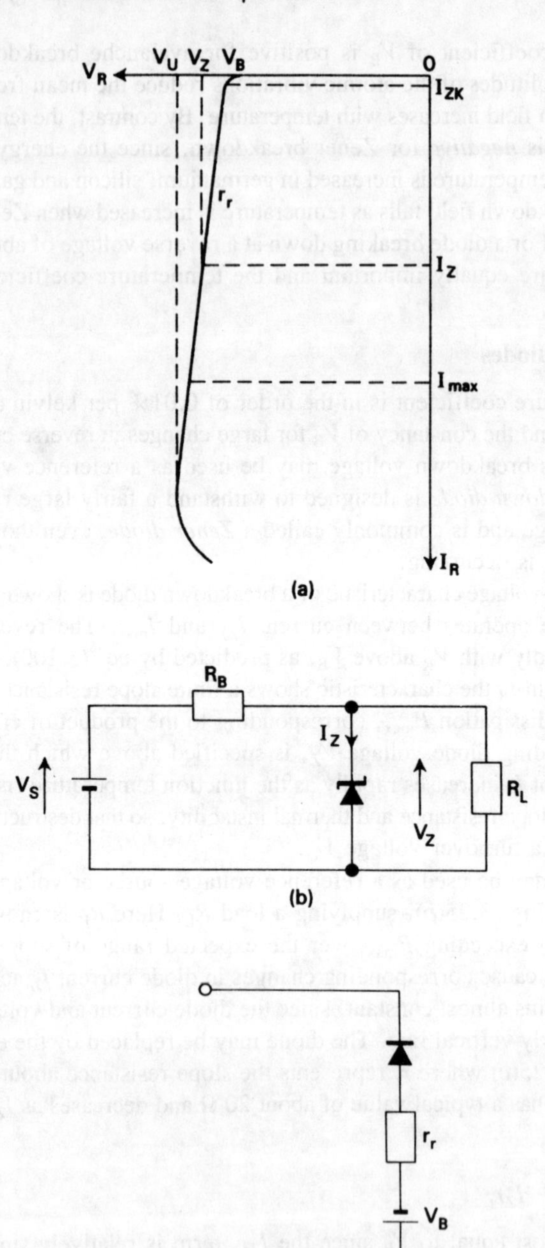

Figure 3.25 **Breakdown diode (*a*) *I–V* characteristic; (*b*) voltage-regulator circuit; (*c*) equivalent circuit.**

$$P_Z = I_Z V_Z$$

so

$$P_{max} = I_{max} V_Z \tag{3.103}$$

Available breakdown diodes have values of V_B from 2 to 200 V, r_r from 0.1 to 1000 Ω and P_{max} from 0.4 to 75 W.

Further increases in doping density, up to the limit of the solubility of impurities in the semiconductor, will cause V_B to move closer to zero, and eventually breakdown occurs for small forward bias voltages. This results in another device, the *tunnel diode*, which is described in Chapter 9.

Characteristics of a practical p-n junction

The theoretical diffusion current-voltage characteristics of eqs (3.5), (3.38) and (3.42) are closely followed by a Ge p-n diode operating at low current densities (*see* Fig. 3.26(*b*)i). At higher current densities with forward bias a practical p-n junction will have a greater voltage drop across it at a given current than predicted due to the ohmic resistance of the semiconductor materials and of their contacts with the external circuit. Thus, if the total resistance outside the junction is R,

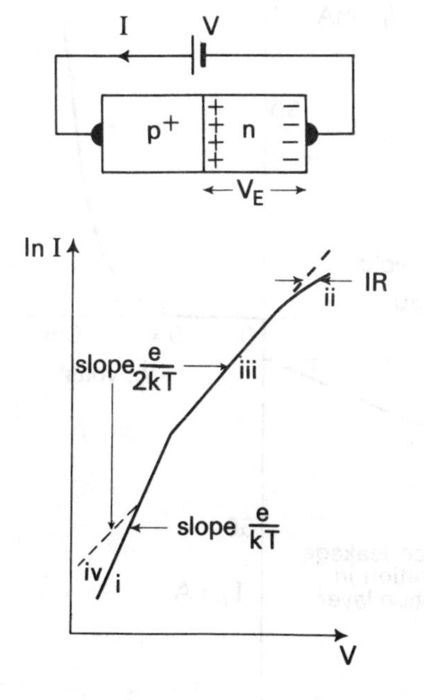

Figure 3.26 (*a*) High level injection in a p-n junction; (*b*) forward characteristic of a p-n junction (i) diffusion current region; (ii) series resistance region; (iii) high level injection region; (iv) recombination region.

the voltage across the junction is $V - IR$, so that a larger voltage must be applied to compensate for the *IR* drop (Fig. 3.26(*b*)ii).

For silicon and gallium arsenide diodes, in addition to the series resistance effect, high level injection and the generation and recombination of carriers in the depletion layer must be considered (Ref. [4]). High level injection occurs when the density of injected carriers is comparable to the density of majority carriers. Thus in a p^+-n diode at high currents both the density of holes injected at the junction and the density of electrons drawn in at the ohmic contact with the n-region are no longer negligible compared with the density of majority electrons in the n-region. This results in potential difference V_E being set up across the n-region in opposition to the applied voltage (Fig. 3.26(*a*)) and leads to a forward current I_F proportional to exp $(eV/2kT)$ (Fig. 3.26(*b*)iii). The slope of the ln *I* against *V* graph is thus halved at high currents, and in practice this may occur at currents below those for which the *IR* drop is significant. Again, under forward bias, extra current occurs to replace excess carriers lost by recombination. This recombination current is also proportional to exp $(eV/2kT)$ so that the forward current becomes

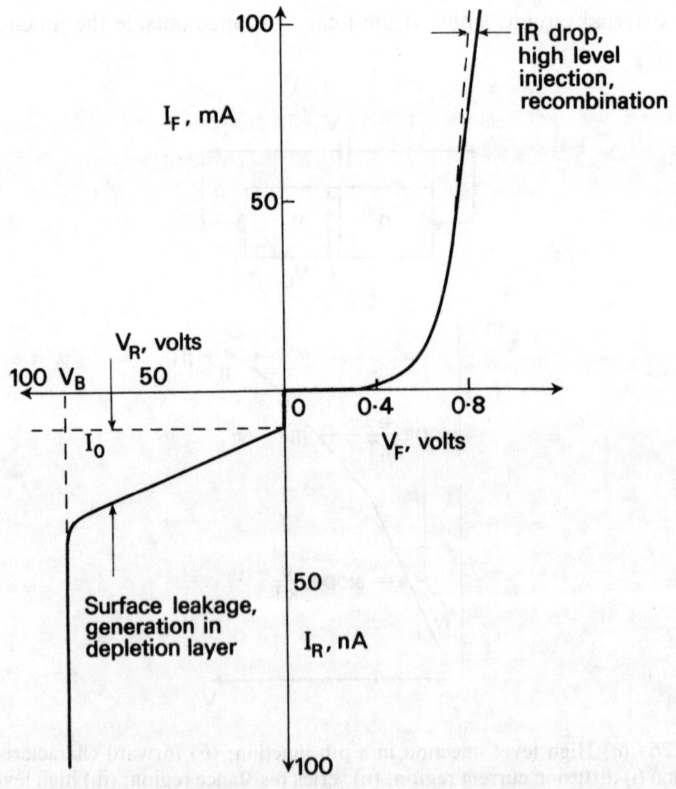

Figure 3.27 **I–V characteristic of a practical p-n junction.**

$$I_F = K_1 \exp(eV/kT) + K_2 \exp(eV/2kT) \tag{3.104}$$

where K_1 and K_2 are constants at constant temperature. A general empirical expression is $I_F \simeq$ const. exp (eV/nkT) where $n = 2$ when the recombination current is the greater and $n = 1$ when the diffusion current predominates (Fig. 3.26(*b*)iv). When both currents are comparable n lies between 1 and 2.

With reverse bias the current I_R is not independent of voltage V_R but shows a slow rise with V_R. This is mainly due to current leakage across the outer surface of the diode, which may be represented by a resistance in parallel with it. It is reduced by chemical treatment of the surface and by hermetically sealing the device. Finally, when $V_R = V_B$, I_R rises sharply at constant voltage and breakdown occurs (Fig. 3.27).

In silicon and gallium arsenide diodes under reverse bias the generation of electron-hole pairs is also important and this leads to an extra current which is proportion to n_i and to the width of the depletion layer d. Thus this current rises with reverse voltage V_R and so depends on $(\psi - V)^{1/2}$ for abrupt junctions and $(\psi - V)^{1/3}$ for linearly graded junctions. It adds to the effect of surface leakage, which may well predominate as shown in Fig. 3.27. For a leakage resistance r_l around a thousand megohms the scale of the reverse current I_R will be in pA, compared with mA for I_F. V_R would be in tens or hundreds of volts and V_F up to one volt or so.

A model to represent a diode with a characteristic of Fig. 3.27 is shown in Fig. 3.28. It combines the models of Figs. 3.17(*c*) and 3.25(*c*) for forward and reverse bias respectively, with a parallel resistance r_l to account for surface leakage and series resistance R for the ohmic resistance.

Computer simulation of a semiconductor diode

The p-n junction and Schottky diodes described in this chapter are important elements in electronic circuits. They may be used alone or as part of other devices, such as the transistors described in later chapters. The integrated circuits of Chapter

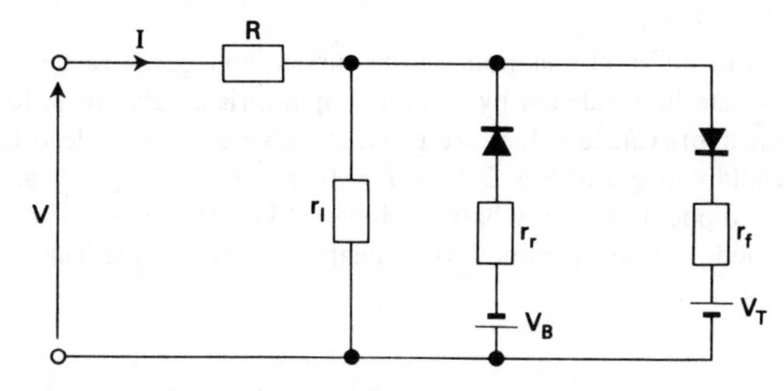

Figure 3.28 **Diode model corresponding to characteristics of Fig. 3.27.**

7 contain many diodes and transistors, together with other components such as resistors and capacitors. In fact, an integrated circuit is so complex that it is normal practice to use a computer to simulate the operation of the circuit, since this ensures that the design is correct before manufacture. It is much more economical to change the simulation rather than a completed circuit and this also applies to circuits made from discrete devices and components.

Computer simulation of an electronic circuit requires that the devices and components used can be described accurately by mathematical equations, which are known as *models*. The circuit nodes are specified and the nodal equations, including the device models, are solved to provide d.c., transient and a.c. analyses.* A widely used general-purpose circuit simulation program developed at the University of California, Berkeley, is SPICE, a Simulation Program with Integrated Circuit Emphasis. Further details about SPICE are given in Appendix 3 but in this chapter we are concerned with the diode model as used in SPICE version 2G.5 and to discuss how parameters derived in this chapter are used in the model.

d.c. analysis

The model used for the d.c. characteristics of a diode is based on eq. (3.5), with the addition of the *emission coefficient* n to give

$$I = I_0 \left[\exp\left(\frac{eV}{nkT}\right) - 1 \right] \tag{3.105}$$

n models the change in slope of $\ln I$ versus V illustrated in Fig. 3.26 and the total resistance outside the junction R of Fig. 3.28 can also be included. It should be noted that this model is more accurate than the one shown in Fig. 3.28 since no linear approximations are made. Consequently it will give much more precise values of current for a given diode voltage, which is one advantage of computer simulation. Equation (3.105) assumes that no reverse breakdown occurs but that the breakdown voltage V_B can be specified. In this case an *exponential* increase in reverse current is assumed and the current I_{BV} at V_B is used to define this variation (Fig. 3.29).

The effect of different materials on the current/voltage characteristic shown in Fig. 3.16 can be simulated by using the appropriate value of I_0 for a small diode. Some approximate values are given in Table 3.1, which lead to the forward threshold voltages of Fig. 3.16 at I = 10 mA from eq. (3.44) and include the Schottky diode or Schottky barrier diode (SBD). The values of I_0 are internal to the diode and are assumed to be unaffected by surface leakage.

* A node is a junction in a network. Nodal equations relate the voltage and current(s) at a node.

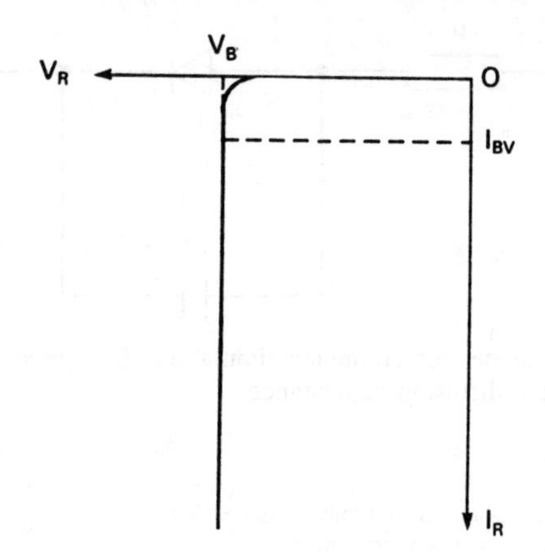

Figure 3.29 Simulated reverse *I/V* characteristic.

Table 3.1 Reverse current and forward voltage values for diodes of various materials

Material	I_0	Forward voltage at I = 10 mA
	A	V
Germanium	3×10^{-6}	0.20
Silicon	1×10^{-14}	0.69
Gallium arsenide	3×10^{-20}	1.04
Schottky diode, silicon device	3×10^{-7}	0.26

a.c. analysis

For a.c. analysis the d.c. model is extended to include capacitance, as shown in Fig. 3.30 where the ideal diode and series resistance R represent the d.c. model, C_j is the depletion layer capacitance given by eq. (3.78) for an abrupt junction or by eq. (3.94) for a graded junction. These equations are generalized into the form

$$C_j = \text{const.}\ (\psi - V)^{-m} \tag{3.106}$$

where m is the *grading coefficient* which is 1/2 for an abrupt junction and 1/3 for a graded junction. The constant is expressed in terms of C_{j0}, the value of C_j at $V = 0$ (Fig. 3.19), so that the constant becomes $C_{j0}\ \psi^m$, with ψ being given by eq. (3.22). Equation (3.106) then becomes

$$C_j = \frac{C_{j0}\ \psi^m}{(\psi - V)^m} \tag{3.107}$$

and typical parameter values given with SPICE for a small silicon diode are given in Table 3.2.

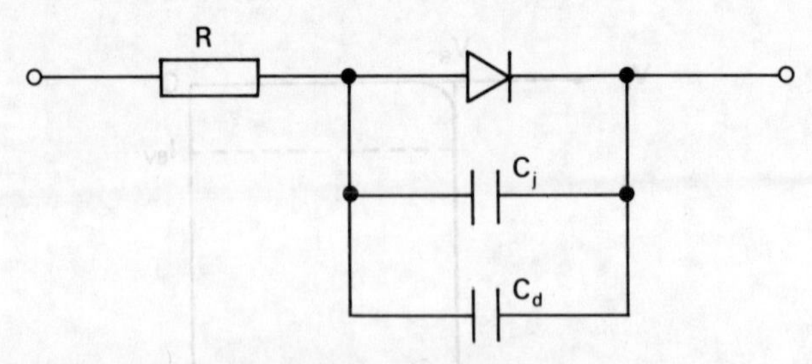

Figure 3.30 Diode model for computer simulation. C_j represents the depletion layer capacitance and C_d the diffusion capacitance.

Table 3.2 Theoretical symbols and SPICE parameters for diode simulation

Symbol	Parameter	Typical value	
I_0	IS	10^{-14} A	(Si)
R	RS	10 Ω	
n	N	1.0	
τ	TT	0.1 ns	
C_{j0}	CJO	2.0 pF	
ψ	VJ	0.6 V	(Si)
m	M	0.5	
W_g	EG	1.11 eV	(Si)
		0.69 eV	(SBD)
		0.67 eV	(Ge)
x	XTI	3.0	(Si)
		2.0	(SBD)
	FC	0.5	
v_B	BV	40 V	
I_{BV}	IBV	10^{-3} A	

A problem arises with eq. (3.107) if it is implemented directly in the simulation program, since C_j tends to infinity as V tends to ψ. In an actual diode the current I increases with V and the IR drop external to the junction prevents the junction voltage V becoming equal to ψ. However, in a simulated diode R may be zero and then the denominator of eq. (3.107) could also become zero, which would cause the simulation to fail. This is prevented by introducing a coefficient, called FC in SPICE, which gives the variation of C_j with V shown in Fig. 3.31. Simulated and theoretical variations are close until V is nearly equal to ψ, when C_j falls instead of rising to infinity. The details of the operation of the coefficient are given in Ref. [5].

C_d in Fig. 3.30 is the diffusion capacitance, which depends on the stored charge q_d supporting the diode current I as described earlier in this chapter under Transient Effects. Then

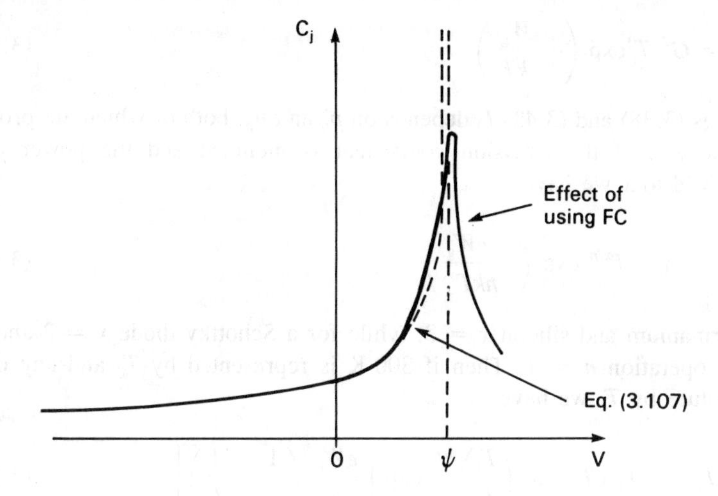

Figure 3.31 Effect of using the SPICE parameter FC for the variation of depletion layer capacitance with diode voltage.

$$I = \frac{q_d}{\tau} \tag{3.108}$$

where τ is the transit time of charge carriers across the diode. For small signals

$$C_d = \frac{dq_d}{dV}$$

$$= \frac{dq_d}{dI} \cdot \frac{dI}{dV}$$

$$= \frac{\tau e I}{kT} \tag{3.109}$$

where dI/dV is obtained from eq. (3.5) by differentiation, and by assuming that $I \gg I_0$ for forward bias. C_d is thus an effect of charge storage and is modelled by defining the transit time τ.

Analysis at different temperatures

All input data for SPICE is assumed to have been measured at 27 °C (300 K) and the simulation also assumes a nominal temperature of 27 °C. However, a very useful feature is that the circuit can be simulated at other temperatures, which is much simpler than building the circuit and operating it in an oven or a refrigerator.

Temperature appears explicitly in the exponential term of eq. (3.5) and, in addition, I_0 has a built-in temperature dependence. From eq. (2.30)

$$n_i^2 = G^2 T^3 \exp\left(-\frac{W_g}{kT}\right) \quad (3.110)$$

From eqs (3.38) and (3.42) I_0 depends on p_n and n_p, both of which are proportional to n_i^2. If the emission coefficient is included and the power of T generalized to x we have

$$I_0 = \text{const. } T^{x/n} \exp\left(\frac{-eW_g}{nkT}\right) \quad (3.111)$$

For germanium and silicon $x = 3$, while for a Schottky diode $x = 2$ and for normal operation $n = 1$. Then if 300 K is represented by T_0 and any other temperature by T_1 we have

$$I_0(T_1) = I_0(T_0) \times \left(\frac{T_1}{T_0}\right)^{x/n} \exp\left[\frac{eW_g}{k}\left(\frac{1}{T_0} - \frac{1}{T_1}\right)\right] \quad (3.112)$$

and eq. (3.112) models the effect of temperature on I_0 through W_g and x.

Temperature also appears explicitly in the value of junction potential ψ (eq. 3.22) for all device models.

The theoretical symbols used in this chapter cannot be used directly in the simulation program and the corresponding SPICE parameters are shown in Table 3.2, together with typical SPICE values. A full list of diode parameters with their names is given in Appendix 3.

Points to remember

- When two different materials are brought into contact a built-in potential appears across the junction, which is known as the contact or diffusion potential.
- Contacts can be either ohmic or rectifying with the p-n junction being the most important rectifying contact. It is also known as the p-n diode.
- The saturation or leakage current flows when a reverse voltage is applied across a diode and is an exponential function of temperature.
- The forward current of a diode is proportional to the saturation current and rises exponentially with forward voltage.
- A p-n junction may be abrupt or graded and has a depletion layer with an equivalent parallel plate capacitance.
- The maximum electric field across the depletion layer depends on the

reverse voltage and may be high enough to cause breakdown of the junction.

* A junction diode can be accurately simulated on a computer using the SPICE program, which has parameters defining the physical properties of the device.

References

[1] Rowe, D.M. and Bhandari, C. *Modern Thermoelectrics*, (Holt Saunders), 1983.
[2] Shockley, W. 'The theory of p-n junctions in semiconductors and p-n junction transistors', *Bell Syst. Tech. J.*, 1949, vol. 28, p. 435.
[3] Chang, C.Y., Chin, S.S. and Hsu, L.P. 'Temperature dependence of breakdown voltage in silicon abrupt p-n junctions', *IEEE Trans. Electron Devices*, 1971, vol. ED-18, p. 391.
[4] Sah, C.T., Noyce, R.N. and Shockley, W. 'Carrier generation and recombination in p-n junction and p-n junction characteristics', *Proc. IRE*, 1957, vol. 45, p. 1228.
[5] Getreu, I.E. *Modelling the bipolar transistor*, (Elsevier), 1978.

Problems

At room temperature $kT/e = 0.025$ eV

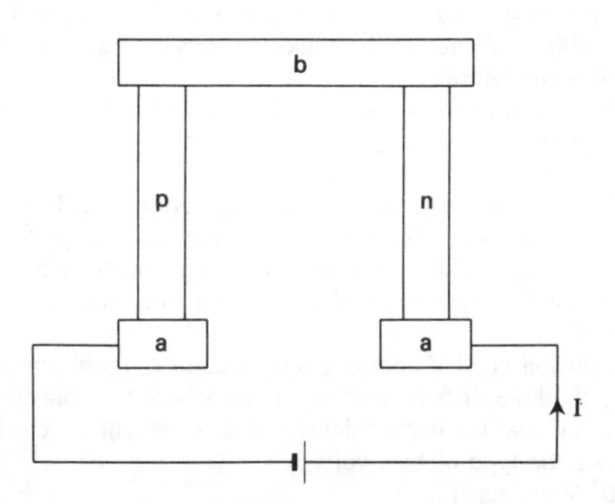

3.1 Ohmic contacts are made between slabs of n- and p-type semiconductor material and metal blocks **a** and **b** as shown above. The couple has a total resistance 7 mΩ, assumed constant, and both blocks **a** are maintained at 20 °C. When a current of 10 A is passed, with all the blocks initially at the same temperature, the initial value of dQ/dt is 0.65 J/s. If the resulting steady-state temperature of **b** is −12 °C, calculate the values of α_p, K and thermal efficiency. Obtain the steady-state temperature of block **b** when the current is increased to 25 A and explain the result.
[0.1 V; 0.03 W/°C; 1.2%; 4.5 °C]

3.2 An ideal Schottky diode is formed between gold and n-type silicon, the work function ϕ_m of gold being 4.8 eV and the electron affinity χ of silicon being 4.05 eV. If N_c is 2.8×10^{25} levels/m^3 and there are 1.5×10^{23} donor atoms/m^3, calculate the barrier heights at room temperature for electrons moving (i) from the metal into the semiconductor and (ii) from the bottom of the semiconductor conduction band into the metal.
[0.75 eV; 0.62 eV]

3.3 A pn junction has a donor density $N_d = 2.0 \times 10^{23}$ atoms/m^3, an acceptor density $N_a = 4.0 \times 10^{22}$ atoms/m^3 and intrinsic density $n_i = 1.5 \times 10^{16}$ electron-hole pairs/m^3 at room temperature. Determine the height of the potential barrier at the junction with (i) a forward bias of 0.6 V and (ii) a reverse bias of 10 V.
[0.18 V, 10.78 V]

3.4 An abrupt silicon p-n junction is formed from p-type material with a resistivity of 1.3×10^{-3} Ω m and n-type material with resistivity of 4.6×10^{-3} Ω m at room temperature. The lifetimes of the p- and n-materials are 100 μs and 150 μs respectively, and the junction area is 1.0 mm^2.

If $\mu_p = 4.8 \times 10^{-2}$ m^2/Vs, $\mu_n = 0.135$ m^3/Vs and $n_i = 6.5 \times 10^{16}$/m^3, calculate the reverse bias leakage current, assuming the p- and n-regions are much longer than the diffusion length.

If a similar junction to the above is formed, except that the lengths of the p- and n-regions are each 50 μm, what is the new leakage current?
[2.7×10^{-13} A; 2.1×10^{-12} A]

3.5 The reverse saturation current of a p-n diode of area S may be expressed as

$$I_0 = eS[D_p\, p_n/L_p + D_n n_p/L_n]$$

where D_n and D_p are the diffusion coefficients and L_n and L_p are the diffusion lengths, assumed to be much less than the diode length. Explain briefly the physical principles underlying the expression.

A silicon p-n diode of area $S = 0.1$ mm^2 has $N_A = N_D = 10^{22}$ atoms/m^3 and a lifetime of 100 μs in each of the p and n materials. Calculate the saturation current at room temperature.

If a similar p-n diode to the one above is formed with the p and n regions each 30 μm long, what is the new saturation current?
[3.26×10^{-15} A, 5.49×10^{-14} A]

3.6 Calculate the forward voltage of an ideal p-n junction for a diode current of 10 mA when I_0 is (i) 1.0 μA, (ii) 1.0 pA at room temperature. Suggest with reasons which of these values of I_0 would correspond to a germanium diode and which to a silicon diode, assuming the same physical dimensions in each case.
[0.23 V; 0.58 V]

3.7 The hole diffusion current across a p-n junction at room temperature is exactly balanced by the hole drift current at a point where the concentration gradient is 10^{18} carriers/m^4 and the carrier density is 2×10^{11}/m^3. Determine the junction field and state the type of bias voltage.
[1.25×10^5 V/m; zero]

3.8 The characteristics of a p-n diode are given by

$$I = I_0 \left[\exp\left(\frac{1.16 \times 10^4\, V}{T} \right) - 1 \right] \text{ amperes}$$

where $I_0 = 10^{-8}$ A at 20 °C and doubles for each 6 °C rise in temperature. If the diode is operated at a constant current of 1 mA estimate the change in forward voltage when the temperature rises from 20 °C to 42 °C.
[–52 mV]

3.9 Assuming that Maxwell–Boltzmann statistics are applicable, show that the current/voltage characteristic of a p-n junction is

$$I = I_0 \left[\exp\left(\frac{eV}{kT}\right) - 1 \right]$$

For a silicon p-n junction, I_0 is proportional to exp $(-1.1 \times e/kT)$ and at 300 K the forward voltage is 600 mV for a current of 10 mA. What would be the voltage at 330 K if the current remained constant at 10 mA?
[0.55 V]

3.10 A certain abrupt p-n junction is made from silicon doped with 5×10^{22} and 1×10^{22} impurity atoms per m^3 on the p- and n-sides respectively. Using Gauss's theorem, or otherwise, obtain the width of the depletion layer when the maximum electric field within the junction has a magnitude of 2×10^7 V/m under reverse-bias conditions.
[1.6 μm]

3.11 A linearly graded silicon p-n junction of area 0.1 mm^2 has a grade constant of 10^{27} atoms/m^4. If the effective reverse bias is $\psi - V$ calculate for a maximum field of 10^7 V/m (i) the depletion layer width; (ii) the effective reverse bias; and (iii) the junction capacitance.
[7.29 μm; 48.58 V; 1.46 pF]

3.12 An abrupt silicon p-n junction is formed from a p-type semiconductor with an acceptor density of $3 \times 10^{23}/m^3$ and an n-type semiconductor with a donor density of $3 \times 10^{22}/m^3$. If the junction area is 1.0 mm^2 and the reverse bias is 10 V calculate (i) the width of the depletion layer, (ii) the maximum field within the depletion layer, (iii) the depletion layer capacitance.

Assuming that a field of 5×10^7 V/m is required for avalanche breakdown, what is (iv) the breakdown voltage of the junction?
[0.7 μm; 2.86×10^7 V/m; 152 pF; 30 V]

3.13 A graded silicon p-n junction is formed by diffusing boron into n-type silicon having a resistivity of 0.015 Ω m. The surface density of boron atoms is $5 \times 10^{23}/m^3$ and the transition from p- to n-type material occurs 1.5 μm below the surface. If the junction area is 1.0 mm^2 and the reverse bias is 10 V, calculate for the depletion layer (i) its width; (ii) its maximum field; (iii) its capacitance; and (iv) its breakdown voltage, assuming the same breakdown field as in Problem 3.12.
[1.6 μm; 9.65×10^6 V/m; 311 pF; 120 V]

3.14 The capacitance C_j of a p-n diode depends on the reverse voltage V_R as shown below:

V_R volts	C_j pF
2.0	8.0
5.0	6.0

Assuming that $C_j = K/V_R^n$ determine K and n and calculate the value of V_R for $C_j = 4.0$ pF.
[10^{-11}; 0.33; 19 V]

3.15 Explain briefly how a depletion layer is formed in a p-n junction. Describe how it leads to a junction capacitance C_j.

Show that C_j may be expressed as

$$C_j = C_{j0}/(1 + V_R/\psi)^n$$

where C_{j0} is the junction capacitance at zero voltage, ψ is the diffusion potential and V_R is the reverse voltage.

Determine the voltage to be applied to a diode so that it resonates at 100 MHz with a 2.0 μH inductor. The diode constants are C_{j0} = 3.0 pF, ψ = 0.78 V and n = 0.33.
[9.59 V]

3.16 Sketch the current-voltage characteristic of a practical p-n diode, explaining carefully how it differs from an ideal diode characteristic.

A certain p-n diode has the following properties at 40 °C:
Reverse current: 30 nA at 20 V, 5 nA saturation value
Forward voltage: 0.44 V at 20 mA

Estimate the reverse resistance across the junction and the series resistance of the diode.
[800 MΩ; 1.5 Ω]

3.17 The following SPICE parameters apply to a junction diode

IS, N, EG, BV, IBV

Explain the meaning of each parameter and whether or not it can be used in the expression $I = I_0\ [\exp\ (eV/kT) - 1]$.

Calculate the forward voltage at I = 10 mA for
(i) a silicon diode with $I_0 = 10^{-14}$ A and W_g = 1.11 eV
(ii) a similar gallium arsenide diode with W_g = 1.43 eV.

What is the value of I_0 for this diode?
[0.69 V; 1.01 V; 2.85 × 10^{-20} A]

4 Bipolar junction transistors and thyristors

The bipolar junction transistor (BJT)

A bipolar junction transistor consists of two p-n junctions formed back-to-back in a single crystal of semiconductor material. Two arrangements are possible, p-n-p and n-p-n. A p-n-p transistor consists of a thin n-region sandwiched between two p-regions, while in the n-p-n transistor the types of semiconductor are reversed. In both types the central region is called the base and the outer regions are called the *emitter* and the *collector* respectively.

In normal operation the emitter-base diode is forward biased and the collector-base diode is reverse biased, and under these conditions current from the emitter flows across the base to the collector. The collector current is carried by holes in the p-n-p transistor and by electrons in the n-p-n transistor, which is generally preferred due to the higher mobility of electrons. The term *bipolar* refers to the fact that *both* types of current carrier are involved in either p-n-p or n-p-n transistors, although holes predominate in p-n-p and electrons predominate in n-p-n, and the device may be referred to as a Bipolar Junction Transistor or BJT.

Practical construction methods are discussed in Chapter 7 and lead to a *graded base* transistor in which the impurity content of the base falls between emitter and collector. The essential theory is more easily derived by considering the limiting case of a transistor with a base having a uniform impurity distribution and this is developed below for the n-p-n transistor. Similar considerations apply to the p-n-p type, with holes substituted for electrons and the polarities of the applied voltages reversed.

When no potentials are applied the potential barriers and energy levels for the uniform-base n-p-n transistor are as shown in Fig. 4.1, while for the normal applied potentials the potential barriers are shown in Fig. 4.2. The emitter-base voltage V_{EB} is normally between 0.2 and 0.7 V, while the collector-base voltage V_{CB} may range from about 2 V to 200 V. Since the collector-base junction is reverse biased the collector current contains a leakage component I_{CBO}, which is mainly due to holes from the collector. I_{CBO} is the current flowing between collector and base when the emitter is open-circuited.

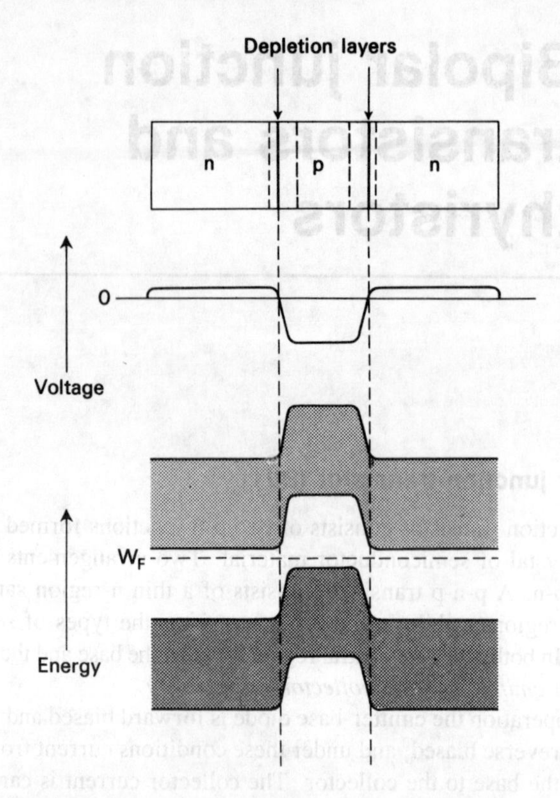

Figure 4.1 **Potential and energy distributions of an unbiased uniform base n-p-n transistor.**

The emitter n-region is much more heavily doped than the p-region, so that as electrons are injected into the base an excess density n_e is set up at the edge of the emitter-base depletion layer (Fig. 4.3). These electrons diffuse across the base towards the collector, and if the effective width of the base region w_b is much less than the diffusion length for electrons very little recombination occurs. At the collector the field across the collector-base junction sweeps all the electrons out of the base, whether they are the original minority carriers in the p-region or excess carriers injected from the emitter. Thus the electron density is zero at the base side of the collector-base depletion layer.

An important feature of the transistor is the fraction of the emitter current that reaches the collector. This can be expressed as the ratio of the direct collector and emitter currents I_C/I_E when V_{CB} is held constant. However, when the transistor is connected, as shown in Fig. 4.2, the measured collector current includes the leakage current of the collector-base diode, I_{CBO}. The leakage current may be excluded by considering the *rate of change* of collector current with emitter current, which is given the symbol h_{fb} expressed here as α_0 where

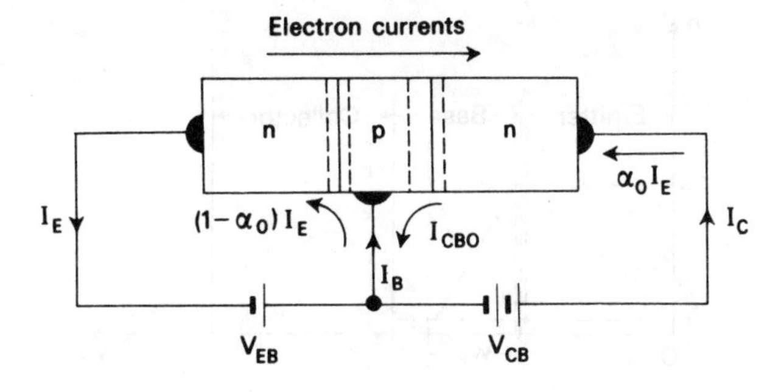

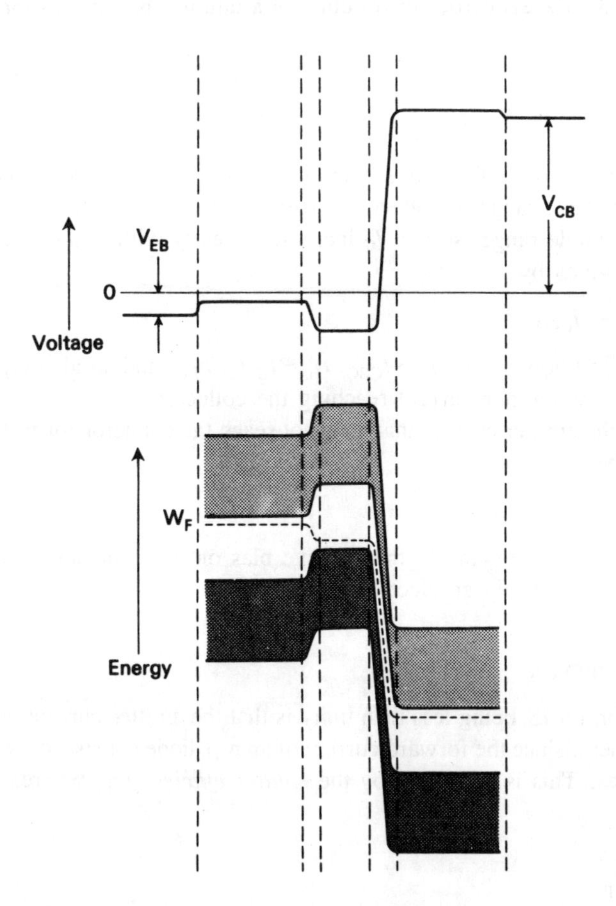

Figure 4.2 Current flow, potential and energy distributions for a normally biased uniform-base n-p-n transistor. The base is the electrode common to both input and output. Electron currents flow in the opposite direction to the conventional currents shown.

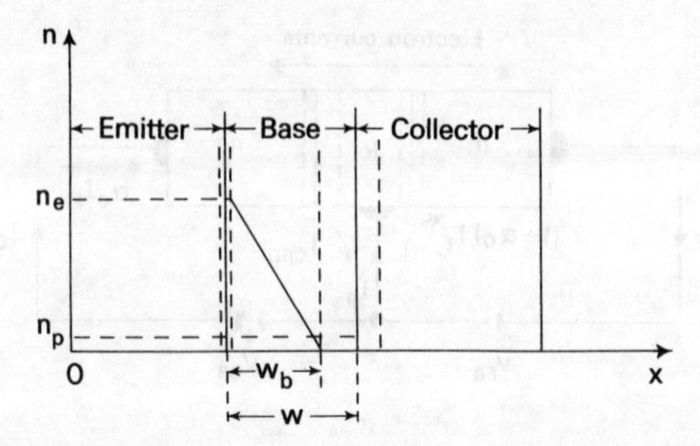

Figure 4.3 Excess carrier distribution in a uniform-base transistor.

$$\alpha_0 = \left.\frac{\partial I_C}{\partial I_E}\right|_{V_{CB}\ \text{constant}} = h_{fb} \tag{4.1}$$

In a practical transistor α_0 lies between about 0.95 and 0.995 and is constant over most of the available range of collector current. In an ideal transistor α_0 is constant over the whole range, so that I_C increases linearly with I_E and the collector current is given by

$$I_C = \alpha_0 I_E + I_{CBO} \tag{4.2}$$

Under these conditions, since $I_C \gg I_{CBO}$, $\alpha_0 \approx I_C/I_E$, h_{FB}, and so also represents the fraction of the emitter current reaching the collector.

The part of the emitter current that does not reach the collector forms the base current I_B, where

$$I_B = (1 - \alpha_0)I_E - I_{CBO} \tag{4.3}$$

as shown in Fig. 4.2. Owing to the positive bias on the collector in an n-p-n transistor, $\alpha_0 I_E$ is entirely an electron current.

Emitter efficiency, γ

The first reason for α_0 being less than unity is that the emitter current will have a hole component, since the forward current of an n-p diode consists of both electrons and holes. This is described by the *emitter efficiency* γ, where

$$\gamma = \frac{I_n}{I_n + I_p} \tag{4.4}$$

I_p is the hole component and I_n the electron component of the current crossing the emitter-base junction. I_p is made very small by doping the n-region much more heavily than the p-region, so forming an n^+-p-n transistor. From eq. (4.4),

$$\gamma = \frac{1}{1 + \dfrac{I_p}{I_n}} = \left(1 + \frac{I_p}{I_n}\right)^{-1}$$

$$\approx 1 - \frac{I_p}{I_n} \tag{4.5}$$

if $I_n \gg I_p$ and using the binomial theorem. I_p may be obtained from eq. (3.32):

$$I_p = \frac{eD_p\, p_n S}{L_{pe}} \left[\exp\left(\frac{eV}{kT}\right) - 1\right] \tag{4.6}$$

S being the effective cross-sectional area of the base and L_{pe} the diffusion length of electrons in the emitter. The width of the base is normally much less than the diffusion length of electrons, so that I_n is obtained from eq. (3.37) with L_n replaced by w_b, which gives

$$I_n = \frac{eD_n n_p S}{w_b} \left[\exp\left(\frac{eV}{kT}\right) - 1\right] \tag{4.7}$$

dividing eq. (4.6) by eq. (4.7) and substituting into eq. (4.5),

$$\gamma = 1 - \frac{D_p\, p_n w_b}{D_n n_p L_{pe}} \tag{4.8}$$

Since $n_p p_p = n_n p_n$ from eq. (2.37) which gives $p_n/n_p = p_p/n_n$, and also $D_p/D_n = \mu_p/\mu_n$ from eqs (2.108) and (2.109), we can write

$$\frac{D_p\, p_n}{D_n n_p} = \frac{\sigma_p}{\sigma_n} \tag{4.9}$$

using eq. (2.66) and where σ_p and σ_n are the conductivities of the p- and n-regions respectively. Then

$$\gamma = 1 - \frac{\sigma_p w_b}{\sigma_n L_{pe}} \tag{4.10}$$

and γ approaches unity when the second term is small. This is achieved by making the base region narrow and also by making $\sigma_n \gg \sigma_p$ which occurs when $n_n \gg p_p$. It is common to make the conductivity of the n-region about 100 times that of the p-region, so that γ is normally greater than 0.99 since w_b in a uniform-base transistor is typically a few μm, which is much less than L_{pe}.

Base transport factor, δ

The second reason for α_0 being less than unity is that some of the electrons diffusing across the base will recombine with the holes present as majority carriers.

This results in a component of base current I_{BF} due to holes replacing those that have recombined and leads to the *base transport factor*, δ, which is the ratio of the electron current entering the collector to the electron current leaving the emitter.

The distribution of excess electrons in the base is shown in Fig. 4.4, and since they move across the base by diffusion, the current at any point is proportional to the slope of the electron distribution curve at that point. Thus the electron current leaving the emitter is

$$\gamma I_E = -eD_n S \left.\frac{dn}{dx}\right|_{x=0} \tag{4.11}$$

and the electron current entering the collector is

$$\delta\gamma I_E = -eD_n S \left.\frac{dn}{dx}\right|_{x=w_b} \tag{4.12}$$

The difference between these two currents constitutes the base current I_{BF} by which electrons which have recombined with holes are replaced from the external circuit, so that

$$I_{BF} = (1-\delta)\gamma I_E \tag{4.13}$$

The component of the base charge due to leakage current is negligible since in a small transistor this current would be a few nA as compared with an emitter current of about 1 mA and a base current of about 10 μA. Also δ is only just less than unity, so that to a first approximation

$$\left.\frac{dn}{dx}\right|_{x=0} \approx \left.\frac{dn}{dx}\right|_{x=w_b} \tag{4.14}$$

and the distribution of electrons in the base may be considered linear, as indicated

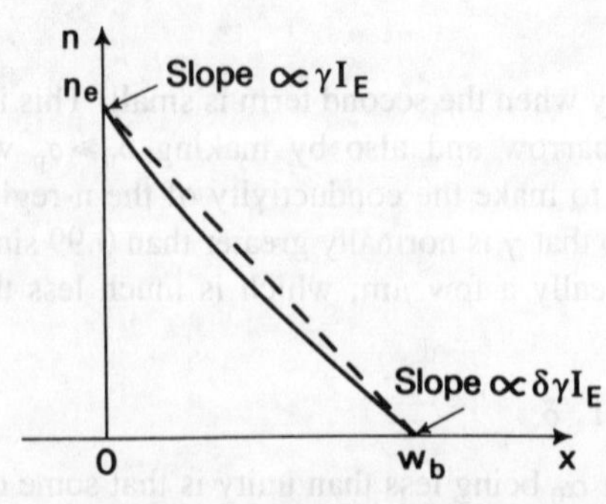

Figure 4.4 **Distribution of excess charge in the uniform base.**

by the dotted line. The effective volume of the base is Sw_b, and the average density of the excess base charge is $\frac{1}{2}\, en_e$, so that the actual excess base charge, q_B, is given by

$$q_B = \frac{en_e w_b S}{2} \tag{4.15}$$

and

$$\gamma I_E = \frac{eD_n n_e S}{w_b} = q_B \frac{2D_n}{w_b^2} \tag{4.16}$$

This equation may be written in the form

$$\frac{q_B}{I_E} = \tau_C \tag{4.17}$$

since $\gamma \approx 1$ and putting $w_b^2/2D_n = \tau_C$.

The excess carrier distribution of Fig. 4.4 is maintained in the base by electrons entering from the emitter and leaving for the collector, so that τ_C represents the average transit time of electrons across the base.

In addition there is always space-charge neutrality in the base to ensure a negligible electric field and allow diffusion processes to dominate, although this does not apply to the graded base, described later. Consequently, there is an equal and opposite excess hole charge q_B related to the base current I_{BF} by the expression

$$I_{BF} = \frac{q_B}{\tau_B} \tag{4.18}$$

Since I_{BF} is due to the process of recombination in the base, we can see that τ_B represents the lifetime of the minority carriers in the base. Combining eqs (4.13), (4.16) and (4.18),

$$1 - \delta = \frac{w_b^2}{2D_n \tau_B} = \frac{w_b^2}{2L_n^2} \tag{4.19}$$

where L_n is the diffusion length of electrons in the base. Hence

$$\delta = 1 - \frac{w_b^2}{2L_n^2} \tag{4.20}$$

and $\delta \to 1$ when $w_b \ll L_n$. This is achieved by making the base width small during manufacture and by making the diffusion length large by means of a long minority carrier lifetime in the base region.

In practice δ is about 0.995 and where τ_B is small as in a high frequency or fast switching transistor the value of δ is again maintained by having a very thin base, (eq. (4.20)).

Base current components

Under normal operating conditions

$$\alpha_0 = \gamma\delta \approx \left(1 - \frac{\sigma_p w_b}{\sigma_n L_{pe}}\right)\left(1 - \frac{w_b^2}{2L_n^2}\right) \tag{4.21}$$

and for typical values of γ and δ given above, α_0 is 0.99.

The fraction of the emitter current reaching the collector depends, not only on the current I_{BF}, but also on a second component, I_{BE}, which is due to the electron component of the emitter current and is given by

$$I_{BE} = (1 - \gamma)I_E \tag{4.22}$$

Thus the total base current is

$$\begin{aligned} I_B &= I_{BF} + I_{BE} - I_{CBO} \\ &= I_E(\gamma - \gamma\delta + 1 - \gamma) - I_{CBO} \\ &= (1 - \alpha_0)I_E - I_{CBO} \end{aligned} \tag{4.23}$$

which is the same as eq. (4.3). The various components of the base current are illustrated in Fig. 4.5.

In a 'good' transistor γ and δ (and so α_0) would be just less than unity, so that I_B is very small and I_C is practically equal to I_E, and such a transistor will also have a large value of h_{FE} (*see* eq. (4.48)).

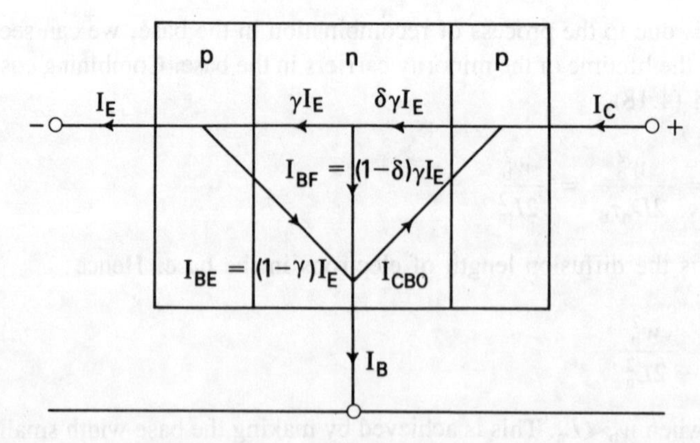

Figure 4.5 Components of the base current. Electron currents flow in the opposite direction to the conventional currents shown.

d.c. characteristics of a transistor

The currents flowing through a transistor are related to the voltages applied between the electrodes, for instance, I_E is an exponential function of V_{EB} because the emitter-base diode is forward biased. The arrangement so far discussed is called the *common-base* configuration, in which the emitter is the input electrode and the collector the output electrode, with the base as the electrode common to both the input and output circuits. Here I_E controls I_C and this process can be easily appreciated from a physical point of view.

However, this is not the only possible arrangement of the transistor. Greater efficiency is obtained if the base is used as the input electrode, so that the small base current controls the much larger collector current. This is known as the *common-emitter* configuration, since the emitter is now the electrode common to the input and output circuits. Although this is the most usual method of operation, it is less easy to appreciate physically, so the common-base characteristics will be discussed first. Two sets of curves are needed to give all the relationships between the voltages and currents of a transistor. These are known as the *input* and the *output* characteristics respectively.

Common base characteristics

Common-base input characteristics are shown in Fig. 4.6. Their general form follows the *I–V* characteristic of a silicon p-n diode under forward bias when V_{CB} is low. However, it can be seen that, if V_{BE} is fixed, I_E rises as V_{CB} is made more positive. This is due to the effect of V_{CB} on the base width w_b which will now be considered in more detail. For a higher emitter efficiency, I_B will be almost entirely due to electrons and the electron distribution in the base region will be as shown in Fig. 4.7(*a*), with the excess density at the edge of the emitter-base depletion layer increasing with V_{EB} as in eq. (3.34) for the diode. The effective base width w_b is less than the actual distance W between the two

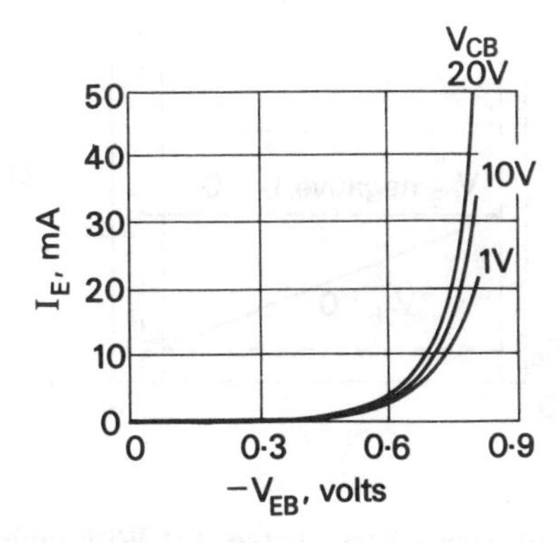

Figure 4.6 Common-base input characteristics (silicon n-p-n transistor).

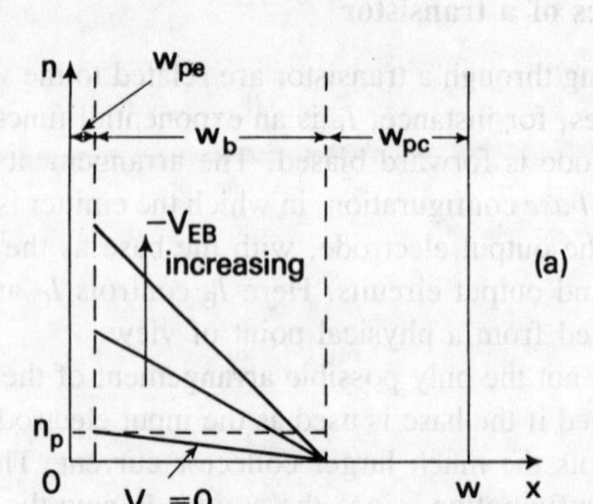

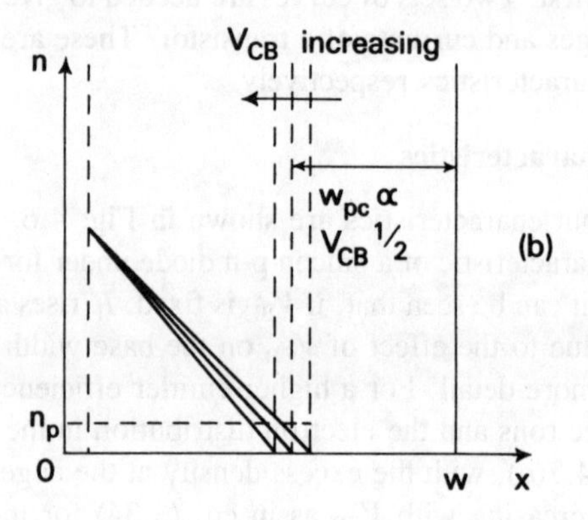

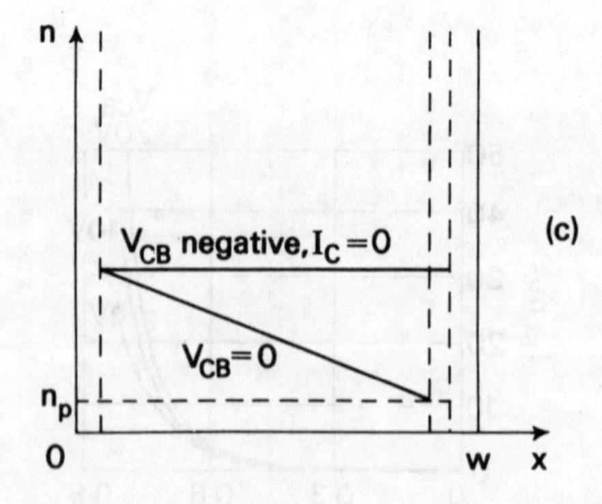

Figure 4.7 **Variation of excess base charge. (*a*) With emitter-base voltage; (*b*) with collector-base voltage; (*c*) in the saturated region.**

metallurgical junctions, since both depletion layers penetrate into the base. The distances penetrated are w_{pe} at the emitter and w_{pc} at the collector, so that

$$w_b = W - w_{pe} - w_{pc} \qquad (4.24)$$

The general expression of w_p, the penetration of the depletion layer into the p-region, is given by eq. (3.72). For an n-p-n transistor $N_d \gg N_a$, so the general expression becomes

$$w_p \approx \left(\frac{2\epsilon}{eN_a}\right)^{1/2} (\psi - V)^{1/2} \qquad (4.25)$$

For the emitter junction V is small and positive and for the collector junction, V is much larger and negative. Hence $w_{pe} \ll w_{pc}$, and is nearly constant over the allowed range of emitter voltages, while w_{pc} varies widely with the much larger changes in collector voltage. Thus we can write

$$w_b \approx W - w_{pc} \qquad (4.26)$$

For a narrow p-region the emitter current may be obtained from eq. (4.7). Putting $w_b = W - w_{pc}$, making S the area of the emitter junction, and neglecting the hole current term since γ is assumed almost unity,

$$I_E = \frac{eSD_n n_p}{(W - w_{pc})} \exp\left[\left(\frac{eV_{EB}}{kT}\right) -1\right] \qquad (4.27)$$

Since $w_{pc} \propto (\psi + V_{CB})^{1/2}$, $W - w_{pc}$ falls as V_{CB} is increased and I_E therefore rises, when V_{EB} is held constant. The effect on the charge distribution is illustrated in Fig. 4.7(*b*). At a sufficiently large value of V_{CB}, $w_{pc} = W$, so that $w_b \to 0$ and the depletion layer extends right through the base to the emitter junction. The emitter current then rises directly with V_{CB} and normal transistor action ceases. This condition is known as *punch-through*, which, like Zener or avalanche breakdown, is destructive unless the current is limited to prevent excessive heat dissipation. It occurs at a collector-base voltage V_{PT}, obtained from eq. (4.25):

$$V_{PT} \approx \frac{eN_a W^2}{2\epsilon} \qquad (4.28)$$

putting $w_p = W$.

The common-base output characteristics of a small transistor (Fig. 4.8) show how I_C varies with V_{CB} for constant values of I_B, I_C being almost equal to I_E and practically independent of V_{CB}. When $I_E = 0$, $I_C = I_{CBO}$, which is the leakage current of the collector-base diode. This is so small that it cannot be seen on a milliampere scale. I_C is then due to a current $\alpha_0 I_E$ added directly to I_{CBO} as in eq. (4.2). Both γ and δ are functions of w_b, so that α_0 and hence I_C increase slowly with V_{CB}. However, for high values of V_{CB} breakdown occurs, as discussed below, and I_C rises rapidly with V_{CB}. I_C flows even when $V_{CB} = 0$, since the

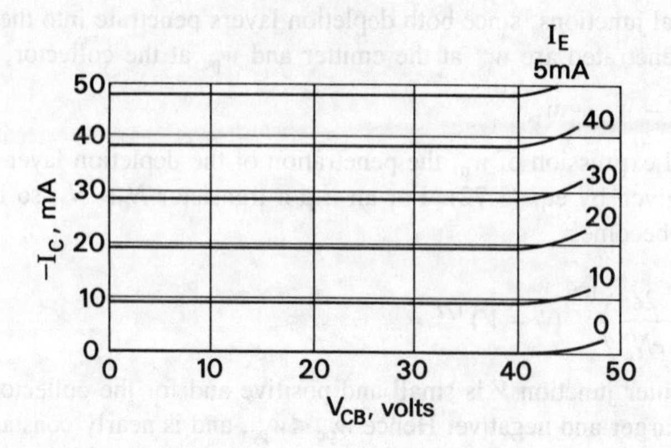

Figure 4.8 **Common-base output characteristics (n-p-n transistor).**

electron distribution in the base still has a gradient towards the collector when V_{EB} is positive. In fact a small forward bias on the collector is required to reduce I_C to zero, as shown in Fig. 4.7(*c*).

The planar junction transistor

In practice single (discrete) junction transistors are manufactured by techniques similar to those described in Chapter 7 for integrated circuits, with the slice being cut up into individual devices. A cross-section for a discrete transistor is shown in Fig. 4.9 and the starting point would be a slice of n^+-silicon. This contrasts with the p-type slice for an integrated circuit which is necessary to provide isolation between devices. An epitaxial n-layer is deposited for the collector, the base region is diffused into this epilayer and the emitter region is diffused into the base. Base and emitter contacts are provided at the top as before but the collector contact is now at the bottom. The n^+-region ensures an ohmic contact and it is mounted on aluminium for the external contact. This construction ensures a lower collector resistance and efficient conduction of heat from the collector. A similar construction is used for discrete p-n-p transistors, starting with a p^+-substrate. The silicon dioxide plays a very important part in protecting the electrodes and their junctions from contamination by the atmosphere and ensures that leakage currents are low and that there are few surface traps. Thus less recombination takes place at low emitter currents and α_0 has a useful value at collector currents down to about 10 μA (*see* Fig. 4.12). Furthermore the characteristics of the transistor remain stable over long periods. The transistor is so small, with a diameter less than 100 μm that a large number can be manufactured at the same time on one wafer of silicon, typically 2,500 on a silicon slice with a diameter of 4 cm. Thus the manufacturing process is very economical and uniform, as in the case of the production of integrated circuits.

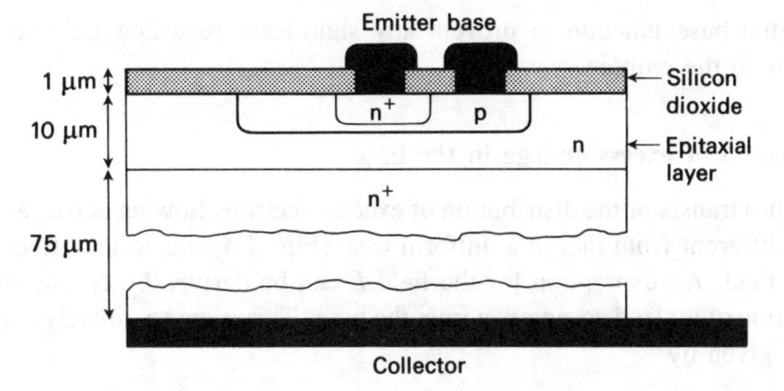

Figure 4.9 Section through a discrete diffused base n-p-n transistor (not to scale).

The base impurities are not uniformly distributed but decrease in concentration from emitter to collector (Fig. 4.10(*a*)) to give a *graded base* about a micrometer wide. The collector-base junction is also graded but the emitter-base junction is effectively abrupt due to the heavy emitter doping. The zero bias potential distribution through the transistor is shown in Fig. 4.10(*b*) and leads to a field E built into the base which, in fact, acts to accelerate electrons from emitter to collector. Since the emitter is heavily doped the potential falls rapidly across

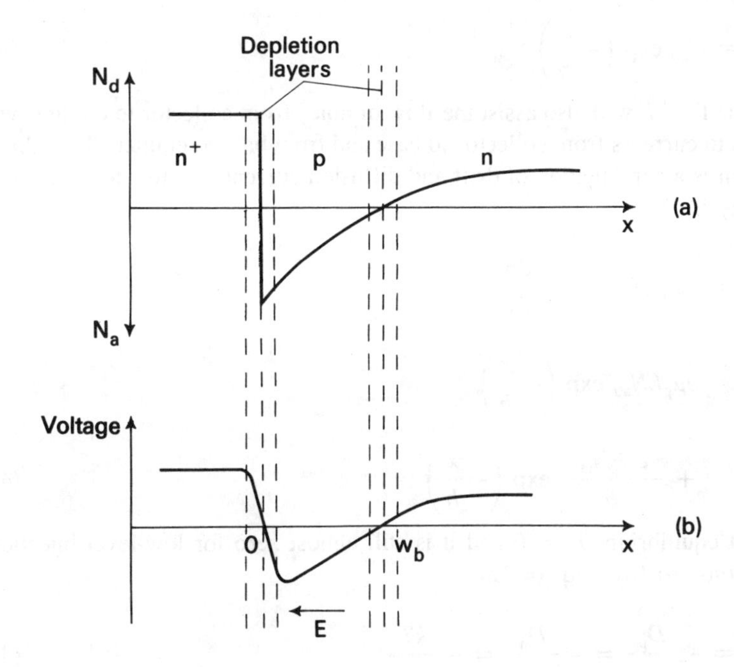

Figure 4.10 n-p-n transistor with graded base. (*a*) Distribution of impurities; (*b*) potential through transistor with zero applied bias.

the emitter-base junction to prevent any significant retarding field returning electrons to the emitter.

Distribution of excess charge in the base

In an n-p-n transistor the distribution of excess electrons flowing across a graded base is different from that in a uniform base (Fig. 4.4) due to the effect of the built-in field. An expression for the field E can be derived by considering the distribution of ionized acceptors across the base. This is approximately exponential and given by

$$N_a = N_{a0} \exp\left(-\frac{x}{b}\right) \quad (4.29)$$

N_{a0} is the acceptor density at the emitter side of the base where $x = 0$ and the length b defines the rate of change of the impurity density across the base. Since at room temperature we can assume that all the acceptors are ionized, there will be a higher density of fixed negative charges at the emitter than at the collector side of the base. Thus the field E assists the flow of minority carriers (electrons) from emitter to collector. The density of majority carriers (holes) at any point in the base is

$$p = N_{a0} \exp\left(-\frac{x}{b}\right) \quad (4.30)$$

and the field E will also assist the flow of holes from collector to emitter, which is due to currents from collector to base and from base to emitter. Then the hole current is a combination of drift and diffusion currents, so that the hole current density is

$$J_p = e\mu_p pE - eD_p \frac{dp}{dx} \quad (4.31)$$

$$= e\mu_p E N_{a0} \exp\left(-\frac{x}{b}\right) + \frac{eD_p N_{a0}}{b} \exp\left(-\frac{x}{b}\right) \quad (4.32)$$

But at equilibrium $J_p = 0$ and it is still almost zero for low-level injection of electrons, so from eq. (4.32),

$$E = -\frac{D_p}{\mu_p b} = -\frac{D_n}{\mu_n b} = -\frac{kT}{eb} \quad (4.33)$$

using eqs (2.108) and (2.109). The minus sign indicates that E opposes hole flow in the positive x-direction, from emitter to collector, but assists electron flow in that direction and eq. (4.33) shows that E is constant for an exponential acceptor distribution across the base.

In order to determine the effect of the built-in field on the distribution of excess charge, q_B, in the base we must consider the electron current density J_n. The equilibrium electron concentration in the base is n_p, which is increased to a new value n_e owing to the injection of excess electrons from the emitter. Then, for a steady current flow, the excess electron density is

$$\Delta n = n_e - n_p \tag{4.34}$$

and the electron current density is

$$J_n = e\mu_n \Delta n E = eD_n \frac{d\Delta n}{dx} \tag{4.35}$$

where the field E is unaffected by low-level electron injection and recombination is neglected. Substituting for E from eq. (4.33) leads to

$$\frac{J_n}{eD_n} = -\frac{\Delta n}{b} + \frac{d\Delta n}{dx} \tag{4.36}$$

which, after rearranging and integrating, gives

$$\ln\left(\frac{J_n b}{eD_n} + \Delta n\right) = \frac{x}{b} + C \tag{4.37}$$

Then, assuming that $\Delta n = 0$ at the collector where $x = w_b$,

$$C = \ln \frac{J_b b}{eD_n} - \frac{w_b}{b} \tag{4.38}$$

so that the excess electron density is

$$\Delta n = -\frac{J_n b}{eD_n}\left[1 - \exp\left(\frac{x - w_b}{b}\right)\right] \tag{4.39}$$

the minus sign indicating negative excess charge. Typically $b \approx w_b/8$, in which case for values of x up to about $0.7w_b$ the exponential term is much less than unity and Δn is constant. This gives the distribution of excess electrons shown in Fig. 4.11, which is compared with the linear distribution for a uniform-base transistor of the same base width carrying the same current.

The total excess charge in the base is obtained by integrating $-e\Delta n$ between the limits of $x = 0$ and $x = w_b$, so that, from eq. (4.39),

$$-\int_0^{w_b} e\Delta n \, dx = \frac{J_n b w_b}{D_n}\left[1 - \frac{b}{w_b}\left(1 - \exp\left(-\frac{w_b}{b}\right)\right)\right] \approx \frac{J_n b w_b}{D_n} \tag{4.40}$$

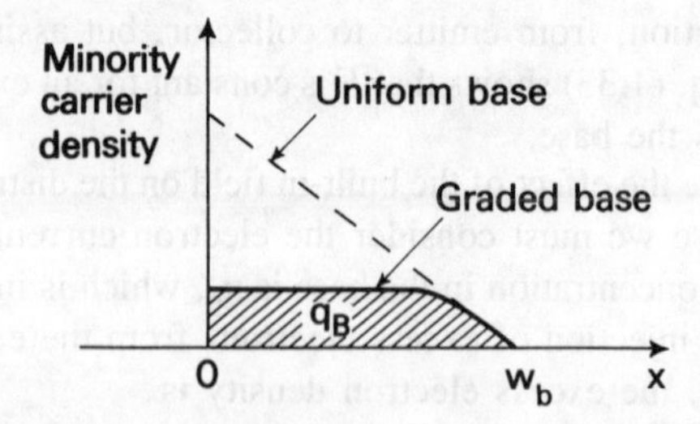

Figure 4.11 Excess base charge distributions in uniform-base and graded-base transistors.

The approximation follows since b is always appreciably less than w_b in order to make the charge distribution constant over as much of the base as possible, which corresponds to the existence of the built-in field E. Multiplying eq. (4.40) by the area of the base and putting $b = w_b/8$ leads to an expression for the excess base charge q_B of

$$q_B = \frac{w_b^2 I_E}{8D_n} \tag{4.41}$$

which may be compared with the expression for the excess base charge in a uniform-base transistor obtained from eq. (4.16) with $\gamma = 1$:

$$q_B = \frac{w_b^2 I_E}{2D_n} \tag{4.42}$$

In a graded-base transistor α_0 is affected by the high doping level of the emitter. From eq. (4.10) in order to make γ approach unity σ_p must be much less than σ_n. But σ_p is high owing to the large value of N_a near the emitter, so that σ_n must be made even larger. This means that the impurity concentration on each side of the emitter-base junction may become so large that the emitter-base breakdown voltage is reduced and this sets a limit on the maximum impurity concentration at the emitter side of the base.

Variation of current gain with collector current

Although the simple theory suggests that α_0 is independent of I_E and I_C (eq. (4.21)), in practice α_0 varies considerably with I_C. Small changes in α_0 produce large changes in the related parameter β_0 (eq. (4.45)) as illustrated in Table 4.1, so the variation of β_0 is shown in Fig. 4.12. At low currents α_0 and β_0 are also low, rising to a flat peak corresponding to the design value and then falling more rapidly at high currents. The letters (*a*), (*b*) and (*c*) in Fig. 4.12 correspond to Figs. 4.13 (*a*), (*b*) and (*c*), respectively.

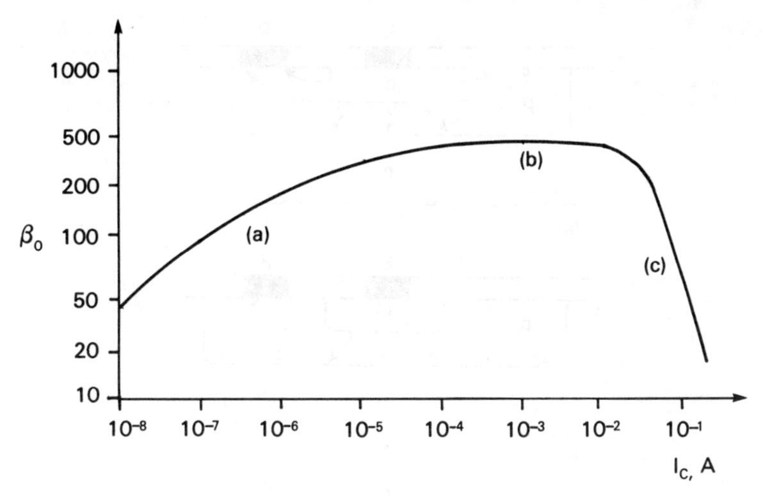

Figure 4.12 Typical variation of current gain β_0 with collector current. The letters refer to Fig. 4.13.

The initial low values of α_0 and β_0 are due to the loss of injected minority carriers by recombination both in the bulk and at the surface of the base (Fig. 4.13(*a*)), as described in Ref. [1]. At low currents a high proportion of injected carriers are lost at trapping centres in the bulk and at the surface, which effectively reduces the base transport factor δ. As the collector current is increased the traps are progressively filled and more minority carriers reach the collector, so that δ increases with the current towards the design value (Fig. 4.13(*b*)).

At high currents the density of the injected carriers is large and requires a correspondingly large base current to replenish the majority carriers which have recombined in the base. The first consequence is that the conductivity of the base is reduced owing to the high density of excess electrons, comparable to the majority carrier density and corresponding to high level injection (Fig. 4.13(*c*)). It may be seen from eq. (4.10) for γ that, if the base conductivity σ_p is reduced, γ also falls, and this is counteracted by making the edge of the emitter junction as long as possible, as shown in Fig. 4.14, where the base and emitter connections are in the form of stripes. In a *power* transistor *current crowding* is a further effect, and this is illustrated in Fig. 4.14(*c*). The base current is high enough to cause a voltage drop along the base region, so that the effective value of V_{BE} is less than the applied value. The voltage between base contact and emitter periphery is thus higher than the voltage between base contact and emitter face. Hence most of the emitter current flows from the periphery and the number of stripes is increased to handle larger currents and distribute the current more uniformly. In a *high-frequency* transistor dimensions of the active areas are reduced by reducing the width of the emitter stripe S and the base width W, typically down to about 0.3 μm and 0.03 μm respectively, which has the effect of reducing capacitance and transit time.

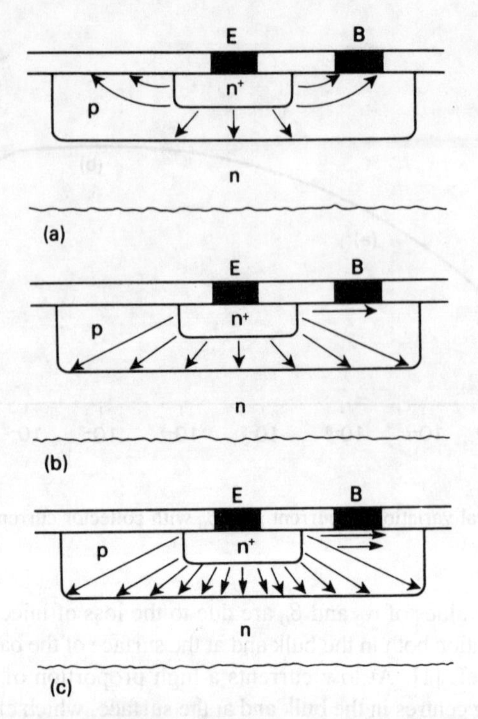

Figure 4.13 Current flow in a diffused base n-p-n transistor. (*a*) Low collector current showing recombination at surface of base with reduction in δ; (*b*) normal current flow; (*c*) high collector current showing high level injection with reduction in γ.

Common-emitter characteristics

The usual method of operating a transistor is with the emitter common to input and output. This connection is shown in Fig. 4.15 and compared with the common-base connection.

In common-emitter operation the base is used as the input electrode. The component of base current $(1 - \alpha_0)I_E$ supplies the losses in emitter current occurring in the base region, so that if the base current I_B is increased both I_E and I_C also increase. I_B is in turn controlled by V_{BE}, but since the essential mechanism is due to I_B the transistor is still considered as a current-controlled device.

$$I_E = \frac{1}{1 - \alpha_0}(I_B + I_{CBO}) \tag{4.43}$$

Substituting in eq. (4.2)

$$I_C = \frac{\alpha_0}{1 - \alpha_0}(I_B + I_{CBO}) + I_{CBO}$$

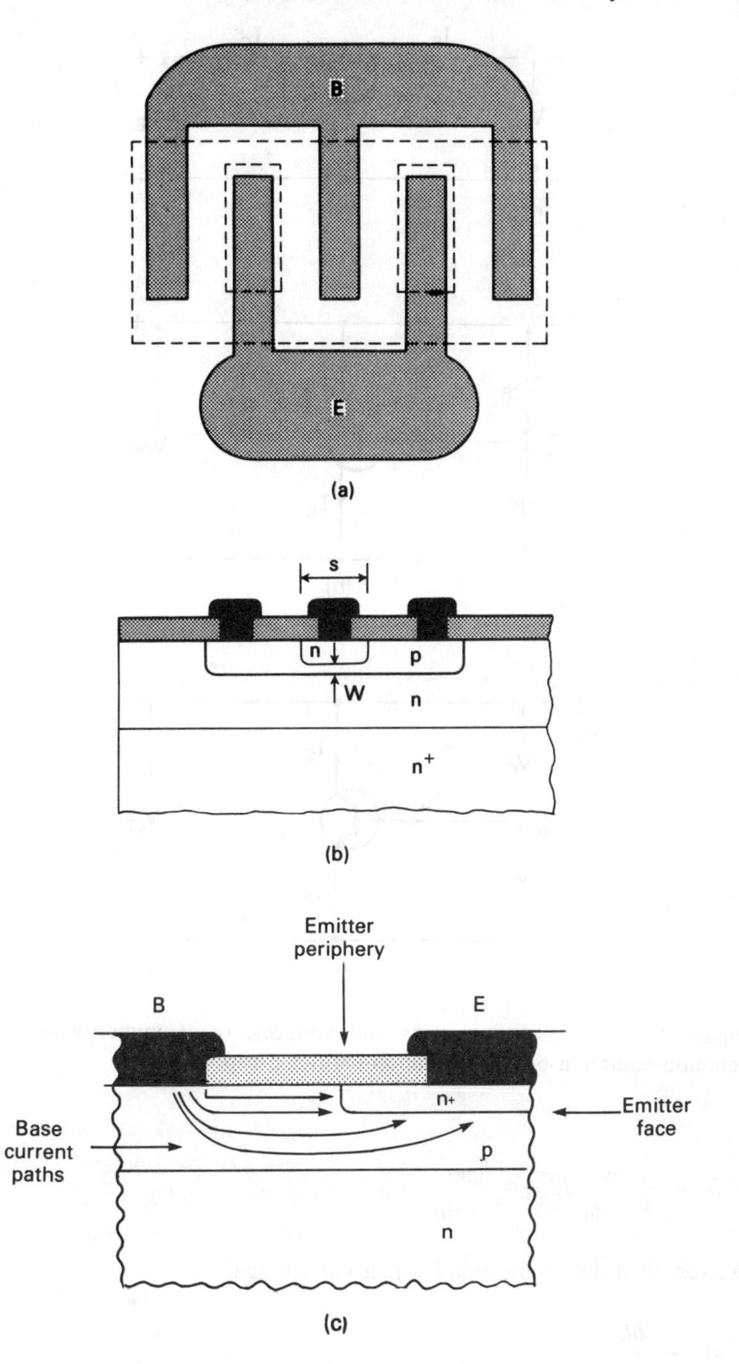

Figure 4.14 Stripe geometry for power and high-frequency transistors. (*a*) Plan view; (*b*) section; (*c*) current crowding effect.

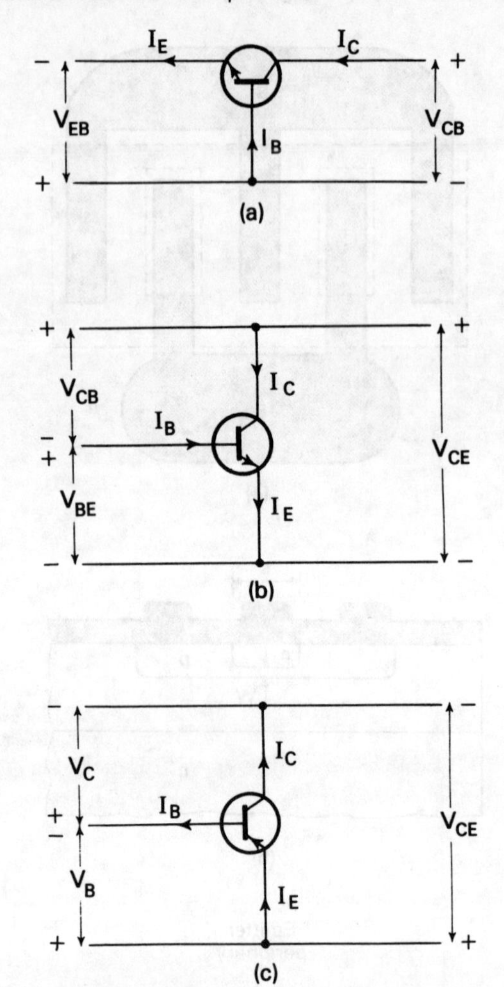

Figure 4.15 Transistor currents and voltages. (*a*) Common-base n-p-n; (*b*) common-emitter n-p-n; (*c*) common-emitter p-n-p.

$$I_C = \frac{\alpha_0}{1 - \alpha_0} I_B + \frac{I_{CBO}}{1 - \alpha_0} \tag{4.44}$$

We can then define the small signal current again

$$\beta_0 = \left.\frac{\delta I_C}{\delta I_B}\right|_{V_{CE} \text{ constant}} \tag{4.45}$$

where

$$\beta_0 = \frac{\alpha_0}{1 - \alpha_0}$$

We can also put

$$I_{CEO} = \frac{I_{CBO}}{1 - \alpha_0}$$

the collector current flowing when $I_B = 0$ (base open-circuited).

β_0 is the small-signal common-emitter current gain and is much greater than unity, as shown in Table 4.1, so that a small base current can control a much larger emitter current.

Table 4.1 Dependence of β_0 and I_{CEO} on α_0

α_0	0.97	0.98	0.99	
β_0	32	49	99	
I_{CEO}	33	50	100	nA

Values of I_{CEO} are given for I_{CBO} = 1 nA

I_{CEO} is the collector leakage current for the common-emitter configuration and is much larger than I_{CBO}, as may be seen from the values given in the table. Both β_0 and I_{CEO} are very sensitive to changes in α_0 since the very small factor $(1 - \alpha_0)$ occurs in the denominator of each of them. The effect of a 2% change in α_0, from 0.97 to 0.99, causes more than a threefold increase in β_0 and I_{CEO}. Thus measurement of β_0 is a very sensitive method of detecting small changes in α_0 and since α_0 depends on manufacturing conditions wide ranges of β_0 (h_{FE}) are quoted for available transistors.

Although I_{CEO} is still small compared with a collector current of a few mA, β_0 is not equal to the large-signal current gain in a practical transistor owing to the variations introduced by small changes in α_0. The large-signal current gain is the *h*-parameter h_{FE}, where

$$h_{FE} = \left.\frac{I_C}{I_B}\right|_{V_{CE}\ \text{constant}} \qquad (4.46)$$

and depends on the *actual* values of I_C and I_B, while β_0 depends on the *changes* in value of I_C and I_B. Equation (4.46) assumes negligible leakage current I_{CEO} and when this is included the expression for I_C becomes

$$I_C = h_{FE}I_B + I_{CEO} \qquad (4.47)$$

For an *ideal* transistor, in which α_0 is assumed constant at all collector currents and I_{CEO} is negligible

$$h_{FE} = \frac{\alpha_0}{1 - \alpha_0} = \beta_0 \tag{4.48}$$

and

$$I_C = h_{FE} I_B + I_{CBO} (1 + h_{FE}) \tag{4.49}$$

The common-emitter output characteristics are shown in Fig. 4.16. The collector-emitter voltage V_{CE} is developed across both junctions, so that

$$V_{CE} = V_{CB} + V_{BE} \tag{4.50}$$

In the *active* region of operation, where V_{CE} exceeds about 0.2 V, the emitter junction has a small forward bias, which is almost constant to maintain the base current constant, and the collector junction has a much larger reverse bias (Fig. 4.17(*a*)). In this region I_C rises slowly with V_{CE} owing to the fall in base width as V_{CE} rises and the corresponding rise in α_0. This effect is much more noticeable than in the common-base characteristics owing to the factor $(1 - \alpha_0)$. For the same reason the increase in I_C due to avalanche breakdown becomes apparent at a lower collector voltage since the change in α_0 is magnified in its effect on I_C, as explained below. The breakdown voltage defines the upper voltage limit of the active region into which the transistor is biased when it is used as an amplifier. The collector dissipation $P_C = V_{CE} I_C$ and is shown as a dotted line in Fig. 4.16. The transistor is operated with P_C below the maximum value, with voltages and currents to the *left* of the line.

For values of V_{CE} below about 0.2 V, I_C falls very rapidly as V_{CE} is reduced, and the transistor is working in the *saturated* region. For a given base current the point at which the rapid fall in current begins is called the 'knee' of the characteristics, and the corresponding value of V_{CE} defines the boundary between

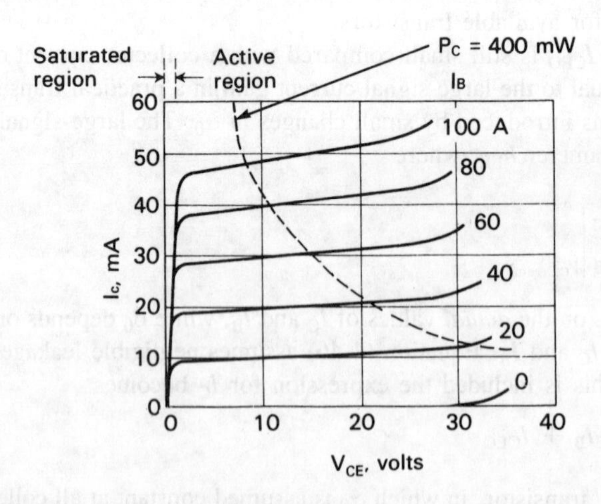

Figure 4.16 Common-emitter output characteristics (n-p-n transistor).

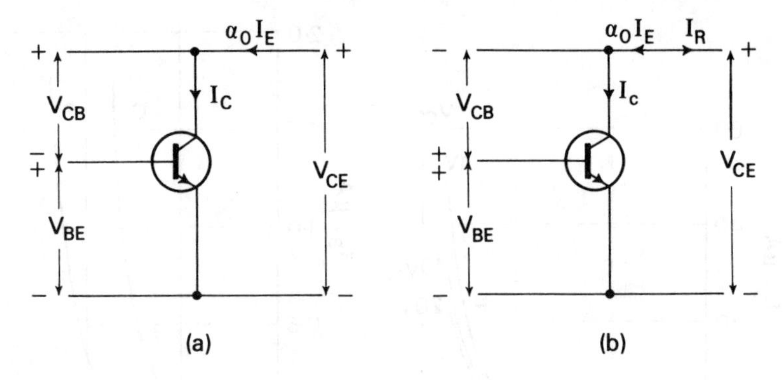

Figure 4.17 n-p-n transistor junction biasing. (*a*) In active region; (*b*) in saturated region.

the saturated and active regions of the transistor. As V_{CE} is reduced in the active region, V_{BE} remains almost constant so that V_{CE} must fall until it is about zero at the knee of the curve. However, electrons continue to flow into the collector since the slope of the density gradient in the base is still towards the collector (Fig. 4.7(*c*)). As V_{CE} is reduced below this voltage, V_{CB} becomes a *forward* bias and current I_R flows from the collector in the reverse direction to the current from the emitter (Fig. 4.17(*b*)). Both junctions are then forward biased so that, in the saturation region,

$$V_{CE} = -V_{CB} + V_{BE}$$

and

$$I_C = \alpha_0 I_E - I_R \tag{4.51}$$

neglecting the leakage current. Thus, as V_{CE} is reduced towards zero, $-V_{CB}$ and I_R both increase and I_C also tends to zero. A detailed analysis, using the Ebers–Moll model given below, shows that when $I_C = 0$, V_{CE} is not zero but has a small value called the *offset* voltage which is normally only a few millivolts. The transistor operates in the saturation region when it is used as a switch. It is particularly suitable for this application owing to the low value of V_{CE} when it is conducting heavily.

The input characteristics (Fig. 4.18(*a*)) depend on eq. (4.27), which shows that I_E increases exponentially with V_{EB}. I_B therefore increases exponentially with V_{BE}, and is also a function of collector voltage since I_E depends on V_{CB} through the base-width term of eq. (4.27). When the base current is zero, $I_E = I_{CBO}/(1 - \alpha_0)$, from eq. (4.3), and a finite value of V_{BE} is required to maintain this value of I_E. The value of V_{BE} for $I_B = 0$ increases with V_{CE}, since the corresponding value of I_E increases with V_{CB}. Thus all the input characteristics cut the voltage axis, and finally, when $V_{BE} = 0$, $I_E = 0$ and $I_B = -I_{CBO}$. Input

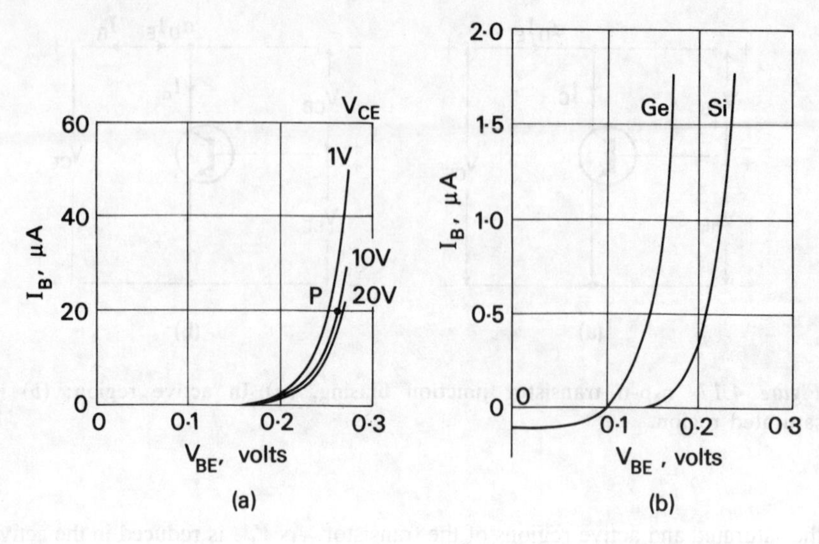

Figure 4.18 Common-emitter input characteristics (n-p-n transistor). (*a*) Germanium transistor; (*b*) comparison of germanium and silicon transistors at low base currents (leakage currents 100 nA and 5 nA respectively).

characteristics at the same value of V_{CE} are compared in Fig. 4.18(*b*) for germanium and silicon transistors having base-to-emitter leakage currents of 100 nA and 5 nA, respectively.

Transistor breakdown voltages

Breakdown can occur at either the emitter or the collector junction. Due to the high level of emitter doping the emitter breakdown voltage is only about 5 V. However, the high doping level ensures that the punch through condition previously described, rarely occurs since the voltage required to extend the collector depletion layer to the emitter would normally be higher than that required for collector breakdown.

Two breakdown voltages are usually specified corresponding to collector-base breakdown with emitter open-circuit, BV_{CBO}, and collector-emitter breakdown with the base open-circuit, BV_{CEO}. They may be considered by introducing another factor, M, the *collector multiplication factor*, so that using eq. (4.21) the fraction of emitter current reaching the collector becomes

$$\alpha = \gamma\delta M$$

$$= \alpha_0 M \qquad (4.52)$$

where

$$M = \frac{1}{1 - (V_{CB}/V_B)^n} \quad \text{(eq. (3.99))}$$

V_{CB} is the actual voltage applied between collector and base, V_B is the breakdown voltage of the collector-base diode, and n is a constant usually about 6 for n-p-n silicon transistors. For normal operation $V_{CB} \ll V_B$ and $M = 1$, so $\alpha = \alpha_0$.

As discussed in Chapter 3 the leakage current of a diode is increased by the factor M, so for a transistor in common base configuration.

$$I_C = M\alpha_0 I_E + MI_{CBO} \tag{4.53}$$

BV_{CBO} is obtained from the circuit shown in Fig. 4.19(*a*) and then from eq. (4.53) for $I_E = 0$

$$I_C = MI_{CBO} \tag{4.54}$$

As V_{CB} approaches V_B, M and I_C tend to infinity, reaching it at $V_{CB} = V_B$, so

$$BV_{CBO} = V_B \tag{4.55}$$

the collector-base diode breakdown voltage.

BV_{CEO} is obtained from the circuit in Fig. 4.19(*b*) and for a common emitter configuration

$$I_C = \frac{M\alpha_0}{1 - M\alpha_0} I_B + \frac{I_{CBO}}{1 - M\alpha_0} \tag{4.56}$$

from eqs (4.44) and (4.52). For $I_B = 0$

$$I_C = \frac{I_{CBO}}{1 - M\alpha_0} \tag{4.57}$$

and I_C tends to infinity as $M\alpha_0$ tends to 1. I_C would become infinite at $M = 1/\alpha_0$ or

$$(V_{CB}/V_B)^n = 1 - \alpha \tag{4.58}$$

using eqs (4.48) and (3.99). This occurs where

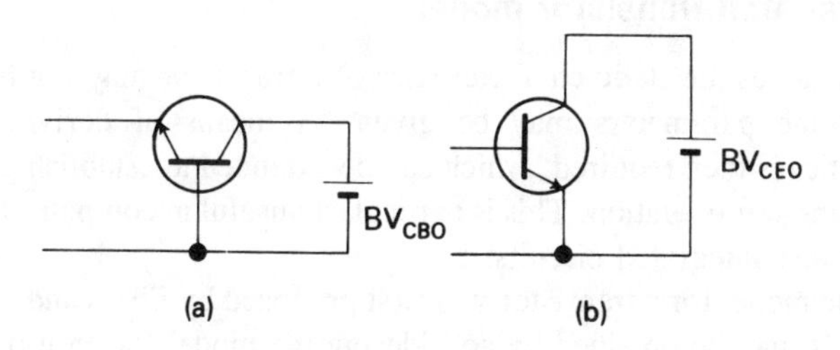

Figure 4.19 Transistor breakdown voltages. (*a*) BV_{CBO}, emitter open-circuit; (*b*) BV_{CEO}, base open-circuit.

$$V_{CB} = BV_{CEO} = V_B(1/h_{FE})^{1/n} \quad (4.59)$$

which is less than BV_{CBO}.

Worked example

At low collector currents in an n-p-n transistor a value of $h_{FE} = 10$ may be assumed and in the collector multiplication factor the constant $n = 6$.

With the emitter open-circuited the breakdown voltage BV_{CBO} is 45 V. Estimate the breakdown voltage with the base open-circuited, BV_{CEO}.

Solution

Using eqs (4.55) and (4.59)

$$V_B = BV_{CBO} = 45 \text{ V}$$

and

$$BV_{CEO} = 45\left(\frac{1}{10}\right)^{1/6}$$
$$= 45 \times 0.68$$
$$= 30.65 \text{ V}$$

BV_{CEO} would apply directly to a transistor in the cut-off condition, as occurs in switching, but as may be seen from the characteristics of Fig. 4.16 the breakdown voltage is reduced when the base current is increased, due to the presence of more electrons. However, when a load resistor is used as described below, the collector voltage falls as collector current is increased, so BV_{CEO} (and BV_{CBO} for common base) indicate the maximum values of breakdown voltage.

The Ebers–Moll transistor model

In some instances the static characteristics of a transistor may not be available, although some parameters may be given. A means of deriving the static characteristics is then required, which can be extended to establish general principles of transistor operation. This is particularly useful in computer-aided design of discrete and integrated circuits.

A suitable model for a transistor was first proposed by Ebers and Moll in 1954 (Ref. [2]). It may be obtained by considering the model for an n-p-n transistor shown in Fig. 4.20(*a*). The terminal currents I_E, I_B and I_C are conventionally shown flowing into the device. For a transistor in the active region (Fig. 4.17(a))

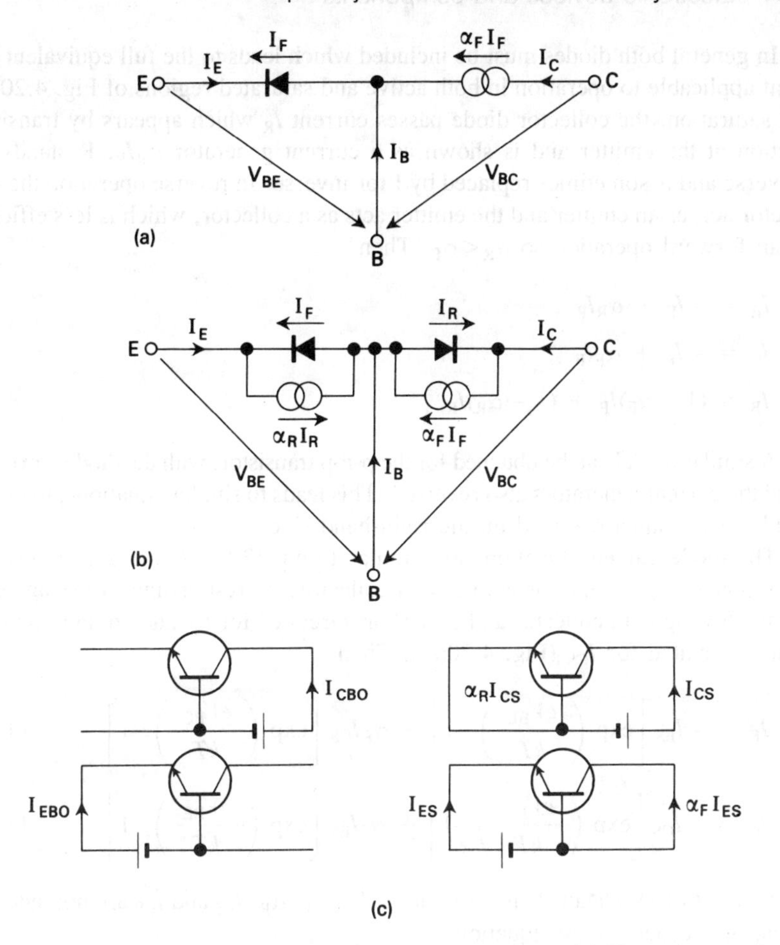

Figure 4.20 The Ebers Moll model of an n-p-n transistor. (*a*) Normal operation; (*b*) general model; (*c*) open- and short-circuit leakage currents.

the emitter diode is forward biased and passes current I_F, which appears at the collector as $\alpha_F I_F$ where α_F corresponds to α_0. The suffix F stands for forward operation and is sometimes replaced by N for normal. The collector diode is reverse biased and so is not included, while the leakage current I_{CBO} is also neglected. Then

$$I_E = -I_F$$

and

$$I_C = \alpha_F I_F \tag{4.60}$$

from eq. (4.2) with $I_{CBO} = 0$.

In general both diodes must be included which leads to the full equivalent circuit applicable to operation in both active and saturated regions of Fig. 4.20(*b*). In saturation, the collector diode passes current I_R which appears by transistor action at the emitter and is shown as a current generator $\alpha_R I_R$. R stands for reverse and is sometimes replaced by I for inverse. In reverse operation the collector acts as an emitter and the emitter acts as a collector, which is less efficient than forward operation so $\alpha_R < \alpha_F$. Then

$$I_E = -I_F + \alpha_R I_R$$

$$I_C = -I_R + \alpha_F I_F$$

$$I_B = (1 - \alpha_F) I_F + (1 - \alpha_R) I_R \tag{4.61}$$

A similar model can be obtained for the p-n-p transistor, with the diodes reversed and the current generators also reversed. This leads to similar equations, to (4.61) with all the signs reversed on the right-hand side.

The diode current equations are similar to eq. (3.5), with leakage currents designated I_{ES} and I_{CS} for emitter and collector, corresponding to leakage currents flowing with collector and base short-circuited for I_{ES} and emitter and base short-circuited for I_{CS} (Fig. 4.20(*c*)). Then

$$I_E = -I_{ES}\left[\exp\left(\frac{eV_{BE}}{kT}\right) - 1\right] + \alpha_R I_{CS}\left[\exp\left(\frac{eV_{BC}}{kT}\right) - 1\right] \tag{4.62}$$

$$I_C = -I_{CS}\left[\exp\left(\frac{eV_{BC}}{kT}\right) - 1\right] + \alpha_F I_{ES}\left[\exp\left(\frac{eV_{BE}}{kT}\right) - 1\right] \tag{4.63}$$

from eq. (4.61). Detailed analysis shows that α_F, α_R, I_{ES} and I_{CS} are not independent but related by the equation

$$\alpha_F I_{ES} = \alpha_R I_{CS} = I_S \tag{4.64}$$

where I_S is the *transport saturation current* used in computer simulation. Also the three terminal currents are related by the expression

$$I_E + I_B + I_C = 0 \tag{4.65}$$

The short-circuit leakage currents may be related to the open-circuit leakage currents as follows. With the collector reverse biased and emitter open-circuited $I_C = I_{CBO}$. When the emitter is short-circuited $I_C = I_{CS}$, so a current $\alpha_R I_{CS}$ can now flow through the emitter (Fig. 4.20(*c*)). In turn this causes current $\alpha_F(\alpha_R I_{CS})$ to flow through the collector in addition to I_{CBO}. Then

$$I_C = I_{CS} = I_{CBO} + \alpha_F \alpha_R I_{CS}$$

so

$$I_{CS} = I_{CBO}/(1 - \alpha_F \alpha_R) \tag{4.66}$$

Similarly, where I_{EBO} is the leakage current with the emitter reverse biased and the collector open-circuited

$$I_{ES} = I_{EBO}/(1 - \alpha_F \alpha_R) \tag{4.67}$$

Equations (4.62) to (4.65) can be combined to describe the operation of the transistor under specified conditions. For instance, in the active region, with emitter forward biased and collector reverse biased, V_{BE} is positive and V_{BC} is negative. Then the base current with short-circuited collector is given by

$$I_B = I_{ES}(1 - \alpha_F) \exp\left[\left(\frac{eV_{BE}}{kT}\right) - 1\right] + I_{CS}(1 - \alpha_R) \tag{4.68}$$

which is another form of eq. (4.3).

The Ebers–Moll equations are particularly useful in describing the characteristics of a transistor in the saturated region, where *both* junctions are forward biased (Fig. 4.17(*b*)). Equations (4.62) and (4.63) then become

$$I_E = - I_{ES} \exp\left(\frac{eV_{BE}}{kT}\right) + \alpha_R I_{CS} \exp\left(\frac{eV_{BC}}{kT}\right) \tag{4.69}$$

$$I_C = - I_{CS} \exp\left(\frac{eV_{BC}}{kT}\right) + \alpha_F I_{ES} \exp\left(\frac{eV_{BE}}{kT}\right) \tag{4.70}$$

The collector-emitter voltage is given by

$$V_{CE} = V_{BE} - V_{BC} \tag{4.71}$$

and eliminating I_E leads to

$$V_{BE} = \frac{kT}{e} \ln \frac{1}{I_{ES}} \left[\frac{I_B + I_C(1 - \alpha_R)}{1 - \alpha_F \alpha_R}\right] \tag{4.72}$$

$$V_{BC} = \frac{kT}{e} \ln \frac{1}{I_{CS}} \left[\frac{\alpha_F I_B - I_C(1 - \alpha_F)}{1 - \alpha_F \alpha_R}\right] \tag{4.73}$$

Eliminating I_{ES} and I_{CS} using eq. (4.64) and putting

$$\frac{\alpha_F}{1 - \alpha_F} = \beta_F, \quad \frac{\alpha_R}{1 - \alpha_R} = \beta_R$$

gives

$$V_{CE} = V_{CEsat} = \frac{kT}{e} \ln \left(\frac{\dfrac{1}{\alpha_R} + \dfrac{I_C}{I_B}\dfrac{1}{\beta_R}}{1 - \dfrac{I_C}{I_B}\dfrac{1}{\beta_F}}\right) \tag{4.74}$$

Using $\alpha_F = 0.98$ and $\alpha_R = 0.80$ as typical values gives $\beta_F = 49$ and $\beta_R = 4$. At room temperature $kT/e = 25$ mV and taking $I_C = 10I_B$, for example, gives $V_{CEsat} = 0.038$ V. When I_B is held constant V_{CE} increases with I_C, as shown in Fig. 4.16. An important point on the $I_C - V_{CE}$ curve occurs where $I_C = 0$ and I_B is finite. Equation (4.74) then reduces to

$$V_{CE} = V_{CEoff} = \frac{kT}{e} \ln \frac{1}{\alpha_R} \tag{4.75}$$

which for the typical values given is 5.6 mV. The $I_C - V_{CE}$ characteristic at constant I_B thus does not pass through the origin but is offset by a small voltage and this may be a disadvantage when the transistor is used as a switch, as described below. The *offset voltage* is reduced by operating the transistor in the *inverted* mode, in which the collector is grounded and acts as the emitter, while the emitter becomes the collector. Under these conditions

$$V_{EC} = V_{ECoff} = \frac{kT}{e} \ln \frac{1}{\alpha_F} \tag{4.76}$$

which is less than V_{CEoff} as $\alpha_F > \alpha_R$.

The operation of a transistor

Consider the circuit of Fig. 4.21(*a*) in which V_{CC} is a fixed direct voltage obtained from a power supply, and R_B and R_L are resistors connected in series with the base and collector respectively. The voltage V_{BB} controls I_B, which in turn controls I_C and hence the *output* voltage V_{CE}. A *change* in the magnitude of any of these direct quantities is known as a *signal*. Such a circuit is the basis of many practical applications of transistors, which fall into two main classes. These involve large and small signals respectively, both of which depend closely on the physical principles which have been described so far.

An n-p-n transistor is shown making both I_B and I_C conventionally positive since they are due to negative charges leaving their respective electrodes. The basic principles, however, apply equally to p-n-p transistors. Then, applying Kirchhoff's law to the base circuit, $V_{BB} = I_B R_B + V_{BE}$ or

$$I_B = \frac{V_{BB} - V_{BE}}{R_B} \tag{4.77}$$

If V_{BB} is a direct voltage which can be varied, I_B can be set at any desired value, which can be predicted by taking $V_{BE} \simeq 0.7$ V for a silicon transistor. Now, I_B is associated with I_C and V_{CE} through the common-emitter output characteristics, but the actual values of collector current and voltage obtained will also depend on the load resistance, R_L. This may be incorporated with the characteristics by means of a *load line*, whose equation is obtained by applying Kirchhoff's law to the collector circuit. This gives $V_{CC} = V_{CE} + I_C R_L$, or

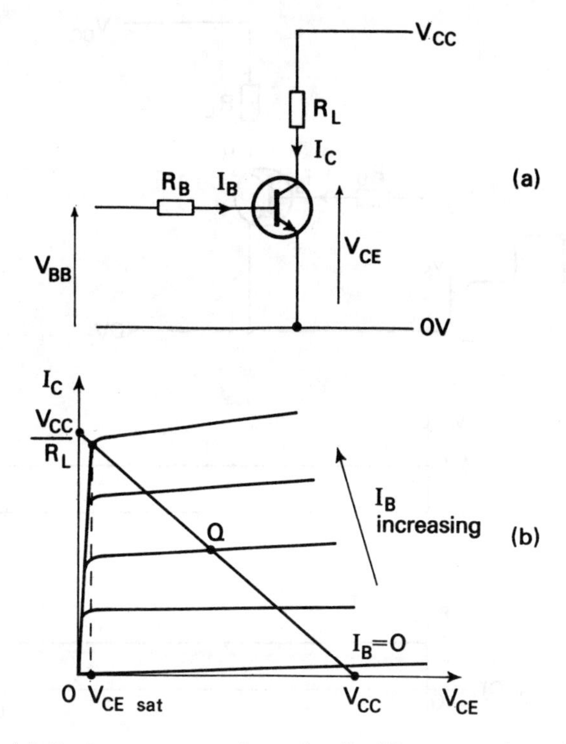

Figure 4.21 (*a*) Basic common-emitter circuit; (*b*) construction of load line.

$$I_C = -\frac{1}{R_L} V_{CE} + \frac{V_{CC}}{R_L} \tag{4.78}$$

If I_C is measured along the y-axis and V_{CE} along the x-axis, eq. (4.78) represents a straight line across the characteristics having a slope of $-1/R_L$ (Fig. 4.21(*b*)), which is the *load line*. It cuts the current axis where $V_{CE} = 0$ and $I_C = V_{CC}/R_L$, and it cuts the voltage axis where $I_C = 0$ and $V_{CC} = V_{CE}$, as may easily be seen from eq. (4.78). The operating values of I_C and V_{CE} are obtained where the load line cuts the characteristic corresponding to a given value of I_B. This occurs at the quiescent point Q, so-called because it refers to direct values without any alternating signals superimposed on them.

Switching operation

Where V_{BB} is replaced by the input signal voltage v_s, which commonly has a rectangular or pulsed waveform, the transistor may be considered to act as a switch (Fig. 4.22(*a*)). v_s is made sufficiently positive for I_B to be large enough to take the transistor into saturation, where the resulting value of V_{CE} is V_{CEsat}, derived in eq. (4.74). This is normally much less than V_{CC} so that the 'switch' is closed

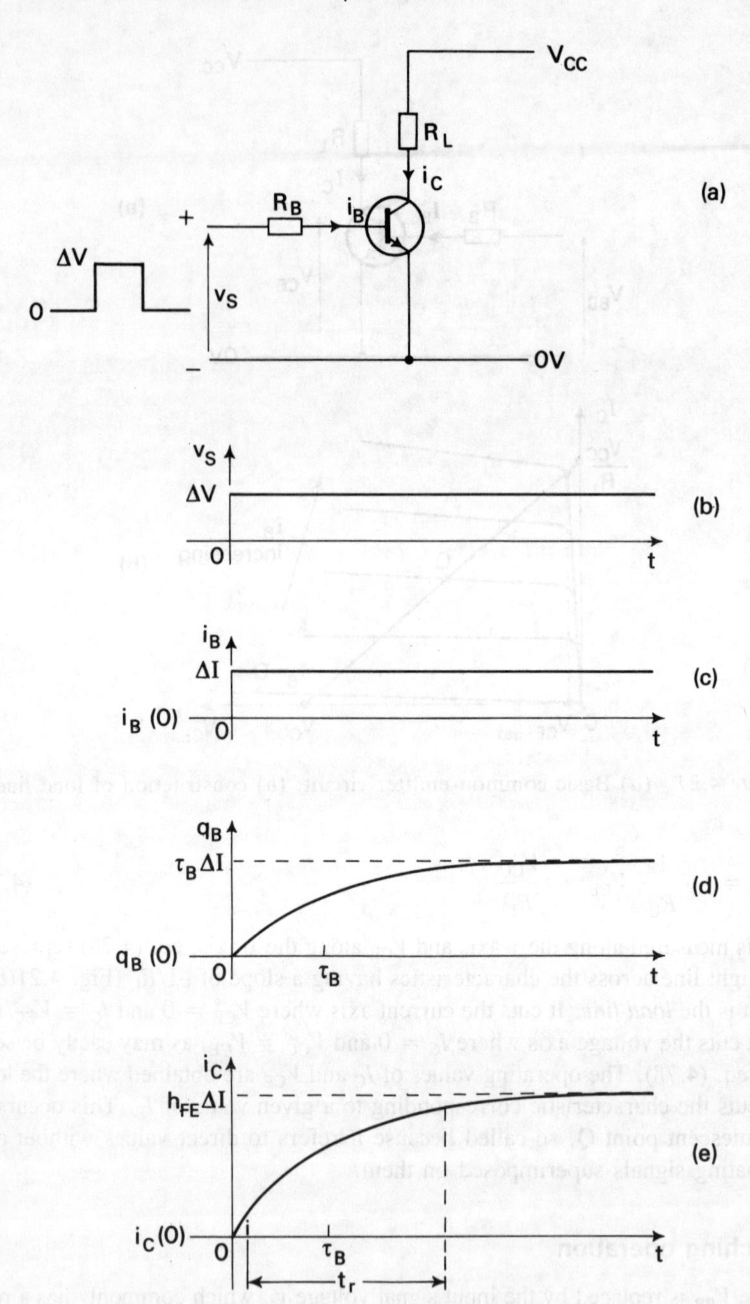

Figure 4.22 Transient response of a transistor to switching on. (*a*) Circuit with signals; (*b*) input-voltage step function; (*c*) base-current step function; (*d*) excess base charge; (*e*) collector current.

with the transistor conducting heavily. When v_s has reached its maximum negative value, I_B is reduced to zero and may be reversed if v_s is large enough. I_C is then reduced to a leakage current which will be I_{CEO} when $I_B = 0$ and can go down to I_{CBO} when I_B is reversed, which will occur when both junctions have become reverse biased, as may be derived from eqs (4.63) and (4.66) with α_F and α_R each taken as zero. The switch is now opened with the transistor cut off.

In this switching operation the active region is rapidly traversed between the 'on' and 'off' conditions. The time taken to change from one of these steady states to the other is of vital importance in many high-speed switching applications, such as the logic circuits in a digital system. This switching time is controlled by the rate at which the excess charge in the base can be established and dispersed, as described below. Here the pulsed waveform is considered as the sum of two step functions, and the switching times are obtained by analysing the response of the transistor to each step function in turn.

Charge control of a transistor

The steady-state operation of a junction transistor depends on the maintenance of an excess charge, q_B, in the base region. The emitter current is then given by eq. (4.17) and

$$I_E = \frac{q_B}{\tau_C} \tag{4.79}$$

where τ_C is the transit time of the minority carriers across the base. A base current I_{BF} due to majority charges is also necessary to make good the loss of charge due to recombination. Since $I_{BF} \approx I_B$, we can write from eq. (4.18) that approximately

$$I_B = \frac{q_B}{\tau_B} \tag{4.80}$$

where τ_B is the lifetime of the minority carriers in the base. Then from eqs (4.2) and (4.46), since α_0 is almost unity and I_{CBO} is very small, we can write approximately that

$$I_E = I_C = h_{FE} I_B \tag{4.81}$$

which gives

$$I_C = \frac{q_B}{\tau_C} \tag{4.82}$$

and

$$\frac{I_C}{I_B} = h_{FE} = \frac{\tau_B}{\tau_C} \tag{4.83}$$

These equations refer to an ideal transistor which can be realized only approximately in practice, but nevertheless they give a useful guide to practical transistor operation. τ_C and τ_B are called *charge control parameters*.

If a signal is applied to the base a varying current is superimposed on the steady current I_B. Thus the charge distribution will change with time so that at any instant the total base current, i_B, may be written as the sum of the steady-state and varying currents:

$$i_B = \frac{q_B}{\tau_B} + \frac{dq_B}{dt} \tag{4.84}$$

a *charge control equation*.

Transient response

Equations (4.83) and (4.84) are particularly useful when considering the response of a transistor to a transient or step function. In the simple common-emitter circuit of Fig. 4.22(*a*) it is assumed that R_B is much larger than the effective input impedance of the transistor, so that a sudden change or step of input voltage ΔV results in a sudden change of base current $\Delta I \approx \Delta V/R_B$. Suppose in the first case that ΔI is small enough to ensure that the transistor remains in the active region of operation. If the initial values of base charge, base current and collector current are $q_B(0)$, $i_B(0)$ and $i_C(0)$ respectively, the step input will increase the base charge by q and the base current by ΔI. Then initially,

$$i_B(0) = \frac{q_B(0)}{\tau_B} \tag{4.85}$$

and after the input has been applied the *change* in base current is

$$\Delta I = \frac{q}{\tau_B} + \frac{dq}{dt} \tag{4.86}$$

Then $\tau_B\, dq/dt = \tau_B \Delta I - q$, or

$$\frac{dq}{\tau_B \Delta I - q} = \frac{dt}{\tau_B} \tag{4.87}$$

so that, integrating,

$$-\ln(\tau_B \Delta I - q) = \frac{t}{\tau_B} + C_1 \tag{4.88}$$

Putting $t = 0$, $q = 0$ initially, $C_1 = -\ln \tau_B \Delta I$, so that

$$\ln\left(1 - \frac{q}{\tau_B \Delta I}\right) = -\frac{t}{\tau_B} \tag{4.89}$$

and

$$q = \tau_B \Delta I \left[1 - \exp\left(-\frac{t}{\tau_B}\right)\right] \tag{4.90}$$

Hence the total base charge as a function of time is given by

$$q_B = q_B(0) + \tau_B \Delta I \left[1 - \exp\left(-\frac{t}{\tau_B}\right)\right] \tag{4.91}$$

From eq. (4.82), $q_B = \tau_C i_C$, and from eq. (4.83), $\tau_B = h_{FE}\tau_C$, so that, substituting into eq. (4.91),

$$i_C = i_C(0) + h_{FE}\Delta I \left[1 - \exp\left(-\frac{t}{\tau_B}\right)\right] \tag{4.92}$$

for the collector current as a function of time. Equations (4.91) and (4.92) are illustrated in Fig. 4.22(*d*) and (*e*). The rise time t_r of the collector current is measured between the times corresponding to 10% and 90% of the total change $h_{FE}\Delta I$, since the time to reach 100% of $h_{FE}\Delta I$ is not well defined. It may be easily found from eq. (4.92) that this gives

$$t_r = 2.2\tau_B \tag{4.93}$$

Typical values are $\tau_C = 10^{-9}$ s, $\tau_B = 10^{-7}$ s and $h_{FE} = 100$, which makes $t_r = 0.22$ μs. This rise time may be unacceptably long for fast switching applications, but it can be shortened by increasing the base current step so that the transistor saturates. Under saturation conditions, from Fig. 4.21,

$$V_{CC} = I_C R_L + V_{CEsat} \approx I_C R_L \tag{4.94}$$

so that

$$I_C \approx \frac{V_{CC}}{R_L} \tag{4.95}$$

and the collector current is determined by the external circuit conditions. Saturation is achieved when $h_{FE}\Delta I > V_{CC}/R_L$, so that when ΔI is large enough for this to occur the collector current is clamped at V_{CC}/R_L before the transient described by eq. (4.92) is completed (Fig. 4.24(*b*)). It is normal to switch the transistor between zero and saturated currents, so that $q_B(0)$, $i_B(0)$ and $i_C(0) = 0$. It is evident from Fig. 4.24(*b*) that the rise time from 10% to 90% of the current V_{CC}/R_L can be made much less than $2.2\tau_B$ as ΔI is increased. Indeed, the rise time can usually be measured over the whole range (0 to 100%) of V_{CC}/R_L since the point at which i_C reaches V_{CC}/R_L is better defined than before.

However, under saturation conditions both the emitter and collector junctions are forward biased and the base charge is supplied through both electrodes. In a uniform base transistor, the total charge in the base is then the result of two triangular distributions as in Fig. 4.23(*a*), which may be more conveniently redrawn as shown at (*b*). Here q_{BO} represents the total base charge needed to take the transistor to the point of saturation and q_S represents the extra base charge needed to hold the transistor firmly in saturation, so that the total charge is

$$q_B = q_{BO} + q_S \tag{4.96}$$

The component of base current associated with q_{BO} is i_{BO}, which is just sufficient to saturate the transistor. Hence

$$i_{BO} = \frac{q_{BO}}{\tau_B} = \frac{V_{CC}}{h_{FE}R_L} \tag{4.97}$$

Similarly the component of current associated with q_S is i_{BS}, which is the part of the base current needed to hold the transistor firmly in saturation, and in the steady state

$$i_{BS} = \frac{q_S}{\tau_S} \tag{4.98}$$

where τ_S represents the lifetime of the extra stored charge. Then, again in the steady state, the total base current is

$$i_B = i_{BO} + i_{BS} = \frac{q_{BO}}{\tau_B} + \frac{q_S}{\tau_S} \tag{4.99}$$

τ_S and τ_B are not equal since τ_S refers to minority carriers originating both in the emitter and the collector, while τ_B refers only to minority carriers from the emitter and τ_S is typically about half τ_B.

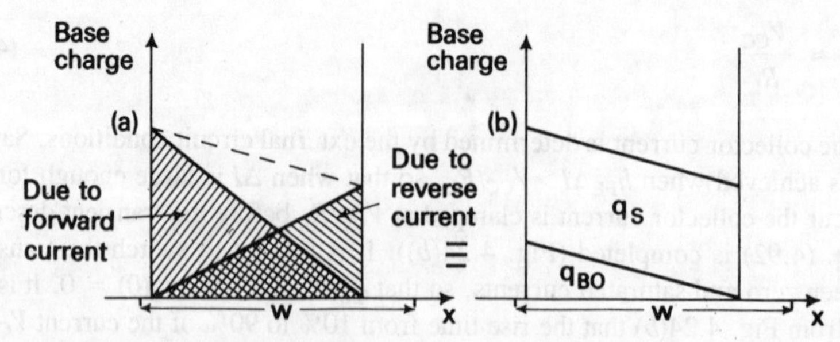

Figure 4.23 Base charge distribution during saturation for a uniform base transistor. (*a*) Excess base charges due to forward and reverse currents; (*b*) saturation base charges.

For any transient change in the saturation region, q_{BO} may be regarded as constant while only q_S changes, so that the base current is given by

$$i_B = \frac{q_{BO}}{\tau_B} + \frac{q_S}{\tau_S} + \frac{dq_S}{dt} \tag{4.100}$$

and after substitution from eq. (4.97),

$$\left(i_B - \frac{V_{CC}}{h_{FE}R_L}\right) = i_{BS} = \frac{q_S}{\tau_S} + \frac{dq_S}{dt} \tag{4.101}$$

The left-hand side of this equation is a current defined by the circuit conditions external to the transistor. Hence, if $i_B = I_{B1}$, the base current used to switch on the transistor, the time for q_S to reach its steady state value can be determined (Fig. 4.24(*c*)).

The transistor is switched off by reversing the input current step (Fig. 4.21(*a*)), but collector current continues to flow and remains constant at V_{CC}/R_L while the charge q_S is removed from the base. If it is assumed that q_S has reached a steady value $\tau_S i_{BS}$ before turn-off, solution of eq. (4.101) shows that it falls exponentially and attempts to reach a terminal value of $-\tau_S I_{B1}$. However, this exponential fall cannot be completed since after a time t_s the point $q_S = 0$ is reached and the transistor then enters the active region (Fig. 4.25(*b*)). t_s is the *storage time* after which the collector current begins to fall as charge q_{BO} is removed with time constant, τ_B. Solution of eq. (4.92) then shows that in this region the collector current is given by

$$i_C = \frac{V_{CC}}{R_L} \exp\left(-\frac{t}{\tau_B}\right) \tag{4.102}$$

where t is measured from the instant at which q_S becomes zero (Fig. 4.25(*c*)). The *fall time*, t_f, measured from 90% to 10% of the current V_{CC}/R_L is given by

$$t_f = 2.2\tau_B \tag{4.103}$$

Again t_f may be unacceptably long, and both t_s and t_f can be reduced if the transistor is turned off with overdrive; i.e. if v_S is decreased below zero to some value V_2 (Fig. 4.26(*a*)). The minority carriers still in the base are then removed through the emitter, so that the emitter junction will behave like a forward-biased diode even though current is flowing in the reverse direction. Hence a negative base current $I_{B2} \approx V_2/R_B$ will also flow until the collector current ceases, corresponding to the bulk of the charge being cleared from the base (Fig. 4.26(*b*)). However, i_B falls to zero only gradually owing to the diffusion of minority carriers from areas of the base remote from the emitter and also from the depletion layers at the emitter and collector junctions.

This reversal of base current increases the rate at which the minority carriers are removed from the base. The total change in the base current is $I_{B1} - I_{B2}$, so

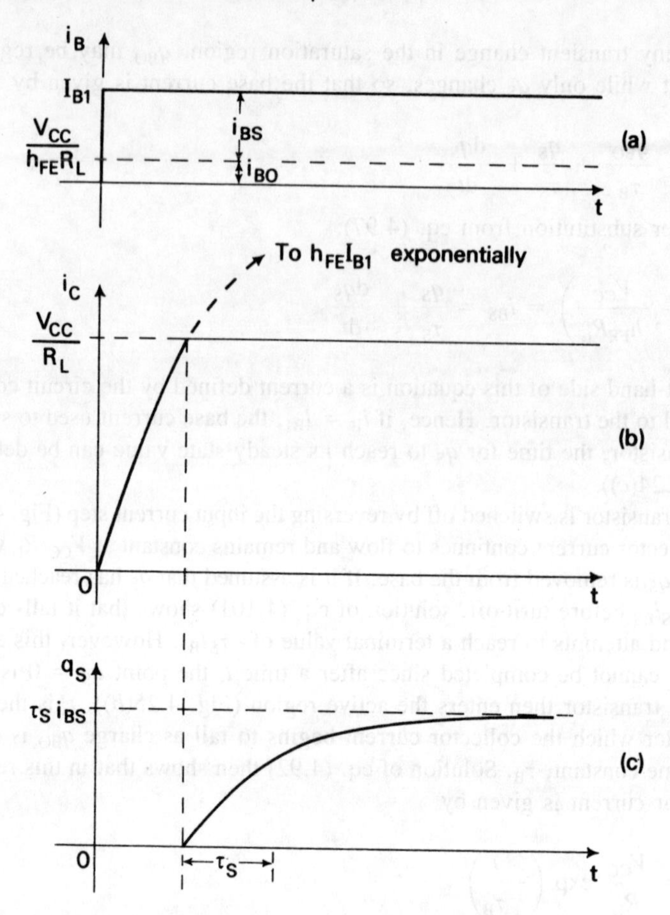

Figure 4.24 **Transient response with overdrive for switching on. (*a*) Base-current step function; (*b*) collector current; (*c*) saturation base charge.**

that q_S will attempt to reach a terminal value of $\tau_S(I_{B1} - I_{B2})$. However, q_S will become zero after time t_s so that

$$q_S = \tau_S(I_{B1} - I_{B2})\left[1 - \exp\left(-\frac{t_s}{\tau_S}\right)\right] \tag{4.104}$$

This equation can be expressed more conveniently in terms of base current since $i_{BS} = I_{B1} - V_{CC}/h_{FE}R_L$ from eq. (4.101), and $q_S = \tau_S i_{BS}$. Hence the storage time t_s will be given by

$$I_{B1} - \frac{V_{CC}}{h_{FE}R_L} = (I_{B1} - I_{B2})\left[1 - \exp\left(\frac{t_s}{\tau_S}\right)\right] \tag{4.105}$$

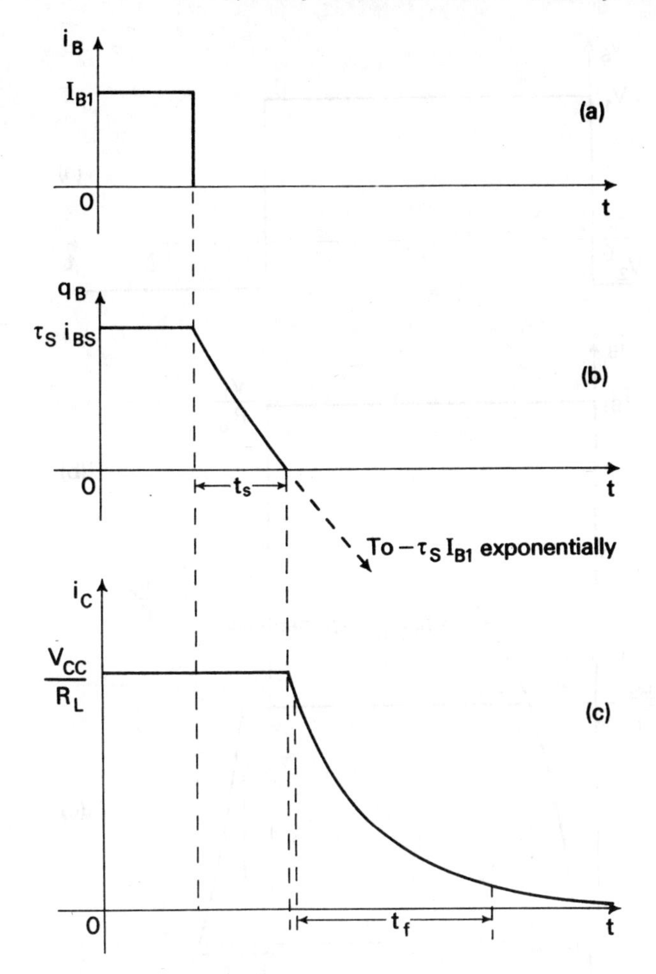

Figure 4.25 Transient response to switching off. (*a*) Base-current step function; (*b*) saturation base charge; (*c*) collector current.

and the more negative I_{B2} is made the smaller will t_s become. After time t_s, q_S has become zero, so that the removal of q_{BO} with time-constant τ_B can begin, with the collector current aiming at a value of $h_{FE}I_{B2}$ instead of zero as in eq. (4.102). Since it starts from a value V_{CC}/R_L the expression for the decay of collector current becomes

$$i_C = \frac{V_{CC}}{R_L} + \left(h_{FE}I_{B2} - \frac{V_{CC}}{R_L}\right)\left[1 - \exp\left(-\frac{t}{\tau_B}\right)\right] \tag{4.106}$$

and the fall time t_f over the whole range of V_{CC}/R_L becomes smaller as I_{B2} is increased, since it corresponds to a smaller part of the exponential decay. Finally

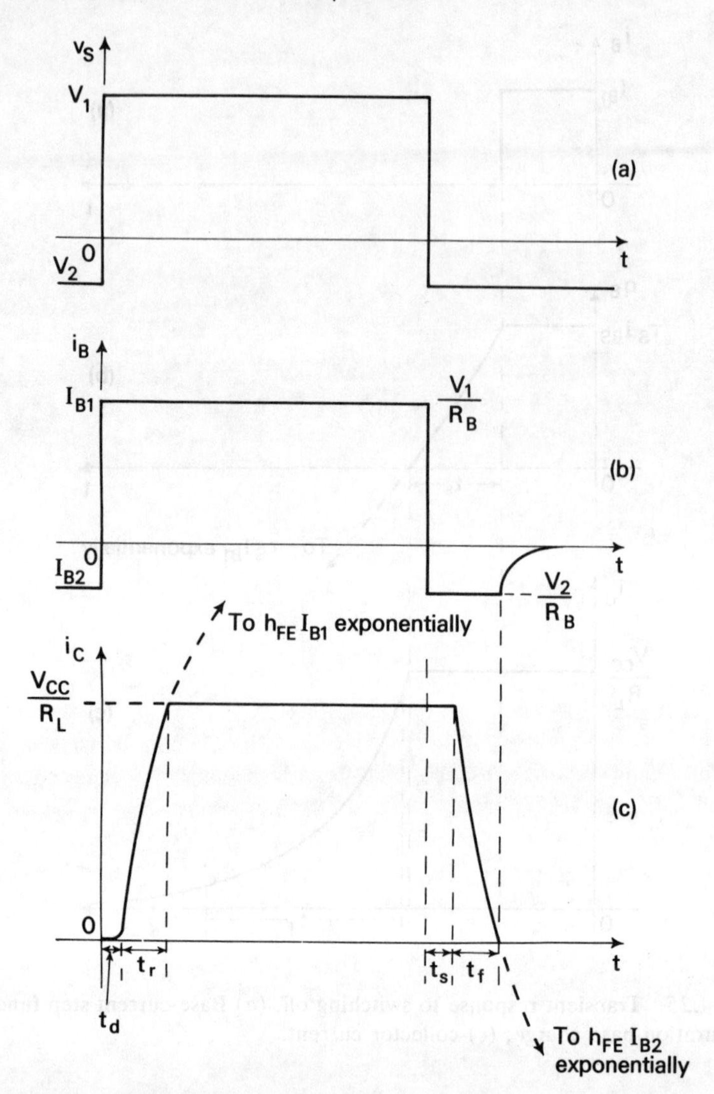

Figure 4.26 Complete transient response with overdrive. (*a*) Input voltage pulse; (*b*) base current; (*c*) collector current.

when $i_C = 0$ all the excess charge in the base has been removed, so that $i_B = 0$ and $t = t_f$. The complete collector current waveform for a transistor turned on and off with overdrive is shown in Fig. 4.26(*c*), which corresponds to normal operation.

For a practical transistor there is also a *delay time* t_d between application of the step input and the time when the current begins to rise to 10% of V_{CC}/R_L. This is due to minority carriers filling the depletion layers in the emitter and collector junctions and so reducing their widths to the values which correspond to forward bias conditions. Until this has been achieved current flow cannot commence, as shown in Fig. 4.26(*c*). (*See also* Problem 4.9.)

Measurement of τ_B and τ_C

In the circuit arrangement of Fig. 4.27(*a*), charge will be transferred from the signal source to the base of the transistor through both the resistor R_B and the capacitor C. For a step input voltage ΔV much larger than the corresponding change in V_{BE}, which in practice means an input of a few volts, the charge transferred initially through the capacitor is $C\Delta V$ (Fig. 4.27(*b*)). This charge then leaks away exponentially through the transistor with time-constant τ_B to give a component of the base charge q_1. The charge q_2 transferred through R_B is given by eq. (4.91) and rises exponentially to a value $\tau_B \Delta I$, where $\Delta I = \Delta V / R_B$, again with time-constants τ_B (Fig. 4.27(*c*)). The total charge transferred is then

$$q_B = q_1 + q_2 \tag{4.107}$$

so that the corresponding change in collector current (Fig. 4.17(*d*)) is given by

$$i_C = \frac{q_B}{\tau_C} = h_{FE} \frac{q_1 + q_2}{\tau_B} \tag{4.108}$$

The parameters τ_B and τ_C can be determined if C and R are adjusted until $q_1 = q_2$. In this condition the fall in q_1 due to recombination in the base is exactly compensated by the rise in q_2, so that q_B rises instantaneously and then remains constant with time, whil i_C undergoes a similar step change (Fig. 4.27(*d*)). Then

$$C\Delta V = \tau_B \Delta I = \tau_B \frac{\Delta V}{R_B}$$

so that

$$\tau_B = CR_B \tag{4.109}$$

If ΔV is made just large enough to switch the transistor from zero collector current to saturation, the required change in base current is

$$\frac{V_{CC}}{h_{FE} R_L} = \frac{\Delta V}{R_B} \tag{4.110}$$

Then

$$h_{FE} = \frac{R_B}{R_L} \frac{V_{CC}}{\Delta V} = \frac{\tau_B}{\tau_C} \tag{4.111}$$

and

$$\tau_C = \frac{\Delta V R_L}{V_{CC} R_B} \tau_B = \frac{\Delta V}{V_{CC}} CR_L \tag{4.112}$$

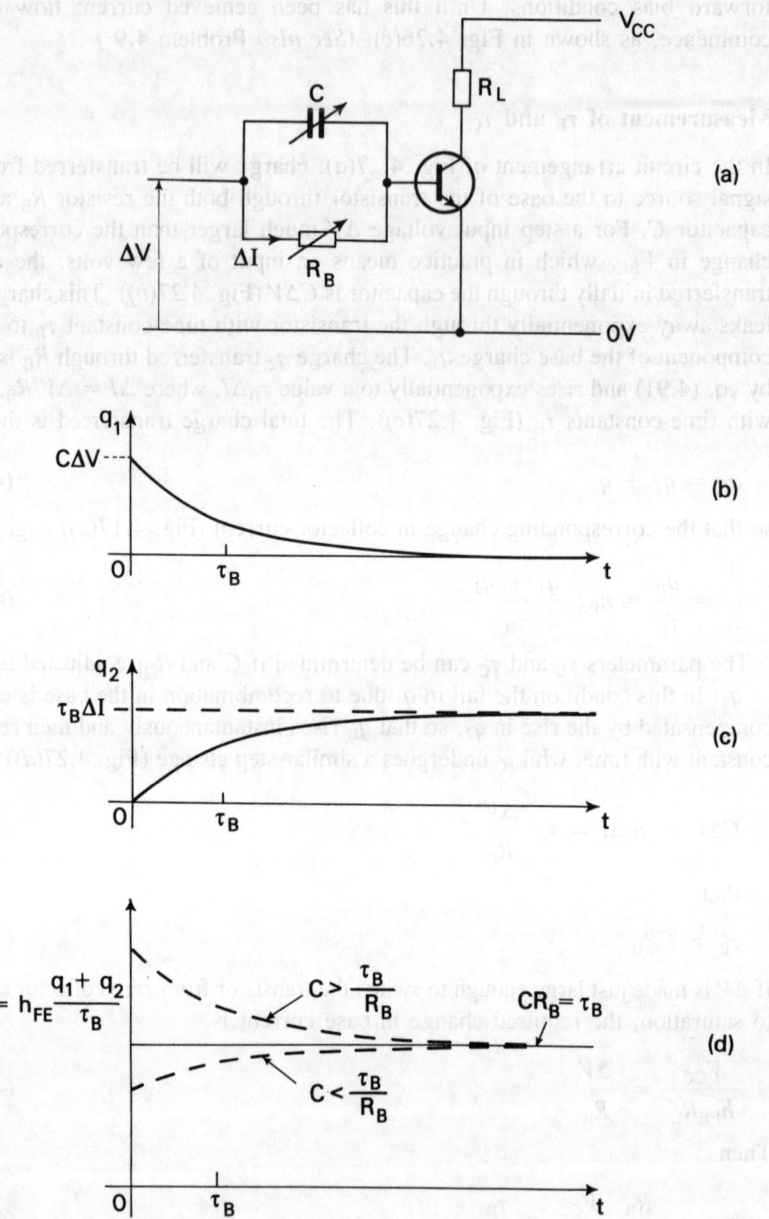

Figure 4.27 Measurement of charge control parameters. (*a*) Measuring circuit; (*b*) variation of charge q_1 introduced through capacitor C; (*c*) variation of charge q_2 introduced through resistor R; (*d*) collector current waveforms.

If R_B is held constant at the value at which the transistor is just saturated in the steady state, C may be adjusted to produce a step change in V_{CE} as observed on an oscilloscope. If C is made too large $q_1 > q_2$ and the recombination of base charge is over-compensated, so that the collector current overshoots its equilibrium value. Similarly, if C is too small, $q_1 < q_2$ and the recombination is insufficiently compensated, giving a rapid initial rise in collector current followed by a slower exponential rise.

Small-signal operation

When the transistor is to be operated as an amplifier, the signals are normally small and the transistor is first biased to a particular operating point such as Q in Fig. 4.21(*b*). The required value of base current may be obtained by connecting the resistor R_B directly to the d.c. supply, as shown in the simple amplifier circuit of Fig. 4.28(*a*), so that $I_B \approx V_{CC}/R_B$ since $V_{BE} \ll V_{CC}$. Corresponding steady values of I_C and V_{CE} are obtained by the intersection of the load line and the characteristics at the point Q.

The input signal to the amplifier is obtained from a generator of voltage V_S and internal resistance R_S, which is connected to the base through a capacitor C. When the waveform of V_S is sinusoidal at frequency f the reactance of C is $1/2\pi fC$, which is made smaller than R_S by choosing a large value for C, say 5 μF, for operating frequencies within the audible range of 20 Hz to 20 kHz. The purpose of the capacitor is to block any direct voltage at the input generator, from the base of the transistor and thus prevent any disturbance of the bias voltage. This is achieved, since the reactance of the capacitor is infinite at zero frequency. The signal waveforms corresponding to sinusoidal changes are illustrated in Fig. 4.28, which shows that a rise in base current causes a rise in collector current and hence a fall in collector voltage. The opposite changes occur for a fall in base current, the changes always occurring with respect to a direct quantity.

In a typical amplifier let us suppose that $V_{CC} = 10$ V, $R_L = 2$ kΩ, $V_{CE} = 5$ V, so that $I_C = 2.5$ mA. Then if $h_{FE} = 100$, $I_B = 25$ μA and $R_B = 400$ kΩ neglecting V_{BE}. The peak values of the signals might be $I_{bm} = 1$ μA, $I_{cm} = 100$ μA and $V_{cm} = 0.2$ V. These are clearly much less than the corresponding direct values and would be very difficult to measure accurately from the characteristics. Hence the performance of a small-signal amplifier is obtained from a circuit which is equivalent to the transistor operating at a particular point Q on its characteristics. This equivalent circuit is composed of resistive and capacitive elements to simulate the effect of high frequencies and also contains one or more generators. All the elements are related to the physical processes described previously so that their values are functions of the operating voltage, current and temperature.

The hybrid-π equivalent circuit for a BJT

The distribution of excess charge for a transistor operating in the active region is shown approximately in Fig. 4.29(*a*), where the effective edge of the charged

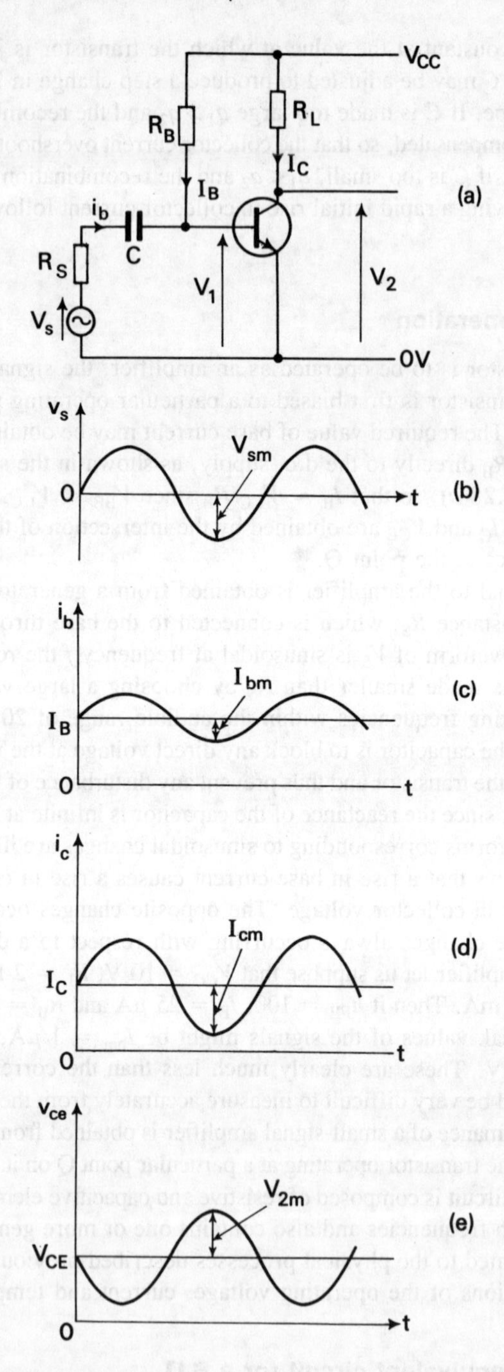

Figure 4.28 **Small-signal operation. (*a*) Amplifier circuit; (*b*) input voltage to base; (*c*) base current; (*d*) collector current; (*e*) collector-emitter voltage.**

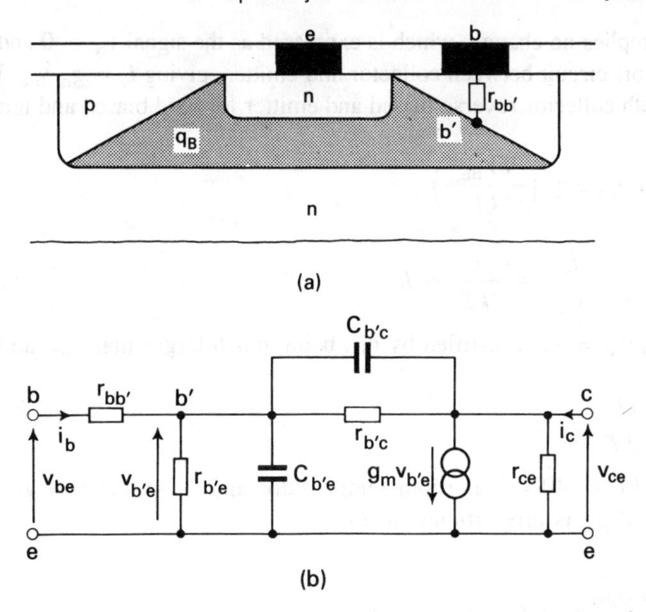

Figure 4.29 Models of a bipolar junction transistor. (*a*) Representation of excess base charge q_B and base spreading resistance $r_{bb'}$; (*b*) hybrid-π equivalent circuit.

region, designated b′ is separated from the external base contact b. The distributed resistance of the base is then represented by $r_{bb'}$, also known as the *base spreading resistance*. This is incorporated in the *hybrid-π equivalent circuit* of Fig. 4.29(*b*), so called because one arm of the π contains a current generator while the other consists of passive components. The elements have suffixes corresponding to the points they join in the circuit in Fig. 4.29(*b*), although other designations are also used (*see* Table 4.2). Each element, except $C_{b'c}$ and $r_{bb'}$, can be related to the collector current and the temperature to allow for variation in these quantities. Provided the signals are small, all the elements are linear and the circuit is then very useful in predicting the performance of the transistor over a wide frequency range. The derivation of the elements is given below.

Transconductance (mutual conductance) g_m

The currents and voltages shown in Fig. 4.29(*b*) represent small changes about the quiescent values defined by operating points such as P and Q (Figs 4.18 and 4.21). g_m is defined by

$$g_m = \left.\frac{\partial I_C}{\partial V_{BE}}\right|_{V_{CE\text{const.}}} = \left.\frac{i_c}{v_{b'e}}\right|_{v_{ce}=0} \tag{4.113}$$

$V_{\mathrm{CEconst.}}$ implies no change, which is expressed as the signal $v_{ce} = 0$ and requires an a.c. short-circuit between collector and emitter, giving $i_c = g_m v_{b'e}$. From eq. (4.63), with collector reverse biased and emitter forward biased and ignoring I_{CS}

$$I_C = \alpha_F I_{ES} \exp\left(\frac{eV_{BE}}{kT}\right)$$

$$\frac{\partial I_C}{\partial V_{BE}} = \frac{i_c}{v_{be}} = \frac{e}{kT}\alpha_F I_C \tag{4.114}$$

Assuming $v_{be} \simeq v_{b'e}$, justified by $r_{b'e}$ being much larger than $r_{bb'}$ and $\alpha_F \simeq 1$

$$g_m = \frac{eI_C}{kT} \tag{4.115}$$

Taking $e/kT = 40\ V^{-1}$ at room temperature and $I_C = 2.5$ mA gives $g_m = 100$ mS and g_m is proportioned to I_C.

Input resistance $r_{b'e}$

The common-emitter current gain (eq. (4.45)) is given by

$$\beta_0 = \left.\frac{\partial I_C}{\partial I_B}\right|_{V_{\mathrm{CEconst.}}} = \left.\frac{i_c}{i_b}\right|_{v_{ce}=0} \tag{4.116}$$

which again requires an a.c. short-circuit between collector and emitter. Then at frequencies low enough for $1/\omega C_{b'e}$ to be much greater than $r_{b'e}$

$$\beta_0 = \frac{i_c r_{b'e}}{v_{b'e}} = g_m r_{b'e} \tag{4.117}$$

so

$$r_{b'e} = \frac{\beta_0}{g_m} = \frac{\beta_0 kT}{eI_C} \tag{4.118}$$

For $\beta_0 = 100$, $g_m = 100$ mS $r_{b'e} = 1$ kΩ at room temperature and $r_{b'e}$ is *inversely* proportional to I_C.

Base-emitter capacitance $C_{b'e}$

$C_{b'e}$ consists of two capacitances in parallel, these being the depletion layer capacitance of the emitter-base junction, C_e, and the *diffusion capacitance*, C_d, so that

$$C_{b'e} = C_d + C_e \tag{4.119}$$

C_e may be significant for a forward biased junction and is obtained from eq. (3.78). C_d arises because the excess charge q_B is related to current I_E (eq. (4.17)), which in turn is a function of V_{EB} (eq. (4.27)) so that

$$C_d = \frac{dq_B}{dV_{EB}} \tag{4.120}$$

Thus C_d links the small signal equivalent circuit with the charge control model. Then assuming $C_d \gg C_e$

$$\begin{aligned} C_d &= \frac{dq_B}{dI_E} \cdot \frac{dI_E}{dV_{EB}} \\ &= \tau_C \frac{eI_E}{kT} \\ &= g_m \tau_C \end{aligned} \tag{4.121}$$

using eqs (4.17) and (4.115) and assuming $I_E \simeq I_C$.

Worked example

A transistor operating at room temperature has an average base transit time of 0.5 ns. Estimate the value of $C_{b'e}$ for collector currents of 2.5 mA and 5.0 mA.

Solution

From eq. (4.121) and for $I_C = 2.5$ mA

$$C_d = \frac{0.5 \times 10^{-9} \times 2.5 \times 10^{-3}}{0.025}$$

$$= 50 \text{ pF}$$

For $I_C = 5.0$ mA

$$C_d = \frac{0.5 \times 10^{-9} \times 5.0 \times 10^{-3}}{0.025}$$

$$= 100 \text{ pF}$$

Since $C_d \simeq C_{b'e}$ these are approximately the values of $C_{b'e}$ at the two currents, so that $C_{b'e}$ is proportional to I_C.

Collector-base resistance $r_{b'c}$

This resistance arises due to the effect of collector voltage on the base width (Fig. 4.7(*b*)) which shows itself as a decrease in V_{EB} when V_{CB} is increased (Fig. 4.6). It may be defined by means of a *voltage feedback factor* μ given by

$$\mu = \left.\frac{\partial V_{EB}}{\partial V_{CB}}\right|_{I_{Econst.}} = \left.\frac{v_{eb}}{v_{cb}}\right|_{i_e=0} \tag{4.122}$$

Again $I_{Econst.}$ implies no change, which is expressed by the small signal $i_e = 0$. Thus μv_{cb} represents the change in the emitter voltage to maintain constant emitter current as the collector voltage is changed. In order to obtain an expression for μ the density of excess electrons, n_e, at the emitter junction is involved (in an n-p-n transistor) so

$$\frac{\partial V_{EB}}{\partial V_{CB}} = \frac{\partial V_{EB}}{\partial n_e}\,\frac{\partial n_e}{\partial w_b}\,\frac{\partial w_b}{\partial V_{CB}} \tag{4.123}$$

From eq. (3.34) for a p-n junction

$$n_e = n_n \exp\left(\frac{eV_{EB}}{kT}\right)$$

so that

$$\frac{\partial n_e}{\partial V_{EB}} = \frac{en_e}{kT} \tag{4.124}$$

Also, from eq. (4.7),

$$I_E \simeq \frac{eD_n n_e S}{w_b}$$

so, for constant I_E,

$$\frac{\partial n_e}{\partial w_b} = \frac{n_e}{w_b} \tag{4.125}$$

Again using eq. (4.26)

$$\frac{\partial w_b}{\partial V_{CB}} = \frac{\partial w_{pc}}{\partial V_{CB}} \tag{4.126}$$

Substituting into eqs (4.122) and (4.123)

$$\mu = -\frac{kT}{ew_b}\frac{\partial w_{pc}}{\partial V_{CB}} \tag{4.127}$$

Since $\partial w_{pc}/\partial V_{CB}$ is positive μ is negative with a value around -10^{-4}. Expressing eq. (4.122) in terms of v_{be} gives

$$-\mu = \left.\frac{v_{be}}{v_{cb}}\right|_{i_e=0} \tag{4.128}$$

and for a base open-circuited to a.c. but with V_{CE} applied

$$-\mu = \left.\frac{v_{b'e}}{v_{cb'}}\right|_{i_e=0}$$

or

$$|\mu| = \frac{r_{b'e}}{r_{b'c}}$$

at low frequencies and

$$r_{b'c} = \frac{r_{b'e}}{|\mu|} = \frac{\beta_0}{|\mu| g_m} \tag{4.129}$$

Taking $r_{b'e} = 1\ \text{k}\Omega$ for $I_C = 2.5$ mA and $|\mu| = 10^{-4}$, $r_{b'c} = 10\ \text{M}\Omega$, and $r_{b'c}$ is proportional to I_C when β_0 and μ are assumed constant.

Output resistance r_{ce}

Again considering low frequencies, where the shunt effects of $C_{b'c}$ and $C_{b'e}$ are negligible, the resistance $r_{b'c} + r_{b'e}$ appears in parallel with r_{ce} when the base is open-circuited. Since $r_{b'c}$ is so large the parallel resistances may be neglected so that when a signal voltage v_{ce} is applied.

$$i_c = g_m v_{b'e} + \frac{v_{ce}}{r_{ce}} \simeq i_e$$

or

$$v_{ce} = -g_m v_{b'e} r_{ce} \tag{4.130}$$

when $i_e = 0$. Then

$$r_{ce} = -\frac{v_{ce}}{v_{b'e}} \cdot \frac{1}{g_m}$$

$$= \frac{1}{|\mu|\ g_m} \tag{4.131}$$

Taking g_m = 100 mS for I_C = 2.5 mA and $|\mu|$ = 10^{-4} gives r_{ce} = 100 kΩ and r_{ce} is also proportional to I_C.

Collector-base capacitance $C_{b'c}$

This is the depletion layer capacitance of the collector-base diode, typically 1 pF or less and depends on V_{CB} according to a relationship similar to eq. (3.94) for a graded base transistor. It becomes important at high frequencies where it is a factor in determining the frequency response as discussed below.

The parameters, and their alternative designations and expressions are given in Table 4.2. The values are typical for a general purpose transistor but would be much changed for a power transistor operating at a collector current of 1.0 A, for example. In this case g_m would rise to 40 S and $r_{b'e}$ would fall to 2.5 Ω, assuming β_0 remains constant.

Table 4.2 Hybrid-π and related parameters for a general purpose planar BJT

Parameter	Expression	Typical value
$r_{bb'}$, $r_{b'}$		10–50 Ω
$r_{b'e}$, r_π	β_0/g_m	1 kΩ
$C_{b'e}$, C_π	$g_m\tau_c + C_e$	60 pF
$r_{b'c}$, r_μ	$\beta_0 r_{ce}$	10 MΩ
$C_{b'c}$, C_μ		1 pF
g_m	$\dfrac{e\|I_c\|}{kT}$	100 mS
r_{ce}, r_0	$1/\|\mu\|g_m$	100 kΩ
β_0, β	$g_m r_{b'e}$	100
f_T	$\dfrac{g_m}{2\pi(C_{b'e}+C_{b'c})} \simeq \dfrac{1}{2\pi\tau_c}$	265 MHz
f_β	f_T/β_0	2.7 MHz
τ_c		0.5 ns
μ		-1×10^{-4}
I_C		2.5 mA
T		293 K

Gain-bandwidth product f_T

One of the most important small-signal parameters of a transistor is the common-emitter short-circuit current gain h_{fe}, also given the symbol β. For sinusoidal signals

$$h_{fe} = \left.\frac{i_c}{i_b}\right|_{v_{ce}=0} \tag{4.132}$$

and it is a function of frequency which can be obtained from the hybrid-π circuit of Fig. 4.29(*b*). It is related to the common-base short-circuit current gain α by the expression

$$h_{fe} = \frac{\alpha}{1 - \alpha} \tag{4.133}$$

so that α is also a function of frequency.

From Fig. 4.29(*b*), if the effect of $r_{b'c}$ is neglected, and the collector and emitter are short-circuited, then

$$i_b = v_{b'e} \frac{1 + jw(C_{b'e} + C_{b'c})r_{b'e}}{r_{b'e}} \tag{4.134}$$

and

$$i_c = g_m v_{b'e} \tag{4.135}$$

so that

$$\left.\frac{i_c}{i_b}\right|_{v_{ce}=0} = h_{fe} = \frac{g_m r_{b'e}}{1 + jw(C_{b'e} + C_{b'c})r_{b'e}} \tag{4.136}$$

At low frequencies the j term is negligible and

$$h_{fe} = g_m r_{b'e} = \frac{\alpha_0}{1 - \alpha_0} = \beta_0 \tag{4.137}$$

As the frequency is increased h_{fe} falls, until at a frequency f_β, $\omega(C_{b'e} + C_{b'c})r_{b'e} = 1$. Here the reactance of $(C_{b'e} + C_{b'c})$ and the resistance $r_{b'e}$ are equal and

$$f_\beta = \frac{1}{2\pi(C_{b'e} + C_{b'c})r_{b'e}} \tag{4.138}$$

Substitution into eq. (4.136) then yields

$$h_{fe} = \frac{\beta_0}{1 + j(f/f_\beta)} \tag{4.139}$$

and f_β is the frequency at which $|h_{fe}|$ has fallen to $\beta_0/\sqrt{2}$ or 70.7% of its low-frequency value β_0, which may be expressed as a fall of 3 dB from the low-frequency value.* At frequencies well above f_β, $\omega(C_{b'e} + C_{b'c}) \ll r_{b'e}$ and i_b may be considered to flow entirely through $(C_{b'e} + C_{b'c})$ (Fig. 4.30(*a*)). Then

* For two powers, P_1 and P_2, the ratio in decibels (dB) is 10 log (P_1/P_2). Where the powers are due to currents I_1 and I_2 flowing through the same value of resistance R the current ratio is 10 log $(I_1/I_2)^2$ = 20 log (I_1/I_2). Thus when I_1 = 0.707 this corresponds to a decibel ratio of 3.01 dB. Similarly for two voltages, V_1 and V_2, developed across the same resistance the voltage ratio is 20 log (V_1/V_2).

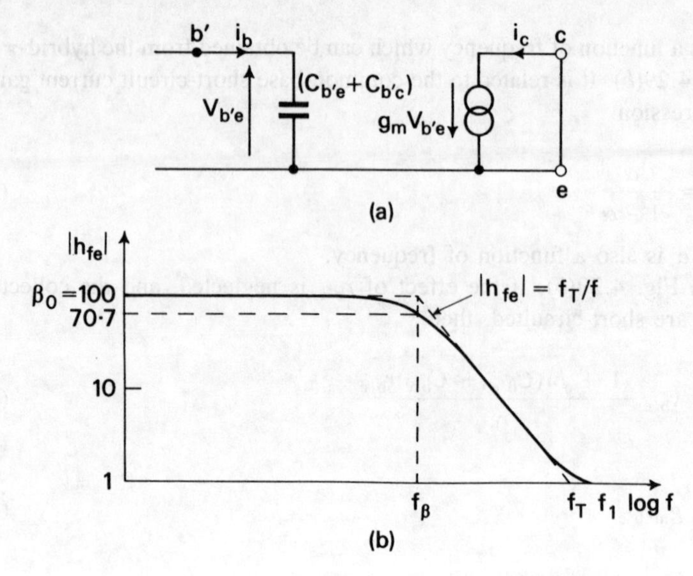

Figure 4.30 **High-frequency response of a junction transistor. (*a*) Equivalent circuit for short-circuit current gain; (*b*) short-circuit current gain as a function of frequency.**

$$i_b = j\omega(C_{b'e} + C_{b'c})v_{b'e} \tag{4.140}$$

and

$$h_{fe} = \frac{g_m}{j\omega(C_{b'e} + C_{b'c})} \tag{4.141}$$

In a practical transistor $|h_{fe}|$ falls as frequency rises in this region until at a frequency f_1, $h_{fe} = 1$. In order to determine f_1 a lower frequency is chosen on the falling part of the curve in Fig. 4.30(*b*) and measured. Then from eq. (4.141) the product of current gain and frequency is

$$|h_{fe}|f = \frac{g_m}{2\pi(C_{b'e} + C_{b'c})} = f_T \tag{4.142}$$

f_T is called the *gain-bandwidth product* or *transition frequency* of the transistor. $|h_{fe}|f$ may be constant in practice, in which case $f_1 = f_T$, but this is not always so. The relationship between f_T and f_β may be obtained from eq. (4.139) since when the j term is much larger than unity,

$$|h_{fe}|\frac{f}{f_\beta} = \beta_0$$

so that

$$f_T = \beta_0 f_\beta \tag{4.143}$$

from eq. (4.142). f_β then occurs at the intersection of the $|h_{fe}| = \beta_0$ and $|h_{fe}| = f_T/f$ parts of Fig. 4.30(*b*).

f_T is quoted by manufacturers as a basic parameters and may be used to estimate $C_{b'e}$, τ_c and τ_B. Rearranging eq. (4.142) gives

$$C_{b'e} + C_{b'c} = \frac{g_m}{2\pi f_T} = \frac{g_m}{\omega_T}$$

or

$$C_{b'e} \simeq \frac{g_m}{\omega_T},$$

neglecting $C_{b'c}$. From eq. (4.121)

$$\tau_c \simeq \frac{C_{b'e}}{g_m} \quad \text{so} \quad \tau_c \simeq \frac{1}{\omega_T}, \; \tau_B \simeq \frac{\beta_0}{\omega_T} \quad \text{and} \quad f_T \simeq \frac{1}{2\pi\tau_C}.$$

Worked example

Using the values of g_m, $C_{b'e}$ and β_0 from Table 4.2 for a general-purpose planar BJT with $I_C = 2.5$ mA, estimate the values of f_T and f_β.

What are the approximate values of f_T and f_β when C_e is neglected?

Solution

$$\tau_C = \frac{C_{b'e}}{g_m} = \frac{6 \times 10^{-11}}{0.1} = 6.0 \times 10^{-10} \text{ s}$$

$$f_T = \frac{1}{2\pi \times 6 \times 10^{-10}} = 265 \text{ MHz}$$

$$f_\beta = \frac{f_T}{\beta_0} = 2.65 \text{ MHz}$$

Neglecting C_e gives $C_{b'e} = 50$ pF from the previous worked example,

$$\tau_C = 5.0 \times 10^{-10} \text{ s}$$

and

$$f_T \simeq 318 \text{ MHz}$$

$$f_\beta \simeq 3.18 \text{ MHz}.$$

Simplified high-frequency equivalent circuit

The hybrid-π equivalent circuit of Fig. 4.29(*b*) is somewhat complicated for the analysis of amplifier performance, and may be simplified by rearrangement. Normally the load resistance R_L in an amplifier as shown in Fig. 4.28(*a*) is much smaller than r_{ce}, since I_C flows through R_L and determines V_{CE}. Thus, considering R_L connected between c and e in Fig. 4.29(*b*), the output voltage V_2 is given by

$$V_2 = -g_m v_{b'e} \frac{r_{ce} R_L}{r_{ce} + R_L} \approx -g_m R_L v_{b'e} \tag{4.144}$$

the minus sign indicating a phase difference of 180° between the output voltage V_2 and $v_{b'e}$. Then the voltage between b′ and c is $v_{b'e}(1+g_m R_L)$. The current through $C_{b'c}$ is

$$j\omega C_{b'c} V_{b'e}(1+g_m R_L)$$

which is the same as the current through a capacitance $C_{b'c}(1+g_m R_L)$ connected between b′ and e, Fig. 4.31(*b*). Similarly, the current through $r_{b'c}$ is

$$\frac{v_{b'e}(1+g_m R_L)}{r_{b'c}}$$

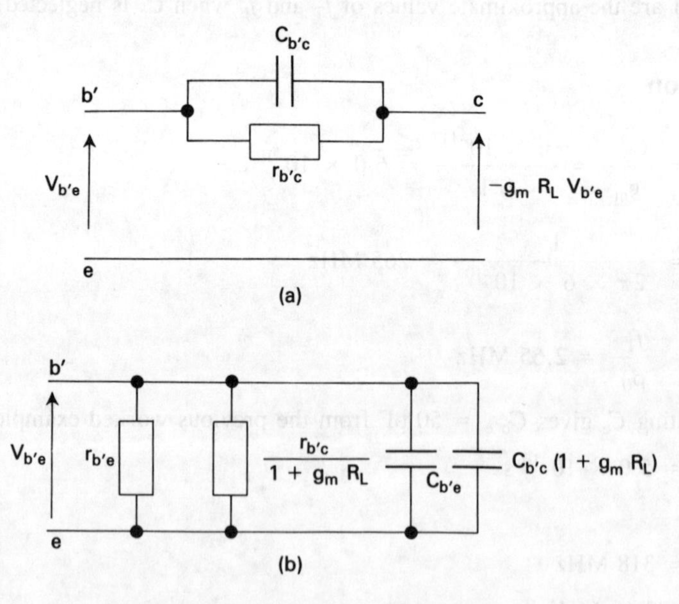

Figure 4.31 Simplification of hybrid-π equivalent circuit. (*a*) Voltages at b′ and c; (*b*) elements joining b′ and c transferred across b′ and e.

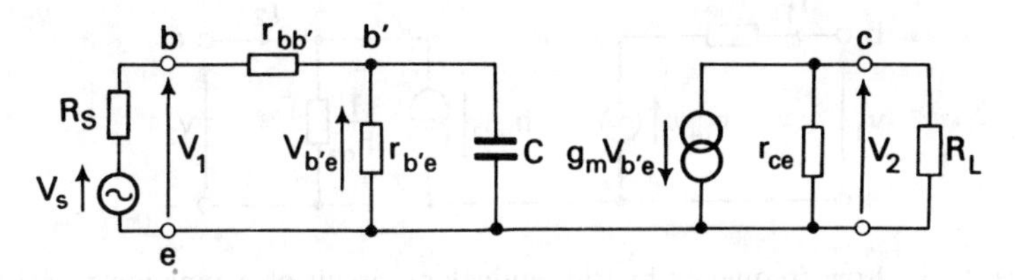

Figure 4.32 **Simplified hybrid-π equivalent circuit.**

which is the same as the current through a resistance $r_{b'c}/(1+g_m R_L)$ connected between b′ and e. Now, for values of $g_m R_L$ up to about 100 this new resistance will still be about 35 kΩ, which is much larger than the value of $r_{b'e}$, so that it may be neglected. Then the approximate equivalent circuit becomes as shown in Fig. 4.32, where

$$C = C_{b'e} + C_{b'c}(1+g_m R_L) \tag{4.145}$$

Thus the frequency response is determined by the input circuit, which depends on R_L and so on the gain of the amplifier.

It is common practice among transistor manufacturers to give a summary of the properties of a transistor in the form of numerical values of I_C, h_{fe} (at low frequencies) and f_T. It may be noted that three of the elements of the circuit of Fig. 4.29(*b*), that is g_m, $r_{b'e}$ and $C_{b'e}$ may be obtained from this information. These are the three most important parameters, since $r_{bb'}$ may well be much less than R_S and the effect of $C_{b'c}$ will be negligible for small values of R_L. While the full data of a transistor should be used wherever possible, the summarized data will nevertheless give some basis for comparing the properties of different types of transistor.

The hybrid parameter equivalent circuit

At low frequencies, up to a few hundred kilohertz for a graded-base transistor, a commonly used equivalent circuit is based on the four *hybrid parameters*, which are purely resistive in this range of frequencies. The transistor is considered as a 'black box' and its internal operation is expressed entirely in terms of the small-signal input voltages and currents (Fig. 4.33). The *h*-parameters can also be defined in terms of the d.c. characteristics and they are given below in terms of both small signals and small changes in the direct voltages and currents of a transistor in the common-emitter configuration.

Input Resistance

$$h_{ie} = \left.\frac{v_1}{i_1}\right|_{v_2=0} = \left.\frac{\partial V_{BE}}{\partial I_B}\right|_{V_{CE\text{const.}}} \tag{4.146}$$

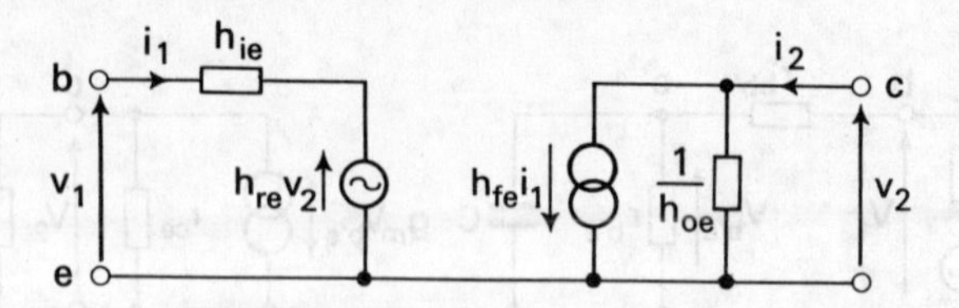

Figure 4.33 Low-frequency hydrid equivalent circuit of a junction transistor.

Reverse Voltage Transfer Ratio

$$h_{re} = \left.\frac{v_1}{v_2}\right|_{i_1=0} = \left.\frac{\partial V_{BE}}{\partial V_{CE}}\right|_{I_{Bconst.}} \tag{4.147}$$

Both of these parameters would be measured with the transistor operating at a point such as P on the input characteristic (Fig. 4.18(*a*)).

The condition $v_2 = 0$ implies that the output terminals are short-circuited to alternating current; and the condition $i_1 = 0$, that the input terminals are open-circuited to alternating current.

Forward Current Transfer Ratio

$$h_{fe} = \left.\frac{i_2}{i_1}\right|_{v_2=0} = \left.\frac{\partial I_C}{\partial I_B}\right|_{V_{CEconst.}} \tag{4.148}$$

Output Conductance

$$h_{oe} = \left.\frac{i_2}{v_2}\right|_{i_1=0} = \left.\frac{\partial I_C}{\partial V_{CE}}\right|_{I_{Bconst.}} \tag{4.149}$$

In practice the parameters could be measured by biasing the transistor to the relevant operating points and applying a 1 kHz signal to the input terminals with the output short-circuited, or to the output terminals with the input open-circuited.

The parameters are called *hybrid* since two of them are dimensionless and the dimensions of the other two are different. The notation of the subscripts is explained by the quantities to which they refer: thus i represents input, r reverse (output to input), f forward (input to output) and o output; e refers to the terminal common to input and output, in this case the emitter. Since a circuit similar to Fig. 4.33 can be drawn for the common-base or the common-collector configuration, the corresponding hybrid parameters would have a subscript ending with b or c respectively, e.g. h_{fb}, h_{fc} and so on.

The *h*-parameters may be expressed in terms of the physical properties of the transistor by relating them to those *hybrid-π parameters* which are important at low frequencies. This leads to the low-frequency hybrid-π equivalent circuit of Fig. 4.34, which may be compared with Fig. 4.33. Then with the output terminals short-circuited the input resistance is

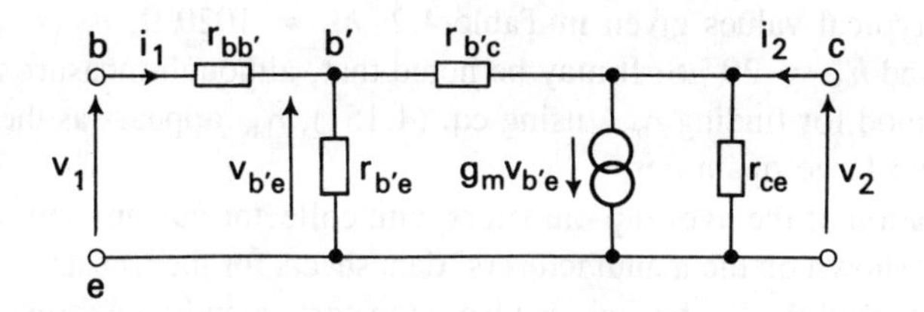

Figure 4.34 **Low-frequency hybrid-π equivalent circuit.**

$$\left.\frac{v_1}{i_1}\right|_{v_2=0} = r_{bb'} + \frac{r_{b'e}r_{b'c}}{r_{b'e} + r_{b'c}} \approx r_{bb'} + r_{b'e} \tag{4.150}$$

or

$$h_{ie} \approx r_{bb'} + r_{b'e} \tag{4.151}$$

Also the forward current transfer ratio is

$$\left.\frac{i_2}{i_1}\right|_{v_2=0} = \frac{g_m v_{b'e}}{v_{b'e}/r_{b'e}} = g_m r_{b'e} \tag{4.152}$$

or

$$h_{fe} = \beta_0 \tag{4.153}$$

at low frequencies. If the input terminals are open-circuited and a voltage v_2 is applied to the output terminals the reverse voltage ratio is

$$\left.\frac{v_1}{v_2}\right|_{i_1=0} = \frac{r_{b'e}}{r_{b'e} + r_{b'c}} = \frac{1}{1 + 1/\mu} \approx \mu \quad \text{since} \quad \mu \ll 1 \tag{4.154}$$

or

$$h_{re} \approx \mu \tag{4.155}$$

The output impedance is obtained from

$$i_2 = \frac{v_2}{r_{ce}} + g_m v_{b'e} \tag{4.156}$$

$$\approx v_2(1/r_{ce} + \mu g_m) \tag{4.157}$$

since $v_{b'e} = v_1 \approx \mu v_2$ with the input open-circuited. From eq. (4.130) $\mu g_m = 1/r_{ce}$, which leads to

$$\left.\frac{i_2}{v_2}\right|_{i_1=0} = h_{oe} \approx \frac{2}{r_{ce}} \tag{4.158}$$

Since the hybrid parameters are expressed in terms of g_m, β_0 and μ their dependence on the transistor properties is obtained from eqs (4.115) and (4.127).

Using the typical values given in Table 4.2, $h_{ie} \approx 1020\ \Omega$, $h_{fe} = 100$, $h_{re} \approx 1 \times 10^{-4}$ and $h_{oe} \approx 20\ \mu S$. It may be noted that, although measurements of h_{ie} gives a method for finding $r_{bb'}$, using eq. (4.151), $r_{bb'}$ appears as the difference between two large quantities.

The variation of the hybrid parameters with collector current and temperature is normally shown on the manufacturers' data sheets for the transistor. However, theoretical variations can be deduced from the corresponding expressions for the hybrid-π parameters. Thus h_{ie} is inversely proportional to I_C in the same way as $r_{b'e}$. h_{oe} is proportional to I_C due to its dependence on $1/r_{ce}$. h_{re} rises with I_C non-linearly in practice. h_{fe} varies with I_C in a similar manner to β_0 (Fig. 4.12) and increases with temperature at approximately 1% per °C rise above room temperature.

The very wide range of available properties is illustrated in Table 4.3 for the three main categories of general-purpose, power and high-frequency transistors. Taking general-purpose transistors as a basis for comparison, power transistors have a high maximum collector current and power dissipation, while high-frequency transistors have a higher gain-bandwidth product for a lower collector current. The applications of discrete bipolar transistors are very numerous, quite apart from the much larger numbers in integrated circuits, but many applications are given in Refs [3] and [4].

Computer simulation of a bipolar junction transistor

Common-emitter model

Simulation of the BJT is simplified by using a common-emitter version of the Ebers–Moll model, in which the transport saturation current I_S is a single parameter applying to both junctions. From eqs (4.61), (4.62) and (4.63) and including emission coefficients n_F and n_R

$$\alpha_F I_F = I_S \left[\exp\left(\frac{eV_{BE}}{n_F kT} \right) - 1 \right] = I_{CC} \tag{4.159}$$

$$\alpha_R I_R = I_S \left[\exp\left(\frac{eV_{BC}}{n_R kT} \right) - 1 \right] = I_{EC} \tag{4.160}$$

Table 4.3 Typical ranges of silicon BJT parameters

	I_{Cmax}	V_{ceo}	P_{Cmax}	h_{fe}	f_T
	A	V	W		MHz
General purpose	0.1–1.0	15–300	0.2–0.6	35–900	50–400
Power	1.0–30	20–1500	6.0–100	2.0–350	7.0–200
High frequency	0.001–0.05	10–45	0.2–1.0	20–300	300–3000

from eqs (4.61) and (4.160)

$$I_C = I_{CC} - \frac{I_{EC}}{\alpha_R}$$

$$= I_{CC} - I_{EC} - I_{EC}\left(\frac{1}{\alpha_R} - 1\right)$$

$$I_C = I_{CT} - \frac{I_{EC}}{\beta_R} \tag{4.161}$$

putting $I_{CT} = I_{CC} - I_{EC}$ and $\beta_R = \dfrac{\alpha_R}{1 - \alpha_R}$,

from eqs (4.61) and (4.159)

$$I_E = I_{EC} - \frac{I_{CC}}{\alpha_F}$$

$$= I_{EC} - I_{CC} - I_{CC}\left(\frac{1}{\alpha_F} - 1\right)$$

$$I_E = -I_{CT} - \frac{I_{CC}}{\beta_F} \tag{4.162}$$

where $\beta_F = \dfrac{\alpha_F}{1 - \alpha_F}$

Equations (4.161) and (4.162) lead to the equivalent circuit of Fig. 4.35 where

$$I_B = \frac{I_{EC}}{\beta_R} + \frac{I_{CC}}{\beta_F} \tag{4.163}$$

This is known as a *transport* model since the current generator represents minority carriers transported across the base and the currents I_{CC} and I_{EC} both depend on the same saturation current I_S.

Early voltage

As illustrated in Fig. 4.7(*b*) the width of the base decreases as V_{CB} increases, which leads to a rise in I_C with V_{CB}. Assuming that V_{BE} remains constant I_C also rises with V_{CE} and the effect is known as basewidth modulation or the Early effect (Ref. [5]). It may be shown that the slope of the $I_C - V_{CE}$ characteristics, corresponding to the output conductance h_{oe}, increases with I_C and that when the characteristics are extrapolated backwards to cut the V_{CE} axis all the lines meet at almost the same point V_A' (Ref. [6] and Fig. 4.36).

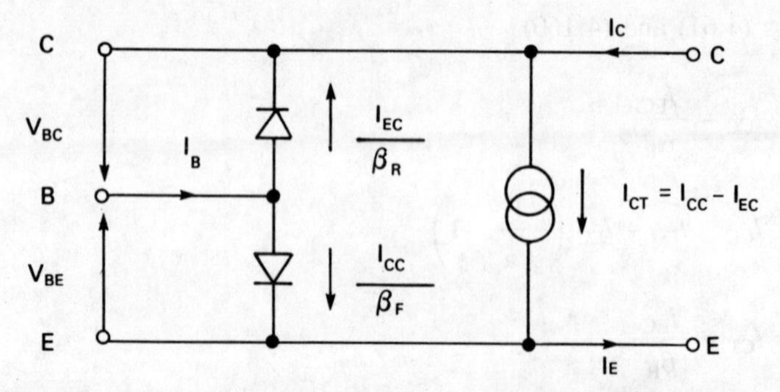

Figure 4.35 Common-emitter transport model for computer simulation of a bipolar junction transistor.

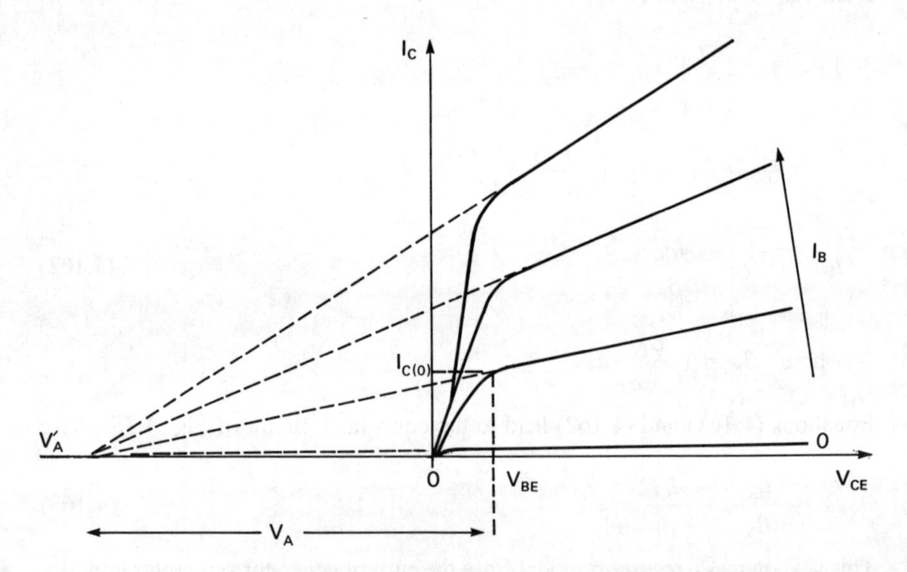

Figure 4.36 Early voltage V_A for a bipolar junction transistor.

The Early voltage V_A is measured from the point where $V_{CE} = V_{BE}$, that is $V_{CB} = 0$, which corresponds to the point at which operation changes from saturated to active (eq. (4.51)) indicated by $I_C = I_C(0)$. Then

$$V_A = V'_A + V_{BE} \tag{4.164}$$

and

$$V_A \simeq V'_A$$

since V'_A is normally at least 50 V, with the output conductance given by

$$h_{oe} = \frac{I_C(0)}{V_A} \tag{4.165}$$

The Early voltage V_A is thus a single parameter which can be used to define output conductance at any value of I_C. A zero output conductance, or a horizontal characteristic, corresponds to an infinite value of V_A.

Knee current I_K

An increase in I_C beyond the normal operating region leads to a fall in β_F. This follows from the corresponding fall in α_0 (or α_F) shown in Fig. 4.12 and occurs above a value of I_C given by the knee current I_K (Fig. 4.37(*a*)). The rate of rise of I_C with V_{BE} is reduced above I_K due to an increase in emission coefficient from 1 to 2, representing the base effects illustrated in Fig. 4.13. β_F is given by the ratio I_C/I_B at a given value of V_{BE} and since the rate of rise of I_B with V_{BE} remains constant this leads to a fall in β_F above I_K, as illustrated in Fig. 4.37(*b*).

Forward and reverse parameters

In general, forward parameters apply to the emitter-base diode and reverse parameters to the collector-base diode, which follows from the Ebers–Moll model. Thus there are forward and reverse current gains, emission coefficients, Early voltages and knee currents. The two depletion layer capacitances are described by equations similar to eq. (3.107) and coefficient FC (Fig. 3.31), while the two diffusion capacitances each require a transit time as in eq. (3.109). Here the transit times refer to minority carriers travelling across the base from emitter to collector (forward) and from collector to emitter (reverse).

The corresponding equations are

$$C_{je} = \frac{C_{joe}}{(1 - V/\psi_e)^{m_e}} \tag{4.166}$$

$$C_{de} = \frac{\tau_f \, eI_E}{kT} \tag{4.167}$$

$$C_{jc} = \frac{C_{joc}}{(1 - V/\psi_c)^{m_c}} \tag{4.168}$$

$$C_{dc} = \frac{\tau_r \, eI_C}{kT} \tag{4.169}$$

Analysis at different temperatures

The saturation current at temperature T_1, $I_S\,(T_1)$, is related to the saturation current at temperature T_0, $I_S\,(T_0)$, where T_0 is normally 300 K, by the expression

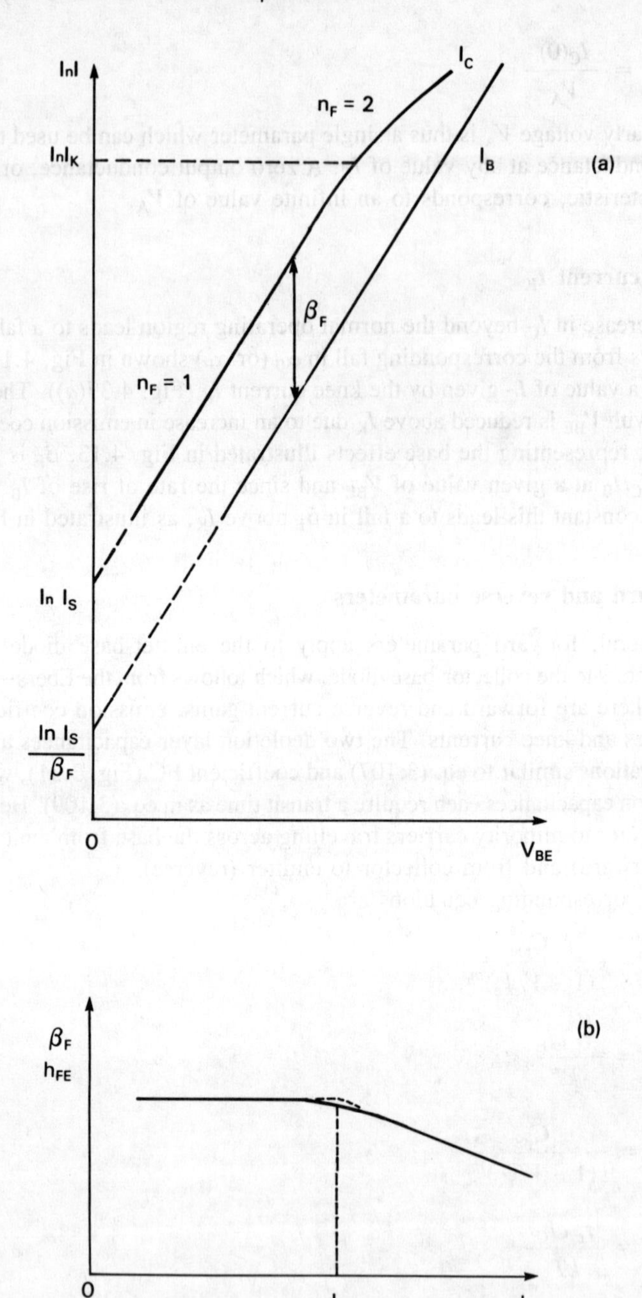

Figure 4.37 (*a*) Variation of ln I_C and ln I_B with V_{BE} in SPICE simulation of a bipolar junction transistor. (*b*) Variation of β_F and h_{FE} with I_C, showing knee current I_K.

$$I(T_1) = I_S(T_0) \times \left(\frac{T_1}{T_0}\right)^x \exp\left[\frac{ew_g}{k}\left(\frac{1}{T_0} - \frac{1}{T_1}\right)\right] \tag{4.170}$$

In addition the temperature dependence of forward and reverse beta is given by

$$\beta(T_1) = \beta(T_0) \times \left(\frac{T_1}{T_0}\right)^x \tag{4.171}$$

The values of both β and the individual diode leakage current are affected here. The values of ψ_e and ψ_c in eqs (4.166) and (4.168) are temperature dependent through eq. (3.22) and so modify the values of C_{je} and C_{jc}.

The theoretical symbols used in this chapter and the corresponding SPICE parameters are shown in Table 4.4, together with typical SPICE values. A full list of bipolar junction transistor parameters with their names is given in Appendix 3. All the values apply to a silicon transistor.

Table 4.4 Theoretical symbols and SPICE parameters for BJT simulation

	Symbol	Parameter	Typical value
	I_S	IS	10^{-15} A
Forward	β_F	BF	100
	n_f	NF	1.0
	V_A	VAF	200 V
	I_K	IKF	10 mA
	C_{jeo}	CJE	2.0 pF
	ψ_e	VJE	0.6 V
	m_e	MJE	0.33
	τ_f	TF	0.1 ns
Reverse	β_R	BR	0.1
	n_r	NR	1.0
	V_A	VAR	200 V
	I_K	IKR	10 mA
	C_{jco}	CJC	2.0 pF
	ψ_c	VJC	0.5 V
	m_c	MC	0.5
	τ_r	TR	10 ns
	x_β	XTB	0
	W_g	EG	1.11 eV
	x	XTI	3
	$r_{bb'}$	RB	100 Ω

The thyristor

The thyristor is a semiconductor device with *three* p-n junctions formed in the same material, which is usually silicon. This gives a p-n-p-n structure (Fig. 4.38(*a*)), with an ohmic contact called the *anode* at the end p-region and a second contact called the *cathode* at the end n-region. A third contact known as the *gate* is usually made to the inside p-region, and this is a control electrode for the two

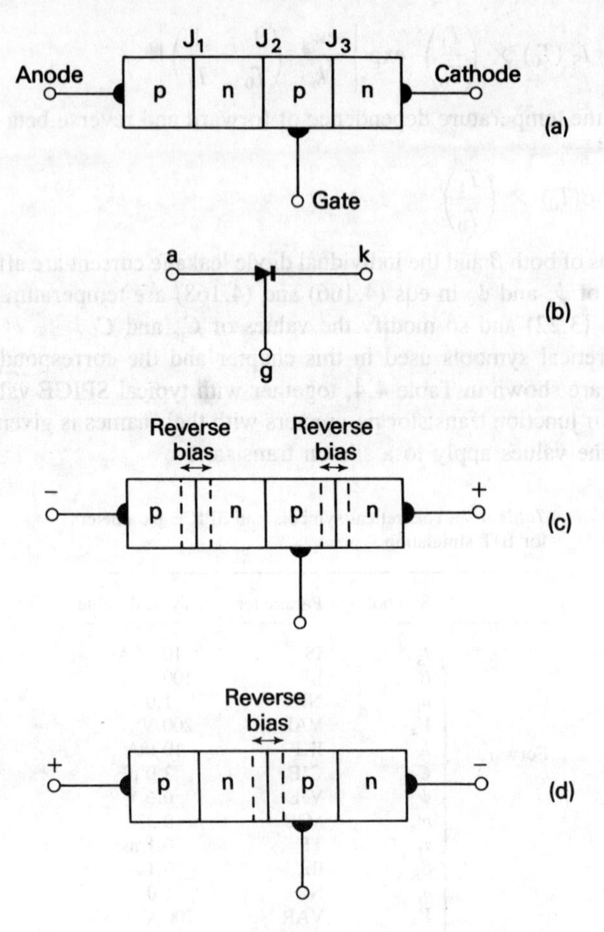

Figure 4.38 The thyristor. (*a*) Arrangement of p-n junctions; (*b*) graphical symbol; (*c*) application of reverse voltage; (*d*) application of forward voltage.

other electrodes through which the main current flow takes place. The thyristor is capable of switching currents up to several hundred amperes at voltages between 1 and 2 kV in the medium power-versions.

Operation of the device depends on the polarity of the voltage applied to the anode with respect to the cathode. When this is negative the two outer junctions, J_1 and J_3, are reverse biased while the inner junction J_2, is forward biased (Fig. 4.38(*c*)). Most of the applied voltage thus appears across J_1 and J_3 and only the leakage current of these junctions can flow through the thyristor. If the voltage is increased sufficiently, avalanche breakdown will occur as in the junction diode and the current will increase rapidly at the breakdown voltage V_{RA} (Fig. 4.39).

If the anode voltage is positive with respect to the cathode J_1 and J_3 are forward biased and only J_2 is reverse biased (Fig. 4.38(*d*)). However, most of the

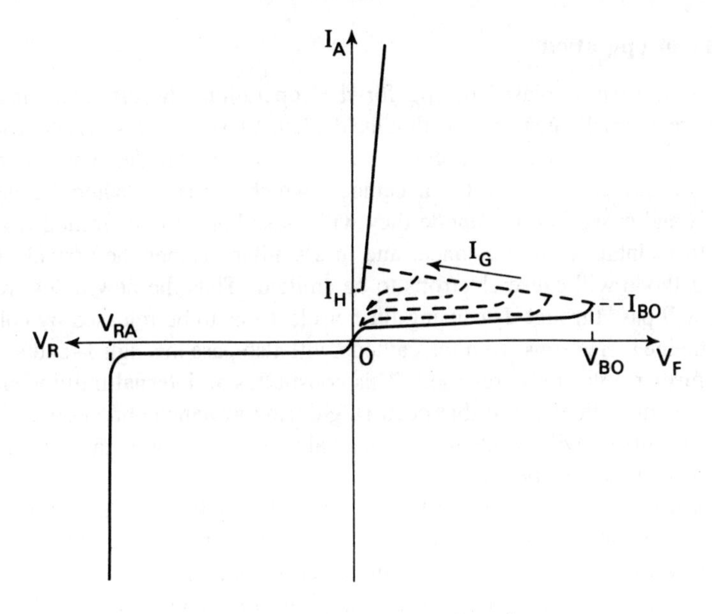

Figure 4.39 Characteristics of thyristor.

voltage will appear across J_2 and again only leakage current will flow; this is known as the *blocking condition*. When the breakdown voltage of this junction is reached the anode current increases rapidly with voltage, until at the *forward breakover voltage,* V_{BO}, and at current I_{BO} the device switches itself into a low-impedance state (Fig. 4.39). The voltage drop then remains at about 1 V up to high values of current, the *I/V* characteristic being similar to that of a silicon rectifier. When the thyristor is operating in this condition the anode voltage remains low until the anode current has been reduced below I_H, the *holding current*, which may range up to about 40 mA.

So far, the gate has been assumed open-circuited, i.e. $I_G = 0$. If the gate is biased positively with respect to the cathode a small current flows between gate and cathode, a part of which is added to the leakage current across J_2. This means that breakdown of the junction can take place at a smaller value of V_{BO} and the larger I_G is made the smaller is the corresponding value of V_{BO}. In fact if I_G is large enough, V_{BO} may be reduced so far that the blocking characteristic in the forward direction disappears and the device behaves as a low impedance for all values of anode voltage. It should be noted that the gate only controls the turning-on of the anode current and has no further effect once the anode voltage has fallen to its low value. The blocking condition is restored only by reducing the anode current below I_H, and in the reverse direction the gate has a negligible effect on the breakdown voltage, although it can cause an increase in reverse current.

Theory of operation

Since J_2 is reverse biased during forward operation, the electrons and holes which are thermally generated within the depletion layer will give rise to a leakage current I_{CO}. Thus an electron current will flow to the anode, which is p-type, and a hole current will flow to the cathode, which is n-type. When the electrons cross J_3 and move into the anode they will cause holes to be emitted from it in order to maintain charge balance, and in a similar manner the arrival of holes at the cathode will cause electrons to be emitted. Thus the new holes from the anode will pass through J_2 causing further electrons to be released by collision, while the new electrons from the cathode will also pass through J_2 and similarly cause further holes to be released. This constitutes an internal multiplication of current, similar in effect to that occurring during avalanche breakdown, so that the anode current will eventually reach a value limited only by the circuit conditions outside the junctions.

The forward operation of the thyristor may be analysed by supposing that it consists of a p-n-p transistor T1 and an n-p-n transistor T2, both sharing junction J_2 (Fig. 4.40). If $I_G = 0$, the two emitter currents and the total current across J_2 are each equal to I_A. The hole current from the anode is then $\alpha_1 I_A$ and the electron current from the cathode is $\alpha_2 I_A$, since these correspond to the two collector currents. Hence the total current flowing across J_2, which includes the reverse leakage current of the junction I_{CO}, is given by

$$I_A = \alpha_1 I_A + \alpha_2 I_A + I_{CO}$$

$$= \frac{I_{CO}}{1 - (\alpha_1 + \alpha_2)} \qquad (4.172)$$

When the reverse voltage across J_2 approaches the breakdown value a factor M must be introduced, given by eq. (3.99). Thus the leakage current becomes MI_{CO} and the two collector currents $M\alpha_1 I_A$ and $M\alpha_2 I_A$ respectively, so that near breakdown

$$I_A = \frac{MI_{CO}}{1 - M(\alpha_1 + \alpha_2)} \qquad (4.173)$$

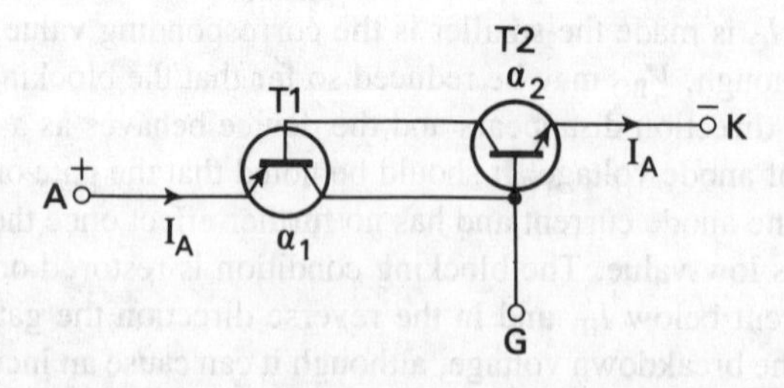

Figure 4.40 Two-transistor model of thyristor.

The thyristor is designed so that at low voltage and currents $\alpha_1 + \alpha_2 \ll 1$, and since $M \approx 1$ under these conditions,

$$I_A \approx I_{CO} \tag{4.174}$$

However, as the voltage is increased, avalanche multiplication occurs at J_2 making $M > 1$, and I_A increases until, at a certain value,

$$M(\alpha_1 + \alpha_2) = 1 \tag{4.175}$$

Thus from eq. (4.173) it can be seen that I_A would become infinite if it were not limited by the applied voltage and circuit resistance, and breakover has occurred corresponding to V_{BO} and I_{BO} in Fig. 4.39. As a result of the increase in current α_1 and α_2 will also have increased until $\alpha_1 + \alpha_2 > 1$, and since the collector current of one transistor is the base current of the other, the base and the collector currents in each transistor will now be of comparable magnitude. Thus the two transistors are in a saturated state and in order to achieve this J_2 must now also be forward biased (Fig. 4.17(*b*)), with its bias having opposite polarity to the biases on J_1 and J_3. Thus the total voltage across the thyristor is the forward bias of one emitter junction plus the saturated voltage of the other transistor, which accounts for the very low voltage across the device after breakover.

The thyristor remains switched on at a low voltage (even though $M = 1$) owing to the increase in $\alpha_1 + \alpha_2$, and it remains in the low-impedance state until the current falls below the value which makes $\alpha_1 + \alpha_2 = 1$. This is the holding current, I_H, below which the device switches off. Equations (4.172) and (4.173) apply to the operation *before* breakover, but are not applicable after breakover since the common collector is forward biased under these conditions.

Finally, if gate current I_G is allowed to flow by closing the switch in Fig. 4.41, a current $\beta_2 I_G$ will be added to the currents flowing across J_2 and at the low currents before breakover β_2 is small so $\beta_2 \approx \alpha_2$. Hence at low voltages

$$I_A = \frac{I_{CO} + \alpha_2 I_G}{1 - (\alpha_1 + \alpha_2)} \tag{4.176}$$

and if I_G is large enough to make $\alpha_1 + \alpha_2 = 1$ the thyristor is switched on at

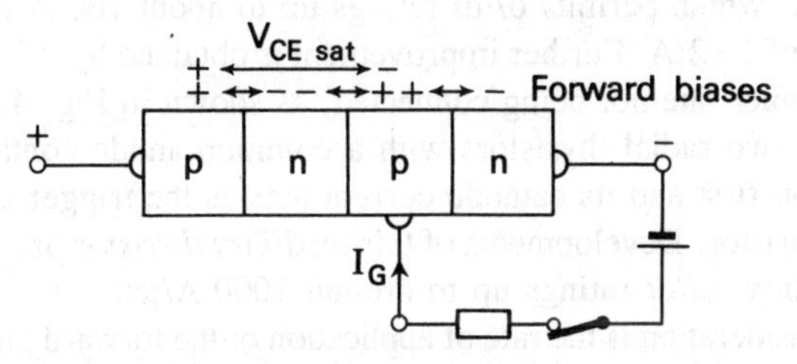

Figure 4.41 Forward voltage and gate voltage of thyristor.

a low value of V_{BO}. In order to trigger a 10 A thyristor into its low-impedance state, a gate current of about 60 mA at a minimum gate voltage of 3 V is required. An average gate dissipation up to 0.5 W is commonly allowed, and if a continuous supply is used to trigger the thyristor this dissipation may be exceeded, resulting in failure of the device. Hence the battery and switch of Fig. 4.41 are replaced by a pulse generator, so that much higher values of I_G and V_G are obtained at the pulse peaks. Thus the average values are much less than the peak values, which ensures that the average dissipation is kept within a safe limit.

The duration of each pulse depends on the time required to *initiate* current multiplication within the thyristor, which is normally a few microseconds, although establishment of the full forward current takes an additional time owing to the spreading velocity, as explained below. Similarly the time needed to turn off the thyristor is determined by the rate at which stored charge can flow out of the device, which can take up to about 30 μs. Gate pulses are not normally applied when the anode is negative as the gate current adds to the leakage current, and this may result in a large current at a high reverse voltage which can easily cause overheating of the device.

Thyristor construction

Thyristors are normally manufactured by successive diffusions of n- and p-type impurities into p-type silicon. A basic construction is shown in Fig. 4.42(*a*), with the anode mounted on a heat sink to keep the temperature of the device below a maximum of about 150 °C. It may be seen that the gate contact is made to a single point, which means that conduction due to internal multiplication is initiated near that point. Thus a high density of electrons and holes is set up in a small area of the cathode and these current carriers then diffuse sideways until the whole cathode is supporting the current. The *spreading velocity* with which this process occurs is constant for a given thyristor and is typically 0.1 mm/μs. Thus, if the cathode has a diameter of 3.5 mm for a 10 A thyristor it will take about 35 μs for complete conduction to be established across the cathode. Hence the rate of rise of current dI/dt must be restricted by the external circuitry, so that a small cathode area will not be required to sustain the whole anode current, which could lead to overheating and failure of the device. The dI/dt rating can be increased by positioning the gate at the centre of an annular cathode, as shown in Fig. 4.42(*b*), which permits dI/dt ratings up to about 100 A/μs when using a trigger pulse of 1–2 A. Further improvement is obtained by using two annular cathodes, the inner one not being connected, as shown in Fig. 4.42(*c*). Effectively there are two radial thyristors with a common anode contact. The inner one is turned on first and its cathode current acts as the trigger current for the main power thyristor. Developments of this *auxiliary thyristor* or *amplifying gate* arrangement allow dI/dt ratings up to around 1000 A/μs.

A second consideration is the rate of application of the forward blocking voltage dV/dt. This is due to C_j, the depletion-layer capacitance of J_2 through which a displacement current flows given by

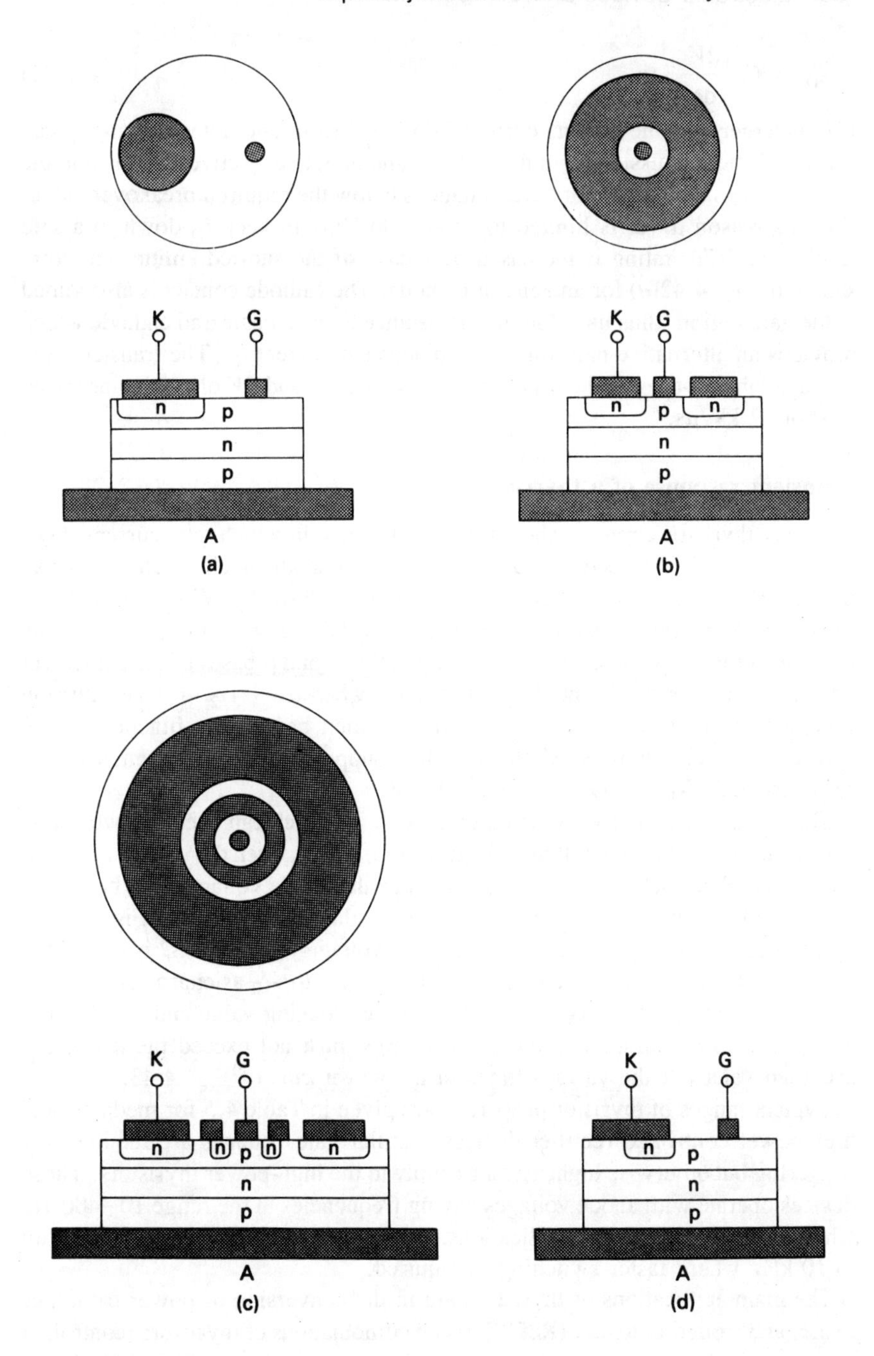

Figure 4.42 Thyristor constructions. (*a*) Basic; (*b*) annular cathod; (*c*) amplifying gate; (*d*) shorted emitter.

$$i_D = C_j \frac{dV}{dt} \tag{4.177}$$

If i_D is larger than the holding current I_H, the electrons and holes forming i_D can cause sufficient emission from the cathode and anode respectively to turn on the thyristor even though the forward voltage is below the required breakover value. For this reason dV/dt is limited to about 100 V/μs to keep i_D down to a safe level. The dV/dt rating is increased by means of the shorted emitter structure shown in Fig. 4.42(*d*) for an annular cathode. The cathode contact is also joined to the gate region, thus introducing a resistance between gate and cathode which provides an alternative path for the displacement current i_D. The transient current available to turn on the thyristor is thus reduced and dV/dt can be increased to about 2 kV/μs.

Transient response of a thyristor

A simply thyristor circuit is shown in Fig. 4.43(*a*), in which gate current flows when switch S1 is closed. With S1 open only a small blocking current flows through R_L so V_{AK} is almost equal to the supply voltage V_S. When S1 is closed there is a *delay time* t_d before I_A rises and V_{AK} falls (Fig. 4.43(*c*)). This is the time taken for sufficient charge to build up in the n- and p-bases of the equivalent transistors to support the holding current I_H, when $\alpha_1 + \alpha_2 = 1$ and turn-on becomes regenerative. The *turn-on* time t_{on} then becomes a function of the spreading velocity until the whole cathode is supporting anode current, when $I_A \simeq V_S/R_L$ and V_{AK} has fallen to around 1 V.

The thyristor remains conducting until $I_A < I_H$ which can occur in two ways. When the supply is an alternating voltage *turn-off* occurs when the voltage reverses. However, when the supply is a direct voltage the anode voltage must be forced to reverse for turn-off. This may be achieved by the simple circuit of Fig. 4.43(*b*) where a capacitor is previously charged to a voltage V_C. When S2 is closed the anode voltage drops to $-V_C$ and as a result I_A drops to $-I_R$ as charge is removed from the n- and p-bases. I_A then returns to the blocking value and V_{AK} returns to V_S. The dI/dt ratings and the dV/dt ratings must not exceed the maximum specified values at the various times in the waveforms of Fig. 4.43.

Typical ranges of thyristor properties are given in Table 4.5 for medium- and high-power controlled rectifier devices. Similar gate voltages are required for triggering but otherwise higher ratings apply to the high-power thyristors. These devices operate with anode voltages having frequencies in the range 10–400 Hz which covers most power applications. Others are available which operate up to 10 kHz where faster switching is required.

The main applications of thyristors are in the conversion of power from a.c. to d.c. and from d.c. to a.c. (Ref [7]) using combinations of thyristors (controlled rectifiers) and diodes (uncontrolled rectifiers). Industrial process control, the control of electrical machines and high-voltage d.c. transmission are among the many areas in which thyristors are applied. They have superseded the thyratron (Chapter 8) in all but the highest-voltage applications.

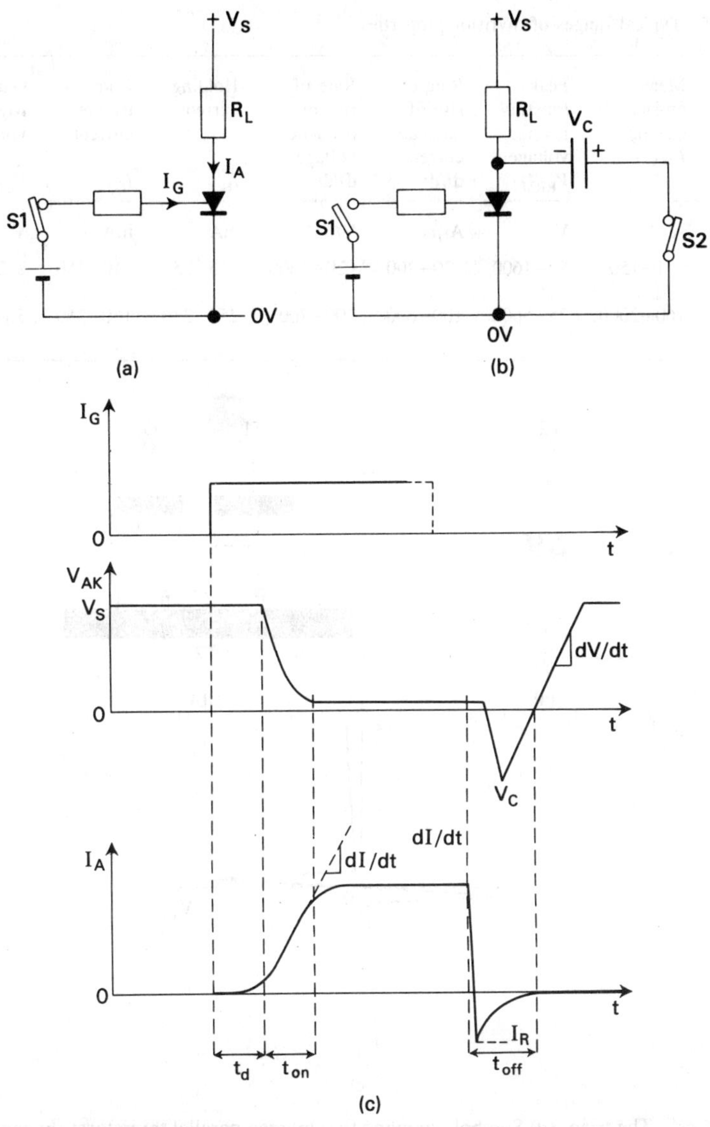

Figure 4.43 Thyristor transient response. (*a*) Turning on a thyristor; (*b*) turning off a thyristor when supply voltage V_s is d.c.; (*c*) gate current, anode voltage and anode current waveforms.

Triac and diac

The *triac* (*tri*ode *a.c.* switch) is a derivative of the thyristor which is a six-layer device and effectively consists of two thyristors in inverse parallel (Fig. 4.44). This provides forward *I/V* characteristics in both the first and third quadrants where terminal T2 acts as an anode. The characteristics can be obtained by

Table 4.5 Typical ranges of thyristor properties

	Mean on-state current I_{TAV}	Peak repetitive reverse voltage V_{RRM}	Rate of rise of on-state current dI/dt	Rate of rise of off-state voltage dV/dt	Holding current I_H	Gate trigger current I_{GT}	Gate trigger voltage V_{GT}
	A	V	A/μs	V/μs	mA	mA	V
Medium power	1–150	50–1600	20–200	20–1600	5–35	10–150	2.5–3.5
High power	160–2000	75–5600	100–800	200–700	250–350	150–350	3.0–3.5

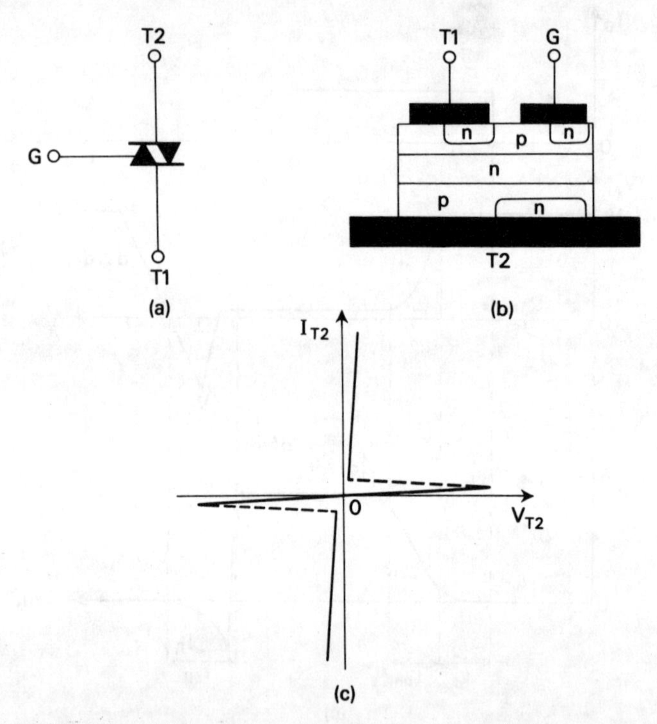

Figure 4.44 The triac. (*a*) Symbol, showing two inverse-parallel thyristors; (*b*) construction; (*c*) *I–V* characteristics, showing conduction for positive and negative voltages on terminal 2.

applying either positive or negative voltages between the gate and terminal T1. This device has many applications in simple control circuits driven from single phase a.c. supplies, in which it provides controlled full wave rectification. Typical ranges of properties are: terminal current 0.45 to 40 A, terminal voltage 25 to 800 V, gate current 5 to 100 mA and d*V*/d*t* 5 to 500 *V*/μs.

The *diac* (*di*ode *a.c.* switch) is normally used to trigger a triac when it is connected in series with the gate. In its simplest form it has a three-layer structure similar to an n-p-n transistor (Fig. 4.45(*b*)). However, the doping concentrations at the two junctions are approximately the same and no contact is made to the p-region. When a voltage is applied across a diac with either polarity one junction is forward biased, the other is reverse biased and only leakage current flows. As the voltage is increased breakdown will occur at a voltage defined by eq. (4.58) to be $BV_B(1 - \alpha)^n$. Then as the current increases after breakdown α also increases and the terminal voltage is reduced. This leads to a symmetrical characteristic with two negative resistance regions, as shown in Fig. 4.45(*c*), where V_{BO} is typically about 30 V.

A more complex form of the diac has a structure like the triac in Fig. 4.44(*b*) but with the gate connection and n-region omitted. Here regenerative action occurs, as in the thyristor, leading to a larger negative resistance and smaller forward voltage drop than with the n-p-n form of the diac.

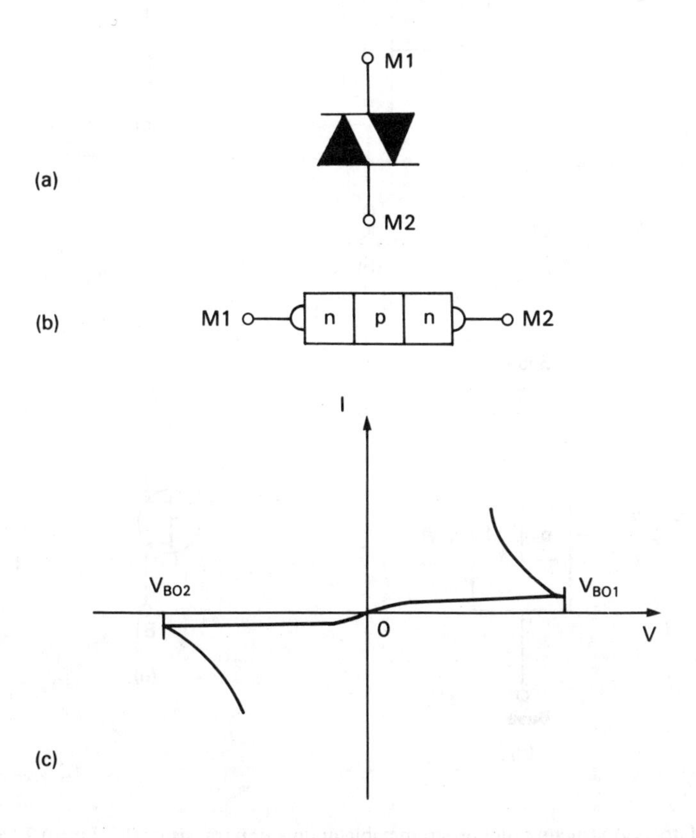

Figure 4.45 The diac. (*a*) Symbol, showing two inverse-parallel diodes; (*b*) construction; (*c*) *I–V* characteristics, showing negative resistance regions after breakdown.

The programmable unijunction transistor (PUT)

The *programmable unijunction transistor* is in fact a four-layer device like a thyristor but with the gate being the middle n-region instead of the middle p-region (Fig. 4.46(*a*)). It is a later version of the unijunction *transistor* (UJT), which is a two-layer device like a junction field-effect transistor (JFET) but having one emitter and two base terminals (Fig. 4.46(*e*)). The PUT and UJT have similar characteristics but in the PUT they may be controlled, or *programmed* by means of the gate voltage and an external resistor in series with the gate.

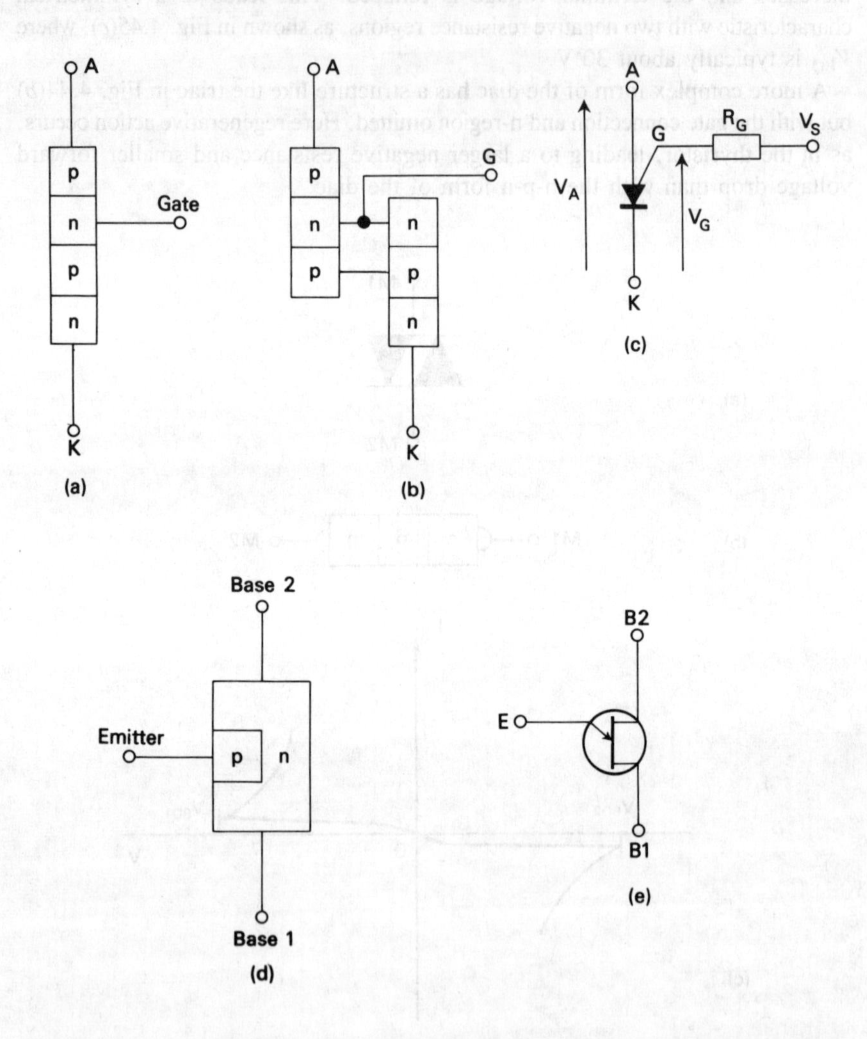

Figure 4.46 (*a*) Structure of a programmable unijunction transistor (PUT); (*b*) 2-transistor model of a PUT; (*c*) symbol and supplies for a PUT; (*d*) structure of a unijunction transistor (UJT); (*e*) symbol for a UJT.

As in the case of the thyristor the PUT may be considered as a p-n-p and an n-p-n transistor sharing the same collector junction (Fig. 4.46(*b*)). The gate supply voltage V_S (Fig. 4.46(*c*)) can lie between about 5 and 35 V and when the anode voltage V_A is less than the gate voltage V_G only a small leakage current I_{GA} can flow (Fig. 4.47). This corresponds to the reverse current of the emitter diode of the p-n-p transistor and since this is in the order of nA, $V_G \simeq V_S$.

The low-current condition continues until $V_A > V_G$ when the emitter diode of the p-n-p transistor becomes forward biased and node current I_A starts to rise. I_A is given by eq. (4.172) where α_1 and α_2 are the current gains of the two equivalent transistors. As I_A rises $\alpha_1 + \alpha_2$ approaches 1 and the PUT switches into a low-voltage state, in the same way as the thyristor. This occurs when the anode voltage V_A has reached a peak point value $V_P = V_S + V_T$. V_T is the *offset voltage*, within the range 0.2 to 0.6 V, and is comparable to the threshold voltage of the emitter diode of the p-n-p transistor.

The peak point current I_P, above which the PUT switches into its low-voltage state, depends on the gate resistance R_G which controls the gate current available at the instance of switching. Just as V_P is programmed by V_S so I_P is programmed by R_G, with currents between 20 and 0.2 μA corresponding to resistances between 200 Ω and 1 MΩ, independent of V_S in a typical device.

When switching occurs V_A falls to the valley voltage V_V and I_A rises to the valley current I_V, which is also programmed by R_G. Typically I_V ranges between 20 mA and 30 μA for resistances between 200 Ω and 1 MΩ at V_S = 20 V.

For anode currents greater than I_V both transistors are saturated and so V_V is about 1 V. V_A rises slowly with I_A in this region and the device has a dynamic resistance $\partial V_A/\partial I_A$ of about 3 Ω above V_V.

The main application of the PUT is in timing circuits and the relaxation oscillator circuit of Fig. 4.48(*a*) will produce waveforms as in Fig. 4.48(*b*). V_A rises as

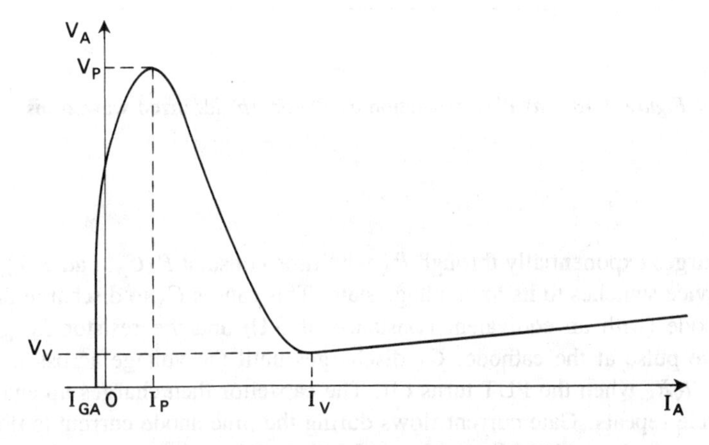

Figure 4.47 I_A–V_A characteristic of a PUT.

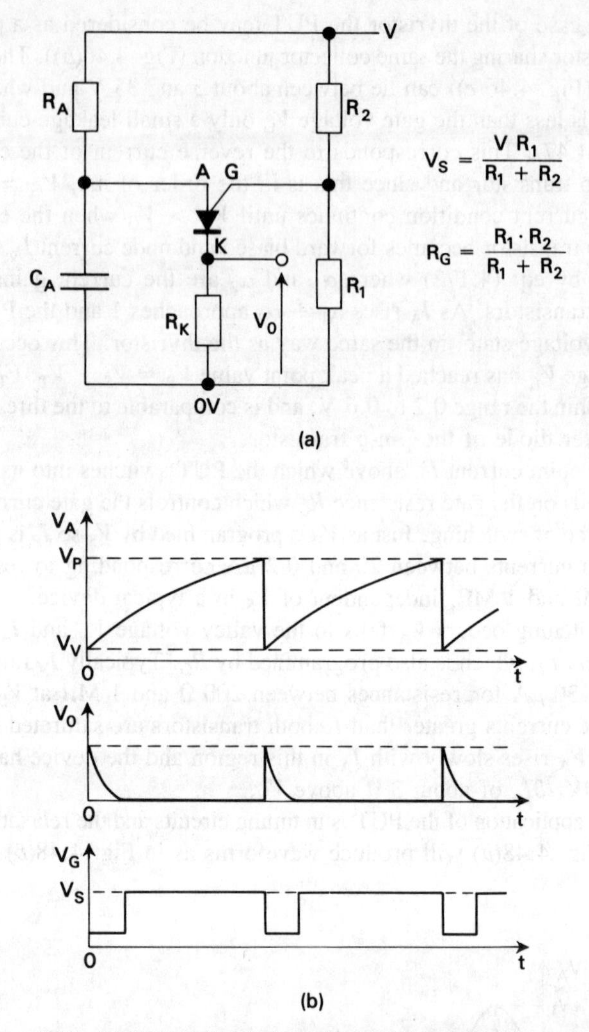

Figure 4.48 (*a*) PUT relaxation oscillator; (*b*) idealized waveforms.

C_A charges exponentially through R_A with time constant $R_A C_A$, and at $V_A = V_P$ the device switches to its low-voltage state. This causes C_A to discharge through the anode (with an equivalent resistance of 3 Ω) and the resistor R_K gives a positive pulse at the cathode. C_A discharges until the voltage across it equals $V_V + I_V R_K$ when the PUT turns off. The capacitor then charges up again and the cycle repeats. Gate current flows during the time anode current is flowing, so the gate voltage also falls to a low value due to the voltage dropped across R_G.

Points to remember

* A bipolar junction transistor or BJT has two p-n junctions, arranged as n-p-n or p-n-p, with the collector current carried by electrons and by holes respectively. The n-p-n transistor is preferred, since the mobility of electrons is higher than that of holes leading to better high-frequency and transient responses.
* The collector current is controlled by the much smaller base current, where the current gain defines the ratio of the two. A high current gain is achieved by heavily doping the emitter and providing a narrow base region. This is achieved by the planar construction which also gives a built-in base field, accelerating minority carriers towards the collector.
* The output characteristics are divided into the active and saturation regions, where the active region is used in amplifier operation and the saturation region is important in switching operation. The characteristics can be defined through the Ebers–Moll model which also forms a basis for computer simulation.
* The transient response can be defined through the charge control model, which includes base and collector time constants. Turn-on time can be reduced by driving the transistor into saturation but the excess base charge has to be dispersed before the transistor can be turned off.
* The frequency response can be defined through the hybrid-π model, with the gain — bandwidth product f_T being the frequency at which the short-circuit current gain has fallen to unity. At low frequencies the hybrid parameter model is used.
* Computer simulation of the BJT can be accurately achieved through the SPICE program, which has parameters defining the physical properties of the device. Of the 40 parameters provided about 15 will give acceptable results in many situations.
* A thyristor has three p-n junctions and conducts only for positive anode voltages, with the instant at which conduction occurs being controlled by a gate. The triac is a derivative which is still gate-controlled but conducts for both positive and negative anode voltages.

References

[1] Werner, W.M. 'The influence of fixed interface charges on current gain fallout of planar n-p-n transistors', *J. Electrochem. Soc.*, 1976, vol. 123, p. 540.
[2] Ebers, J.J. and Moll, J.L. 'Large signal behaviour of junction transistors', *Proc. IRE*, 1954, vol. 42, No. 2, p. 1761.

[3] Millman, J. and Grabel, A. *Microelectronics*, 2nd ed., (McGraw-Hill), 1987.
[4] Taub, H. and Schilling, D. *Digital Integrated Electronics*, (McGraw-Hill), 1977.
[5] Early, J.M. 'Effects of space-charge layer widening in junction transistors', *Proc. I.R.E.*, 1952, vol. 40, p. 1401.
[6] Getreu, I.E. *Modelling the Bipolar Transistor*, (Elsevier), 1978.
[7] Davis, R.M. 'Power diode and thyristor circuits', *IEE Monograph Series* No. 7, 1979.

Problems

4.1 An n-p-n transistor with equal junction areas of 1 mm^2 has an excess electron density of 10^{20} per m^3 maintained at the emitter-base junction. If the effective base width is 2×10^{-5} m and the electron mobility is 0.39 m^2/V s at room temperature, sketch the approximate distribution of electrons in the base region and estimate the collector current.
[7.8 mA]

4.2 Explain the terms *emitter efficiency* γ and *base transport factor* δ and relate them to the current gain h_{FE} of an n-p-n transistor. Discuss the variation of h_{FE} with collector current.

A transistor has $\gamma = 0.995$ and $\delta = 0.985$ at $I_C = 2$ mA. The minority carriers flowing across the base occupy an effective volume of 10^{-13} m^3 and their mean transit time is 1 ns. Determine (i) the number density of minority carriers and (ii) the recombination time in the base.
[1.25×10^{20} carriers/m^3; 49 ns]

4.3 A planar transistor with a base 3 μm wide and of mean area 10^{-2} mm^2 is operated at a collector current of 2 mA. The minority carrier distribution in the base has a constant value of 2×10^{20} carriers/m^3 up to 2.4 μm from the emitter side and it may then be assumed to fall linearly to zero at the collector side of the base. Estimate: (i) the excess base charge; (ii) the mean transit time of carriers across the base; and (iii) the diffusion capacitance.
[8.64×10^{-13} C; 4.32×10^{-10} s; 346 pF]

4.4 The base of an n-p-n silicon planar transistor is fabricated by diffusing boron into n-type silicon which has a resistivity of 10^{-2} Ωm. The density of boron atoms at the emitter side of the base is 2×10^{24} m^3, and the base width is 2 μm. Determine the field within the base and the transit time of electrons across it.
[7.6×10^4 V/m; 0.2 ns]

4.5 An n-p-n transistor is operated at a collector current of 1 mA, the base current being 20 μA and the collector-base leakage current 1 nA. If the recombination time of holes and electrons in the base, τ_B, is 0.2 μs, estimate: (i) the current gain h_{FE}; (ii) the leakage current I_{CEO}; (iii) the transit time of electrons across the base, τ_C; and (iv) the excess base charge q_B.
[50; 51 nA; 4 ns; 4 pC]

4.6 Show how the current gain h_{FE} and its variation with collector current are each controlled by the constructional features of a bipolar transistor. The leakage current I_{CBO} of a p-n-p transistor is related to the collector-base voltage V_{CB} by the expression

$$I_{CBO} = \frac{10^{-8}}{1 - (V_{CB}/50)^6} \text{ amperes where } V_{CB} \text{ is in volts.}$$

If $h_{FE} = 40$, estimate the collector breakdown voltage (i) with the base open-circuited and (ii) with the emitter open-circuited.
[27 V; 50 V]

4.7 The base of a silicon p-n-p transistor has a uniform donor density of 3×10^{22} m^{-3} and at room temperature the diffusion potential of the emitter-base diode is 0.7 V. If it is operated with V_{EB} = 0.5 V and V_{CB} = −10 V, estimate the distances penetrated by the depletion layers into the base at the emitter and the collector. If the metallurgical base width is 10 μm what is the value of the punch-through voltage?
[0.094 μm; 0.69 μm; 2.35 kV]

4.8 The transistor shown below has the following parameters: α_F = 0.98. α_R = 0.10, I_{CS} = 157 pA, I_{ES} = 0.016 pA. Estimate the three terminal currents and voltages for each of the two base input voltage levels at room temperature.
(Hint: $V_{CE} \sim$ 0 V, $V_{BE} \sim$ 0.7 V may be taken as initial assumptions in obtaining I_C and I_B.)
[Input +10 V: I_C = 5 mA, V_{CE} = 78 mV, I_B = 4.65 mA, V_{BE} = 706 mV, I_E = 9.65 mA; Input −10 V: I_C = 0.141 pA, V_{CE} = 10 V, I_B = 0, V_{BE} = −10 V, I_E = -3×10^{-16} A]

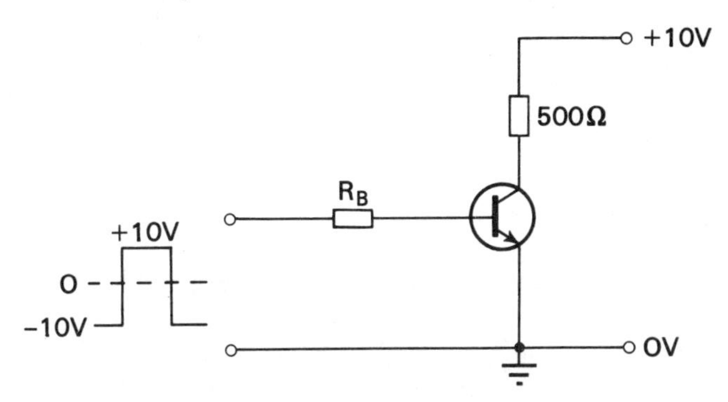

4.9 Sketch and explain the collector current waveform of a saturating inverter using a single bipolar transistor when the input is a pulse with both positive and negative excursions.

An inverter uses an n-p-n transistor with h_{FE} = 60, τ_s = 30 ns and base-emitter capacitance assumed constant at 2 pF. The base resistor is 10 kΩ, the collector resistor 1 kΩ and the collector supply 5 V. If the input is a ±5 V pulse, calculate (i) the delay time, (ii) the storage time, assuming V_{BE} and V_{CE} are both zero in saturation. Explain briefly how a diode connected between base and collector would affect the storage time (Schottky diode).
(Hint: take a 10 V step applied to input and assume V_{BE} stops rising when it has reached 0 V).
[13.8 ns; 16.1 ns]

4.10 The transistor shown in the circuit of Problem 4.8 has current gain h_{FE} = 80, base time-constant τ_B = 0.8 μs and storage time-constant τ_s = 0.4 μs. Making the same assumptions as in Problem 4.8, calculate the value of R_B which will ensure that V_{CE} changes from 10 V to zero in 0.1 μs. Determine also the corresponding storage time.
[4.7 kΩ; 230 ns]

4.11 A certain transistor has f_T = 100 MHz, low frequency current gain h_{fe} = 90, $r_{bb'}$ = 50 Ω and $r_{b'e}$ = 1.2 kΩ. It is used as an amplifier with a collector load of 500 Ω and is supplied from a source of 1mV e.m.f. with internal resistance 500 Ω. Determine the output voltage at low frequencies and the bandwidth of the amplifier. The effects of r_{ce}, $C_{b'c}$, $r_{b'c}$ and the coupling components may be neglected.
[26 mV; 3.5 MHz]

4.12 Obtain the common-emitter hybrid parameters of a transistor operating at room temperature with $I_C = 2$ mA, $\beta_0 = 70$, $\mu = 4 \times 10^{-4}$ and $r_{bb'} = 50\ \Omega$.
[$h_{ie} = 925\ \Omega$; $h_{re} = 4 \times 10^{-4}$; $h_{fe} = 70$; $h_{oe} = 64\ \mu$s]

4.13 The SPICE parameters for the BJT model given in Appendix 3 include forward and reverse sets. Identify these sets of parameters and describe the operating conditions of the transistor for which *both* sets would be necessary.

Determine the value of the base-collector junction capacitance at a reverse bias of 3.0 V when the typical SPICE values are used.
[0.53 pF]

4.14 Determine the forward breakover voltage of a thyristor when the break-down voltage of the middle junction is 520 V, $\alpha_1 + \alpha_2 = 0.2$ in the two-transistor model and the factor n is 6 in eq. (3.99).
[500 V]

5 Optoelectronic devices

Optoelectronics covers a wide variety of devices in which there is interaction between light energy and electrons, and two main classes occur, corresponding to the absorption and generation of light. Light is absorbed to produce an electrical otuput in the photocell, photodiode, phototransistor and solar cell. Light is generated due to an electrical input in the light emitting diode, semiconductor laser, gas and ruby lasers. This is also true of plasma displays, considered in Chapter 8, and the phosphor coating of a cathode ray tube. Liquid crystal displays however, use electrical energy to operate as a shutter on an existing light source and so are passive devices.

In this chapter we are concerned with wavelengths ranging from the near infrared, ~ 10 μm, to the near ultra-violet, ~ 0.2 μm. The range of wavelengths visible to the human eye is between about 0.7 and 0.4 μm, with a peak at about 0.5 μm. This corresponds to a photon energy range between 1.77 and 3.10 eV (Fig. 5.1) which is also the range of energy gaps of common semiconductors.

Units of illumination

In a light source such as a tungsten filament at high temperature, or a fluorescent tube, photons will be emitted covering a continuous range of frequencies. In the case of the filament this is due to the energy levels in the conduction band of a metal being continuous, which allows a very large number of transitions from the top of the valence band.

We are concerned with the visible part of the power radiated by the source, the *luminous flux*, which is the total visible light energy emitted by a source in one second and is measured in *lumens* (lm). It is found experimentally that a source of power 1 W radiating at a wavelength of 555 nm gives a luminous flux of 680 lm. At any other wavelength this value is scaled by the relative response at that wavelength, so where the response is 0.5, for example, the luminous efficiency is 340 lm/W. Thus, since practical power sources contain a range of wavelengths, their luminous efficiencies are much lower, being about 100 lm/W for the sun and about 40 lm/W for a daylight fluorescent tube.

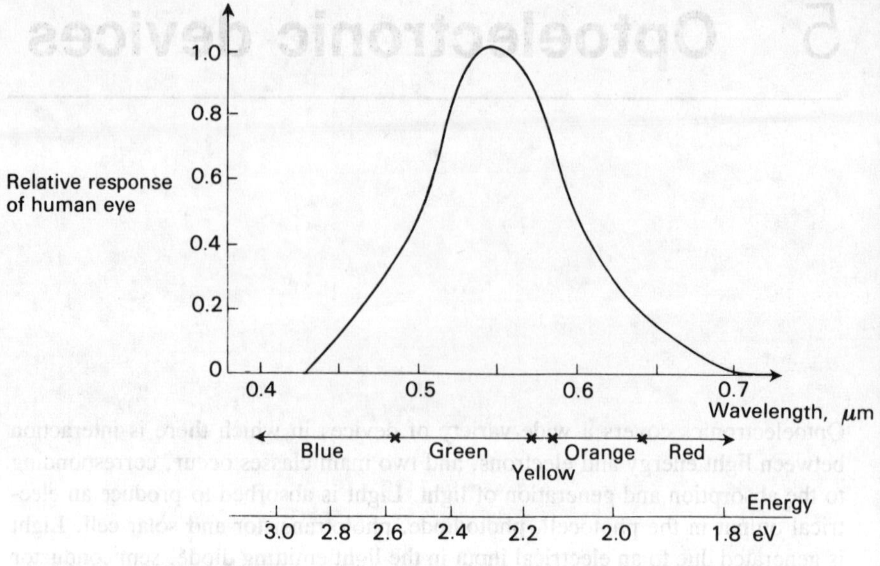

Figure 5.1 Spectral response of the human eye correlated with the energy gaps of semiconductors.

Only a part of the visible light energy radiated by a source is intercepted so that the illumination of a given surface is measured in lumens per square metre or *lux* (lx). Thus 1 lux = 1 lm/m^2 and a detector receiving 1000 lux from a source of efficiency 20 lm/W, for example, would experience a power density of 50 W/m^2 or 5 mW/cm^2.

Light absorption in a semiconductor

Let us now consider in more detail the result of illuminating a semiconductor, an effect that was introduced in Chapter 2. The minimum frequency of radiation at which electron-hole pair generation can occur is given by

$$hf_0 = W_g \tag{5.1}$$

as shown in Fig. 5.2. In terms of wavelength,

$$hc/\lambda_0 = W_g \tag{5.2}$$

and

$$\lambda_0 = \frac{hc}{W_g} = \frac{1.24}{W_g} \text{ micrometres} \tag{5.3}$$

where W_g is in electronvolts and λ_0 is the maximum or *threshold wavelength*. Thus for germanium with $W_g = 0.66$ eV, λ_0 is 1.88 μm; while for silicon with

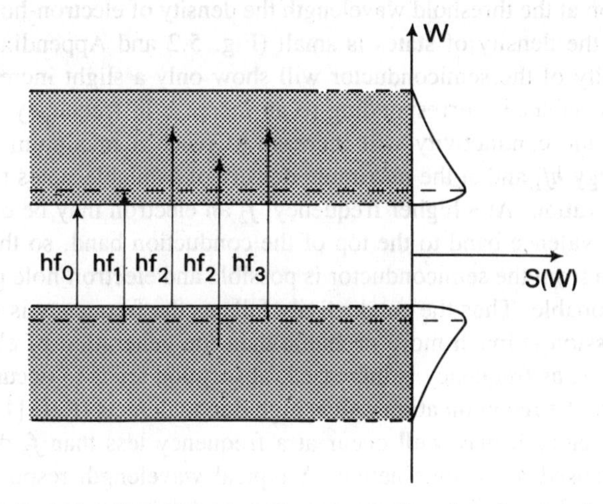

Figure 5.2 Relationship between the energy bands of a semiconductor and radiation of various frequencies.

$W_g = 1.1$ eV, λ_0 is 1.13 μm. In both materials electron-hole pair generation occurs for wavelengths well into the infra-red region, but for wavelengths greater than λ_0 the effect cannot take place.

Worked example

Determine the threshold wavelength for gallium arsenide which has an energy gap of 1.43 eV.

Solution

Expressing the energy gap in joules leads to

$$\lambda_0 = \frac{hc}{W_g} = \frac{6.63 \times 10^{-34} \times 3 \times 10^8}{1.6 \times 10^{-19} \times 1.43}$$

$$= 0.87\ \mu\text{m}$$

This means that in gallium arsenide electron-hole pair generation would occur for wavelengths just into the infra-red. However, in this material as opposed to germanium and silicon, direct *recombination* occurs between electrons and holes. This leads to the application of gallium arsenide in light-emitting diodes, described later in this chapter.

For radiation at the threshold wavelength the density of electron-hole pairs will be low since the density of states is small (Fig. 5.2 and Appendix 2), so that the conductivity of the semiconductor will show only a slight increase. As the wavelength is reduced, corresponding to an increase in frequency, the density of states and the conductivity will increase to reach a maximum at λ_1. This occurs at energy hf_1 and at the maximum density of available states for electron-hole pair generation. At a higher frequency, f_2 an electron may be excited from the top of the valence band to the top of the conduction band, so that emission of the electron from the semiconductor is possible and electron-hole pair generation is less probable. Thus the conductivity falls as the frequency is raised to f_3, at which emission is much more probable than the generation of electron-hole pairs. However, as frequency is increased, absorption tends to occur nearer the surface, where the recombination rate is high due to defects (Ref. [1]). Thus the minimum wavelength may well occur at a frequency less than f_3 due to a fall in response caused by recombination. A typical wavelength response curve is shown in Fig. 5.3 for a silicon photodiode, where the maximum wavelength does correspond to λ_0 for silicon.

Photodetectors

When absorption of light energy produces a corresponding electrical output the device is called a *photodetector*. Two important properties of photodetectors are the *gain*, defined as the number of charge carriers passing between the contact electrodes per second for each photon absorbed per second and the *response time* to a step change of light input. These properties are summarized in Table 5.1 for the photodetectors considered in this chapter.

Photoconductive cell, light dependent resistor (LDR)

The principle of a photoconductive cell is illustrated in Fig. 5.4(*a*). A constant voltage V applied across a semiconductor slab of conductivity σ causes current

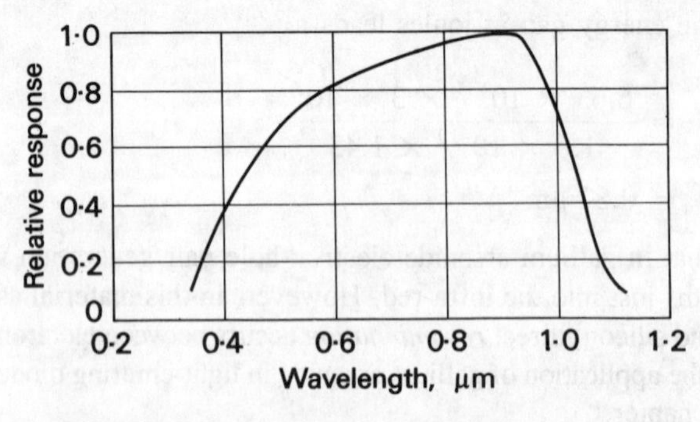

Figure 5.3 Spectral response of a silicon photodiode

Table 5.1 Typical gain and response times for common photodetectors

	Gain	Response time
		s
Photoconductor	1 to 10^6	10^{-1} to 10^{-3}
p-n photodiode	1	10^{-11}
p-i-n photodiode	1	10^{-8} to 10^{-12}
Metal-semiconductor photodiode	1	10^{-11}
Avalanche photodiode	10^2 to 10^4	10^{-10}
Bipolar phototransistor	10^2	10^{-8}
Field effect phototransistor	10^2	10^{-7}

I to flow in the absence of illumination. When the material is illuminated the conductivity increases by $\Delta\sigma$, which is proportional to Δn_i, the number of electron-hole pairs released by the light. Then I increases by ΔI so

$$\sigma = \frac{KI}{V},$$

where K is a constant depending on the cell dimensions,

$$\sigma + \Delta\sigma = K\left(\frac{I + \Delta I}{V}\right),$$

and

$$\Delta\sigma = \frac{K\Delta I}{V} \tag{5.4}$$

Thus the current change ΔI is proportional to Δn_i and so is also proportional to the light intensity.

A common semiconductor for a photoconductive cell is cadmium sulphide, since it has a spectral response similar to that of the human eye. Such a cell is thus suitable for a light meter where direct measurement of ΔI would give an indication for camera exposure settings. A typical construction for a cadmium sulphide cell is shown in Fig. 5.4(*b*), the whole being protected by a transparent cover. In the dark the resistance is about 10 MΩ at room temperature for a 1.4 cm diameter cell, and the resistance falls with increasing illumination as shown in Fig. 5.4(*c*). The resistive properties of the cell would be used in a circuit as shown in Fig. 5.4(*d*), where V_0 rises with light intensity and could operate a control circuit when a certain level is exceeded.

The photoconductive cell has a relatively slow response time since this is mainly

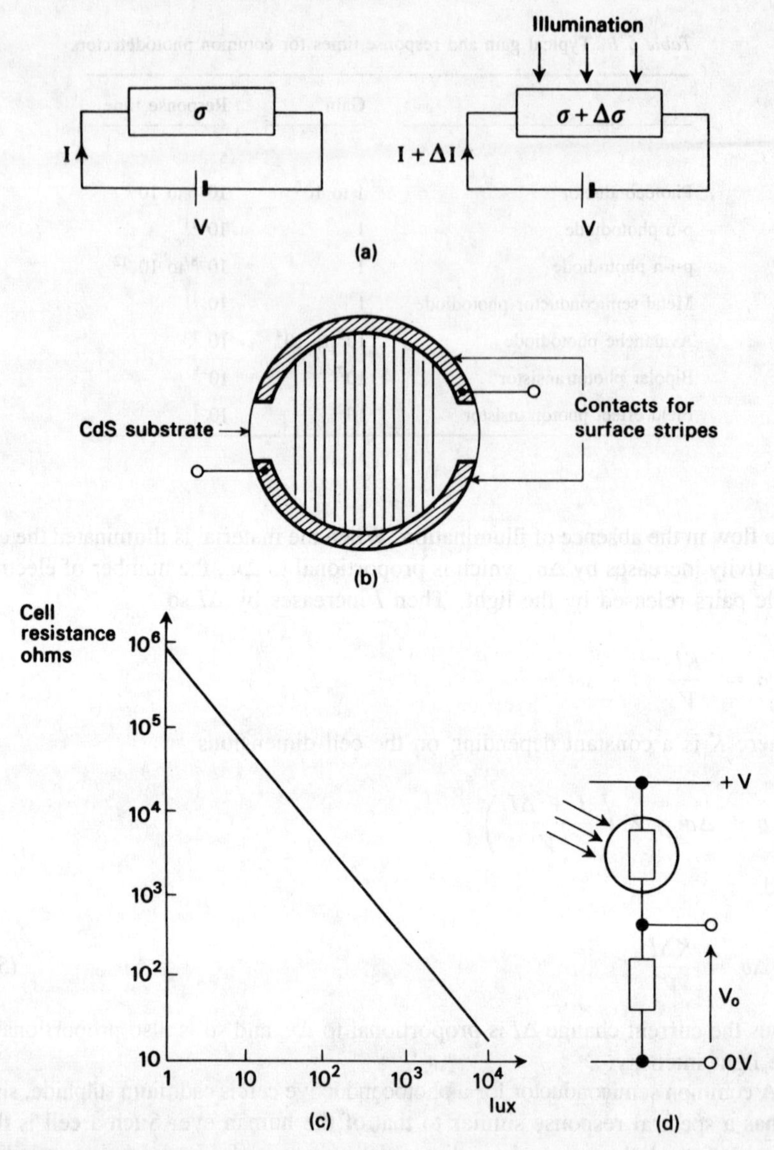

Figure 5.4 Photoconductive cell. (*a*) Principle of operation; (*b*) construction of cadmium sulphide cell; (*c*) variation in cell resistance with illumination; (*d*) circuit for operation.

dependent on the carrier recombination time. This is important when the light intensity flickers due to modulation and typically the resistance rise time is 75 ms and fall time 350 ms.

The *photoconductive gain* may be defined as the ratio of recombination time

τ to transit time τ_t through the device. A typical gain of 10^5 and a recombination time of 100 ms would thus correspond to a transit time of 1 μs. Then where the current released by absorption of photons is I_p, $\Delta I = I_p\ \tau/\tau_t$.

Cadmium sulphide has an energy gap of 2.42 eV, so λ_0 is 0.52 μm. The response can be extended to longer wavelengths by means of impurities which insert levels in the energy gap and CdS is the most commonly used material for the visible range. For the infra-red, lead sulphide with W_g = 0.41 eV and λ_0 = 3.02 μm provides a high impedance detector and indium antimonide with W_g = 0.17 eV and λ_0 = 7.29 μm provides a low impedance.

p-n junction photodiode and phototransistor

If the encapsulation of a diode is made transparent then radiation can penetrate to the junction region and electron-hole pairs will be generated both near the junction and away from it. Electrons released near the junction will be swept into the *n-region* by the junction field and holes released near the junction will be similarly swept into the *p-region* (Fig. 5.5(*a*)). Electron-hole pairs formed away from the junction may recombine before they can be separated by the junction field. Thus the excess carriers near the junction provide a photocurrent I_P which adds directly to the current generated thermally, the leakage current I_0. In a reverse-biased photodiode with zero illumination I_0 is known as the *dark current*, and after illumination the reverse current rises from I_0 to $I_0 + I_P$.

I_P is related to the power P of the incident radiation, since if there are n_P photons incident per second on the diode,

$$P = n_P hf = n_P hc/\lambda \tag{5.5}$$

and if there are n_{iP} electron-hole pairs created per second by the radiation

$$I_P = n_{iP} e \tag{5.6}$$

Dividing eq. (5.6) by eq. (5.5)

$$\frac{I_P}{P} = \frac{e\lambda n_{iP}}{hcn_P} = \frac{e\lambda\zeta}{hc} \tag{5.7}$$

where I_P/P is the responsivity R and $\zeta = n_{iP}/n_P$ the *quantum yield*.* In practice the quantum yield is a few per cent and is a measure of the efficiency of the diode. Thus, from eq. (5.7), I_P is proportional to the power of the incident radiation at a particular wavelength. The illumination is also related linearly to the number of photons passing through one square metre of surface normal to the beam in

* The percentage quantum efficiency QE of a photodiode at a wavelength λ in mm can be defined in terms of the responsivity R in A/W as QE = $\dfrac{1.24 \times 10^5 \times R}{\lambda}$ %.

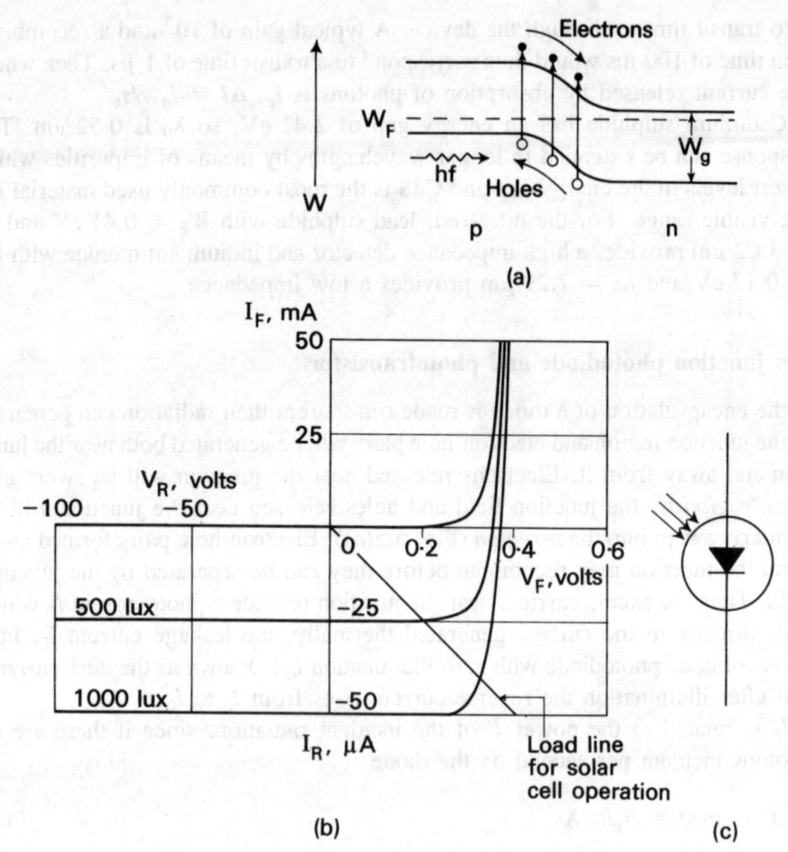

Figure 5.5 p-n photodiode. (*a*) Energy diagram for zero bias; (*b*) characteristics of a silicon photodiode; (*c*) symbol.

one second, so the photocurrent is still proportional to the illumination with λ replaced by a term covering the visible range of wavelengths.

The *I/V* characteristic of a silicon p-n photodiode is shown in Fig. 5.5(*b*). The reverse current, $I_R = I_0 + I_P$ is almost independent of the reverse voltage, since the field across the junction is sufficiently strong to extract all the current carriers created at the junction, even for low values of applied voltage. The reverse current is proportional to the illumination and can be measured directly; alternatively a resistor can be connected in series with the diode and supply and current changes observed as voltage changes across the resistor. A typical sensitivity is 0.7 A mW^{-1} cm^{-2}. Such a diode is useful as a detector of visible or infra-red radiation and is suitable for use in photometers, high speed counting or punched card/tape readers.

Greater sensitivity is obtained if the junction is the collector junction of a *phototransistor*. This can be a normal transistor in a transparent encapsulation,

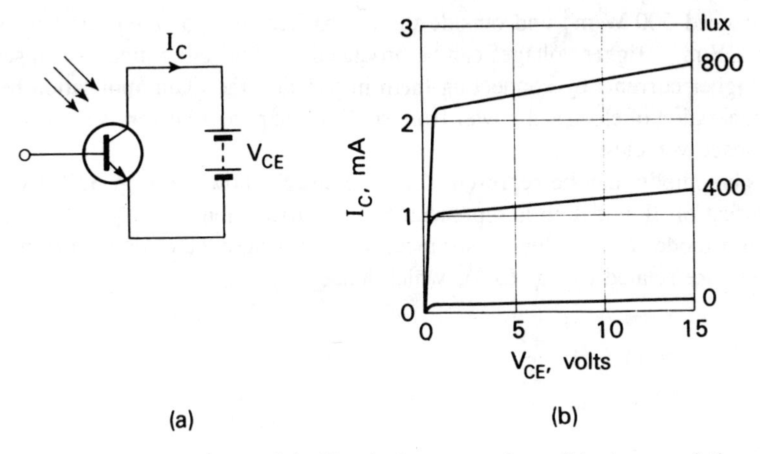

Figure 5.6 Phototransistor. (*a*) Circuit for operation; (*b*) characteristics of a silicon phototransistor.

which is connected in the common-emitter configuration with the base open-circuited. In the absence of radiation the collector current is then $(h_{FE} + 1)I_{CBO}$ (eq. (4.49)) or approximately $h_{FE}I_{CBO}$. When radiation falls on the collector-base junction (Fig. 5.6(*a*)) the leakage current I_{CBO} is increased by the photocurrent I_P, so that the collector current now becomes $I_C = h_{FE}(I_{CBO} + I_P)$. Thus, owing to the current gain of the transistor, the photocurrent is greater than the diode photocurrent and leads to the characteristics of Fig. 5.6(*b*). The base may be connected to the emitter through a resistor in order to reduce the dark current $h_{FE}I_{CBO}$ and hence to improve the ratio I_P/I_{CBO} at high temperatures.

Transparent encapsulations are also available for the MOST and the thyristor, with light energy falling on the gate region in each case. In the *photoFET*, extra charges are released on the surface of the substrate, so that the drain current is controlled by the illumination rather than the gate voltage. In the *photothyristor* the extra charges cause triggering of the thyristor by illumination rather than gate current.

Solar cell

It may be seen from Fig. 5.5(*b*) not only that the reverse current has increased, but also that the characteristic passes through a forward voltage at zero current. Thus an open-circuited photodiode generates an e.m.f. when the junction is illuminated, since extra holes move into the p-region and extra electrons move into the n-region owing to the junction field. This means that the junction is converting light energy directly into electrical energy, and such a device is called a *solar cell*. A silicon solar cell can produce an open-circuit voltage up to about 0.5 V and a short-circuit photocurrent of about 200 μA per square millimetre of junction area at a light energy of 1 kW/m^2. Even in the UK the light energy can exceed 500 W/m^2 and outside the atmosphere the solar power density is 1.36 kW/m^2. Higher voltages can be produced by connecting junctions in series and higher currents by connecting them in parallel, the main application being the generation of electrical power from sunlight, in particular for use in satellites and space vehicles.

A photodiode may be represented by the circuit shown in Fig. 5.7(*a*) where the effect of illumination is represented by a current generator I_P across a normal p-n diode. R_L is a load resistance, and, in general, the diode current and voltage are related by eq. (3.5), which leads to

$$\ln(I+I_0) = \ln I_0 + \frac{eV}{kT}$$

and

$$V = \frac{kT}{e} \ln\left(1 + \frac{I}{I_0}\right) \tag{5.8}$$

When the diode is open-circuited by disconnecting R_L illumination of the diode makes $I = I_P$, so that the open-circuited voltage is

$$V_{oc} = \frac{kT}{e} \ln\left(1 + \frac{I_P}{I_0}\right) \tag{5.9}$$

V_{oc} increases with the illumination as shown in Fig. 5.5(*b*).

With R_L connected across the diode, a load line can be drawn as shown on Fig. 5.5(*b*), so that

$$I = I_P - I_L \quad \text{where} \quad I_L = V/R_L \tag{5.10}$$

and

$$V = \frac{kT}{e} \ln\left(1 + \frac{I_P - I_L}{I_0}\right) \tag{5.11}$$

which is less than V_{oc}. Finally for $R_L = 0$ the diode behaves as a current generator with $I_L = I_P$ and $V = 0$.

A typical construction for a solar cell is shown in Fig. 5.7(*b*) and both p-on-n and n-on-p solar cells have been made, having series resistances of about 0.4 Ω and 0.7 Ω respectively. Silicon or gallium arsenide is used, with the gallium arsenide cell having a higher efficiency. At room temperature the predicted efficiencies are 22 to 26%, but in practice values of 10 to 15% are obtained. This is mainly due to energy loss by reflection and the effect of series resistance. The efficiency falls as temperature rises, being 13% at 200 °C instead of 26% at 20 °C on predicted values.

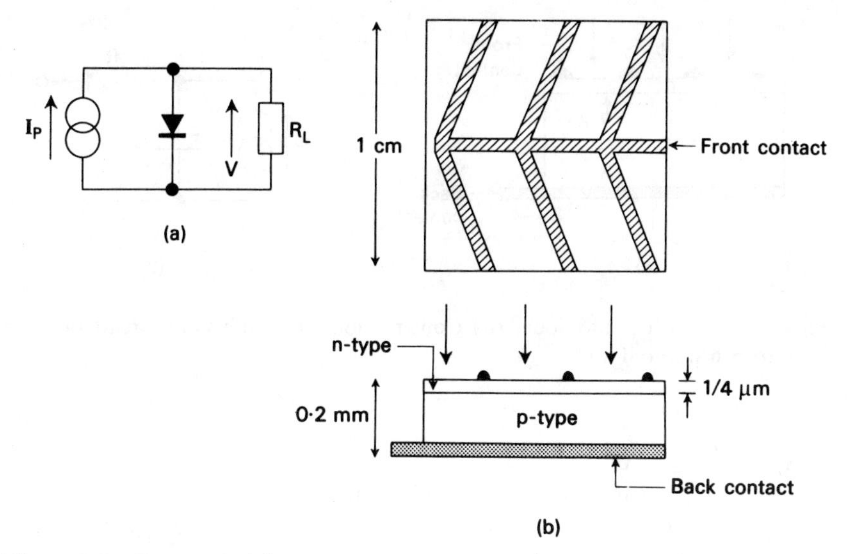

Figure 5.7 Solar cell. (*a*) Equivalent circuit and load; (*b*) top and side views of construction (not to scale).

p-i-n photodiode

In this device an intrinsic region (or lightly doped p- or n-region) is sandwiched between a p- and an n-region, as shown in Fig. 5.8(*a*). When the diode is reverse biased, depletion regions occur at the junctions of both p- and n-regions and the effective depletion layer width is increased by the effect of the i-region. Thus the depletion layer capacitance is much less than that of a normal p-n diode, which gives a much faster response to modulated light (Table 5.1). An a.c. equivalent circuit for both types of diode is shown in Fig. 5.8(*b*), where C represents the depletion layer capacitance and R the series resistance.

Electron-hole pairs are released within the i-region when radiation is absorbed, while electrons drift to the n-region and holes to the p-region under reverse bias, adding to the reverse current. The width of the i-region can be adjusted for optimum sensitivity and frequency response, with diodes having sensitivities up to 4 μA/mW/cm^2 being available. The diode can be used in optical distance measurement, star tracking and fibre optic termination. p-i-n diodes are also used with microwaves (Chapter 9).

Avalanche photodiode (APD)

If the reverse bias voltage is increased to a value near the breakdown voltage, avalanche multiplication occurs, giving a much larger gain than for other photodiodes (Table 5.1). Both the leakage and photocurrents are subjected to avalanche multiplication and the combined multiplication factor is given by

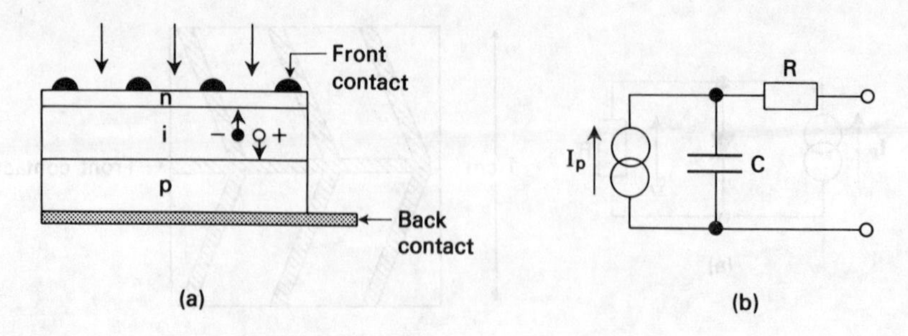

Figure 5.8 p-i-n photodiode. (*a*) Construction; (*b*) equivalent circuit (also applies to p-n photodiode).

$$M = \frac{I}{I_p} = \frac{I_{ph} + I_d}{I_{ph0} + I_{d0}} = \frac{1}{1 - (V_R/V_B)^n} \tag{5.12}$$

where I_{ph} and I_d are the multiplied photo- and dark currents, I_{ph0} and I_{d0} are the low voltage photo- and dark currents, V_R is the reverse bias voltage, V_B is the breakdown voltage and n is a constant. The form of M is the same as for a normal p-n diode (eq. (3.99)).

The equivalent circuit of an avalanche photodiode is shown in Fig. 5.9 where the multiplier M can have a value between 1 and 100 controlled by V_R. The avalanche process, being random, gives rise to a noise current which rises more rapidly with M than the optical current does and this restricts the values of V and M which are used. The diode has low noise performance over wide bandwidths, with a typical gain-bandwidth product of 80 GHz, and is used with lasers and in fibre optic communication systems.

Display devices (Ref. [2])

The general principle of an active display device is that radiation is emitted from a solid when it is supplied with energy in some form. Electrons then move from a higher energy level W_2 to a lower level W_1 with emission of a photon of energy $hc/\lambda = W_2 - W_1$, as discussed in Chapter 1. A band of wavelengths usually occurs since W_1 and W_2 belong to two groups of energy levels. The emission

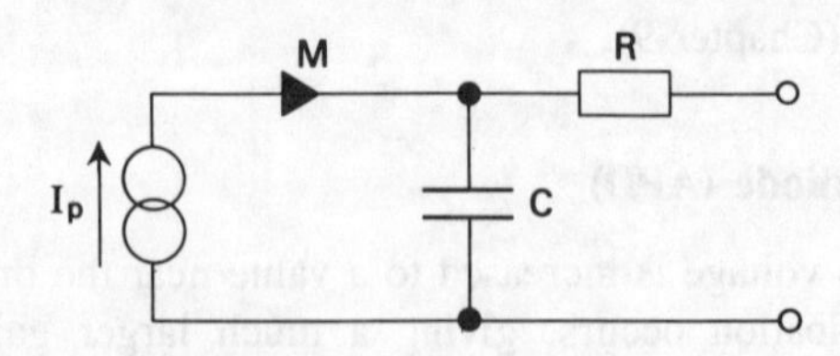

Figure 5.9 Avalanche photodiode equivalent circuit.

of radiation is known as *luminescence*, unless it is the result of thermal stimulation, when it is called *incandescence*.

When the excitation is removed the luminescence will persist for a time equal to the lifetime of the transition between the two energy levels; this phenomenon is known as *fluorescence*. However, if there are metastable or long-lifetime states between W_2 and W_1 electrons can remain trapped in them until they are released later by thermal excitation. In this case the luminescence persists for much longer than before; this is known as *phosphorescence*. Materials with this property are known as phosphors, most of which are based on zinc sulphide, ZnS.

The method of excitation takes various forms. For example, a very common active display device is the cathode ray tube, described in Chapter 8. In this device the excitation is due to electron bombardment of a phosphor-coated screen, which provides *cathodoluminescence*, as well as phosphorescence and fluorescence. Another use of a phosphor is in the *electroluminescent* display, which is formed when a phosphor powder in a transparent insulating binding is sandwiched between two electrodes, one of which is transparent. Luminescence is then obtained when an a.c. voltage up to 200 V or a d.c. voltage up to 100 V is applied.

Plasma displays rely on the glow produced in a cold-cathode discharge, normally using neon gas as described in Chapter 8. An a.c. excitation is normally applied with the transparent electrodes outside the discharge tube. This has a peak value of about 150 V to start the discharge and falls to about 90 V to maintain it. Both plasma and electroluminescent displays are suitable for large area displays, such as travel indicators and road signs.

A second form of electroluminescent display occurs in the light emitting diode, which depends on injection electroluminescence in a p-n junction and is described below.

Light emitting diode (LED)

When an electron-hole pair has been generated by external excitation the recombination of electron and hole may cause the radiation of a photon. This does not occur with equal probability in all semiconductors and the suitability of a semiconductor as a light source depends on the variation of energy with momentum in the crystal. This is introduced in Appendix 1, where it is shown that the energy W rises with the square of the momentum vector k, except near a forbidden band. The situation is more complicated in a real crystal, and the dependence of the energy on the momentum vector is shown diagrammatically for germanium and silicon in Fig. 5.10(*a*), and for a material such as gallium arsenide in Fig. 5.10(*b*). The energy gap W_g corresponds to the difference between the maximum energy in the valence band and the minimum energy in the conduction band. In gallium arsenide these energies both occur at the same value of k, which allows a direct transition with the emission of a photon, so that gallium arsenide is known as a *direct-gap* semiconductor. However, in germanium and silicon the maximum and minimum energies occur at different values of k, so that an electron must

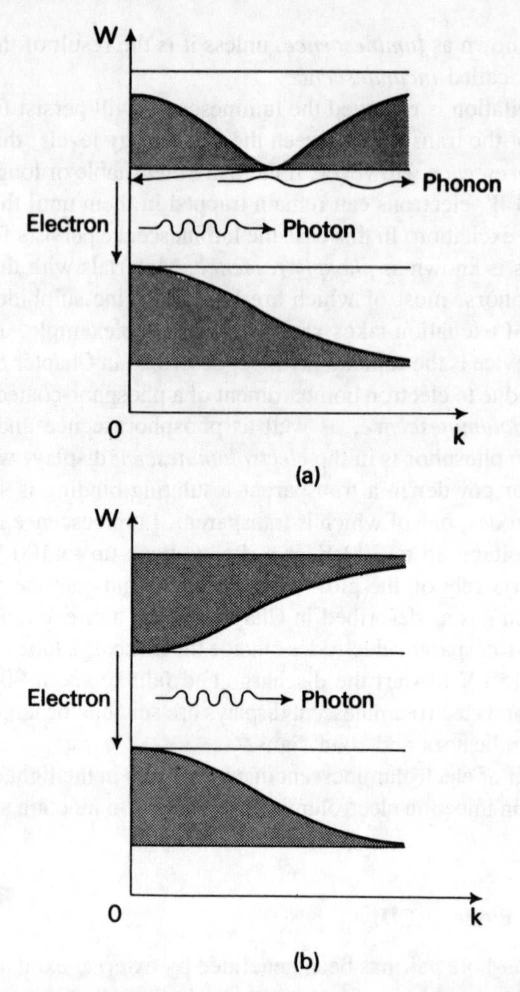

Figure 5.10 Energy-momentum vector diagrams. (*a*) Indirect gap semiconductor; (*b*) direct gap semiconductor.

lose momentum in order to have the value of k corresponding to the maximum energy of the valence band, after which a transition back to the valence band can take place.

The momentum is lost through the release of a phonon, which has a very much higher momentum than does a photon of higher energy.*

The probability of a phonon having the correct momentum being released at

* For both photon and phonon the momentum is h/λ and the energy is hc/λ. For a photon λ is about 10^{-6} m while for a phonon λ is about 10^{-10} m, giving a momentum about 10 000 times that of a photon. For a photon c is 3×10^8 m/s, while for a phonon c is the velocity of sound in a solid, typically 3×10^3 m/s. Thus the energy of a phonon is only about 1/10 that of a photon.

the same time as a photon of the correct energy is very small, so that this type of transition is very unlikely to happen. Germanium and silicon are known as *indirect-gap* semiconductors, and in these materials recombination takes place mainly through traps, which give rise to one or more intermediate levels in the forbidden band. An electron can fall from the conduction band into a trapping level and thence return to the valence band either directly or via other trapping levels to recombine with a hole.

Commercial LED materials

Gallium arsenide

As explained below a p-n diode is required which can be formed by using zinc as an acceptor and diffusing it into n-type GaAs. Alternatively, silicon can be used both as a donor, replacing gallium in a high-temperature diffusion and as an acceptor, replacing arsenic in a low-temperature diffusion. The wavelength of the radiation depends on the energy gap, as given by eq. (5.3) so the theoretical wavelength for GaAs is 0.87 μm from an energy gap of 1.43 eV. The practical values are slightly larger, as shown in Table 5.2 but are in the near infra-red.

Table 5.2 Characteristics of commercial LED materials (after Ref. [2])

Material	Dopant	Wavelength of peak emission	Colour
		μm	
GaAs	Zn	0.90	Infra-red
GaAs	Si	0.90 to 1.02	Infra-red
GaP	N, light doping	0.57	Green
GaP	N, heavy doping	0.59	Yellow
GaP	Zn, 0	0.70	Red
$GaAs_{0.6}P_{0.4}$		0.65	Red
$GaAs_{0.35}P_{0.65}$	N	0.63	Orange
$GaAs_{0.15}P_{0.85}$	N	0.59	Yellow

Gallium Arsenide Phosphide

The wavelength can be reduced into the visible region by adding phosphorus, and gallium arsenide phosphide is used in sources of visible red light. The wavelength corresponding to proportions of phosphorus between 0 and 100% are

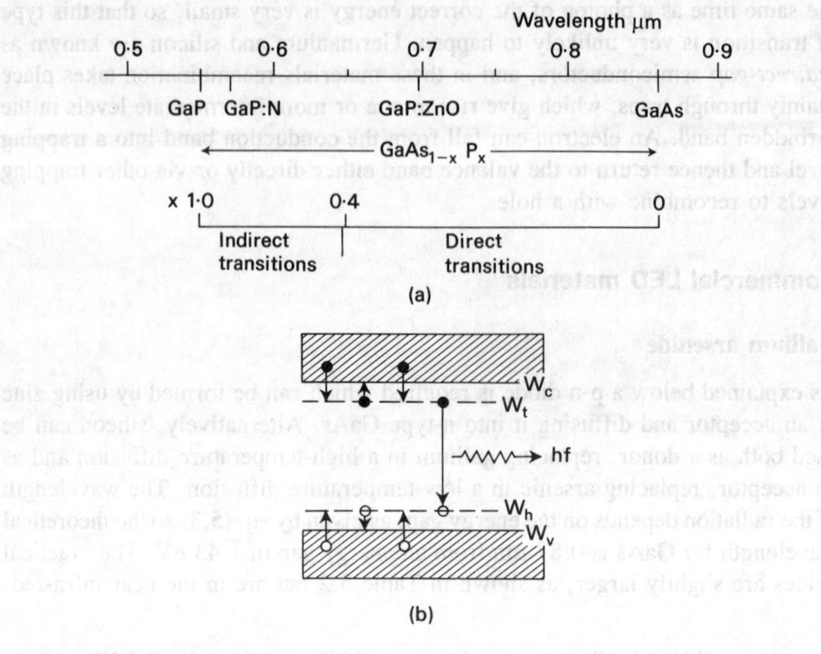

Figure 5.11 Gallium arsenide phosphide light-emitting diodes. (*a*) Wavelengths of radiation; (*b*) energy diagram for nitrogen doping.

illustrated in Fig. 5.11(*a*). Increasing the phosphorus content ultimately leads to gallium phosphide with $W_g = 2.24$ eV, corresponding to the green part of the spectrum, but when the proportion of phosphorus is increased above about 43%, indirect transitions occur. However, proportions of 65% and 85% will provide direct transitions when nitrogen doping is used, giving wavelengths at 0.63 μm (orange light) and 0.59 μm (yellow light).

Most commercial LED displays use material with the composition $GaAs_{0.6}P_{0.4}$, emitting at about 0.65 μm, although GaP : ZnO is also available. Even though only about 0.1% of the electrical input is converted into light output the application of LEDs as small, convenient light sources is widespread.

Gallium phosphide

One way of producing direct transitions for green light is to dope gallium phosphide with nitrogen (Ref. [3]). The effect of replacing phosphorus atoms with nitrogen atoms is to introduce extra levels W_t and W_h in the energy gap (Fig. 5.11(*b*)). W_t represents an electron trap and W_h a hole trap and direct transitions can occur from W_t to W_h.

For light doping W_t and W_h are close to their respective band edges and a wavelength of 0.57 μm (green light) is obtained. Heavier doping leads to W_t and W_h moving slightly away from their band edges and the wavelength rises to 0.59 μm (yellow light). These values compare with 0.56 μm obtained from the energy gap of 2.24 eV.

When zinc is used as a donor and oxygen as an acceptor W_t and W_h move nearer the middle of the energy gap and red light is emitted at a wavelength of 0.69 μm.

Even in a direct-gap semiconductor, significant light output is obtained only when large numbers of electrons and holes recombine in unit time. For a single crystal with no junctions a high temperature would be required to increase the density of electron hole pairs (eq. (2.30)), and doping introduces one type of carrier at the expense of the other type. However, if a p-n junction is formed using degenerate semiconductors, as in a tunnel diode but with slightly lower doping, the energy bands are as shown in Fig. 5.12(*a*) at zero bias. There is a high concentration of electrons in the conduction band of the n-region and a high concentration of holes in the valence band of the p-region. When forward bias is applied (Fig. 5.12(*b*)), the electrons at the edge of the conduction band occupy energy levels directly above the levels of holes at the edge of the valence band. Direct recombination takes place with one photon emitted for each electron transition, and since the carrier concentrations are high a useful light output is obtained.

A section through one form of light emitting diode is shown in Fig. 5.13(*a*). Not all the light generated at the junction will be emitted from the diode due to two main factors: (1) light absorption in the anode, and (2) internal reflection at the top anode surface. (1) is reduced by making the anode very thin and in gallium arsenide diodes it can be n-type since the absorption in p-type is considerably larger. Even a ray such as *A*, normal to the surface, will be partially reflected back into the diode due to (2). At a critical angle θ_c, *all* the light will be internally reflected (ray *B*) and θ_c is given by

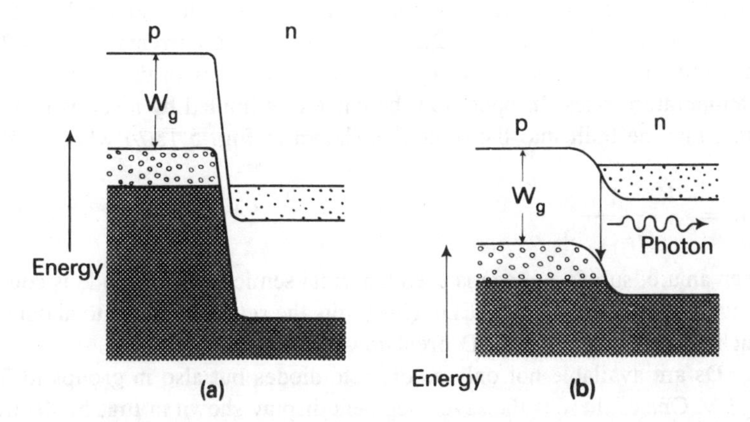

Figure 5.12 Energy diagrams for a light emitting diode or diode laser. (*a*) Zero bias; (*b*) forward bias.

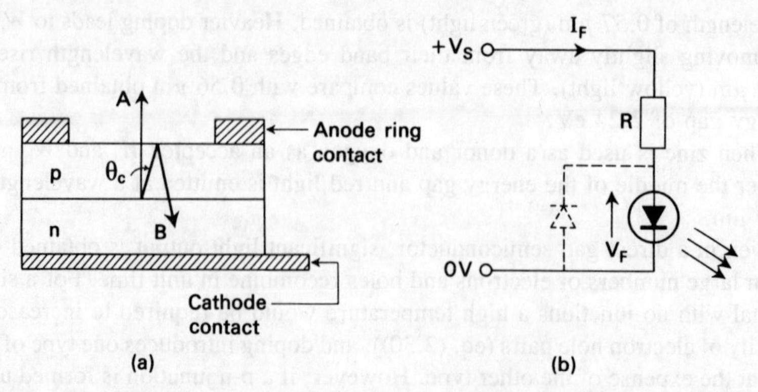

Figure 5.13 Light-emitting diode. (*a*) Construction; (*b*) circuit for operation.

$$\sin \theta_c = \frac{n_2}{n_2} \tag{5.13}$$

where n_1 is the refractive index of the medium in contact with the anode, which is 1.0 for air, n_2 is the refractive index of the anode which is 3.6 for gallium arsenide. In this case the critical angle is 16° but if a transparent epoxy dome is mounted on the diode for which $n_1 = 1.5$ then the critical angle is increased to 24.6°. Suitable choice of lens material and design allows the viewing angle to be increased into the range 30° to 60°. This is an advantage where the LED is used as an indicator lamp but still means that the luminous intensity varies widely with viewing angle outside this range. However, where the LED is used as a source in a fibre optic system narrower angles between 8° and 11° are provided in order to concentrate the light energy.

The values of forward voltage V_F and current for a reasonable light output depend on the colour. Typically for red LEDs V_F is around 1.8 V, while for green or yellow LEDs around 2.2 V is required and currents of 10 to 20 mA are common in all types. The light output increases with the current but falls as temperature rises. In operation the current is limited by a series resistance, which may be built into the diode, as shown in Fig. 5.13(*b*) where

$$R = \frac{V_S - V_F}{I_F} \tag{5.14}$$

When an a.c. supply voltage is used a normal semiconductor diode is connected in inverse parallel with the LED. This limits the reverse voltage to about 0.6 V, which is well below the LED breakdown voltage.

LEDs are available not only as discrete diodes but also in groups to form a *display*. One of these is the seven-segment display shown in Fig. 5.14(*a*) where each segment can be supplied from one LED through a plastic lens, thus magnifying the diode surface (Fig. 5.14(*b*)). The decimal numbers 0 to 9 can then be

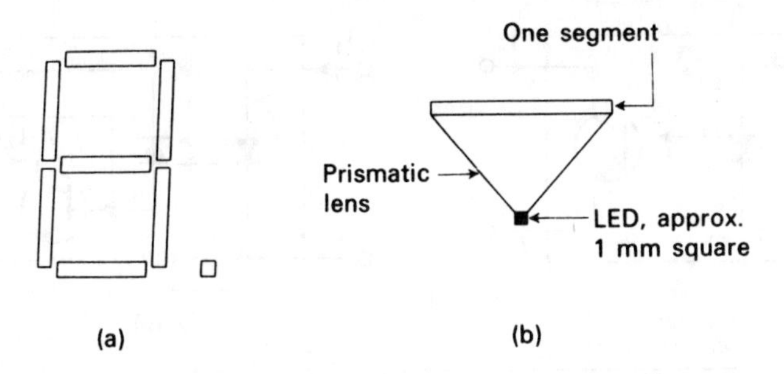

Figure 5.14 Seven-segment display. (*a*) Layout of segments; (*b*) lens system.

generated by energizing the appropriate segments and an extra LED can provide the decimal point. Such displays are very popular in mains-operated microprocessor systems and digital multimeters, where a polarity indication is also provided.

Optically coupled isolator

An LED and a photodiode or phototransistor mounted in the same encapsulation provide optical coupling but electrical isolation up to 3 kV. An amplifier or a logic driver can be included as shown in Fig. 5.15. The current transfer ratio I_2/I_1 is about 0.02 for a photodiode and about 7.0 for a logic driver. The complete electrical isolation allows two circuits with widely different ground voltages to be coupled. Logic circuits of different families such as TTL and CMOS can also be coupled with ground isolation. Pulse transformers can be replaced and signals from a line can be coupled into a computer at data rates up to 10 M bit/s.

Liquid crystal displays (LCD) Refs [4,5]

Liquid crystals are liquid organic compounds with cigar-shaped polar molecules. The polarizability of the molecules causes them to associate in the liquid with all the molecular axes pointing in the same direction (Fig. 5.16(*a*) and (*b*)).

In both the cases shown, the long axes are parallel and the short axes randomly oriented but in the *smectic* structure (Fig. 5.16(*a*)) the molecules are aligned to form layers, while in the *nematic* structure (Fig. 5.16(*b*)) the molecules are placed at random. In the *cholesteric* structure (not shown) the molecules are in sheets with a similar structure to the nematic but each sheet is slightly rotated with respect to its neighbours. At present nematic liquids are used for devices and the liquid is contained between glass plates to form a *nematic liquid crystal* (Fig. 5.16(*c*)).

The plate spacing is typically 10 μm and each plate has transparent conducting electrodes of indium or tin oxide, so that an electric field can be applied across the liquid. Each segment of the display is a liquid crystal which can have any shape but again a seven-segment display is common.

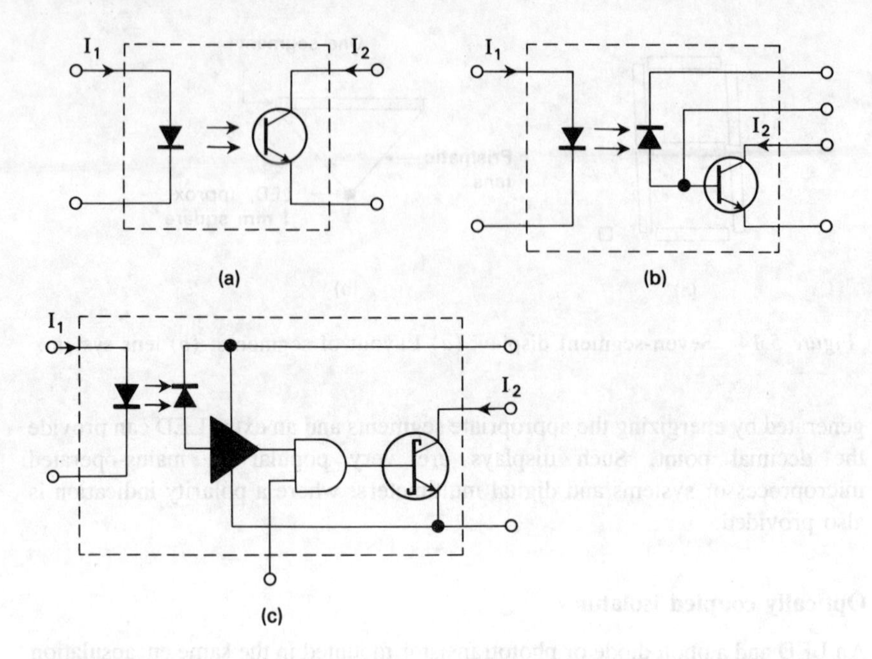

Figure 5.15 Optically coupled isolator. (*a*) Direct coupling; (*b*) coupling through a photodiode; (*c*) coupling through a logic driver with Schottky output transistor.

In early LCDs a direct field was applied which caused the molecules to become efficient scatterers of white light, while without a field the crystal was transparent. In this case an energized segment appears white and a high value of incident illumination produces a bright display. This is a *dynamic scattering display* which has the two disadvantages of a relatively short life and poor contrast due to reflection.

It has been superseded by the *twisted nematic* or *field effect* display (Fig. 5.17) using a positive nematic liquid in which the molecules align themselves parallel with a field. The plates are treated by unidirectional rubbing or angular evaporation of a dielectric film to ensure that the molecules align themselves in the plane of each plate in the absence of a field (Fig. 5.17(*c*)). The display device is then assembled with the aligned directions at 90° and filled with the liquid. The effect of the twist in the molecules is to rotate the plane of polarization of the light wave through 90° across the crystal and the device is viewed through crossed polarizing filters, the principle of which is illustrated in Figs 5.17(*a*) and (*b*).

With no applied field the device appears transparent, as the plane of polarization of the emergent light is parallel with that of the filter (Fig. 5.17(*c*)). When a field is applied the molecules align themselves across the crystal and so have virtually no effect on the light (Fig. 5.17(*d*)). However, the planes of polarization are now crossed, so the device appears dark. When the field is removed the

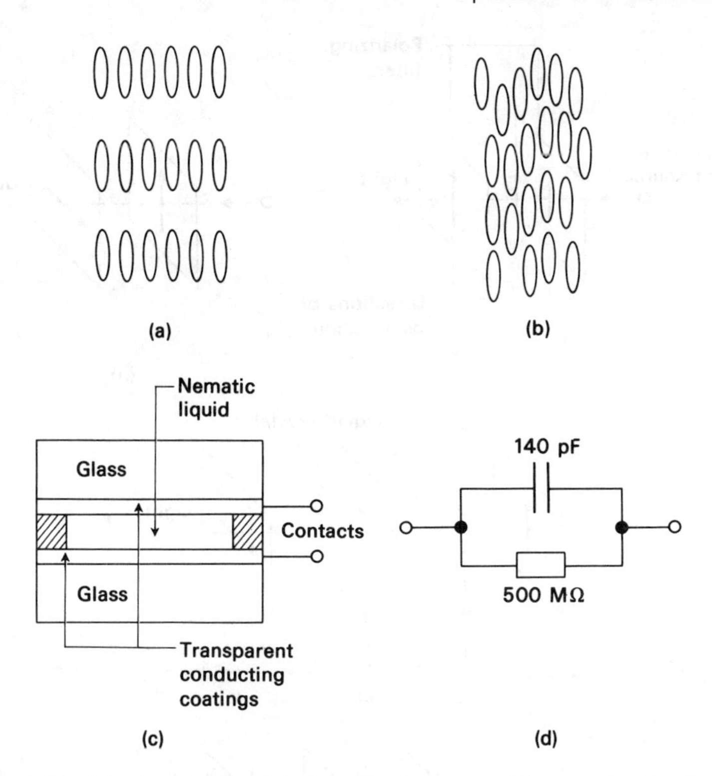

Figure 5.16 Liquid crystals. (*a*) Smectic structure; (*b*) nematic structure; (*c*) construction of liquid crystal cell; (*d*) equivalent circuit.

original twisted configuration is restored by the action of elastic forces from the surfaces.

The left-hand plate can be made reflective so that ambient lighting is returned through the crystal. The other plate then appears dark against a lighter background when a field is applied. Alternatively the display can be lit from the back as shown in Fig. 5.17(*c*) and (*d*). An a.c. drive is used to avoid electrolytic dissociation of the liquid and typically 4 to 17 V r.m.s. is required, with a square waveform. An equivalent circuit for a single segment of a display with 12 mm high characters is shown in Fig. 5.16(*d*). This illustrates the capacitive nature of the cell and the current may be calculated assuming that the resistive component is negligible. For a voltage of 6 V r.m.s. at 50 Hz the current is about 0.3 μA and rises linearly with frequency. A threshold of about 2 V d.c. must be added and a frequency range between 30 and 100 Hz is typical, with turn-on and off times of about 100 ms.

Due to their low power requirements LCDs can be driven by CMOS logic circuits and at present they are mainly applied in digital watches and clocks and in battery-operated calculators and multimeters. Compared with light emitting

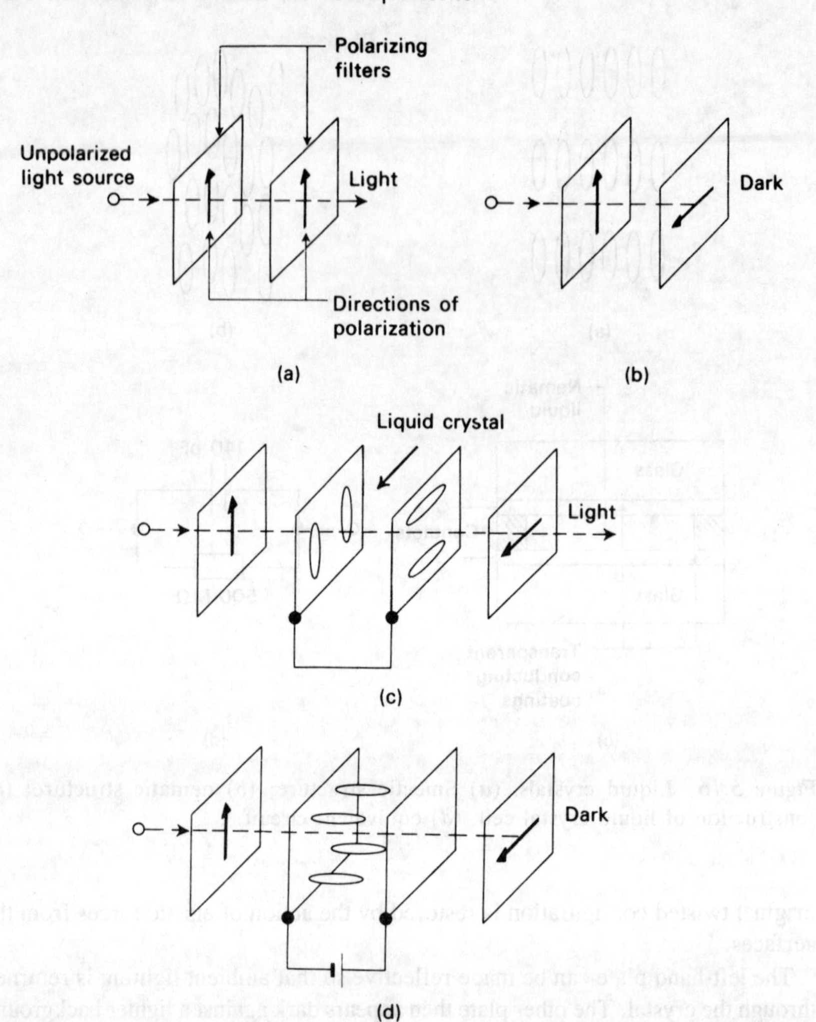

Figure 5.17 Twisted nematic (field effect) display. (*a*) Parallel polarizing filters; (*b*) crossed polarizing filters; (*c*) no applied field; (*d*) field applied.

diodes they have a much lower current consumption and can be read more easily in bright light giving a reflective contrast of about 70:1. Larger displays can also be made in LCDs.

Lasers

When electromagnetic radiation interacts with matter, transitions occur between the atomic energy levels, as first suggested by Bohr in 1913. However, it was not until 1917 that Einstein considered the *probabilities* of absorption and emission of radiation by an atom, and he showed that a second emission process, *stimulated emission*, must also occur. This process is more significant at microwave frequencies than at optical frequencies, and in 1954 a device was introduced called the *maser*. The principle of the maser was extended to optical frequencies with the advent of the *laser* in 1960, the word standing for Light Amplification by Stimulated Emission of Radiation.

The light energy produced by a laser is concentrated into a very narrow beam which forms an intense source of heat. In addition all the waves in the beam are in phase, which is unique in a light source. These two properties have led to many applications of the laser since its discovery. Drilling small holes even in diamonds, fine welding and production of integrated circuits all use its *heating* properties, while surveying, metrology, inspection and non-destructive testing, visual display systems and communications all use its unique *optical* properties. The statement that the laser is a solution in search of a problem no longer holds true.

Laser fundamentals

In 1901 Planck introduced the quantum theory to account for the observed spectrum of the energy radiated from the black body. He showed that the energy could not change continuously but only in units of one quantum, hf joules. For a black body radiating energy within a frequency range f to $f+df$ this leads to an expression for the energy radiated per unit volume, $E_f\,df$. E_f is the energy density and is given by

$$E_f = \frac{8\pi f^2}{c^3} \frac{hf}{\exp(hf/kT) - 1} \tag{5.15}$$

This equation is illustrated in Fig. 5.18 for a black body at 1500 K and is in agreement with the observed experimental values of E_f.

Now suppose that the radiation from a black body at temperature T is in thermal equilibrium with atoms having energy levels W_1, W_2, and so on. The energy of an atom can be increased if an electron is excited from a level W_1 to a higher level W_2 when a photon of radiation of frequency f_{21} is absorbed, where

$$hf_{21} = W_2 - W_1$$

(from eq. (1.11)). The photon is re-radiated after a short time when the atom returns to level W_1, a process known as *spontaneous emission*. Hence at any instant the atoms will be distributed between the available energy levels, and if N_1 atoms have energy W_1 the number of atoms with any other energy level W_n is determined by Boltzmann statistics, so that

$$N_n = N_1 \exp\left(-\frac{W_n - W_1}{kT}\right) \tag{5.16}$$

For atoms in the levels W_1 and W_2, then,

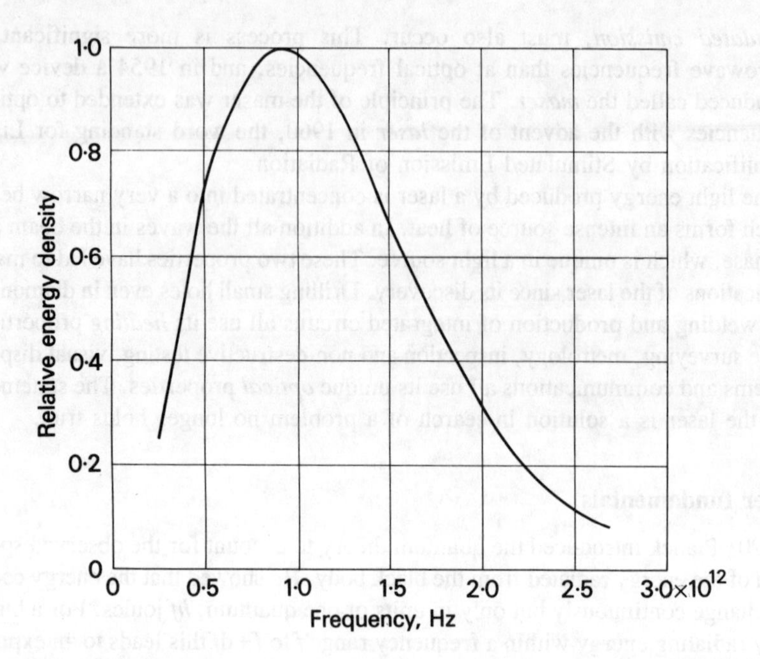

Figure 5.18 Energy density and radiation frequency for a black body at a temperature of 1500 K.

$$N_2 = N_1 \exp\left(-\frac{hf_{21}}{kT}\right) \tag{5.17}$$

which means that N_2 is less than N_1 when the exponential term is less than unity. A dynamic equilibrium exists between the two levels such that in the same time interval the number of transitions from W_1 to W_2 equals the number of transitions from W_2 to W_1.

Einstein's three radiation coefficients

The number of atoms raised from W_1 to W_2 in a time interval dt by absorption of photons will be proportional both to the number of atoms of energy W_1 and to the energy density of the radiation at frequency f_{21}. If the constant of proportionality is B_{12}, the rate of change of the number of atoms in level W_2 is

$$\frac{\mathrm{d}N_2}{\mathrm{d}t} = B_{12}N_1E_{f21} \tag{5.18}$$

B_{12} is called the *coefficient for absorption of radiation*.

The number of atoms falling from W_2 to W_1 by radiation of a photon is similarly proportional to the number of atoms in level W_2. Then if the constant of proportionality is A_{21} the rate of change of the number of atoms in level W_1 is

$$\frac{dN_1}{dt} = A_{21}N_2 \tag{5.19}$$

A_{21} is called the *coefficient of spontaneous emission*. Then, if only these two processes exist, in equilibrium

$$\frac{dN_1}{dt} = \frac{dN_2}{dt} \tag{5.20}$$

Hence

$$E_{f21} = \frac{A_{21}}{B_{12}}\frac{N_2}{N_1} \tag{5.21}$$

$$= \frac{A_{21}}{B_{12}} \exp\left(-\frac{h_{f21}}{kT}\right) \tag{5.22}$$

using eq. (5.17). However, it is clear that the form of eq. (5.22) does not correspond to eq. (5.15), which has been derived theoretically and confirmed experimentally. Einstein therefore suggested a second emission process which is *stimulated* by the presence of radiation of frequency f_{21}. The number of transitions from W_2 to W_1 by this process is then proportional both to the number of atoms in W_2 and the energy density of radiation at frequency f_{21}. Thus an additional term is required in eq. (5.19) which gives, for transitions from W_2 to W_1,

$$\frac{dN_1}{dt} = A_{21}N_2 + B_{21}N_2E_{f21} \tag{5.23}$$

where B_{21} is the *coefficient of simulated emission*. Again at equilibrium $dN_1/dt = dN_2/dt$, so that, from eq. (5.18) and (5.23),

$$E_{f21}(B_{12}N_1 - B_{21}N_2) = A_{21}N_2 \tag{5.24}$$

Substituting for N_1 from eq. (5.17),

$$E_{f21}\left[B_{12} \exp\left(\frac{hf_{21}}{kT}\right) - B_{21}\right] = A_{21}$$

and

$$E_{f21} = \frac{A_{21}}{B_{12} \exp\left(\frac{hf_{21}}{kT}\right) - B_{21}} \tag{5.25}$$

This equation then agrees with Planck's equation (5.15) if

$$B_{12} = B_{21} \tag{5.26}$$

and

$$\frac{A_{21}}{B_{21}} = \frac{8\pi h f_{21}^3}{c^3} = E_{f21}\left[\exp\left(\frac{hf_{21}}{kT}\right) - 1\right] \tag{5.27}$$

Equation (5.26) shows that the probabilities of absorption and stimulated emission of a photon are equal; Einstein's three coefficients are illustrated in Fig. 5.19.

The application of stimulated emission

A photon of energy hf_{21} which interacts with an atom of energy W_2 will reduce its energy to W_1 and cause the stimulated emission of a second photon, also of energy hf_{21}. Thus the number of photons has doubled and amplification of the original signal has occurred. However, photons of energy hf_{21} will also be emitted at random time intervals by spontaneous emission, and these constitute a source of electrical noise which interferes with the signal. The relative powers of the signal and noise depend on the relative number of photons emitted per second. Thus from eq. (5.27) the ratio of stimulated to spontaneous emission under conditions of thermal equilibrium is

$$\frac{B_{21}E_{f21}}{A_{21}} = \frac{1}{\exp\left(\dfrac{hf_{21}}{kT}\right) - 1} \tag{5.28}$$

Now consider a stream of photons passing through a material which has a difference in energy levels corresponding to the energy of the photons. If the photons

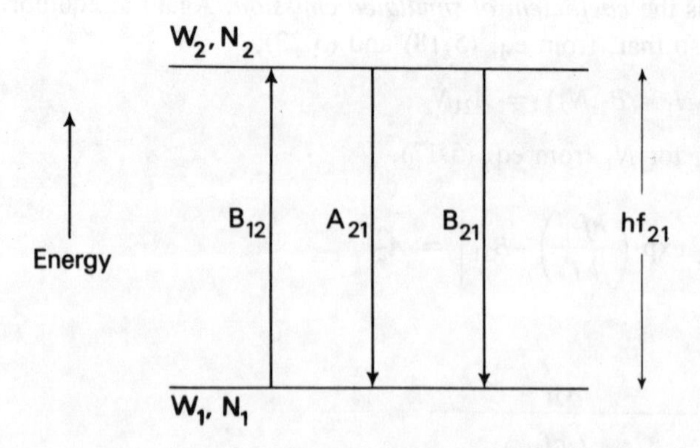

Figure 5.19 Einstein's radiation coefficients.

strike unexcited atoms they may be absorbed and removed from the stream, which thus loses energy. However, if the photons strike excited atoms, more photons can be produced which are added to the stream and increase its energy. Since the probabilities of absorption and stimulated emission are the same (eq. (5.25)), both attenuation and amplification of the stream occur simultaneously and amplification can only predominate if there are more atoms in the higher level than in the lower level.

Population inversion

Under equilibrium conditions the lower level has more atoms than the higher level (eq. (5.16)). Thus for amplification to be possible the population of atoms, N_2, in the upper level, must exceed the population, N_1, in the lower level, the process leading to $N_2 > N_1$ being known as *population inversion*. This can be achieved in a material where the lifetime of the upper level exceeds that of the lower level, for example when the upper level is metastable.

If there is a third normal level, W_3, above the metastable level, W_2 (Fig. (5.20)), a strong microwave signal of frequency f_{31}, where

$$hf_{31} = W_3 - W_1 \tag{5.29}$$

can sometimes be used to raise the energy of electrons in some of the atoms from W_1 to W_3 (Fig. 5.21)). This will continue until the populations of W_3 and W_1 are equal, since then the absorption of radiation is just balanced by the stimulated emission. Since W_3 is a normal level, atoms will remain in it only for a very short time, and will then return to W_1, either directly or by way of W_2, which

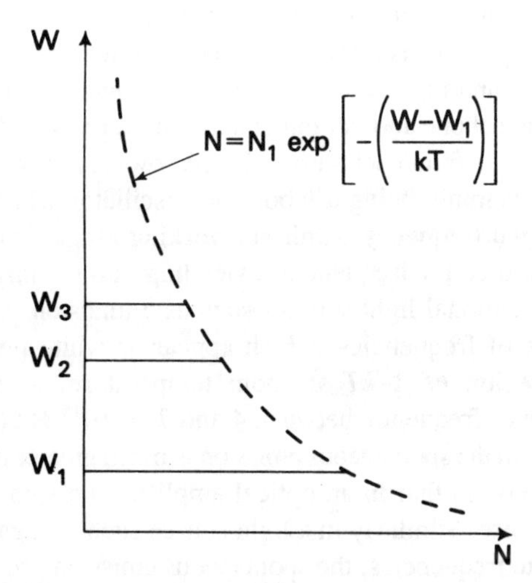

Figure 5.20 Distribution of atoms between energy levels.

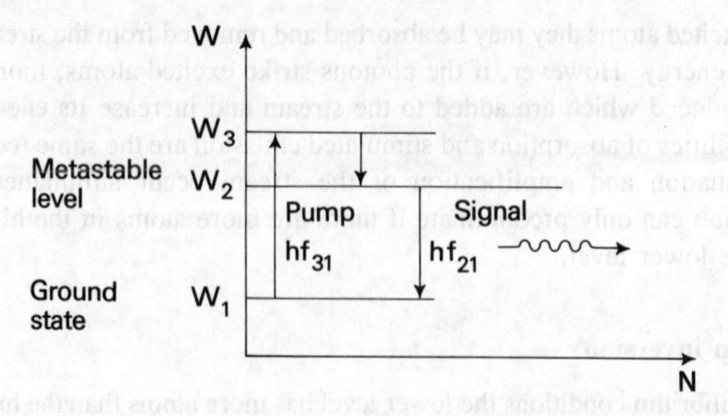

Figure 5.21 Population inversion.

is chosen to be a metastable level with a much longer lifetime. Thus the number of atoms in W_2 can be arranged to exceed the number in W_1 because a 'head' of atoms is maintained in W_3. f_{31} is known as the *pump* frequency, and the amplified signal has a frequency f_{21}, typical values being 10 and 4 GHz respectively. In fact, avalanche multiplication of the photons is occurring since one photon colliding with an atom releases a second photon and both of these release two more. Thus the number of photons multiplies in a similar manner to the electron avalanche.

Coherence of emission

Since a photon incident on an atom causes the emission of another photon, their associated electromagnetic waves are in phase (Fig. (5.22(*a*)). Thus the waves reinforce each other and the emission is said to be *coherent*. However, the spontaneous emission of photons is a random process, so that the phase relationships between their waves and the incident wave is also random (Fig. 5.22(*b*)). Thus the waves tend to cancel one another out and the emission is said to be *incoherent*.

Coherent sources with frequencies up to the microwave region are comparatively easy to obtain, an example being a laboratory oscillator which can provide an output at any selected frequency within its working range. It is far less easy to obtain a coherent source for frequencies extending into the infra-red and visible regions, since conventional light sources such as a tungsten filament lamp emit a continuous range of frequencies, which appear as white light.

In the visible region $hf_{21} \gg kT$ at room temperature, as may be seen by substituting a value of frequency between 4 and 7×10^{14} Hz (Table 1.1). Thus in thermal equilibrium the spontaneous emission is much greater than the stimulated emission (eq. (5.28)), so that in an optical amplifier the signal would be completely lost in the noise. Similarly in a light source such as a gas discharge tube, which emits discrete frequencies, the spontaneous emission predominates and the light output is incoherent. However, in a laser the stimulated emission is amplified

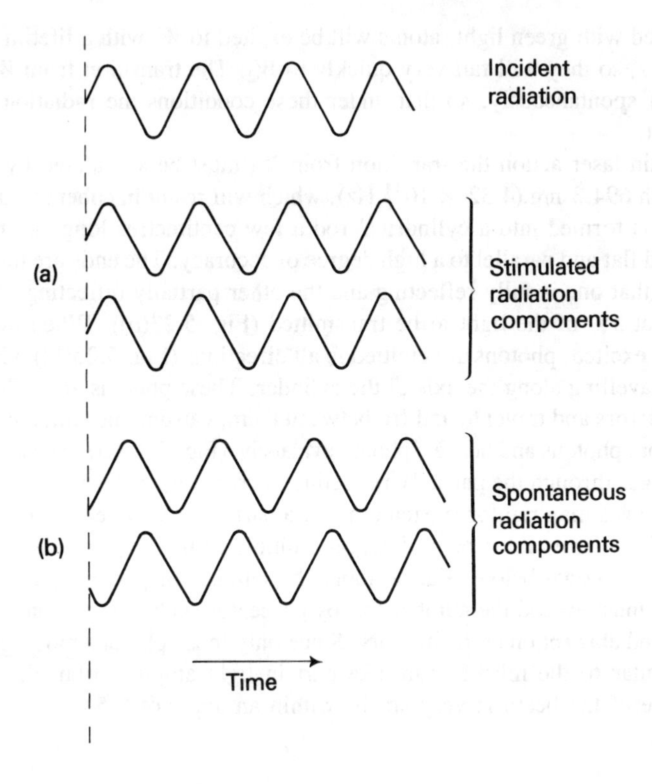

Figure 5.22 Phase relationships for (*a*) coherent and (*b*) incoherent emission.

and separated from the spontaneous emission to give an optical oscillator of coherent radiation at high intensity.

Solid state lasers

Ruby laser

Suitable energy levels for laser action are found in chromium ions present as an impurity in an aluminium oxide crystal, about 1 in 5000 of the aluminium ions having been replaced by chromium. This combination is a version of the precious stone, ruby, being pink instead of deep red in colour. The chromium ion in a ruby crystal provides another suitable set of three energy levels in the ruby laser. The energy difference $W_3 - W_1$ corresponds to a band of wavelengths between 520 and 600 nm. This represents *green* light, which is used for pumping, and since $hf_{31} \gg kT$, $N_1 \gg N_2$ at room temperature. The metastable level W_2, which has a lifetime of 3 ms at room temperature, gives rise to transitions to W_1 corresponding to a wavelength of 694.3 nm, which represents *red* light. If the crystal is irradiated with green light, atoms will be excited to W_3 with a lifetime of less than 10^{-9} s, so they will fall very quickly to W_2. The transition from W_2 to W_1 will occur spontaneously, so that under these conditions the radiation will be incoherent.

To obtain laser action the transition from W_2 must be stimulated by light of wavelength 694.3 nm (4.32×10^{14} Hz), which will result in coherent radiation. The ruby is formed into a cylindrical rod a few centimetres long and the ends are ground flat and parallel to a high degree of accuracy. The ends are then silver coated so that one is fully reflecting and the other partially reflecting, allowing up to about 5% of the light to be transmitted (Fig. 5.23(*a*)). When the atoms have been excited, photons are emitted in all directions (Fig. 5.23(*b*)) with a few of them travelling along the axis of the cylinder. These photons are reflected by the end mirrors and travel to and fro between them, causing the stimulated emission of more photons and hence a photon avalanche (Fig. 5.23(*c*)). The laser beam then emerges through the partially reflecting mirror (Fig. 5.23(*d*)). In order that the light waves may reinforce each other at a particular wavelength the distance between the mirrors must be an integral multiple of the *half-wavelength* of the radiation, as shown below. Furthermore the number of photons gained in the avalanche must exceed the number lost by processes such as emission out of the cylinder and absorption at the mirrors. Since only those photons moving exactly perpendicular to the mirrors can take part in light amplification, the angular divergence of the beam is very small, within an angle of 0.5°.

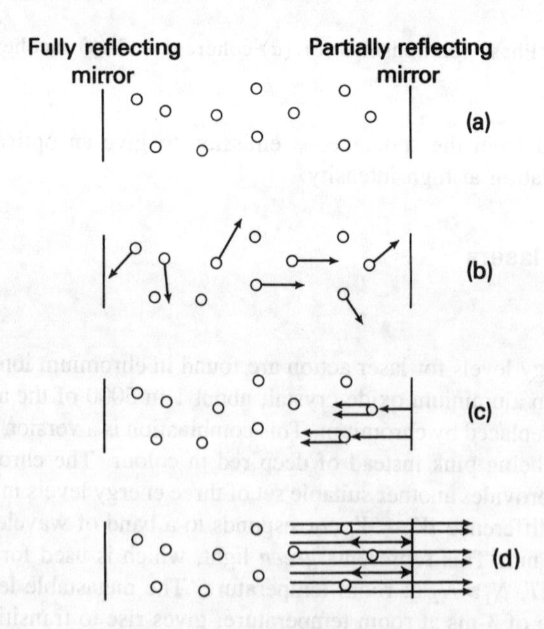

Figure 5.23 Formation of a photon avalanche in a laser.

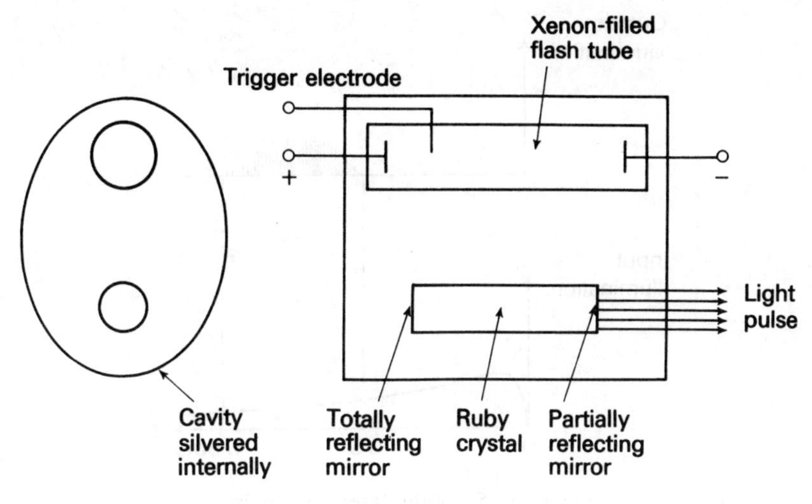

Figure 5.24 Ruby laser.

The initial excitation of green light can be obtained from a xenon flash tube (Fig. 5.24) in which an arc is set up by discharging a bank of capacitors. The ruby crystal and the flash tube are mounted at the foci of an elliptical internally reflecting cavity to ensure an efficient transfer of energy to the ruby crystal. Only a small proportion of the output from the tube is of the correct wavelength; the remainder causes heating of the crystal. Thus the laser is normally operated with a pulsed output to prevent overheating and destruction of the rod. The output energy from the flash tube must be above a threshold value which is required to maintain the population of W_2 above that of W_1, and typical values are between 500 and 1000 J per pulse. The corresponding light output from the laser is between 0.1 and 1.5 J for a 4 cm × 1 cm crystal, with the output occurring in a burst of irregular pulses (Fig. 5.25). These occur because the photon density builds up so rapidly in the crystal that stimulated transitions occur faster than the rate at which population inversion is maintained. Thus at the end of each small pulse the population of W_2 has fallen below the threshold value for sustained emission and the next pulse only appears after population inversion has been restored.

Laser action has also been obtained in neodymium doped materials. The commonest host materials are yttrium aluminium garnet and glass and the light output has a wavelength of 1.06 μm. The energy level W_1 in the laser transition (Fig. 5.21) is about 0.25 eV above the ground state and so has a negligible population at normal operating temperatures. This four-level transition therefore allows population inversion at a much lower pumping power than for the three-level ruby transition. A xenon flash tube can be used for pulsed output and tungsten-halide incandescent lamps are used for continuous wave (CW) operation.

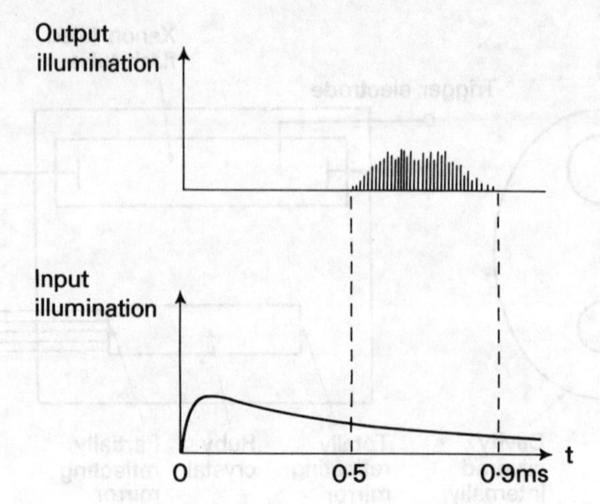

Figure 5.25 Ruby laser light pulses.

Threshold pumping levels of about 3 J at 3% conversion efficiency can be achieved, and continuous output powers of over 1 kW are obtainable with YAG lasers. The glass lasers are less suitable for CW operation but peak pulsed powers over 10^9 W can be produced.

Q-switching

The energy of all the small pulses shown in Fig. 5.25 may be concentrated into a single high-power pulse by operating in the Q-switched mode. Here the laser is first excited without feedback by preventing reflection from one of the end mirrors, thus increasing the optical losses, since the mirror system can be considered as a cavity resonator whose Q-factor depends on the losses within it (eq. (9.9)). Stimulated emission cannot occur if an end mirror is not allowed to reflect into the laser rod and the excitation greatly increases the population of W_2. If the end mirror is then suddenly allowed to reflect, the feedback of light energy causes a rapid inversion of the excited level. The photon avalanche builds up very quickly and an intense burst of radiation, known as a *giant pulse*, occurs in a very short time. The Q-switch is opened about 0.5 ms after the initiation of the exciting flash and a single output pulse is obtained lasting about 0.1 μs. The peak output power can be increased by up to 100 times by Q-switching, and peak output powers up to 100 MW have been reported using large crystals.

The Q-switch can take several forms as shown in Fig. 5.26. Physical rotation of one of the mirrors ensures that exact parallelism only occurs for a short time, typically 10 to 30 ms, during which time a giant pulse lasting about 30 ns is generated. A passive Q-switch uses a dye which absorbs low-intensity radiation but becomes transparent when the intensity has reached a high value. The transmission coefficient of such a bleachable absorber can increase from about 10^{-6} to

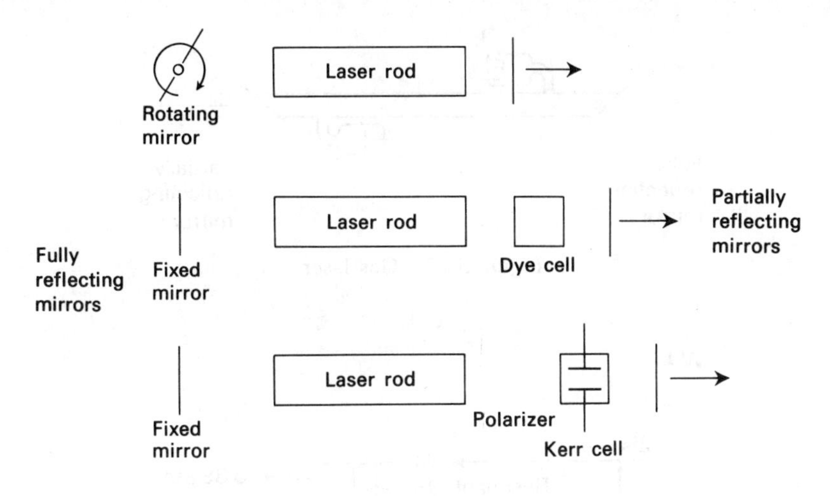

Figure 5.26 Q-switching by rotating mirror, dye cell and Kerr cell.

0.3 in a few nanoseconds, and the effect is reversible so that the material recovers in time for the next pulse. For ruby lasers the dye kryptocyanine has been used. An electro-optic Q-switch is the Kerr cell for which a material such as KDP (potassium dihydrogen phosphate) can be used.This rotates the plane of polarization when the voltage is applied so that the polarizer absorbs light that makes a double pass through the cell. When the voltage is removed light is rcturned to the laser rod and the giant pulse is produced.

Gas lasers

Helium-neon laser

A continuous light output can be obtained from a discharge in a gas or a mixture of gases, laser action being achieved by mounting the discharge tube between two mirrors. A commonly used mixture is that of the inert gases helium and neon, approximately in the proportion of 7 to 1 by volume. The discharge can be maintained by a direct voltage of a few hundred volts between an anode and a cathode, as shown in Fig. 5.27. An earlier arrangement uses an alternating voltage of frequency about 30 MHz applied through electrodes on the outside of the tube (electrode-less discharge).

Helium atoms are excited in the discharge to the two $2s$ levels, which are metastable (Fig. 5.28). Neon also has two metastable levels, $2s$ and $3s$, whose energies are close to the two helium levels. Thus, when a helium atom collides with a neon atom, energy can be transferred to the neon atom by a process known as a *resonant transfer*. Spontaneous emission then occurs due to neon atoms dropping from the metastable levels to the lower levels. The strongest spectral lines

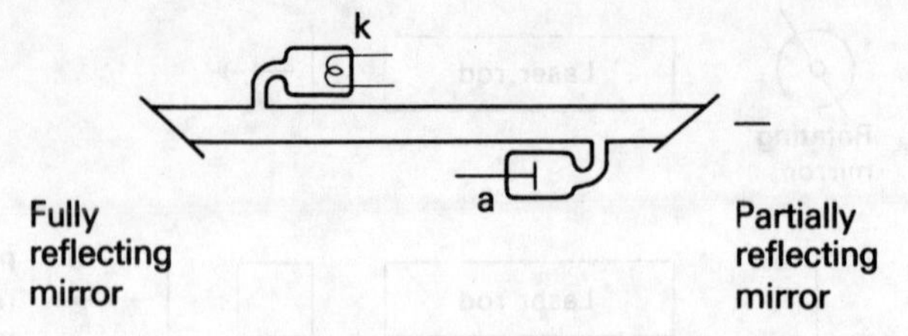

Figure 5.27 Gas laser.

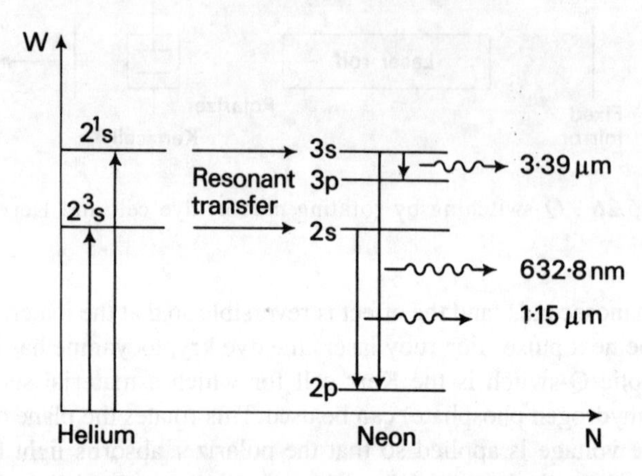

Figure 5.28 Transitions in helium and neon.

occur for the transitions shown in Fig. 5.28, at wavelengths of 632.8 nm in the visible red region and at 1.153 μm and 3.39 μm in the infra-red.

The discharge tube is sealed at each end by optically flat fused-quartz windows mounted at an angle of 55.5° to the tube axis (Fig. 5.27). This is the *Brewster angle* at which reflection of the 632.8 nm (4.74×10^{14} Hz) light is a minimum and transmission of this wavelength is a maximum, which helps to select it as the main output wavelength of the laser.

Of the many gases in which an inverted population of atoms can be established, argon and carbon dioxide have emerged as among the most important for laser applications. Argon provides coherent radiation in the visible spectrum at 458 nm and 515 nm, and an output of 1 W is obtainable at an efficiency of about 0.1%. Where high continuous power is required the carbon dioxide laser is used. This can provide output powers of many kilowatts at 10.6 μm in the far infra-red. The efficiency is increased by the addition of helium and nitrogen, with which a value of up to 30% is obtainable, and it is thought that resonant transfer occurs between metastable levels in the nitrogen and the upper CO_2 levels. The efficient fluorescence of many organic compounds makes them suitable for yet another system known as the dye laser.

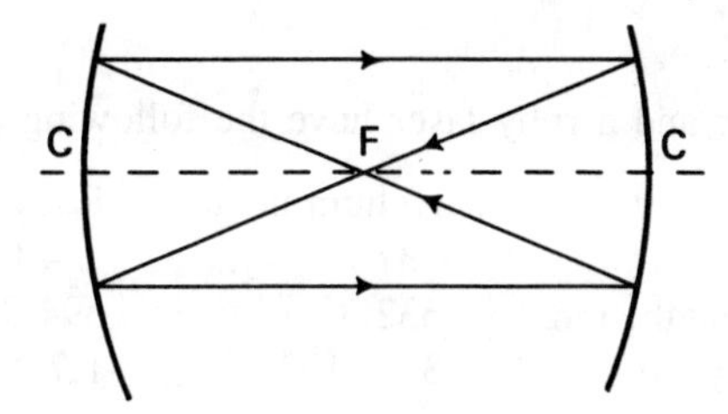

Figure 5.29 Confocal mirror system.

Laser modes

The two mirrors required for laser action may be spherical, with the centre of curvature, C, of each lying in the surface of the other, so that their focal points coincide in the centre of the tube at F (Fig. 5.29). The distance between the mirrors has to be an exact multiple of a half-wavelength for reinforcement of reflected waves to occur. The system is termed *confocal*, and even if the mirrors are slightly tilted a regenerative path similar to the one shown is still possible. Thus the adjustment of spherical mirrors when the laser is set up is much less critical than for a plane mirror system, or *Fabry–Perot interferometer*. Again one of the mirrors allows up to about 5% of the light to be transmitted, so that a light output is obtained at one end. The mirrors are coated with alternate layers of high- and low-refractive-index materials, which have a thickness of $\lambda/4$ at 632.8 nm. These reflect 99% of the light at this wavelength but much less at other wavelengths.

In effect the mirrors form a resonant cavity, in both the gas and solid-state lasers. Standing waves are set up in the cavity when the optical spacing between the mirrors is an integral number of half-wavelengths, the situation being analogous to the standing waves for an electron in a potential well, discussed in Appendix 1. If the optical spacing between the mirrors is L then resonance occurs when

$$L = \frac{n\lambda}{2} = n\frac{c}{2f} \tag{5.30}$$

or the resonant frequency is

$$f = n\frac{c}{2L} \tag{5.31}$$

where n is an integer.

Each integral value of n corresponds to a frequency at which oscillations may occur and constitutes a resonance or *mode*. A number of values of L are possible, depending on the path followed by the light between the mirrors. Where L is the optical spacing along the longitudinal axis of the discharge tube the *longitudinal* or *axial modes* are given by eq. (5.30).

Worked example

A helium-neon laser and a ruby laser have the following properties.

	Helium-neon	Ruby
Length, cm	37	4.0
Radiation wavelength, nm	632.8	694.3
Radiation velocity, m/s	3×10^8	1.7×10^8

In each case determine the number of modes and the frequency of the radiation.

Solution

Helium-neon laser

From eq. (5.30)

$$n = \frac{2L}{\lambda}$$

$$= \frac{2 \times 37 \times 10^{-2}}{632.8 \times 10^{-9}}$$

$$= 1.16 \times 10^6 \text{ modes}$$

$$f = \frac{c}{\lambda}$$

$$= \frac{3 \times 10^8}{632.8 \times 10^{-9}}$$

$$= 4.74 \times 10^{14} \text{ Hz}$$

Ruby laser

$$n = \frac{2 \times 4 \times 10^{-2}}{694.3 \times 10^{-9}} = 1.15 \times 10^5 \text{ modes}$$

$$f = \frac{1.7 \times 10^8}{694.3 \times 10^{-9}} = 2.45 \times 10^{14} \text{ Hz}$$

The laser will only generate those modes which correspond to optical paths very close to the axis, but even then a very large number are possible and the cavity has to be designed with only a few resonances having a high Q-factor.

These modes are separated by a frequency $c/2L$ hertz, which is 406 MHz for a laser of length 37 cm. The energy levels of a gas have a finite width, which leads to a broadening of the spectral lines, and for the 632.8 nm line this broadening corresponds to a frequency spread of about 1.5 GHz. Thus output frequencies are possible centred on 4.74×10^{14} Hz, but within a band 1.5 GHz wide which will include about three modes.

This is illustrated in Fig. 5.30, which shows a typical variation of the power gain of the laser as a function of frequency. The gain curve and the spectral line curve coincide, and the bandwidth is taken where the gain has dropped to half its maximum value, which includes three modes in the diagram. Each mode contains a number of resonances corresponding to optical paths between the mirrors, whose lengths satisfy eq. (5.31). The mode separation is increased by reducing L, and using the example given above, if the spacing L is reduced to about 10 cm the mode separation rises to 1.5 GHz, so that only one mode is possible within the spectral linewidth.

The interaction of the various modes causes the output beam to form a characteristic pattern instead of a single spot. However, careful cavity design and the use of a short tube have resulted in lasers being commercially available with only one mode amplified, which will provide a spot source.

Laser applications

A detailed description of the applications of lasers is beyond the scope of this book, but their main features and applications are listed in Table 5.3 and more detailed descriptions are given in Ref. [5]. One of the earliest uses of a ruby laser was in range-finding with a Q-switched pulse, in which the time delay between transmission and reception of a pulse is measured. The ruby laser is also used as a high-intensity source in cutting, drilling and welding operations, including the manufacture of integrated circuits. In medicine a particular application occurs in eye surgery, where a laser beam can be used to re-attach a detached retina. Another application of the laser is holography, in which a three-dimensional photograph of an object can be obtained and for which a coherent source is essential.

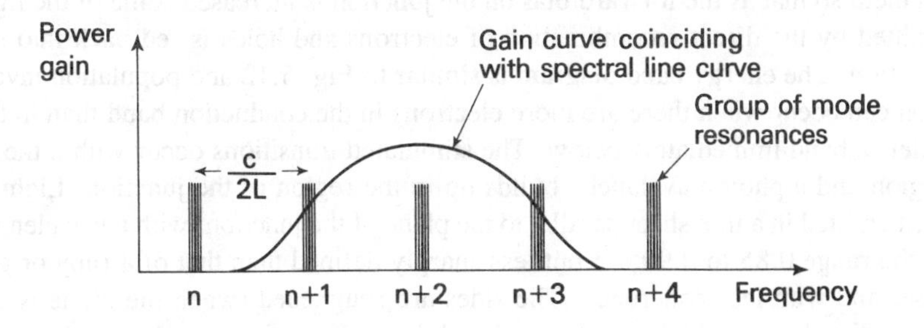

Figure 5.30 Modes of a typical cavity compared with laser power-gain curve.

Table 5.3 Characteristics and applications of some common lasers

Laser	Principal output wavelength	Output	Maximum power	Normal application
	μm		W	
Ruby	0.6943	Pulsed	10^8	Drilling, opthalmology, holography
Nd^{3+}:YAG/glass	1.06	Pulsed CW	10^9 10^3	Drilling, welding, rangefinding, holography
Helium-neon	0.6328 1.1152 3.39	CW CW CW	10^{-1} 10^{-3} 10^{-3}	Surveying, metrology, holography, velocity measurement
Argon	0.458 0.515	CW	10	Opthalmology, holography
Carbon dioxide	9.2–10.8	Pulsed CW	10^7 10^4	Cutting, welding
Nitrogen	0.3371	Pulsed	10^{-2}	Pumping dye lasers
Dye	0.265–0.960	Pulsed	10^8	Spectroscopy, pollution detection
Gallium arsenide	0.85–0.90	Pulsed CW	10 0.01	Communications, rangefinding, pollution
GaAs/GaAlAs	0.85	Pulsed	0.01	Compact discs, video discs, optical printers

The diode laser (injection laser)

The production of a light output from a heavily doped p-n junction has been described in connection with the light-emitting diode. The diode laser, or *injection* laser, is a development of the LED in which a reflecting cavity is formed simply by polishing the ends of the crystal perpendicular to the plane of the p-n junction (Fig. 5.31). These end surfaces reflect about 30% of the light falling on them so that as the forward bias on the junction is increased some of the light emitted by the direct recombination of electrons and holes is fed back into the junction. The energy band diagram is similar to Fig. 5.12 and population inversion can occur when there are more electrons in the conduction band than in the valence band immediately below. The stimulated transitions occur within the p-region and a photon avalanche builds up in the region of the junction. Light is then emitted in a thin sheet parallel to the plane of the junction, with a wavelength in the range 0.85 to 0.90 μm but less sharply defined than that of a ruby or gas laser and with less coherence. The sides are roughened (when the diode is cut out of the slice in which it is formed) and this reduces the loss of optical energy from the structure, known as a *Fabry–Perot cavity*.

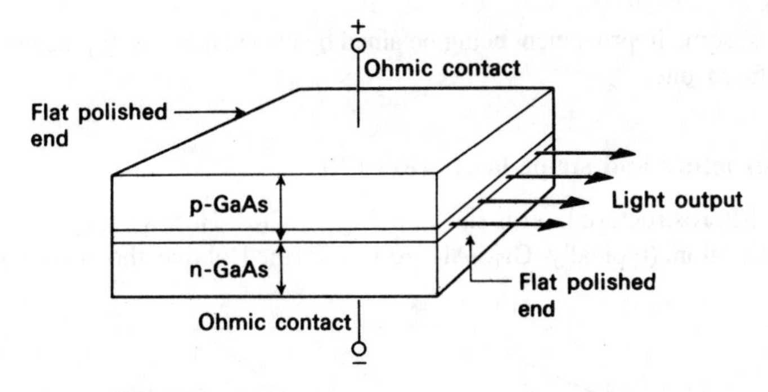

Figure 5.31 Diffused diode laser (not to scale).

Population inversion occurs in an active region a few micrometers thick, near the metallurgical junction, only when the forward current has exceeded a threshold value. At lower currents there is spontaneous emission in all directions and absorption of radiation as in an LED, but the optical gain increases with current due to feedback of light from the mirrors. Lasing occurs at a threshold current density J_t (Fig. 5.33(*b*)) where the gain per unit length g and the loss per unit length α (mainly due to free carrier absorption and scattering) satisfy the expression (Ref. [1]).

$$R \exp (g - \alpha)L = 1$$

or

$$g = \alpha + (1/L) \ln (1/R) \tag{5.32}$$

Here L is the length of the cavity, typically a few hundred micrometers, and R is the reflectivity of the ends of the cavity. $R = \sqrt{(R_1R_2)}$ if the reflectivities R_1 and R_2 at each end are different. g is related to J_t by the expression

$$g = \beta J_t \tag{5.33}$$

where β is the gain factor and substitution in eq. (5.32) leads to

$$\beta J_t = \alpha + (1/L) \ln (1/R)$$

and

$$J_t = \frac{1}{\beta}\left[\alpha + \frac{1}{2L} \ln \frac{1}{R_1R_2}\right] \tag{5.34}$$

α and β depend on the manufacturing technique of the p-n junction, (Ref. [1]) and are modified to reduce the threshold current, which is nearly 10^5 A/cm^2 at room temperature for a simple diffused laser of the type shown in Fig. 5.31. This can be reduced to about 10^3 A/cm^2 by operating at a very low temperature (77 K) but reduction in the room temperature threshold current requires different structures, some improvement being obtained by using an epitaxial junction instead of a diffused one.

Heterostructure and stripe lasers (Ref [7])

A single heterostructure laser is shown in Fig. 5.32(*a*). Here a gallium aluminium arsenide region (typically $Ga_{0.7}Al_{0.3}As$) is formed above the p-type gallium

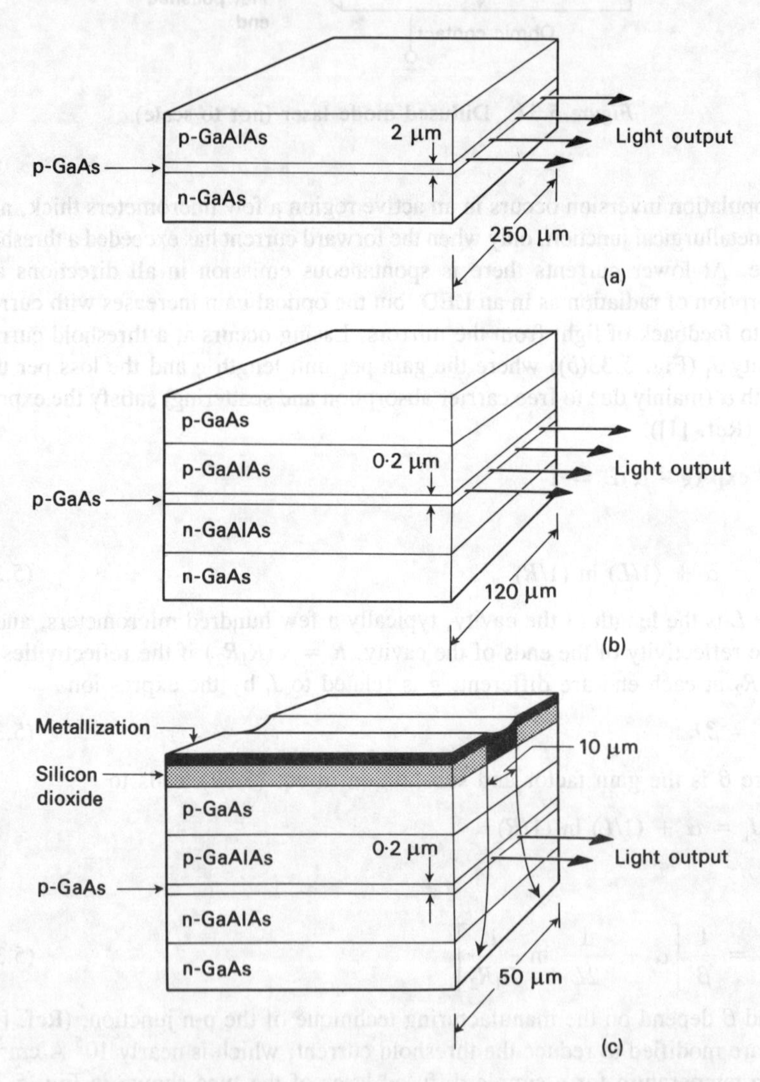

Figure 5.32 Heterojunction diode lasers. (*a*) Single heterostructure; (*b*) double heterostructure; (*c*) double heterostructure (DH) with stripe. Each layer approximately 1 μm thick except where indicated (not to scale).

arsenide region which introduces a discontinuity in the refractive index at the interface. This partially confines the optical flux to the active region and lowers α. A potential barrier is also introduced at the interface which confines the electrons injected across the junction to a narrow region and so increases β. The resultant value of J_t is then about 10^4 A/cm^2 at room temperature or about 10 A in a practical device, which is normally operated in a pulsed mode with a peak output power of about 10 W. The wavelength at peak intensity is around 0.91 μm with a width at 50% intensity of about 4.5 nm (spectral width, *see* Fig. 5.33(*a*)).

A double heterostructure (DH) laser, shown in Fig. 5.32(*b*), has an active region of gallium arsenide bounded on *both* sides by gallium aluminium arsenide regions. This restricts the optical flux to an even narrower region, increases β and reduces J_t further to about 10^3 A/cm^2. A practical device has a threshold current of about 0.5 A and a peak pulsed output power of around 200 mW. The wavelength at peak intensity is about 0.85 μm with a spectral width of about 4.5 nm.

A stripe laser, shown in Fig. 5.32(*c*) again has a double heterostructure but the electron current is confined to a very narrow operating region of about 10 μm compared with over 100 μm in the device shown in Fig. 5.32(*b*). This is achieved by depositing a silicon dioxide insulating layer so that current flows only in the stripe region and the stripe laser is particularly suitable for continuous wave operation, with a practical threshold current in the region of 20 to 100 mA. The device is biased at this current and a modulating current superimposed for communication purposes, giving an output power of about 10 mW. Diode DH lasers are now being applied not only in optical communications systems as described below but also for digital audio discs (compact discs), video discs and optical printers.

Optical communications

Since the frequency of optical radiation is around 10^{14} Hz an extremely high bandwidth should be possible in an optical communication system. However, the available modulation techniques are limited to much lower frequencies, depending as they do on modulating the forward current of an LED or diode laser.

For LEDs the optical output rises linearly with current until the period of the modulating frequency approaches the recombination time of carriers in the active region (Ref. [7]). The maximum frequency is several hundred MHz and the output power is a few mW for about 100 mA of current. The range of wavelengths at peak intensity is about 0.7 to 1.2 μm with a spectral width of 20 to 30 nm, much larger than for a diode laser.

For diode lasers the optical output rises very steeply with current, so that small current variations lead to large amplitude modulations of the output. In double heterostructure stripe lasers, the linearity of the light output-current characteristic above threshold is improved by reducing the stripe width and devices with a stripe only a few μm wide are being investigated. The modulation frequency is limited to several hundred MHz by a strong resonance which occurs between the carrier population inversion and the photons in the optical cavity. The output power is

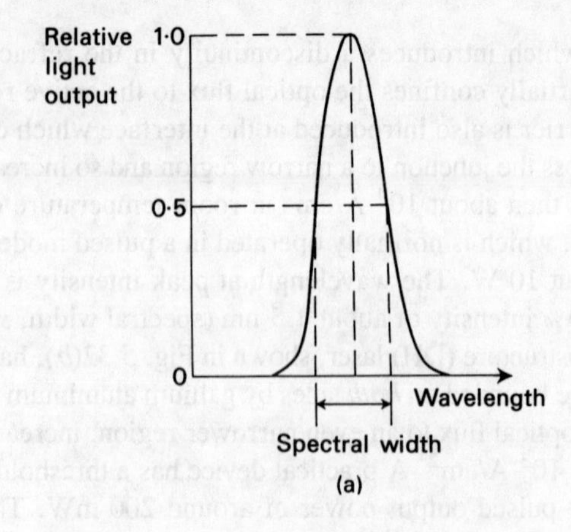

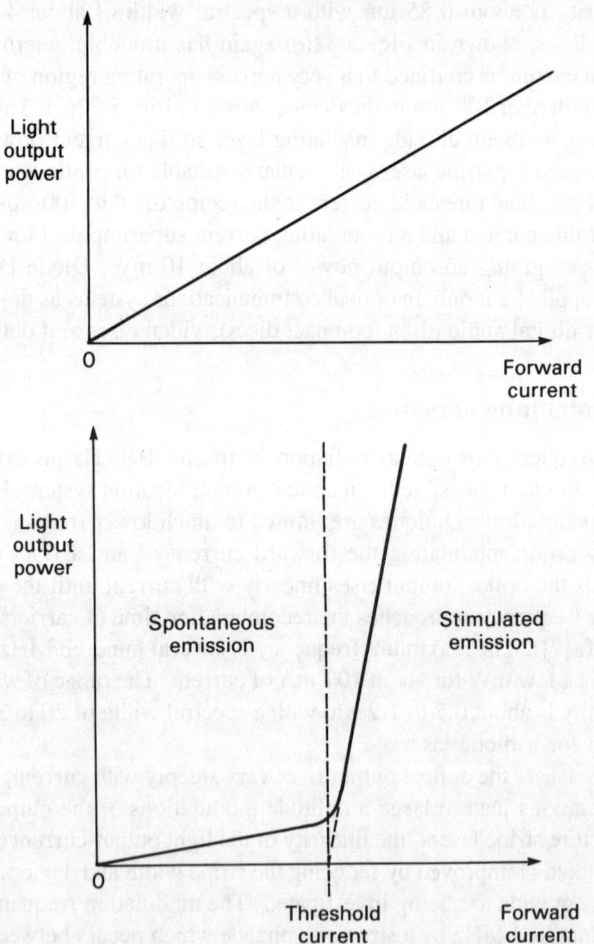

Figure 5.33 Semiconductor light-source properties. (*a*) Spectral response; (*b*) ideal light output–current characteristic for a diode laser; (*c*) ideal light output–current characteristic for an LED.

around 10 mW for about 100 mA of current. Pulsed modulation can be used, preferably with the diode biased near the threshold to avoid oscillations, and rates up to 1000 M bit/s have been attained.

The threshold current of a diode laser increases by about 1% per °C rise in junction temperature, which may cause loss of lasing and a sharp reduction in light output power. For LEDs a rise in junction temperature leads to a fall in light output power of about −1% per °C, which is significant but not critical to the device operation since there is no threshold. The output wavelength is a strong function of temperature, typically rising at 0.4 nm/°C, due to shifts in the band-gap leading to changes in the peak optical gain.

In general, diode lasers with their higher output power and narrow spectral width are preferred for high-frequency transmission over long distances, while the lower power LEDs with a wider spectral width are used for shorter, lower-frequency systems. A lens is required to convert the shape of the output from an edge-emitting device to the circular profile of an optical fibre.

Optical fibres (Ref. [9])

The transmission path for any communications system must have constant properties for reliable operation, which makes the open atmosphere unacceptable for optical communication during adverse weather conditions. An alternative is now available since the development of glass fibres with losses less than 1 dB/km, compared with some thousands of dB/km before 1970.

Three main types of optical fibre are used, all made of high purity fused silica glass and involving multimode or single mode propagation (Fig. 5.34). In the first of these (Fig. 5.34(*a*)) radiation is transmitted through a glass fibre with a diameter between 50 and 300 μm, which is much larger than the transmitted wavelength. This core is clad with a glass of refractive index a few per cent lower with an abrupt boundary between core and cladding. This adds mechanical strength, reduces scattering loss and protects the core from surface contamination. Typical values of refractive index are 1.48 for the core and 1.45 for the cladding which ensures total internal reflection according to eq. (5.13). Total internal reflection confines the radiation to the fibre but a large number of possible paths exist, each with a different transmission time. In pulsed operation this limits the fibre length and repetition frequency that can be used before pulses overlap due to spreading, which is typically 30–50 ns/km.

A single-mode optical fibre has a core diameter of 3 to 8 μm, which represents a few wavelengths (Fig. 5.34(*b*)). Typical values of refractive index are 1.460 for the core and 1.456 for the cladding. Only a single path now exists due to suppression of other modes, while the pulse spreading is around 1 ns/km per per cent relative spectral width. For the lasers described above the spectral width is about 0.5% of the wavelength at peak intensity so a pulse spreading of about 0.5 ns/km could be expected, which becomes 3–4 ns/km for an LED. However, a single mode diode laser source has a relative spectral width around 0.001% so the pulse spreading is about 10^{-3} ns/km, indicating that dispersion is less important than attenuation of the fibre.

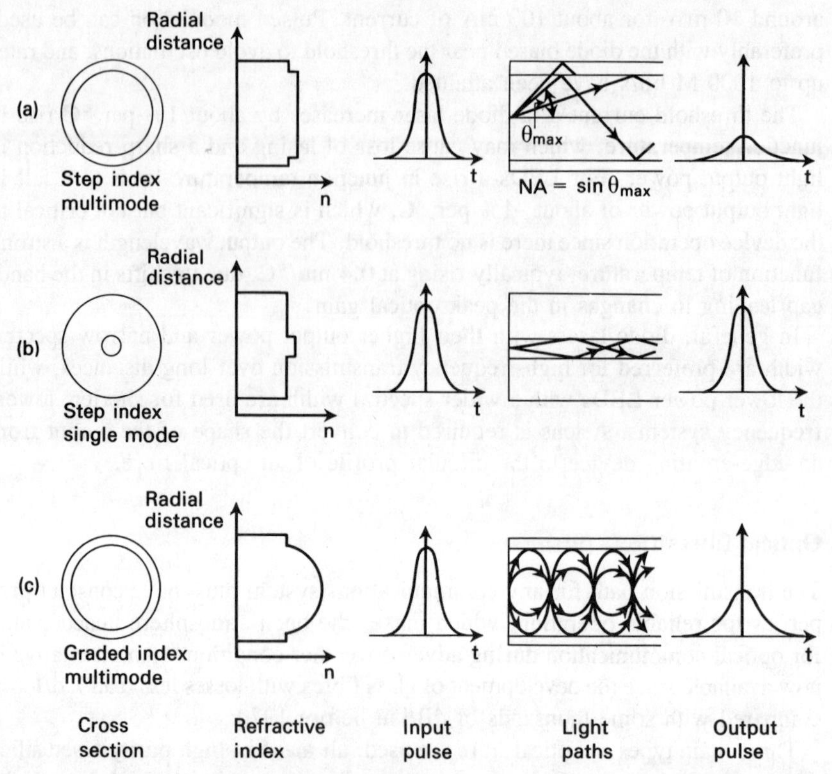

Figure 5.34 Optical fibres (after Ref. [8]). (*a*) Step index multimode; (*b*) step index single mode; (*c*) graded index multimode.

The numerical aperture *NA* is related to the maximum angle at which a light ray will be accepted by the fibre, which is determined by the angle of total internal reflection θ_{max} (Fig. 5.34(*a*)). $NA = \sin \theta_{max}$ for the step index multimode fibre and in general it is in the range 0.13 to 0.50. This is important when considering the coupling of a source to a fibre, since for a low-loss fibre with *NA* = 0.14 the coupling loss with a diode laser is only around 3 dB but for an edge emitting LED the coupling loss is around 14 dB.

The third type of fibre, which is again multimode, has a core with a diameter from 40 to 300 μm and a refractive index that is graded, rather than introducing a step change as in the other fibres (Fig. 5.34(*c*)). A typical maximum value of refractive index for the core is 1.48, with a cladding value of 1.46. A graded index fibre continually refocuses the beam along its length, which results in a higher fibre bandwidth than with a multimode step index fibre. Typically the refractive index varies parabolically, considered from the centre of the core and the diameter is 50 μm.

Plastic is also used for optical fibres in two ways
(1) as the cladding with a glass core, and
(2) as both cladding and core.
Both are multimode and (1) can be step or graded index while (2) is only step index. These fibres have lower cost than all-glass fibres but generally the attenuation is increased and the bandwidth is reduced. The all-plastic fibre has a large diameter which makes interconnection easier. Characteristics and applications of optical fibres are shown in Table 5.4.

Optical fibre systems

A very common requirement in a communications system is for joining or splicing together lengths of optical fibre. Methods for a permanent joint include fusing the two ends with an electric arc, splicing with a glass tube and V-groove splicing. Here the two fibres are mounted in a V-groove cut in a supporting plate with a retaining plate glued over it (Fig. 5.35). This gives a joint with a loss below 0.2 dB and can be extended to multiple fibres. A wide range of demountable connections is also available, which are more difficult to achieve than optical fibre splices. This is because they must allow for repeated connection and disconnection with perfect alignment of the fibres to avoid introducing high attenuation.

Sources and detectors

The attenuation–wavelength characteristic for an ultra-low loss single mode silica-based fibre is shown in Fig. 5.36. This shows that attenuation is at a minimum around 0.8, 1.3 and 1.5 μm which are thus very important wavelengths for optical transmission.

For sources in the 0.8–0.9 μm range the GaAs/AlGaAs DH system on a GaAs substrate is the most developed and is used for both lasers and LEDs. The energy gap is adjusted to cover this range by changing the AlGa composition. For the 1.0–1.7 μm range the InGaAsP/InP system on an InP substrate is the most favourable at present.

At the receiving end of the fibre avalanche photodiodes or p-i-n photodiodes are the preferred types of detector. Both are capable of responding to modulation

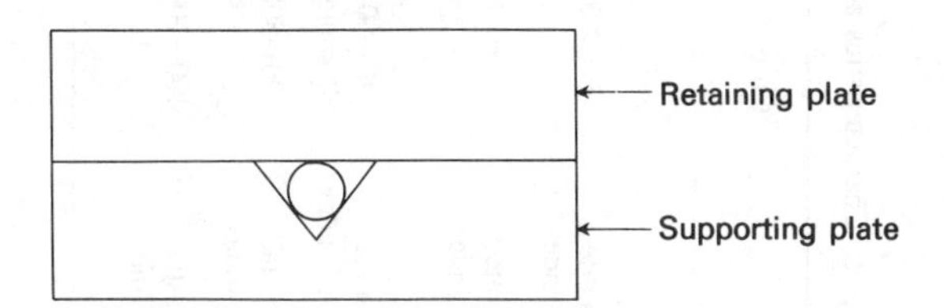

Figure 5.35 V-groove for splicing optical fibres.

Table 5.4 Characteristics and applications of optical fibres (after Ref. [9])

	Core dia.	Cladding dia.	NA	Attenuation	Bandwidth	Application
	μm	μm		dB/km	MHz/km	
Multimode step index	50–400	125–500	0.16–0.5	4–50	Up to 100	Short distance, limited bandwidth, low cost
Single mode step index	3–10	50–125	0.08–0.15	2–5 1 at 0.85 μm 0.6 at 1.3 μm	Up to 40 000	Long distance, high bandwidth, require single-mode diode laser source
Multimode graded index	30–60 50 standard	100–150 125 standard	0.2–0.3	2–10	150–2000	Medium distance, medium/high bandwidth, multimode LED or diode laser source
Plastic clad step index	100–500	300–880	0.2–0.5	5–50	5–25	Short distance, low bandwidth, low cost, relatively easy termination
All plastic step index	200–1000	450–2250	0.5–0.6	350–1000 at 0.65 μm	—	Very short distance, 10s of metres so bandwidth not specified, low cost, easy coupling and termination

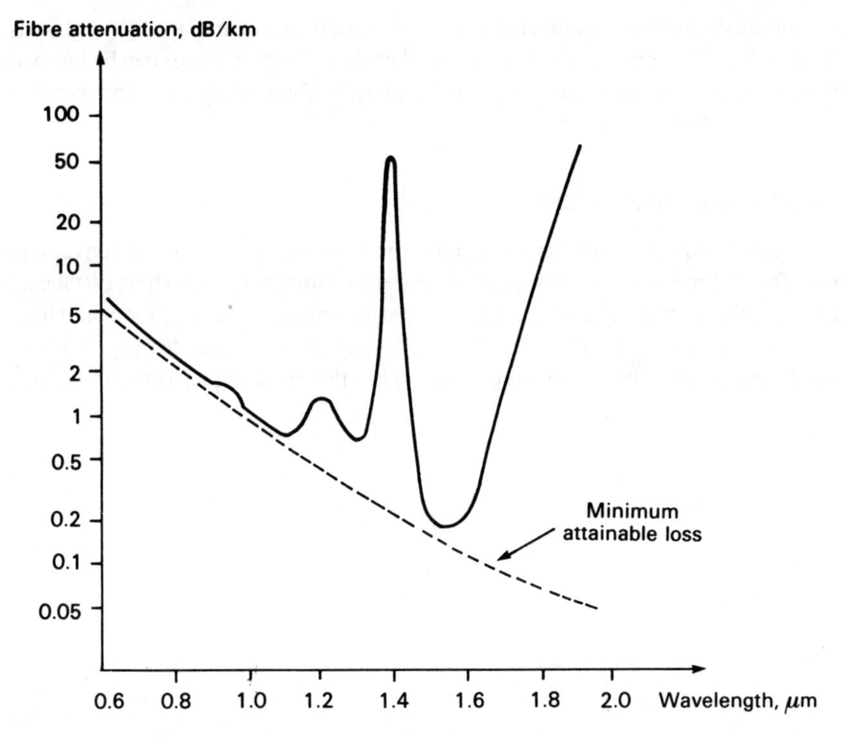

Figure 5.36 Attenuation versus wavelength for an ultra-low loss single-mode fibre based on silica. The peaks are due to absorption by hydroxyl (-OH) ion impurities. The dotted line shows the minimum loss, controlled by Rayleigh scattering (after Ref. [2]).

frequencies in the hundreds of MHz, with the higher internal gain of the avalanche photodiode giving it a 10 dB greater power sensitivity. For detection, germanium can be used, operating as a direct-gap semiconductor with an energy gap of 0.81 eV. This corresponds to a threshold wavelength λ_0 of 1.53 μm so Ge can be used over the whole band of useful transmission wavelengths. However, the relatively large dark current of a Ge photodiode is a major disadvantage and has led to the development of detectors based on III–V alloys. Both InGaAs and InGaAsP can be used over the 1.0–1.7 μm range, with the latter providing photodiodes with fewer crystal defects and lower dark current. Here the energy gap of 1.35 eV (λ_0 0.92 μm) for InP is reduced by adjusting the relative proportions of In and Ga and of As and P.

At present, optical communication systems have the advantage of a greater distance between repeaters than in electrical cable systems. Due to the attenuation and distortion introduced, a repeater is required every 2 km in an electrical system to restore the signal quality, but this distance is increased to 8 to 12 km in a fibre optic system. A further advantage is that there is no interference from external electrical signals, which makes optical fibres particularly suitable for communications along an electrical railway system for example. Again glass is cheaper than the copper used in an electrical cable and optical fibres can be inserted in existing ducts to increase their capacity, 6 to 8 fibres being mounted together to form an optical fibre cable.

Integral optoelectronics

The optical fibre is a circular waveguide for light which is the most convenient form for all long-distance transmission systems. However, over short distances, for example in an integrated circuit, a planar waveguide is more practical where the core is replaced by a thin film. This is deposited on a substrate (Fig. 5.37(*a*)) and the other side may be covered by air. The film has a higher refractive index

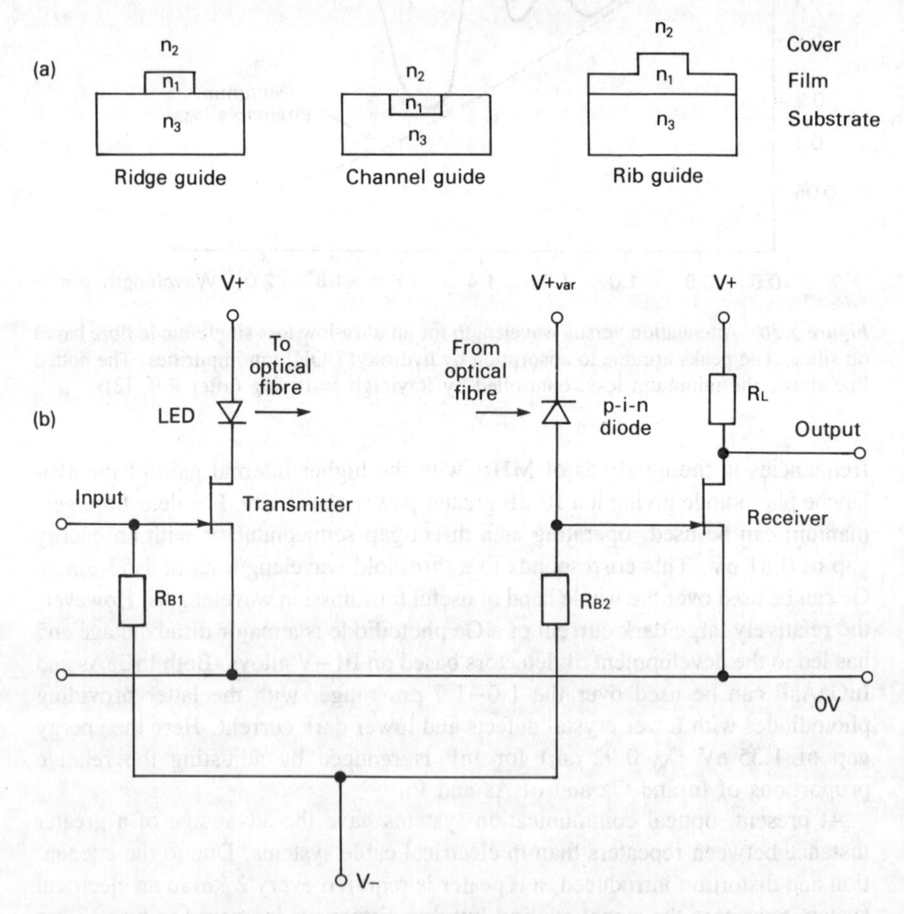

Figure 5.37 (*a*) End views of planar waveguides, $n_1 > n_2 \gg n_3$. Cross-sectional area typically 3–50 μm^2. (*b*) Optoelectronic integrated circuit elements to provide a transmitter, a receiver or a repeater for an optical fibre system.

than either the substrate or the cover, which ensures that the light is confined within the film.

In addition to waveguides other passive components can be produced such as beam splitters and filters. Frequency-selective rejection filters are based on a one-dimensional diffraction grating where light is reflected back at a wavelength dependent on the grating spacing. Active components are also available using electro-optic materials in which the refractive index is changed by an applied electric field, such as lithium niobate. These include directional couplers, switches and bistable devices. A faster alternative to current modulation of a semiconductor injection laser is provided by electro-optic modulators, which can be simple on-off devices or depend on optical phase-shifting. These are all applications of *integrated optics* and their operating speed is much greater than that of integrated circuits due to the absence of parasitic elements. They should allow the exploitation of the very large potential bandwidth of single-mode optical fibres in practical systems.

The integration of optoelectronic devices, optical components and transistors on the same substrate has also been achieved, being based on III–V semiconductors. Optical sources and detectors are readily available in GaAlAs/GaAs on a semi-insulating GaAs substrate. GaAs can be used as the film and GaAlAs as the cover of a planar waveguide, which is coupled to an optical fibre. Typically, an optical transmitter consists of an LED with an FET driver and an optical receiver has a p-i-n diode with an FET amplifier (Fig. 5.37(*b*)). Each is separately coupled through a waveguide on the integrated circuit to an optical fibre, so that the chip could be used as a receiver or a transmitter. An optical repeater is used to restore the signal at intervals along the transmission path. This is formed by connecting the receiver output to the transmitter input, normally through pulse shaping circuits.

The wide range of integrated optical components reviewed above are being actively developed and will become part of the rapidly advancing technology of optical fibre communication systems.

Points to remember

* Light can be absorbed by a semiconductor at wavelengths below a threshold value which is inversely proportional to the energy gap.
* Photodetectors can be in the form of a resistance, a p-n diode, a transistor or a thyristor.
* The solar cell is a photodiode designed to convert light into electrical energy, with many solar cells being grouped into an array in practical converters.

* Light is emitted in display devices at a wavelength inversely proportional to the difference between two electron energy levels.
* In the light-emitting diode or LED the energy gap is controlled by the doping level to produce light of different colours.
* The liquid crystal display LCD operates as an electronically controlled shutter on an external light source.
* Lasers provide a coherent light source at a particular wavelength and may be based on a gas, a solid or a p-n junction.
* In an optical communications system the transmitter is a diode heterostructure laser, joined by a glass fibre to a photodetector as a receiver.
* Optical components are being integrated on to a single substrate, as in the integrated circuit, to provide a means of optical communication in future commercial systems.

References

[1] Sze, S.M. *Physics of semiconductor devices*, 2nd ed., (Wiley), 1981.
[2] Wilson, J. and Hawkes, J.F.B. *Optoelectronics an introduction*, (Prentice-Hall International), 1983.
[3] Bergh, A.A. and Dean, P.J. *Light-emitting diodes*, (Clarendon Press), 1976.
[4] Chandrasekhar, S. *Liquid crystals*, (Cambridge University Press), 1980.
[5] Ready, J.F. *Industrial applications of lasers*, (McGraw-Hill), 1978.
[6] Gooch, C.H. *Injection electroluminescent devices*, (Wiley), 1973.
[7] Selway, P.R., Goodwin, A.R. and Kirby, P.A. 'Semiconductor laser light sources for optical fiber communications' in Sandbank, C.P. (Ed.), *Optical Fiber Communication Systems*, (Wiley), 1980.
[8] Giallorenzi, T.G. 'Optical communications research and technology: fibre optics', Proc. IEEE, 1978, vol. 66, no. 7, pp. 744–80.
[9] Senior, J.M. *Optical fibre communications principles and practice*, (Prentice-Hall International), 1985.

Problems

5.1 (*a*) Explain briefly two mechanisms by which electron-hole pairs are generated in a semiconductor. Define the terms *recombination time* and *threshold wavelength*. What are the conditions for a constant electron-hole pair concentration?

(*b*) Describe the construction and give the characteristics of a photoconductive cell suitable for detecting visible light. The light source illuminating such a cell is suddenly switched off. Deduce the time for the photocurrent to fall to 10% of its initial value, if the recombination time is 250 ms.

(*c*) Cadmium sulphide with an energy gap of 2.42 eV is often used to form a photoconductive cell. Calculate the corresponding threshold wavelength and explain how this could be extended in practice to cover the visible spectrum from about 0.4 μm to 0.7 μm.

[575 ms, 0.51 μm]

5.2 A photodiode has a quantum yield of 5% and uses a semiconductor with an energy gap of 1.1 eV. Estimate the maximum wavelength λ_0 that can be absorbed. If the diode current under reverse bias rises by 200 μA when light of wavelength λ_0 is absorbed, estimate the absorbed light power.
[1.12 μm; 4.4 mW]

5.3 A single solar cell has a short-circuit current of 50 μA at room temperature when the illumination is 800 lux. The dark current is 8 nA and a minimum open circuit voltage of 1.4 V is required from an array of cells, when the illumination is 100 lux. How many cells are required and how should they be connected?
[9 cells, connected in series]

5.4 A solar cell with a dark current of 5 nA at room temperature delivers a short-circuit current of 15 mA when exposed to a certain illumination. A solar power supply is required with 5 V open-circuit voltage and 2 A short-circuit current when the illumination level is increased by 100%. How many solar cells would be needed and how would they be connected?
[871, 13 in series $\times$ 67 in parallel]

5.5 Sketch those elements of a seven-segment display which are illuminated for each of the numerals from 0 to 9. How many letters of the alphabet can be displayed?

5.6 The optical spacing between the mirrors of a gas laser is 37 cm and the bandwidth of their spectral gain is 1.5 GHz. Determine (*a*) the frequency separation of the axial modes, and (*b*) the number of modes in the laser.
[(*a*) 406 MHz; (*b*) 3]

5.7 The spacing between the mirrors of a ruby laser is 4.0 cm. Estimate the frequency spacing of the axial modes, given the velocity of light in ruby is 1.7×10^8 m/s.
[2.12 GHz]

5.8 A double heterojunction laser has the following properties at room temperature: cavity length 500 μm, cavity width 200 μm, mirror reflectivity 0.25, threshold current 2 A, loss factor 20 cm^{-1}. Estimate the gain factor β.
[24×10^{-3} cm/A]

6 Field-effect transistors and charge transfer devices

The junction-gate field-effect transistor

The field-effect transistor, or FET, was proposed by Shockley in 1952, but it was not possible to manufacture it in large numbers until semiconductor techniques were sufficiently advanced. It has been commercially available since 1960 and has unique properties, some of which are complementary to those of a bipolar transistor, such as a very high input impedance.

One form of the device, known as a junction-gate field-effect transistor (JFET), normally consists of a bar of n-type silicon with an ohmic contact at each end known as the *source* and the *drain* respectively. The bar is enclosed for part of its length by heavily doped p-type silicon, known as the *gate*, so that a p-n junction is formed along it (Fig. 6.1(*a*)). If the voltage between gate and source, V_{GS}, is initially zero and the drain-to-source voltage, V_{DS}, is made positive, an electron current I_D will flow from source to drain, whose magnitude depends on the effective resistance of the bar (Fig. 6.2(*a*)). The junction is subjected to a reverse bias which increases from zero at the source to a maximum at the drain, owing to the voltage drop along the bar. Consequently depletion layers whose width increases with the reverse bias extend into the n-region, since its resistivity is higher than that of the p-region (Chapter 3) and I_D flows through the tapering channel between the depletion layers. This is called an *n-channel* device; the opposite arrangement with a p-type bar and an n-type gate would be a *p-channel* device. A practical form is shown in Fig. 6.1(*b*) with a single gate and could be manufactured by the processes used for planar transistors (Chapter 4).

As V_{DS} is increased the channel becomes narrower and the resistance between source and drain increases. Eventually, when V_{DS} reaches a critical value V_P, known as the *pinch-off voltage* the two depletion layers almost meet at the drain and the channel is said to be *pinched off*. Thus I_D increases non-linearly with V_{DS} until pinch-off has been reached (Fig. 6.2(*b*)). At all values of V_{DS} the value of I_D at any point in the channel may be obtained by combining eqs (2.60) and (2.49), so that

$$I_D = N_d e \mu E_D S \tag{6.1}$$

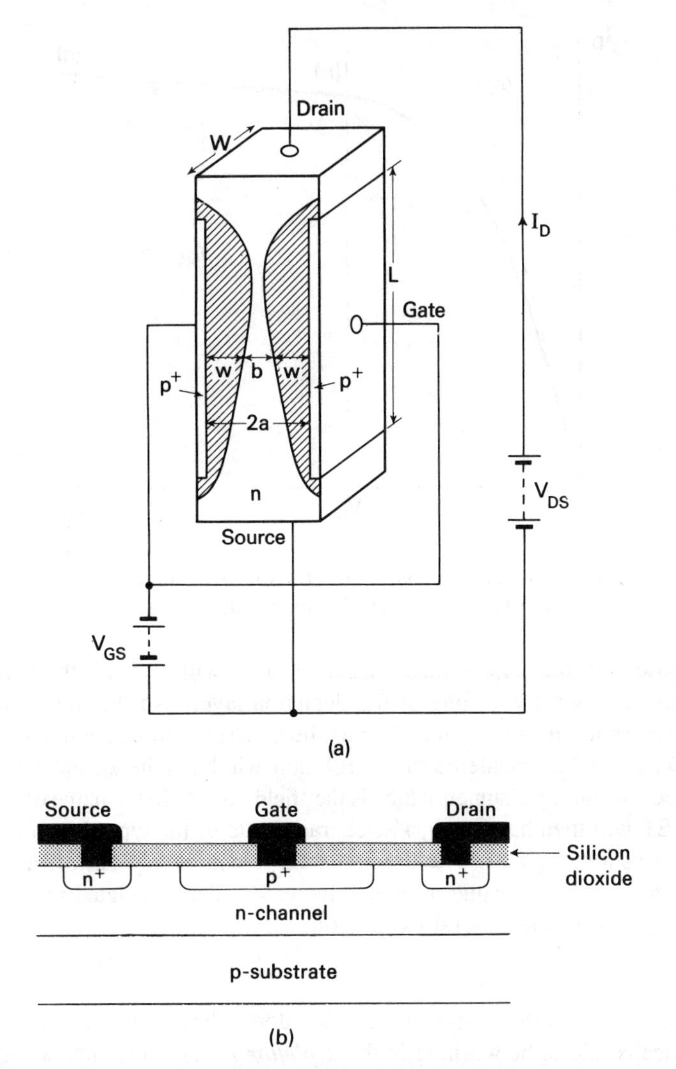

Figure 6.1 Junction-gate field-effect transistor. (*a*) Diagram and applied voltages; (*b*) practical JFET.

where E_D is the field accelerating electrons from source to drain. At pinch-off the cross-sectional area of the channel, S, is very small near the drain, so that E_D becomes very large to maintain the flow of current and thus prevents the depletion layers quite meeting each other. The current *density* is very high at this point and the electrons approach their maximum drift velocity (Chapter 2) as they shoot through the very narrow gap. Thus for values of V_{DS} above pinch-off the drain current is almost independent of voltage (Fig. 6.2(*c*)).

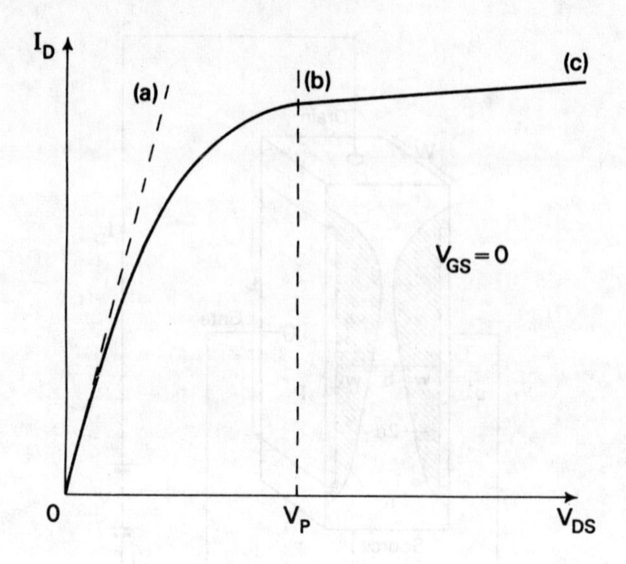

Figure 6.2 Drain-current/voltage characteristic of a field-effect transistor, with zero gate voltage. (*a*) Ohmic region; (*b*) pinch-off; (*c*) saturation region.

Suppose now that V_{GS} is made negative. This will increase the reverse bias and cause a general widening of the depletion layers, so that pinch-off occurs at a lower value of drain voltage. Thus the electric field in the depletion layers due to V_{GS} may be considered to control their width and hence the current flowing in the conducting channel, which is the 'field effect', in the name of the device. The JFET will then have an I_D/V_{DS} characteristic of the type shown in Fig. 6.3 with V_{GS} as a parameter. The drain current is completely cut off at any drain voltage for a negative value of gate voltage, V_P. This is related to the gate and drain voltages by the general expression

$$V_{DS} = V_{GS} + V_P \tag{6.2}$$

which is shown on the characteristics as a dotted line. When biased negatively the device is said to be working in the *depletion mode*, since increasing the bias depletes the channel of charge carriers. The input impedance is then virtually that of a reverse-biased silicon p-n junction, which is about $10^{10}\ \Omega$ at room temperature and far higher than for a conventional transistor. If the bias were made positive, the junction would be forward biased and the flow of carriers enhanced since the channel would become wider. The input impedance in this *enhancement mode* becomes that of a forward-biased p-n junction, r_e, which is very low so that operation with positive values of V_{GS} is not normally recommended.

The current through a field-effect transistor consists only of majority carriers and so it is a *unipolar* device, while in bipolar transistors minority carriers diffuse across the base and their space charge is neutralized by an equal number

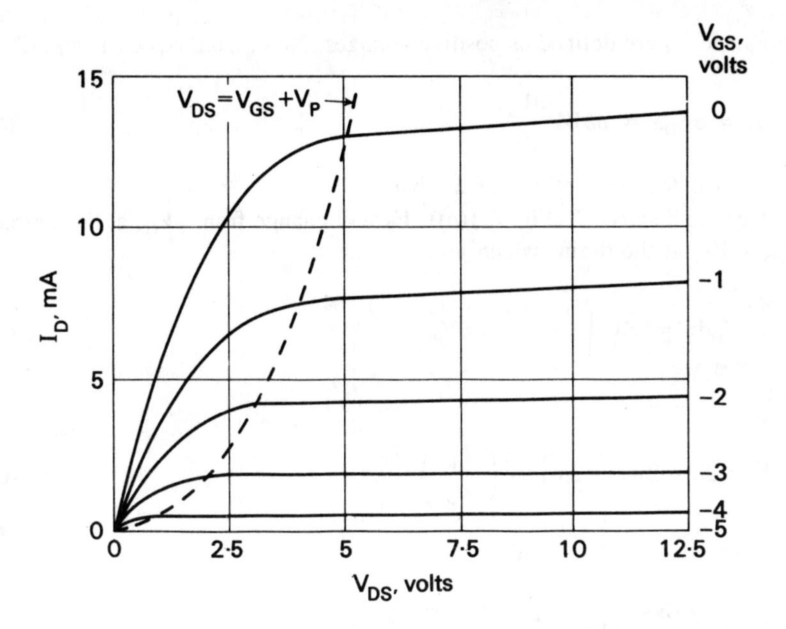

Figure 6.3 Output characteristics of an n-channel JFET for $V_P = 5$ V.

of majority carriers. Also the leakage-current effects present in bipolar transistors and discussed in Chapter 4, are not met in field-effect transistors, which are preferred in applications where a high input impedance is required.

Theoretical Characteristics of a JFET

An approximate expression for the characteristics shown in Fig. 6.3 may be obtained by considering the device shown in Fig. 6.1(*a*). The width w of each depletion layer is given approximately by eq. (3.74) for a p^+-n diode, so that

$$w = \left(\frac{2\epsilon V_R}{eN_d}\right)^{1/2} \tag{6.3}$$

V_R is the junction reverse bias voltage which is a function of distance x along the bar, and reaches a maximum value of $V_{DS} - V_{GS}$ at $x = L$. Under pinch-off conditions $w \approx a$, so that, from eq. (6.3),

$$V_P = \frac{eN_d a^2}{2\epsilon} \tag{6.4}$$

and

$$V_R = V_P \frac{w^2}{a^2} \tag{6.5}$$

so V_P and V_R are defined as positive voltages. Now, from eqs (6.1) and (2.66),

$$I_D = \sigma E_D S = \sigma b W \frac{dV_R}{dx} \tag{6.6}$$

where σ is the conductivity of the bar.

Over the distance L (Fig. 6.1(*a*)), V_R will change from $-V_{GS}$ at the source to $V_{DS} - V_{GS}$ at the drain, which gives

$$\int_0^L I_D dx = \sigma W \int_{-V_{GS}}^{V_{DS} - V_{GS}} b \, dV_R \tag{6.7}$$

But

$$b = 2a - 2w = 2a \left[1 - \left(\frac{V_R}{V_P} \right)^{1/2} \right] \tag{6.8}$$

using eq. (6.5), which leads to

$$I_D = \frac{2aW\sigma}{L} \left[V_R - \frac{2}{3} \frac{V_R^{3/2}}{V_P^{1/2}} \right]_{-V_{GS}}^{V_{DS} - V_{GS}} \tag{6.9}$$

after integration

$$I_D = \frac{2aW\sigma}{L} \left[V_{DS} - \frac{2}{3} \frac{(V_{DS} - V_{GS})^{3/2}}{V_P^{1/2}} + \frac{2}{3} \frac{(-V_{GS})^{3/2}}{V_P^{1/2}} \right] \tag{6.10}$$

which is an approximate expression for the characteristics *below* pinch-off.

In amplifier applications the JFET is normally operated in the *saturation* region *above* pinch-off. An expression for the saturation current, I_{Dsat}, is obtained by putting $V_{DS} - V_{GS} = V_P$ in eq. (6.10), which gives

$$I_{Dsat} = \frac{2aW\sigma}{L} \left[\frac{V_P}{3} + V_{GS} + \frac{2}{3} \frac{(-V_{GS})^{3/2}}{V_P^{1/2}} \right] \tag{6.11}$$

This expression is independent of V_{DS}, and when $-V_{GS} = V_P$, $I_{Dsat} = 0$, confirming that the drain current is cut off for a gate voltage more negative than the pinch-off voltage. A useful relationship between saturation current and gate voltage is obtained by considering the value of I_D for $V_{GS} = 0$. This is the current I_{DSS}, given by

$$I_{DSS} = \frac{2aW\sigma}{L} \frac{V_P}{3} = \frac{2aV_P Ne\mu}{3} \frac{W}{L} \tag{6.12}$$

obtained from eq. (6.11), and is a constant since V_P depends on the electrical properties of the bar (eq. (6.4)). Substitution into eq. (6.11) gives

$$I_{\text{Dsat}} = I_{\text{DSS}}\left[1+3\frac{V_{\text{GS}}}{V_{\text{P}}}+2\left(\frac{-V_{\text{GS}}}{V_{\text{P}}}\right)^{3/2}\right] \tag{6.13}$$

The I_{Dsat} vs V_{GS} curve is the transfer characteristic of the device, relating input current for a device with alloyed junctions. If diffused junctions are formed, as in the planar JFET illustrated in Fig. 6.1(*b*), it is found experimentally (Ref. [1]) that the transfer characteristic obeys the equation

$$I_{\text{Dsat}} = I_{\text{DSS}}\left(1+\frac{V_{\text{GS}}}{V_{\text{P}}}\right)^2 \tag{6.14}$$

Equations (6.13) and (6.14) are compared in Fig. 6.4 the average difference being about 10%, which suggests that a square law is quite a good approximation to eq. (6.13) also. A typical value of I_{DSS} is 13 mA at $V_{\text{DS}} = 15$ V and a typical value of V_{P} is 5 V. Similar considerations apply to the p-channel JFET but with V_{DS} negative, V_{GS} positive and V_{P} defined as negative.

The FET is used as an amplifier by connecting a load resistance in series with the drain, just as the conventional transistor amplifier has a collector load resistance. A similar load-line construction determines the operating point, while the input parameter is V_{GS} and not input current. For small-signal operation an equivalent circuit is required, and again the mutual conductance, g_{m}, is defined from the slope of the imput characteristic, eq. (6.14). Then

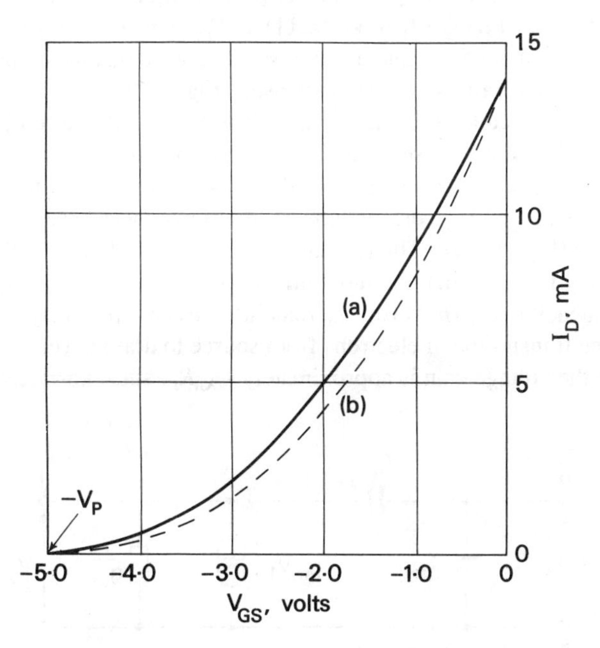

Figure 6.4 Input characteristics of JFETs. (*a*) Diffused junction, square law; (*b*) alloyed junction, eq. (6.13).

$$\left.\frac{\partial I_{\text{Dsat}}}{\partial V_{\text{GS}}}\right|_{V_{\text{DS const.}}} = g_{\text{m}} = \frac{2I_{\text{DSS}}}{V_{\text{P}}}\left(1+\frac{V_{\text{GS}}}{V_{\text{P}}}\right) \tag{6.15}$$

or

$$g_{\text{m}} = \frac{2}{V_{\text{P}}}\left(I_{\text{Dsat}}I_{\text{DSS}}\right)^{1/2} \tag{6.16}$$

g_{m} is not constant, but rises linearly with V_{GS} and is proportional to $I_{\text{D sat}}^{1/2}$, I_{DSS} and V_{P} being constants depending on the physical properties of the device (eqs (6.12) and (6.4)). It may be noted that the value of g_{m} at $V_{\text{GS}} = 0$ is

$$g_{\text{m0}} = \frac{2I_{\text{DSS}}}{V_{\text{P}}} = \frac{4}{3}\frac{aW\sigma}{L} = \frac{4aNe\mu}{3}\frac{W}{L} \tag{6.17}$$

using eq. (6.12), so that g_{m0} has a value $1\frac{1}{3}$ times the conductance of the bar, known as the *open-channel* conductance. Theoretically the output characteristics are horizontal straight lines, since eq. (6.11) is independent of V_{DS}, but in practice I_{Dsat} rises slowly with V_{DS} and the output conductance g_{ds} is given by

$$\left.\frac{\partial I_{\text{Dsat}}}{\partial V_{\text{DS}}}\right|_{V_{\text{GS const.}}} = g_{\text{ds}} \tag{6.18}$$

This is due to a progressive lengthening of the pinched-off region, which decreases the effective channel length L in eq. (6.11) as V_{DS} is increased. Practical values of g_{m} and g_{ds} are about 5 mS and 50 μS respectively, so that the output resistance is about 20 kΩ and falls as I_{Dsat} is increased (Fig. 6.3).

An equivalent circuit for the device is shown in Fig. 6.5. Here C_{gs} represents the capacitance of the reverse-biased gate-to-channel diode close to the source, while C_{gd} represents the capacitance of the same diode close to the drain. Thus C_{gs} is larger than C_{gd} since the depletion layer is narrower near the source than near the drain (Fig. 6.1(*a*)). The resistance of the reverse-biased diode is omitted since it is very large, so that the input impedance of the device is capacitive and falls as frequency rises. However, g_{m} remains constant up to very high frequencies since the transit time of electrons from source to drain is very short. At low frequencies the voltage gain is approximately $-g_{\text{m}}R_{\text{L}}$, since normally $R_{\text{L}} \ll 1/g_{\text{ds}}$.

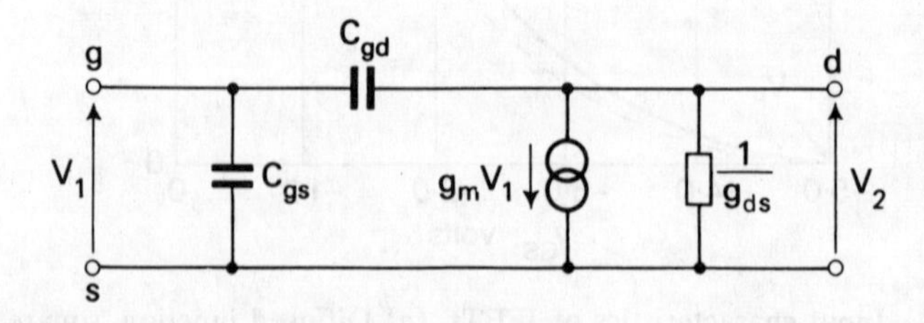

Figure 6.5 **Small-signal equivalent circuit of a field-effect transistor.**

The equivalent circuit of Fig. 6.5 is very similar to the hybrid-π equivalent circuit of a bipolar transistor (Fig. 4.31). Its analysis is simplified by the transformation of C_{gd} to a capacitance $C_{gd}(1+A)$ in parallel with C_{gs} by a method similar to that considered previously. Also a unity gain-bandwidth product may be defined, similar to eq. (4.142), since at very high frequencies the short-circuit current gain of a JFET will also be inversely proportional to frequency. The gain-bandwidth product f_T is then obtained by equating the current through C_{gs} to the short-circuit output current $g_m V_1$ which leads to

$$f_T = \frac{g_m}{2\pi C_{gs}}$$

Short channel effects in JFETs

The JFET static characteristics defined by eqs (6.13) and (6.14) depend on the ratios a/L and W/L. They also assume a constant mobility μ and a uniform charge distribution. For a given W/L ratio a long channel device may be defined as one having $a/L \ll 1$ or typically $L \geq 10a$. However, as L is reduced the transit time of carriers between source and drain falls, which is important for a high-frequency device. Ultimately, the value of L approaches that of a, which leads to a short-channel JFET in which the electric field between drain and source is higher than that of a long-channel JFET.

Instead of the mobility being constant it now becomes field-dependent, as discussed in Chapter 2 for silicon and in Chapter 8 for gallium arsenide. In each case the drift velocity finally becomes field-independent at a constant saturation level u_s of about 10^5 m/s. This occurs at a field of about 10^7 V/m for Si and 5×10^6 V/m for GaAs. Additionally in GaAs the drift velocity reaches a peak of about 2×10^5 m/s at about 3×10^5 V/m before falling to the saturation velocity u/s.

The effect of field-dependent mobility on the static characteristics is to reduce the drain current for given gate and drain voltages, as compared with the constant mobility equation (6.13). Also, when the drift velocity is saturated under the gate, the doping profile of the gate diode has a large effect on the transfer characteristic, I_D vs V_{GS}. It is found that a graded junction leads to a linear characteristic in this case, while for constant mobility the doping profile has negligible effect (Ref. [3]).

High-frequency operation

At high frequencies the finite time for carriers to travel from source to drain, the transit time τ, becomes a limiting factor. For the constant mobility case the transit time is given by

$$\tau = \frac{L}{u} = \frac{L}{\mu E_D} \simeq \frac{L^2}{\mu V_{DS}} \tag{6.19}$$

where E_D is the field between drain and source. For the saturated velocity case we have

$$\tau = \frac{L}{u_s} \qquad (6.20)$$

and in each case τ falls as L is reduced. Typical values of τ lie between about 10 and 1000 ps.

In general, eq. (6.19) is applicable to long JFETs, with $L \geq 20$ μm, say, and eq. (6.20) is applicable to short JFETs with L at a few μm. In silicon devices the mobility gradually falls as L is reduced and E_D increases, until the velocity saturates at u_s. In gallium arsenide devices, where the u/E_D characteristic is more complicated, the mobility is considered to be constant in the source region and the velocity to be constant in the drain region. Since electrons have a higher mobility than holes in both Si and GaAs only n-channel JFETs are used at high frequencies.

A cut-off frequency f_T can be obtained in terms of τ in a similar manner to the bipolar transistor, which leads to $f_T = 1/2\pi\tau$. This f_T is expected to be higher in GaAs than in Si, since the low-field mobility is higher in GaAs.

As gate length L is reduced τ falls and f_T rises. However, L/a must be greater than 1 to ensure that the gate can control carrier flow adequately, so the channel depth a must also be reduced. For a given drain current this is compensated for by a higher doping level (eq. (6.12)) which is limited to about 5×10^{23} atoms/m^3 to avoid breakdown. In turn, this leads to a minimum gate length of about 0.1 μm and a maximum f_T near 100 GHz.

At the present time, in microwave JFETs L lies between 1 and 0.5 μm giving a maximum f_T around 30 GHz (Ref. [2]). A gallium arsenide or cadmium phosphide substrate is used with an aluminium gate deposited directly on it, to form a Schottky diode. This is a MESFET (MEtal Semiconductor FET) which is further discussed in Chapter 9.

Worked example

A JFET has length $L = 20$ μm and is operated at $V_{DS} = 5$ V. Confirm that the carriers have constant mobility for a device manufactured in either Si or GaAs.

Calculate the transit time and gain–bandwidth product for devices in both of these materials. Also obtain these properties for a GaAs FET with $L = 1$ μm.

Solution

$$\text{Field } E_D = \frac{V_{DS}}{L} = \frac{5}{20 \times 10^{-6}} = 2.5 \times 10^5 \text{ V/m}$$

This is well below the saturation field for Si and below the peak field for GaAs, so the device will operate with carriers of constant mobility.

Transit time

$$\tau = \frac{L^2}{\mu V}$$

and

$$\mu = 0.15 \text{ m}^2/\text{V s for Si so } \tau = \frac{(20 \times 10^{-6})^2}{0.15 \times 5}$$

$$= 533 \text{ ps}$$

$$f_T = \frac{1}{2\pi\tau} = \frac{1}{2 \times \pi \times 533 \times 10^{-12}}$$

$$= 0.30 \text{ GHz}$$

$$\mu = 0.85 \text{ m}^2/\text{V s for GaAs}$$

so

$$\tau = \frac{(20 \times 10^{-6})^2}{0.85 \times 5} = 94.1 \text{ ps}$$

$$f_T = \frac{1}{2 \times \pi \times 94.1 \times 10^{-12}}$$

$$= 1.69 \text{ GHz}$$

For a device with $L = 1$ μm, $E_D = 5 \times 10^6$ V/m and the drift velocity is saturated with a value of 10^5 m/s, approximately.

$$\tau = \frac{u_s}{L} = \frac{10^5}{10^{-6}} = 10^{-11} = 10 \text{ ps}$$

$$f_T = \frac{1}{2 \times \pi \times 10^{-11}} = 15.9 \text{ GHz}$$

JFET operation at low drain voltage

In addition to its application as a high-impedance wide-band amplifier, the FET is very useful as a switch. Since it is basically a device whose resistance is controlled by the gate voltage, all the characteristics pass through the point $I_D = 0$, $V_{DS} = 0$, so that the offset voltage is negligible compared with 0.1–2 mV for the conventional transistor. Another unique property is that for positive or negative values of V_{DS} less than about 0.1 V, I_D varies linearly with V_{DS} (Fig. 6.6).

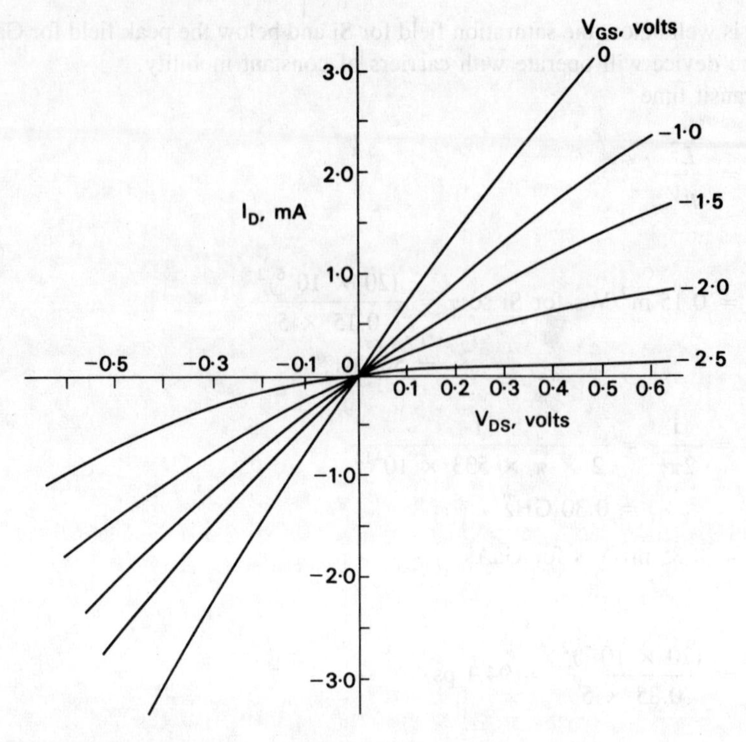

Figure 6.6 I_D–V_{DS} characteristics of a JFET near the origin.

Such low values of drain voltage have little effect on the widths of the depletion layers, which increase only with increasing negative gate voltage and so lead to an increase of drain-to-source voltage. The device operates as a voltage-dependent resistor (VDR), with a typical variation from 150 Ω to 450 Ω as V_{GS} is changed from 0 V to –2.0 V. The range of V_{DS} over which I_D rises linearly may be increased to about ± 1.0 V using negative feedback. The VDR can be used, for example, in a voltage-controlled filter or in an amplifier with voltage-controlled gain.

Temperature and voltage limitations

One effect of temperature may be deduced from eq. (6.12) which shows that I_{DSS} is proportional to conductivity σ. This falls as temperature rises due to the fall in mobility (eq. (2.55)). Thus at a given gate voltage I_{Dsat} also falls as temperature rises (eq. (6.14)), typically at –25 μA/°C at 20 °C. In addition, since the gate current is the saturation current of a p-n junction this will increase rapidly with temperature (eq. (3.45)) and reduce the input resistance of the JFET.

If the drain voltage is increased sufficiently a value will be reached at which drain current increases sharply. This is due to breakdown of the drain-gate junction and the combined effect of V_{DS} and V_{GS}. If V_{DSO} is the breakdown voltage when V_{GS} = 0 V then at any other gate voltage breakdown occurs at V_{DSO} + V_{GS}, which will always be less than V_{DSO} since V_{GS} carries an opposite sign. Thus, for example, if V_{DSO} = 28 V and V_{GS} = –3 V breakdown would occur at V_{DS} = 25 V.

The metal-oxide-silicon field-effect transistor

In another form of field-effect transistor the gate junction is replaced by an insulating layer and metallization. Owing to the method of construction it is commonly called the metal-oxide-semiconductor field-effect transistor (MOSFET), the MOS transistor (MOST), or the insulated gate FET (IGFET).

n-channel enhancement mode MOST

One form is illustrated in Fig. 6.7(*a*), where the substrate material is lightly doped, p-type silicon, nearly intrinsic and therefore of high resistivity. This is coated with a layer of silicon dioxide, which is an insulator, and then two heavily doped n-regions are diffused into the substrate through holes etched in the layer. Aluminium ohmic contacts are made to each of the n^+-regions, again called the source and the drain respectively. Finally, a metal film, also of aluminium, is deposited on the oxide between source and drain to produce the gate.

The gate and substrate now form a parallel-plate capacitor, with the silicon dioxide as the dielectric, so that a charge of one sign on the gate electrode induces a charge of the opposite sign on the substrate. Thus for a negative gate voltage V_{GS} holes will be attracted to the surface of the substrate, which becomes more strongly p-type due to the *accumulation* of holes. Since V_{DS}, the drain voltage, is positive as before, two reverse-biased p-n junctions are formed in series between source and drain and no drain current can flow. However, for a positive gate voltage, a depletion region is formed because holes in the p-type substrate are repelled from the interface region between substrate and oxide. At a positive *threshold voltage* V_T, sufficient depletion has occurred to allow any additional gate voltage to attract free electrons to the interface to form an n-type *inversion layer*. Thus a conducting channel is formed between the two n^+-regions and drain current flows, both the thickness of the layer and the current increasing with the voltage $V_{GS} - V_T$. Since the channel is conductive only when the gate voltage is positive, this is an enhancement mode MOST, with the characteristics shown in Figs 6.8(*b*).

The drain current does not suddenly begin to flow when $V_{GS} = V_T$ as there is *weak inversion* for V_{GS} just below V_T. This corresponds to the electron concentration in the inversion layer being equal to the hole concentration in the substrate, which allows a small, subthreshold drain current to flow. When $V_{GS} \geq V_T$ there is *strong inversion* and the drain current rises with $V_{GS} - V_T$ as shown

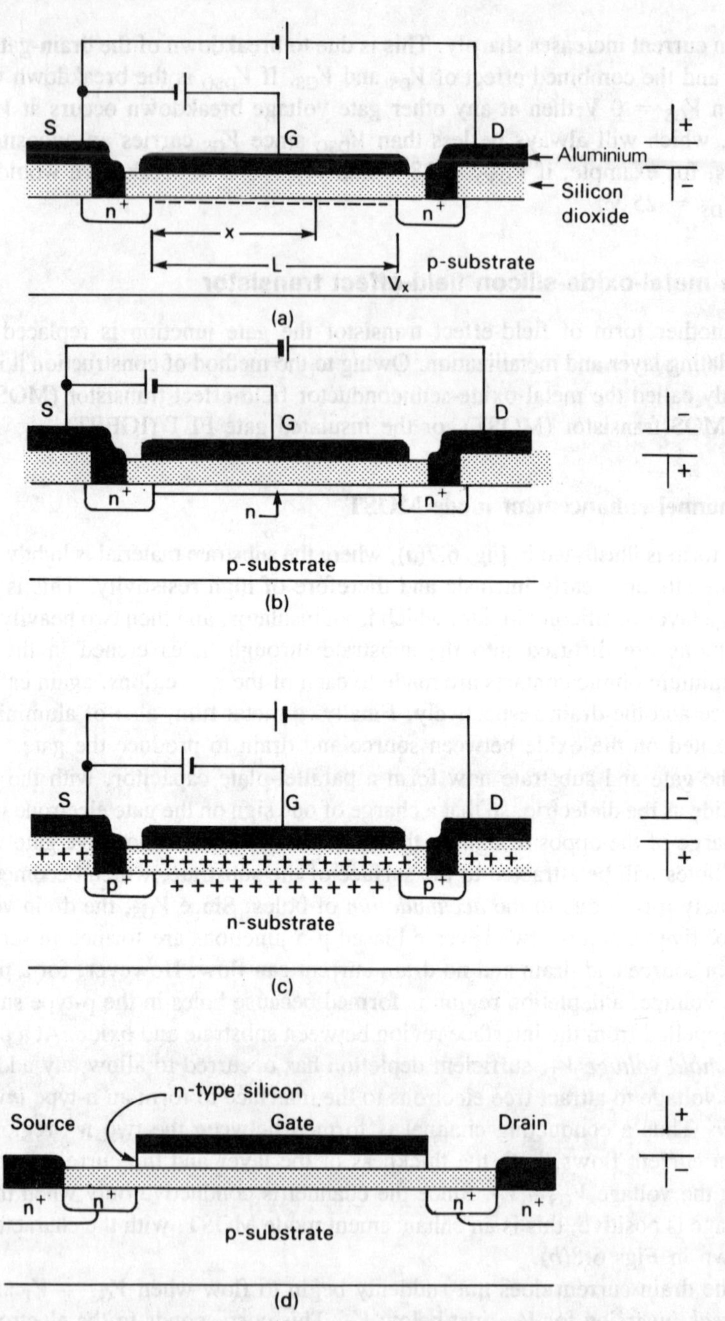

Figure 6.7 Metal-oxide-silicon field-effect transistor with MOS capacitor showing gate voltage applied to top plate and charge induced in the channel on the bottom plate. (*a*) n-channel enhancement mode MOST; (*b*) n-channel depletion mode MOST; (*c*) p-channel enhancement mode MOST; (*d*) n-channel enhancement mode silicon gate MOST.

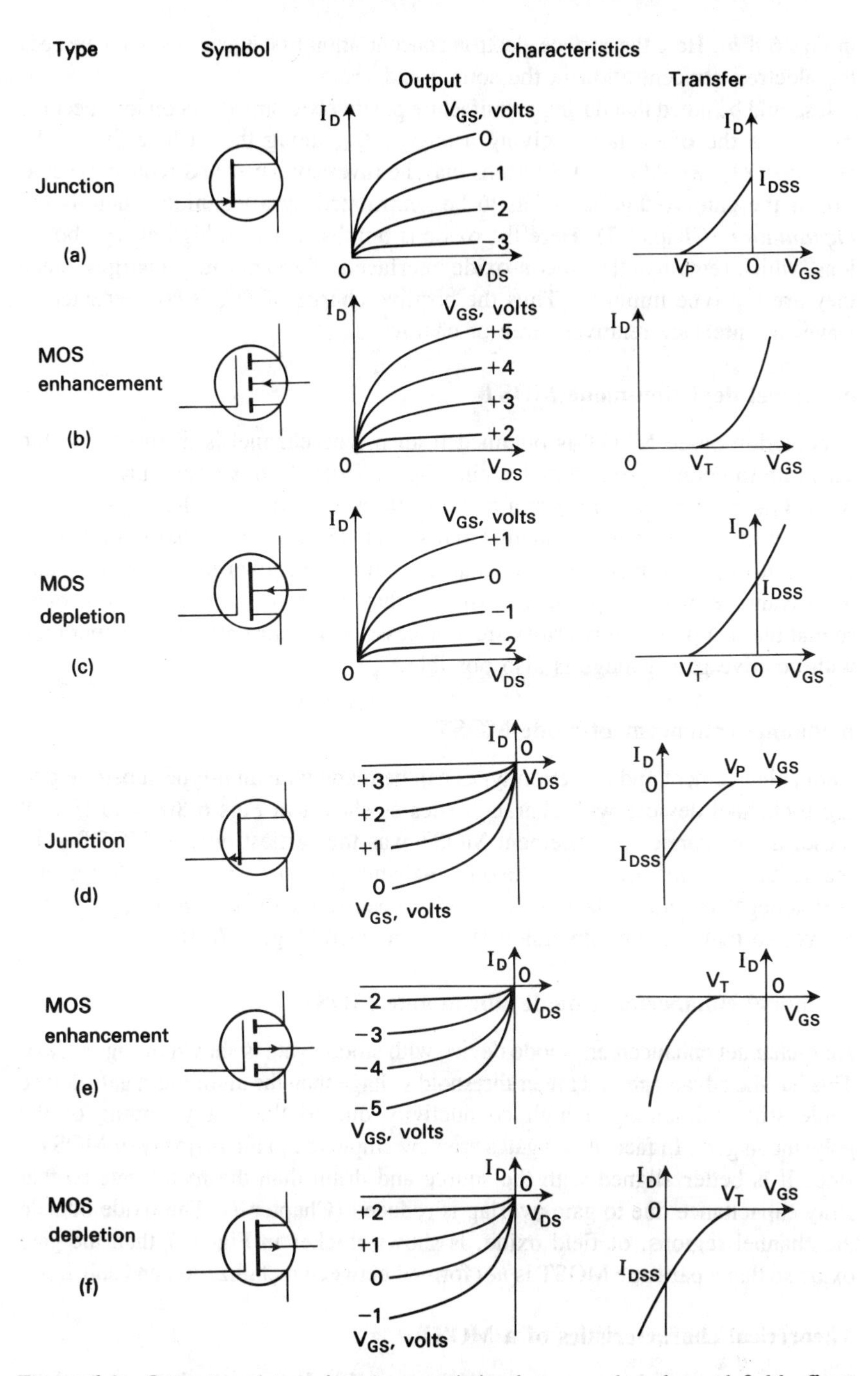

Figure 6.8 **Output and transfer characteristics for n- and p-channel field-effect transistors, (a), (b), (c) and (d), (e), (f) respectively.**

in Figs 6.8(*b*). Here the surface electron concentration first equals and then exceeds the electron concentration in the source and drain.

It should be noted that during manufacture positively charged silicon ions become trapped in the oxide layer, giving a charge Q_{SS} along the surface facing the substrate. Q_{SS} would attract electrons into the inversion layer and reduce the control of the gate voltage so it has to be neutralized by a technique such as *ion implantation* (Chapter 7). Here the oxide is bombarded with high energy boron ions, which remain at the silicon-oxide interface as fixed negative charges, since they are a p-type impurity. Thus the positive charge of Q_{SS} is compensated to leave the interface relatively free of charge.

n-channel depletion-mode MOST

A depletion-mode MOST is obtained if an n-type channel is diffused into the substrate to connect source and drain (Fig. 6.7(*b*)). In this case, current flows when $V_{GS} = 0$ and is cut off at a negative threshold voltage. The negative gate repels electrons from the n-channel and so induces a positive charge in it, converting it to p-type material. The characteristics are shown in Figs 6.8(*c*), and the advantage of this type of transistor is that it can be operated at zero bias, so that bias supplies and resistors are unnecessary. Enhancement-mode operation with positive gate voltage is also possible.

p-channel enhancement-mode MOST

Both enhancement and depletion types can be made with an n-type substrate giving p-channel devices with characteristics as shown in Figs 6.8(*e*) and (*f*). Of these, the p-channel enhancement MOST was the earliest type of MOST to be maufactured in quantity and is the only p-channel type now available. It operates with a negative gate voltage which tends to neutralize the charge Q_{SS} described above, so that a p-type inversion layer is formed (Fig. 6.7(*c*)).

n-channel enhancement-mode silicon gate MOST

An n-channel enhancement-mode device with silicon gate is shown in Fig. 6.7(*d*). This has the advantage of a lower threshold voltage than the aluminium gate device while still maintaining a high conductivity due to the heavy doping of the polysilicon gate. In fact silicon gates are now employed in the majority of MOSTs, since it is better aligned with the source and drain than the metal gate so that stray capacitance due to gate overlap is reduced (Chapter 7). The oxide outside the channel regions, or field oxide, is shown thicker in Fig. 6.7 than the gate oxide so that a parasitic MOST is *not* formed between metallization and substrate.

Theoretical characteristics of a MOST

In an n-channel enhancement-mode device with $V_{GS} = 0$, the gate will be charged negatively with respect to the p-type substrate owing to the contact potential difference between them. Current will begin to flow only when $V_{GS} > 0$, so that the threshold voltage V_T is positive. Similarly in a depletion-mode device the metal will be positive with respect to the n-type semiconductor when $V_{GS} = 0$ so that V_T is negative. Thus, V_T can be regarded as a voltage built into the device which is controlled by the gate material, aluminium or silicon, and manufacturing processes as described in Chapter 7. V_T determines the mode of operation and, like V_P for a JFET, is the gate voltage at which current begins to flow.

An expression for the I_D–V_{DS} characteristics can then be derived if it is assumed that the oxide is thicker than the conducting channel. The electric field in the insulator is given by Gauss's law

$$E = \frac{D}{\epsilon_{ins}\epsilon_0} \tag{6.22}$$

where D is the charge per unit area on the gate electrode or in the channel. If the thickness of the insulator is d and the potential of the channel with respect to the source is V_x at a distance x from the source,

$$D = \frac{\epsilon_{ins}\epsilon_0(V_{GS} - V_x - V_T)}{d} \tag{6.23}$$

which includes the effect of the contact potential.

Then the current in the channel is obtained from eq. (6.6) so that

$$I_D = DW\mu \frac{dV_x}{dx} \tag{6.24}$$

where W is the width of the device as before (Fig. 6.1).

Substituting from eq. (6.23),

$$\frac{dV_x}{dx} = \frac{I_D\, d}{\epsilon_{ins}\epsilon_0\, W\mu(V_{GS} - V_x - V_T)} \tag{6.25}$$

Inserting the limits $V_x = 0$ at $x = 0$ and $V_x = V_{DS}$ at $x = L$, leads to the integral

$$\int_0^{V_{DS}} (V_{GS} - V_x - V_T)dV_x = \int_0^L \frac{I_D\, d}{\epsilon_{ins}\epsilon_0\, W\mu}\, dx \tag{6.26}$$

and, after integration,

$$I_D = \frac{\epsilon_{ins}\epsilon_0\, W\mu}{L\, d} V_{DS}\left(V_{GS} - \frac{V_{DS}}{2} - V_T\right) \tag{6.27}$$

It may be noted that the capacitance between gate and substrate is controlled by the oxide and given by

$$C_{OX} = \frac{\epsilon_{ins}\epsilon_0\, WL}{d} \tag{6.28}$$

and that the capacitance per unit area of oxide is given by

$$C_0 = \frac{\epsilon_{ins}\epsilon_0}{d} \tag{6.29}$$

which has units of F/m^2. Then, bringing V_{DS} inside the brackets leads to

$$I_D = \mu C_0 \frac{W}{L}\left[(V_{GS} - V_T)V_{DS} - \frac{V_{DS}^2}{2}\right] \tag{6.30}$$

which may be written in the form

$$I_D = K\left[(V_{GS} - V_T)\, V_{DS} - \frac{V_{DS}^2}{2}\right] \tag{6.31}$$

where

$$K = \mu C_0 \frac{W}{L} \tag{6.32}$$

in units of A/V^2. K is known as the *transistor gain factor* and may be further defined as

$$K = K'\frac{W}{L} \tag{6.33}$$

where

$$K' = \mu C_0 \tag{6.34}$$

K' is known as the *process gain factor* since it depends on the oxide thickness through C_0 and on whether the transistor is n- or p-channel through μ. W/L is the aspect ratio of the transistor and is an important design feature as discussed below. Equation (6.31) is an expression for the characteristics below saturation. In the MOST this occurs at the drain voltage for which $E = 0$ at the drain end of the gate. From eqs (6.22) and (6.23) with $V_x = V_{DS}$, saturation occurs when $V_{DS} = V_{GS} - V_T$, and insertion of this condition into eq. (6.30) leads to

$$I_{Dsat} = \frac{K}{2}(V_{GS} - V_T)^2 \tag{6.35}$$

Then eq. (6.35) applies only to drain currents for which $V_{GS} > V_T$ and again shows that I_{Dsat} rises with V_{GS} (compare eq. (6.14)).

Equation (6.31) and the square-law relationship of eq. (6.35) hold for all MOSTs, with K being positive for n-channel and negative for p-channel devices as μ is effectively negative for holes. The sign of V_T depends on the type of channel and whether it is a depletion- or an enhancement-mode device. This may be seen from Fig. 6.8 and from Table 6.1 where the polarities of K and the voltages are summarized for the available types of MOST.

Table 6.1 Polarities of MOST quantities

MOST type	K	V_{DS}	V_{GS}	V_T
n-channel depletion	+	+	–/+	–
n-channel enhancement	+	+	+	+
p-channel enhancement	–	–	–	–

For a typical MOST in an integrated circuit, where d is about 0.1 μm and ϵ_{ins} about 4.0, the value of C_0 is about 3.5×10^{-4} F/m^2. The surface mobility μ is about 0.06 m^2/V s for the electrons and –0.02 m^2/V s for holes. These values lead to $K_n' \simeq 2.1 \times 10^{-5}$ A/V^2 for n-channel and $K_p' \simeq -7 \times 10^{-6}$ A/V^2 for p-channel devices.

For a depletion device I_{DSS} is given by eq. (6.35) with $V_{GS} = 0$ which leads to

$$I_{DSS} = \frac{K}{2} V_T^2 \tag{6.36}$$

This is important in nMOS integrated circuits with depletion loads as discussed in Chapter 7, and for typical values of $V_T = -2.0$ V, $K' = 2 \times 10^{-5}$ A/V^2 and $W/L = 0.5$, $I_{DSS} = 20$ μA. These values are subject to variation since V_T is dependent on the manufacturing process.

Drain resistance

Another important property of a MOST in a switching circuit is the drain resistance, particularly at low values of V_{DS}. All the I_D/V_{DS} characteristics pass through the point $I_D = 0$, $V_{DS} = 0$, as the device is ohmic in this region, like the junction FET. Then the drain resistance r_D is given by the inverse of the slope of the characteristics.

Using eq. (6.31) the required slope is

$$\left.\frac{\partial I_D}{\partial V_{DS}}\right|_{V_{GS}\ \text{constant}} = K\,(V_{GS} - V_T - V_{DS}) \tag{6.37}$$

$$r_D = \frac{1}{K(V_{GS} - V_T - V_{DS})}$$

$$= \frac{1}{K' \dfrac{W}{L} (V_{GS} - V_T - V_{DS})} \tag{6.38}$$

Equation (6.38) is plotted as a function of V_{DS}, for a fixed value of V_{GS} in Fig. 6.9(*a*) which shows r_D tending to infinity as V_{GS} approaches $V_{GS} - V_T$. An

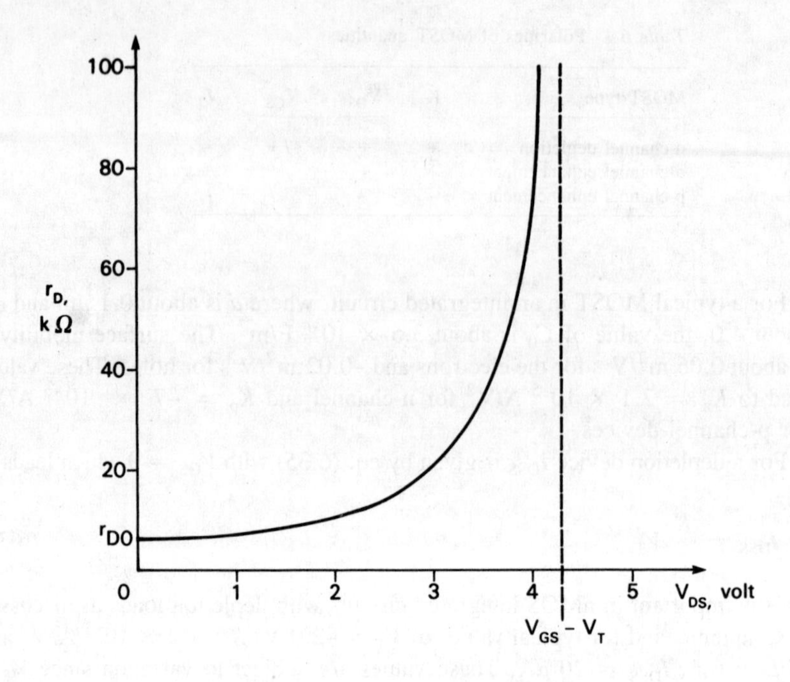

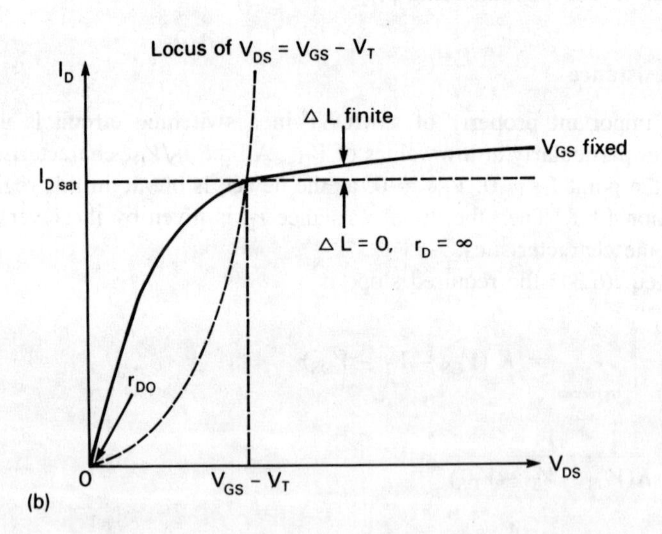

Figure 6.9 (*a*) Variation of drain resistance with drain voltage for a MOST in the ohmic region. $V_{DS} = 5$ V, $V_T = 0.7$ V, $W/L = 2$ and $r_{DO} = 5.8$ kΩ. r_D tends to infinity at $V_{DS} = V_{GS} - V_T$. (*b*) I_D/V_{DS} characteristic for a MOST with fixed V_{GS}. Channel length modulation and threshold voltage reduction lead to the finite value of r_D in saturation.

infinite value of r_D corresponds to zero slope on the I_D/V_{DS} curve in saturation, that is the characteristics are horizontal. The lowest value of r_D, r_{DO}, occurs at $V_{DS} = 0$ and represents the drain resistance at the origin of the characteristics. Thus

$$r_{DO} = \frac{1}{K' \dfrac{W}{L} (V_{GS} - V_T)} \tag{6.39}$$

r_{DO} also represents the on-resistance of a device working with low V_{DS} in a switching circuit, for example.

A MOS switch is made more ideal by reducing r_{DO}, which is achieved by increasing the aspect ratio *W/L*. Using typical values for an n-channel enhancement MOST in a logic circuit we have $K' = 2 \times 10^{-5}$ A/V^2, $W/L = 2$, $V_{GS} = 5$ V, $V_T = 0.7$ V, which leads to $r_{DO} = 5.8$ kΩ. The importance of this result to the switching time is further discussed under nMOS and CMOS inverters later in this chapter. The actual value of r_{DO} is dependent on the manufacturing process through V_T.

Channel length modulation

As in junction FETs, in the saturation region the drain current rises slowly with drain voltage (Fig. 6.9(*b*)) and as before this is due mainly to a reduction in channel length as V_{DS} is increased. In a MOST the drain-substrate junction is reverse biased and, since the drain is doped much more heavily than is the substrate, the depletion layer penetrates into the substrate. The depletion layer width increases with V_{DS} which leads to a shortening of the conducting channel ΔL. An expression for ΔL can be obtained from eq. (3.74) by making N_j equal to N_a, the acceptor concentration of the substrate. Taking the substrate potential as $V_{GS} - V_T$ leads to

$$\Delta L = \left[\frac{2\epsilon_r\epsilon_0}{eN_a}\right]^{\frac{1}{2}} [V_{DS} - (V_{GS} - V_T)]^{\frac{1}{2}} \tag{6.40}$$

An additional effect occurs in MOSTs with short channel length L (Ref. [3]). When the depletion layer widths at source and drain become appreciable some of the charge from the gate appears in the source and drain. This leads to a reduction in V_T so that V_{GS} is more effective and I_D is increased.

Channel shortening or channel length modulation is included in eq. (6.35) to give

$$I_D = \frac{K'W}{L - \Delta L} (V_{GS} - V_T)^2 \tag{6.41}$$

in the saturation region and ΔL increases with V_{DS} according to eq. (6.40). When ΔL is negligible the current is I_{Dsat}, which is independent of V_{DS} (Fig. 6.9(*b*)).

Substrate voltage

In many applications, the substrate is connected to the source, but in nMOS circuits, for example, the substrate is normally at ground voltage and the source of the load transistor is more positive (Fig. 6.16(a)). The substrate is thus reverse biased with respect to the source and it acts as a second gate with a voltage opposing that of the normal gate. This has the effect of raising the threshold voltage V_T.

The threshold voltage can be expressed in terms of the substrate-source or bulk-source, voltage V_{BS} (Ref. [3]) so that

$$V_T = V_{TO} + \gamma[(V_{BS} + \phi)^{1/2} - \phi^{1/2}] \tag{6.42}$$

Here V_{TO} is the threshold voltage for $V_{BS} = 0$,

$$\gamma = \left(\frac{2e\,\epsilon_r\epsilon_0\,N_a}{C_0}\right)^{1/2} \tag{6.43}$$

$$\phi = \frac{2kT}{e}\ln\frac{N_a}{n_i} \tag{6.44}$$

with ϕ being the surface potential for strong inversion ($V_{GS} > V_T$). Typical values for γ and ϕ are given later in Table 6.4.

MOS device scaling

The size of a single silicon chip is limited to about 1 cm square as the manufacturing yield is reduced above this, so the number of logic functions that can be provided depends ultimately on the size of individual transistors. Thus the development of larger memories and more powerful microprocessors, for example, is a consequence of scaling down device dimensions. However, reducing the channel length L alone leads to some undesirable effects and further changes are required, as discussed below.

As L is reduced the depletion layer widths of the source and drain junctions become comparable to L. The potential distribution in the channel is then influenced by the drain voltage as well as the gate voltage and so becomes two-dimensional. This in turn leads to a decrease in threshold voltage V_T with channel length (Ref. [3]).

In addition, for constant drain voltage V_{DS}, the electric field between drain and source, V_{DS}/L, is increased by reducing L. The channel mobility becomes field-dependent and eventually the carrier velocity saturates (Fig. 2.15), while further reduction in L would lead to breakdown of the drain-substrate diode.

These short-channel effects are undesirable and one approach is to avoid them by scaling down all the dimensions and voltages of the MOST. This maintains the same electric fields within the device, so that the simple theory of a long channel can still be applied. The doping level of the substrate N_a is increased to scale down the widths of the depletion layers (eq. (3.74)). Also the oxide thickness d is reduced to maintain the gate field at reduced gate voltage.

If a factor κ is used, which is greater than 1, a dimensional reduction is achieved by multiplying by $1/\kappa$ as illustrated in Table 6.2. Since C_0 is inversely proportional to d it is increased by κ and since I_D is proportional to C_0, W/L and V^2 (eq. (6.30)) it is reduced by $1/\kappa$. The parameter τ is the propagation delay due to transit across the channel, or transit time. It corresponds to L/u, where u is the carrier velocity given by V_{DS}/L, so that $\tau = L^2/\mu V_{DS}$ and is reduced by $1/\kappa$.

Table 6.2 Effects of scaling

Parameter	Scaling factor
L	$1/\kappa$
W	$1/\kappa$
d	$1/\kappa$
N_a	κ
V_{DS}	$1/\kappa$
V_{GS}	$1/\kappa$
C_0	κ
I_D	$1/\kappa$
τ	$1/\kappa$
Device area	$1/\kappa^2$
Power dissipation per gate $I_D \cdot V_{DS}$	$1/\kappa^2$
Power dissipation per unit area	1
Number of devices per unit area	κ^2

The number of devices per unit area of silicon does increase rapidly with scaling but while reducing the power supply below 5 V minimizes short-channel effects it can only be applied where the electronic system stands alone. Examples are a digital watch and a pocket calculator, where the supplies are typically 1.5 V and 3.0 V, respectively. Most other systems require a 5 V supply and in order for a scaled integrated circuit to be compatible, the applied and threshold voltages must be maintained. Since the threshold voltage V_T theoretically increases with $N_a^{1/2}$ (Ref. [3]) the substrate doping is further increased to restore V_T to its previous level. Also, power dissipation is reduced by less than $1/\kappa^2$ due to maintaining the supply at 5 V. Simulated examples of the effects of scaling are given after the sections on nMOS and CMOS logic circuits and illustrate the performance improvements obtained by area reduction.

Transconductance

The mutual conductance in the saturated region (transconductance) is obtained from eq. (6.35):

$$\left.\frac{\partial I_{Dsat}}{\partial V_{GS}}\right|_{V_{DS}\ \text{const}} = g_m = K(V_{GS} - V_T) \tag{6.45}$$

so that once again g_m rises linearly with V_{GS}, with practical values similar to those of the junction-gate device. An equivalent circuit like that in Fig. 6.5 may be used, the various components having comparable values, and a gain-bandwidth product defined as $g_m/2\pi C_{gs}$.

The MOST is also very useful as a switch and as a variable resistance. Its high input impedance is retained for positive values of V_{GS}, unlike the junction-gate device. A practical consequence is that a comparatively low resistance should always be connected between the gate and the source electrodes, since excessive accumulation of charge on the gate can lead to breakdown of the dielectric, where E exceeds the breakdown value in eq. (6.22).

Temperature and voltage limitations

The main temperature effect is that I_D falls as temperature is increased for a given gate voltage. This is because K is proportional to mobility μ from eq. (6.29) and μ falls as temperature T rises. K is found to change at about –0.3 per cent/°C. The threshold voltage also falls as temperature rises at about –3 mV/°C and this would tend to cause an increase of drain current (Ref. [3]).

If the voltage between drain and gate exceeds a certain value then a sharp rise in drain current occurs. This is again caused by breakdown of the oxide layer under the gate metallization which is irreversible, unlike the breakdown of a JFET. Additionally, the source and drain depletion layers can become so wide that they meet, a condition known as *punch through* and which is followed by high drain current.

Computer simulation of field-effect transistors

Both JFETs and MOSFETs can be simulated using the SPICE program. The JFET model is derived from the FET model of Schichman and Hodges which, in its simplest form, is similar to eq. (6.31) for the MOSFET, with V_T replaced by V_P expressed as a negative quantity for an n-channel JFET, so that $V_{DS} = V_{GS} - V_P$.

Channel length modulation is included by defining ΔL through a parameter λ so that

$$\Delta L = \lambda L\ V_{DS} \tag{6.46}$$

At constant V_{GS} eq. (6.41) reduces to

$$I_D = \frac{\text{const.}}{L - \Delta L} = \frac{\text{const.}}{L(1 - \lambda\ V_{DS})} \tag{6.47}$$

in the saturation region and I_D rises with V_{DS} as shown in Fig. 6.9(*b*). λ is a measure of the output conductance in saturation, which can be obtained from eq. (6.47). Applying the binomial theorem with $\lambda V_{DS} \ll 1$ gives I_D = const./L $(1+\lambda V_{DS})$. Differentiating I_D with respect to V_{DS} and substituting for the constant from eqs (6.35) and (6.41) leads to $g_{ds} = \lambda I_{D\,sat}$ for the output conductance and $r_D = (g_{ds})^{-1}$ in saturation.

The theoretical symbols and corresponding SPICE parameters for the JFET are shown in Table 6.3, together with typical SPICE values. A full list of parameters with their names is given in Appendix 3.

Table 6.3 Theoretical symbols and SPICE parameters for JFET simulation

	Symbol	Parameter	Typical value
	V_P	VTO	-2.0 V
	K	BETA	10^{-3} A/V^2
	λ	LAMBDA	10^{-4} V^{-1}
	C_{gs}	CGS	5.0 pF
	C_{gd}	CGD	1.0 pF
Gate diode	ψ	PB	0.6 V
Gate diode	I_0	IS	10^{-14} A

Three MOSFET models are provided in SPICE which differ in the formulation of the I/V characteristic and are described in detail in Ref. [4]. At level 1 is the Schichman–Hodges model based on eq. (6.31) and which is useful for simulating relatively large devices. The level 2 model incorporates most of the second-order effects of small-sized devices. The level 3 model is semi-empirical and is described by a set of parameters defined by curve fitting rather than by a physical background. A total of 42 parameters are available but reasonable results can be obtained with a subset of these, taken from levels 1 and 2 and valid down to about 2 μm length.

The theoretical symbols and corresponding SPICE parameters appropriate to a level 1 simulation are shown in Table 6.4 (over), together with typical SPICE values. A list of parameters with their names is given in Appendix 3.

MOST capacitances

One category of capacitance is the area contribution due to the depletion layers of the source-substrate and drain-substrate diodes. These are calculated from an equation such as eq. (3.107), except that in the MOST simulation the zero bias capacitance CJ is expressed per unit area. Hence the junction areas of source and drain, AS and AD respectively, are also required (Fig. 6.10). The substrate doping level NSUB corresponds to $N_a = N_j$ in eq. (3.107).

A second category is the periphery contribution due to the overlap between

Table 6.4 Theoretical symbols and SPICE parameters for MOST simulation

Symbol	Parameter	Typical value	
K'	KP	2×10^{-5} A/V^2	n-channel
		6.67×10^{-6} A/V^2	p-channel
μ	UO	6×10^{-2} m^2/V s	n-channel
		2×10^{-2} m^2/V s	p-channel
V_T	VTO	−1.0 to −4.0 V	n-channel depletion
		0.4 to 1.0 V	n-channel enhancement
		−0.4 to −1.0 V	p-channel enhancement
d	TOX	10^{-7} m	
γ	GAMMA	0.37 V$^{1/2}$	
ϕ	PHI	0.65 V	
λ	LAMBDA	0.02 V^{-1}	
N_a	NSUB	4×10^{21} m^{-3}	
C_{jo}	CJ	2×10^{-4} F/m^2	junction area
ψ	PB	0.8 V	
m	MJ	0.5	
x_j	XJ	10^{-6} m	
	CGSO	3×10^{-10} F/m	channel width
	CGDO	3×10^{-10} F/m	channel width

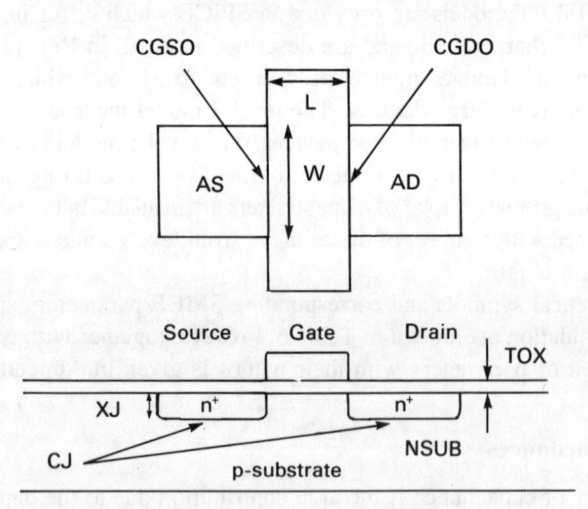

Figure 6.10 SPICE parameters for MOST simulation.

the gate edge and the source or drain. This is defined by CGSO and CGDO which are specified per unit width of channel and so depend on W (Fig. 6.10). The sidewall capacitance per metre of drain and source periphery may also be defined if required.

The VMOS (vertical MOS) transistor

The MOSTs described so far have been small-signal, low-power devices. This is due to the small channel area, low breakdown voltage and the square-law transfer characteristic giving rise to high harmonic content for large signals.

In the VMOS transistor these difficulties are overcome and devices with characteristics that are linear between 0.4 and 2.0 A, breakdown voltages up to 100 V and a dissipation of 25 W are available. A typical section through the device is shown in Fig. 6.11(*a*), where an n^- epilayer is deposited on an n^+ substrate to form the drain. Channel and source regions are then diffused in a manner similar to the production of biplar transistors (Chapter 7). A V-shaped groove is etched through the channel and source regions and precise dimensions are maintained. Silicon dioxide is grown over the V-groove gate and aluminium metallization deposited, with a permanent connection between source and channel.

The current flows vertically through a very short inversion layer, produced in the p-region as for a conventional MOST. The channel length is only about 1.5 μm, compared with about 5 μm previously and two current paths are available, one each side of the groove. The drain has a large area since it is on the back of the chip and all these features contribute to the high current capacity.

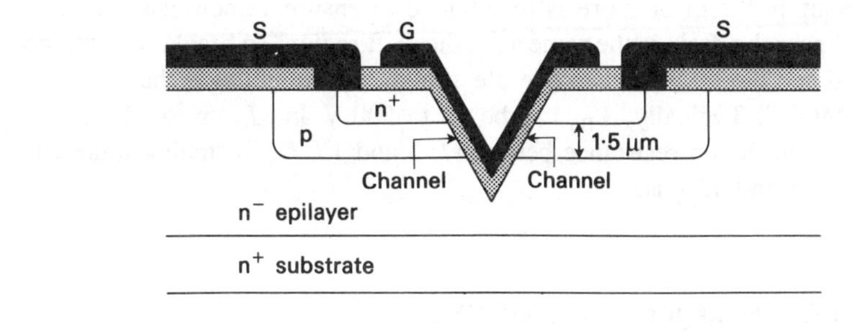

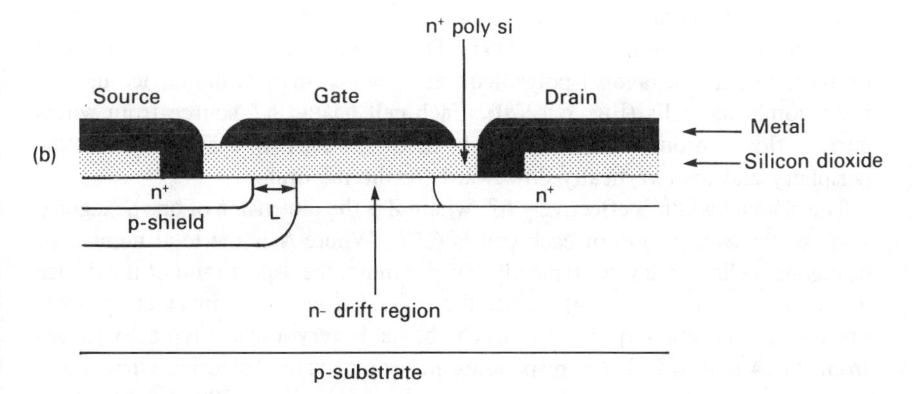

Figure 6.11 (*a*) Vertical (V-channel) MOS transistor, (*b*) Section through a DMOS (double-diffused MOS) transistor.

The drain current I_D depends on the product of mobile channel charge and velocity (eq. (6.24)) and in the saturation region of the VMOST characteristics the short channel length means that the velocity also saturates (Fig. 2.15). Thus I_D varies with channel charge which is proportional to $V_{GS} - V_T$, so that I_D rises linearly with V_{GS} rather than as a square law in normal MOSTs.

The extra n^- region leads to a wide depletion layer spreading into the drain at high voltages, thus reducing the maximum field across the p-n junction and so increasing the breakdown voltage.

The VMOST can be used as a high-current switch with an on-resistance around 1 Ω or as a wideband amplifier with g_m about 0.27 S. Typically, V_{DS} can be up to 80 V and I_D up to 2 A. Switching times can lie between 2 and 10 ns.

The DMOS (double-diffused MOS) transistor

The DMOST like the VMOST provides a short inversion layer and a linear $I_D - V_{GS}$ characteristic from an enhancement mode device. An extra p-diffusion shields the source (Fig. 6.11(*b*)) and the channel length is determined by this p-shield, on which an n-type inversion layer is formed when $V_{GS} \geq V_T$. Electrons leaving the inversion layer pass through the drift region in which a uniform field of about 10^6 V/m or more is maintained to ensure velocity saturation.

The field near the drain is the same as in the drift region, so breakdown effects are reduced and the DMOST can operate at higher drain voltages than the conventional MOST. Typically, V_{DS} can be up to 500 V and I_D up to 12 A, with V_T 3 to 4 V and the on-resistance between 0.3 and 1.0 Ω. Switching times can lie between 25 and 150 ns.

The HEXFET (Hexagonal-gate MOSFET)

The HEXFET operates in a similar manner to the DMOST but the drain is on the back of the chip like the VMOST. The effective device width is extended by means of the hexagonal polysilicon gate, which divides the device up into honeycomb-like cells (Fig. 6.12(*a*)). Each cell has an n^+ source from which current flows through the inversion layer of the narrow channel around the periphery and then vertically downwards to the n^+ drain.

The channel width is effectively $6Z$, where Z is the dimension of one hexagonal side, so the aspect ratio of each cell is $6Z/L$. Where N is the total number of hexagonal cells per device, typically 10^3 per mm^2, the aspect ratio of the device is $6NZ/L$, which is very high. Since the on-resistance r_D, is inversely proportional to aspect ratio (eq. (6.39)) this can be made very low and typically ranges from 0.014 Ω to 0.2 Ω. Corresponding maximum values of drain current and voltage are 145 A and 100 V for the low r_D, and 22 A and 500 V for the high r_D. Switching times lie between about 45 and 300 ns.

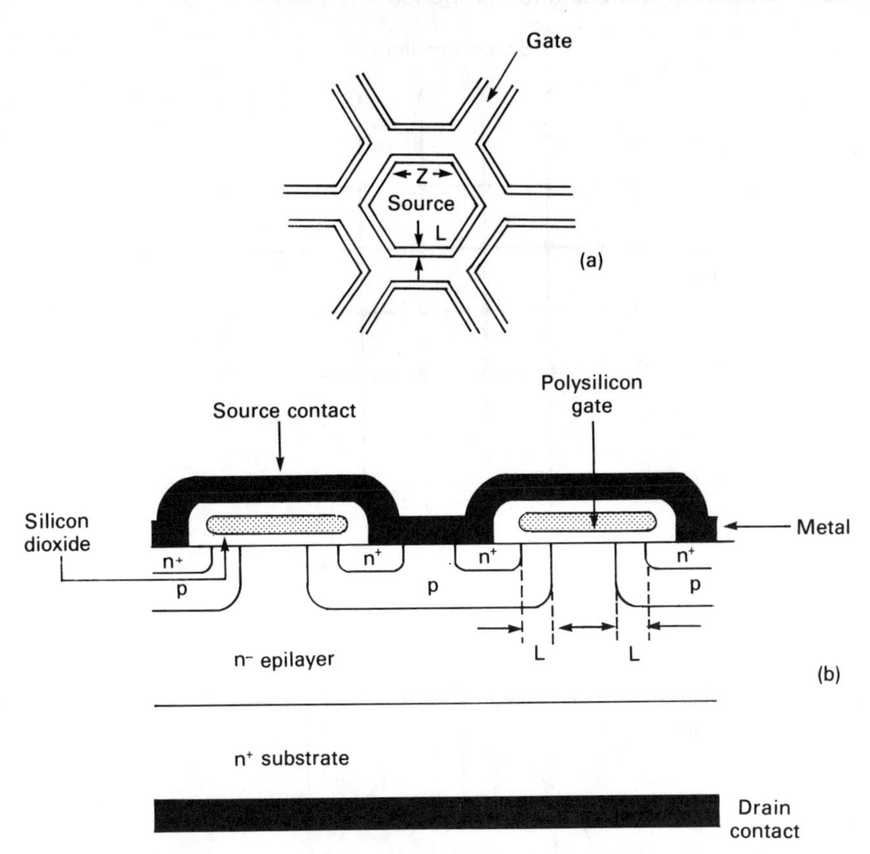

Figure 6.12 (*a*) Plan view of a HEXFET (hexagonal gate MOSFET). (*b*) Section through a HEXFET.

MOST memory cells

A semiconductor random-access memory consists of a symmetrical matrix of 'wires' with a memory cell at each intersection (Fig. 6.13(*a*)). Each cell can store one bit (binary digit) corresponding to 0 or 1 and normally represented by voltage levels of approximately 0 V and 5 V respectively. A cell is *addressed* when it receives a signal on both its *X* and *Y* lines, also called *row* and *column* lines, or *select* and *bit* lines.

Dynamic read/write memory cell

The largest read/write or random-access memories (RAMs) use one n-channel enhancement MOS transistor T1 and an MOS capacitor C_s (Fig. 6.13(*b*)). A 1 is stored when C_s is charged and 0 when it is discharged, while the MOST provides access for reading or writing. C_B is the stray capacitance between the bit line and 0 V and is up to 10 times C_s in value.

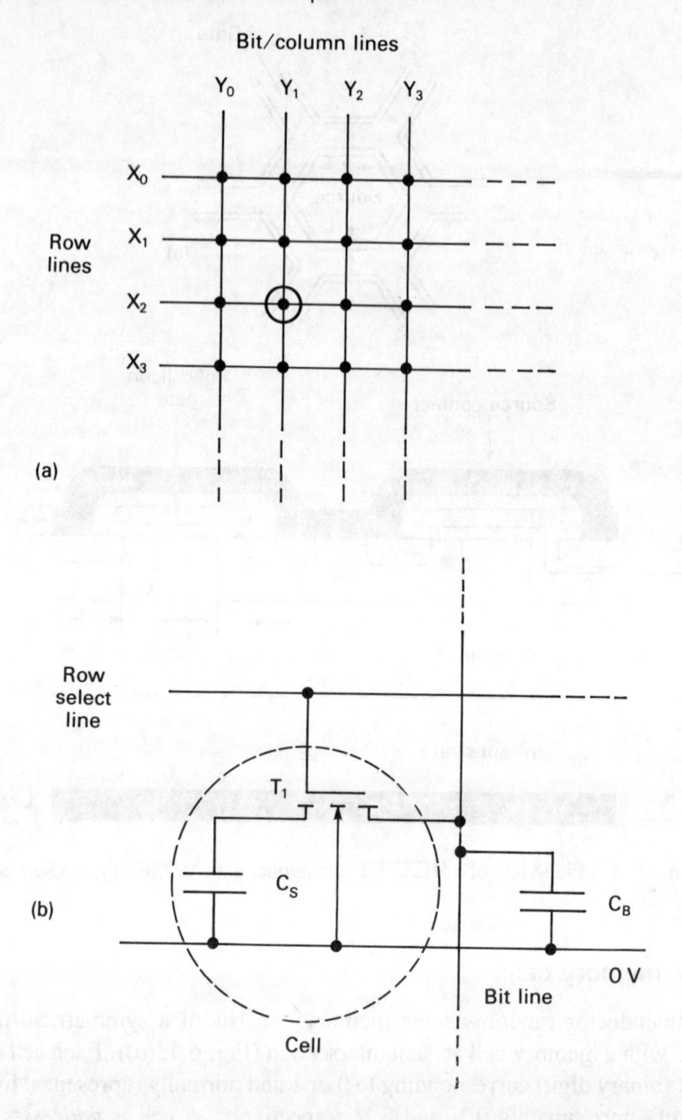

Figure 6.13 (*a*) Random-access memory matrix, represents a memory cell. (*b*) Dynamic memory cell. A regenerative sense amplifier is connected to the bit line for the read/refresh operation.

C_s has a capacitance of about 0.1 pF and the silicon dioxide dielectric has a leakage current i_l of about 50 pA. Consequently, the charge on C_s has to be refreshed at regular intervals for its voltage V_s to be maintained at the desired level. An estimate of the discharge time t can be obtained by assuming a linear change in voltage from a maximum of 5 V. Then

$$t = \frac{C_s V_s}{i_l} = \frac{10^{-13} \times 5}{5 \times 10^{-11}} = 10 \text{ ms.}$$

For a drop down to 3 V this leads to a practical refresh interval of 4 ms, which is likely to increase as cell design is improved.

For writing, the bit line is taken to the desired logic level, 0 or 1, which charges C_B. The cell is addressed by taking the select line positive, T1 conducts and the charge on C_B is shared with C_s.

For reading the bit line is connected to a regenerative sense amplifier and the select line is taken positive. T1 conducts, the charge on C_s is shared with C_B and the level on C_B detected by the sense amplifier. Since $C_B \gg C_s$ the reading is destructive, so the level on C_s is restored by the sense amplifier immediately after reading. When a logic 1 has been stored, the C_s voltage is V_{DD} and the charge is $V_{DD}C_s$. When the cell is read by connecting C_B in parallel the voltage is V_{out} and the charge on the combined capacitance is V_{out} $(C_s + C_B)$. Since the charge is constant

$$V_{out} = \frac{V_{DD}C_s}{C_s + C_B}$$

and for $C_B = 10\ C_s$, $V_{out} = 0.09\ V_{DD}$. The same bit line is used for both write and read/refresh operations, with current flowing into the cell for writing and out of it for reading.

Dynamic memories require special controllers to provide refresh cycles as well as reading and writing but 256 Kbit single chip devices are available and larger sizes are sure to follow. Note that 1 K = 1024 = 2^{10} and is the unit of memory size.

Read-only memory cells

The read/write memory cell described above is called *volatile* because when the power supply is turned off the data is lost due to lack of refreshing. A *non-volatile* memory retains its data when the supply is turned off and an example of this is the read-only memory (ROM). Here, data is programmed into each location by masking during manufacture. Alternatively, the gate of the memory cell transistor is modified so that charge can be stored within it and retained over a period of at least 10 years, if necessary. This allows users to program their own read-only memories and this leads to two types, the erasable-programmable ROM (EPROM) with ultra-violet erasure and the electrically erasable and programmable ROM (EEPROM or E^2PROM). Each of these has memory cells with a floating-gate avalanche-injection MOS (FAMOS) transistor, the first having one floating gate and the second with an additional control gate.

EPROM memory cell

The polysilicon gate of a FAMOS transistor is buried in the oxide and so disconnected or floating, with a thin oxide layer separating it from the substrate (Fig.

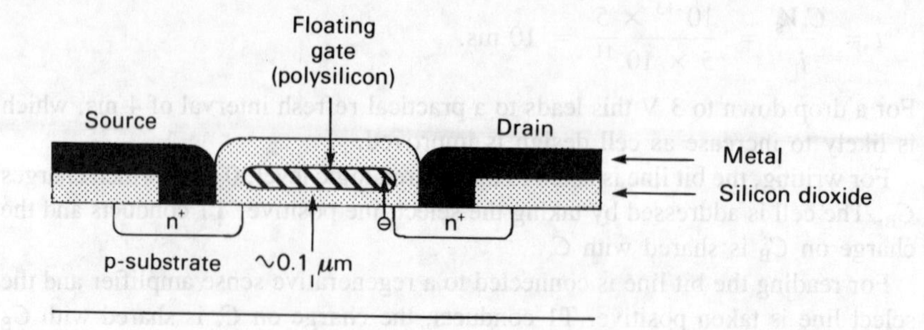

Figure 6.14 Section through a FAMOS (floating-gate avalanche-injection MOS) transistor.

6.14). The threshold voltage is too high to allow formation of an n-channel on the substrate and in the memory this allows the drain voltage to rise to the 1 level without current flow.

When the voltage is increased in pulses during programming to cause avalanche breakdown of the drain diode a large number of electrons cross the drain junction, some of which penetrate the oxide and accumulate in the floating gate. These injected electrons reduce V_T by about 10 V to a value that allows the n-channel to form and the transistor becomes conducting, which corresponds to a 0 level in the memory. The negative charges in the gate are retained even when the 1 level voltage is applied and are only removed when the chip is irradiated with ultra-violet light, which causes the oxide to become conducting and the charges to leak away. Normally, the memory is covered with an opaque label to prevent erasure by normal daylight, which takes about 3 years, or sunlight, requiring about 1 week. However, when erasure is required, a high intensity ultra-violet source is used which completes the operation within about one hour.

EEPROM memory cell

In this type of memory cell an external control gate is added to the FAMOS transistor, which allows electrical erasure (Fig. 6.15). Initially, the drain current is zero due to the high threshold voltage of the floating gate and the device is in the 1 state as before. During programming the threshold is reduced by applying positive voltage pulses to the control gate and drain. This causes avalanche injection into the floating gate to put the device into the 0 state. During erasure positive pulses are applied to the control gate with drain and source grounded, which causes stored electrons to pass from the floating gate to the control gate and out of the device. The high threshold voltage is restored and the device returns to the 1 state.

The MOST inverter

Due to the small area required for MOSTs they are widely used as switches in large-scale integrated circuits, such as memories and micro-processors.

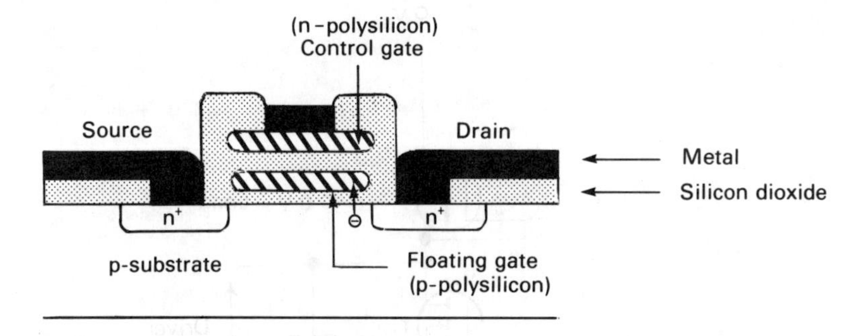

Figure 6.15 Section through an electrically-erasable FAMOS transistor.

The relative sizes of a bipolar transistor, a diffused resistor and a MOS transistor are shown in Fig. 7.11, which shows that while the dimensions of a bipolar transistor and resistor are similar, the MOS is much smaller. Hence it would be unreasonable to form an inverter by using a resistor as the drain load of a driver MOST. Instead chip area is conserved by using a second MOST and the inverter can take two forms, depending on whether the load and driver MOSTs have the same type of channel or opposite types (complementary MOSTs).

nMOS inverter

The first form is shown in Fig. 6.16 for two n-channel devices. The transistor T2 should provide a large resistance to limit the current when the driver T1 is conducting. T1 should have a lower resistance so that the output voltage V_0 is low when it is conducting. For a given gate thickness the drain current of a MOST is controlled by the *aspect ratio* of the channel, W/L (eq. (6.27)). Then for T2 the aspect ratio should be lower than for T1 and as can be seen from Fig. 6.16(*b*) a MOST can be wider than it is long.

The circuit is called a *ratioed* inverter since the value of V_0 when T1 is conducting depends on the ratio of the effective resistances of T1 and T2. For a given manufacturing process this will depend on the aspect ratios of T1 and T2 which are combined in the *inverter ratio* $(W/L)_{T1}/(W/L)_{T2}$. This is actually a conductance ratio for the two transistors. From Fig. 6.16(*b*) $(W/L)_{T1}$ is 2:1 and $(W/L)_{T2}$ is 1:2, giving an inverter ratio of 4:1 in this case, with values up to 8:1 also being used, particularly in logic gates.

Transfer characteristic

Since T2 is a depletion mode transistor with $V_{GS2} = 0$ it will always be in a conducting state. The I_D - V_{DS} characteristic is shown in Fig. 6.17(*a*) which

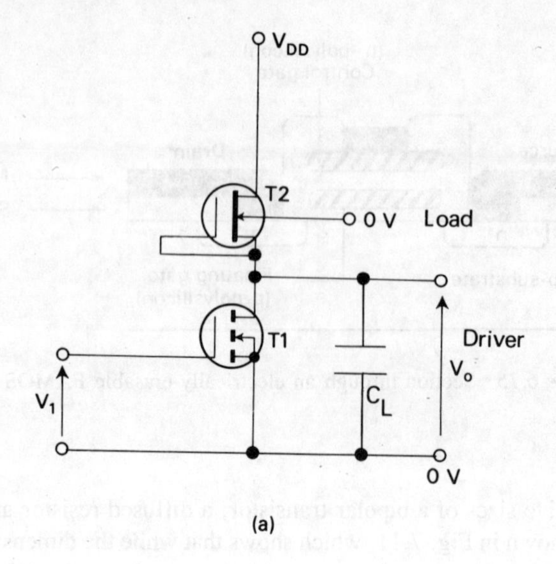

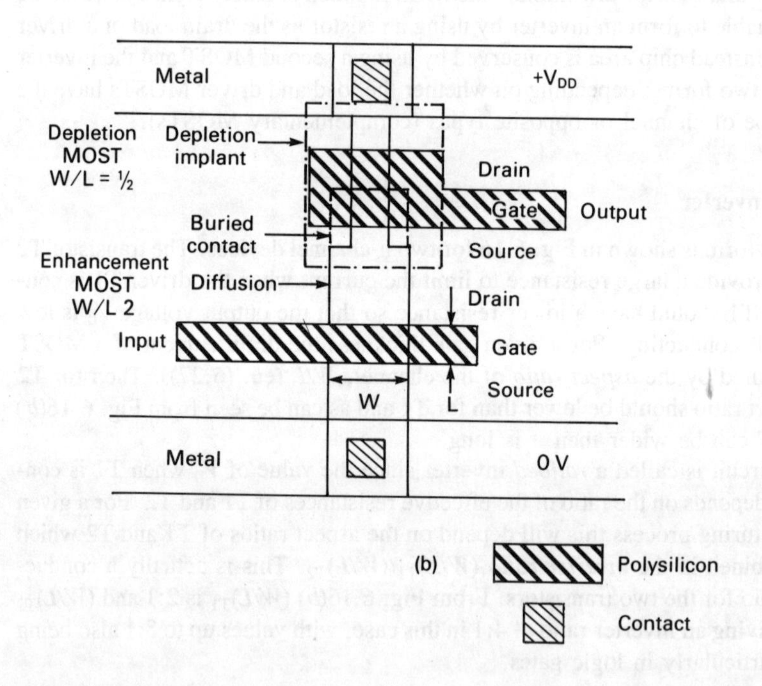

Figure 6.16 **nMOS inverter. (*a*) Circuit; T1 is an n-channel enhancement driver or pull-down transistor, T2 is an n-channel depletion load or pull-up transistor, (*b*) plan view of a possible layout. The formation of the implant, diffusion, contact and polysilicon layers is explained in Chapter 7.**

represents a constant current generator rather than an ohmic resistance. In the saturated region the load current is derived from eqs. (6.35) and (6.36), so that

$$I_{D2} = \frac{K_2}{2} V_{T2}^2 \tag{6.48}$$

Then for the whole inverter

$$\begin{aligned} I_{D1} &= I_{D2} \\ V_1 &= V_{GS1} \\ V_0 &= V_{DS1} \\ V_{DS2} &= V_{DD} - V_{DS1} \end{aligned} \tag{6.49}$$

In the steady states V_1 will be either near 0 V (and so less than V_{T1}), for a logic 0 (low) input or near V_{DD} for a logic 1 (high) input. The I_D - V_{DS} characteristic for T1 is shown in Figs 6.17(*b*) and (*c*) superimposed on the characteristic for T2, with V_{DS2} corresponding to $-V_{DS1}$ according to eq. (6.49). The operating point of the inverter is always given by the intersection of the two characteristics.

For a low input (Fig. 6.17(*b*)) T1 is cut off, with I_{D1} very low since it is the drain-source leakage current, while T2 operates in the ohmic region. Then from eq. (6.31), putting

$$V_{DS2} = V_{DD} - V_0$$

$$I_{D2} = K_2 \left[(V_{DD} - V_0)(-V_{T2}) - \frac{(V_{DD} - V_0)^2}{2} \right]$$

and

$$I_{D1} = I_{D2} \tag{6.50}$$

Hence V_0 can be obtained from eq. (6.50) and will be just below V_{DD} due to the very low current flow.

For a high input (Fig. 6.17(*c*)) T1 conducts, since $V_{GS} > V_{T1}$ and T2 operates in the saturated region. Then, from eq. (6.31), putting $V_{GS1} = V_{DD}$

$$I_{D1} = K_1 \left[V_0 (V_{DD} - V_{T1}) - \frac{V_0{}^2}{2} \right]$$

$$I_{D2} = \frac{K_2}{2} V_{T2}{}^2$$

$$I_{D1} = I_{D2} \tag{6.51}$$

and again V_0 can be obtained from eqs (6.51) being a few hundred millivolts.

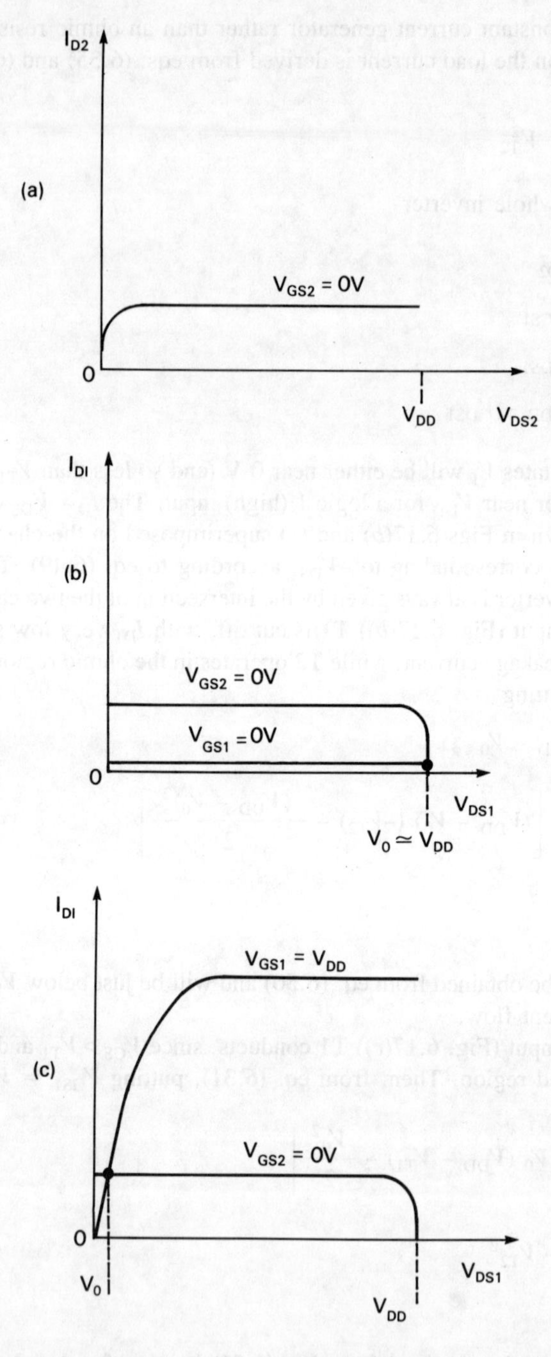

Figure 6.17 nMOS inverter. (*a*) I_D/V_{DS} characteristic for T2; (*b*) combined I_D/V_{DS} characteristic, $V_{GS1} = 0$ V; (*c*) combined I_D/V_{DS} characteristic, $V_{GS1} = V_{DD}$.

Thus when the input is low the output is high, and vice versa, so the circuit behaves as a logic inverter.

The output voltage will change between the static levels as the input voltage changes during switching, which is shown on the *transfer characteristics* of Fig. 6.18. This has been obtained by SPICE simulation with the aspect ratios of Fig. 6.16 and the typical values given in Table 6.3 for KP, CGSO, CGDO, C J and TOX.

VTO = +1 V for T1 and –3 V for T2 and AS = AD = 60 μm^2 for each transistor. W_1 = 8 μm, L_1 = 4 μm, W_2 = 4 μm and L_2 = 8 μm. The complete program is given in Appendix 3.

As V_1 is increased from 0 V V_O will not begin to change until V_1 is just above V_{T1}. V_O then falls rapidly as T1 approaches saturation which occurs when $V_{GS1} = V_{DS} + V_{T1}$, corresponding to $V_1 = V_O + V_{T1}$ (Fig. 6.18). V_O then falls much more slowly as V_1 is increased to V_{DD}, which in this case is 5 V, and reaches a value of 0.29 V for V_1 = 5 V.

It should be noted that static power is dissipated in the inverter only when T1 is on. The current is then given by eq. (6.48) and substituting for K_2 gives

$$I_{D2} = K' \frac{W_2}{L_2} \frac{{V_{T2}}^2}{2} = I_{D1} \tag{6.52}$$

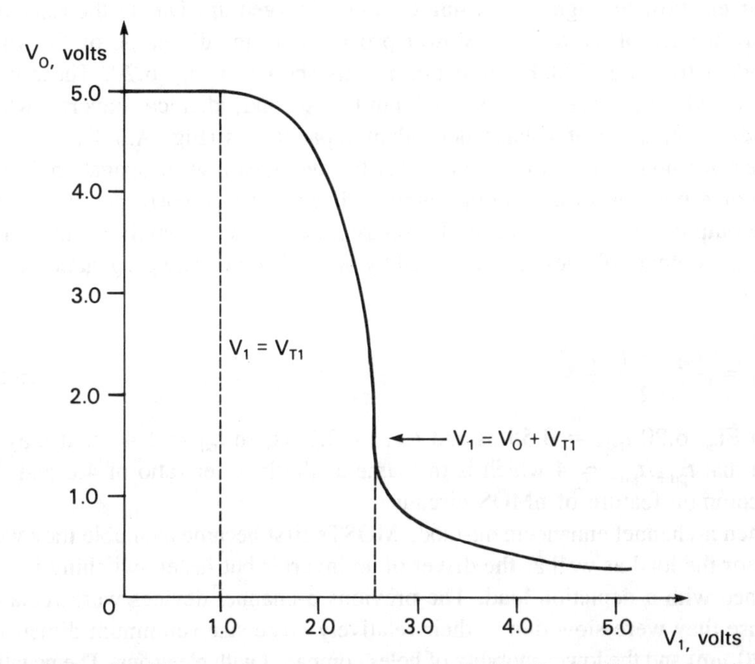

Figure 6.18 nMOS inverter transfer characteristic obtained by SPICE simulation for the circuit of Fig. 6.16. V_0 = 0.29 V for V_1 = 5 V, V_0 = 1.55 V for V_1 = 2.55 V, i.e. $V_1 = V_0 + V_{T1}$.

and static power dissipation

$$P_S = V_{DD} I_{D2} \tag{6.53}$$

Using the values $K' = 2 \times 10^{-5}$ A/V^2, $W_2/L_2 = 0.5$, $V_{T2} = -3$ V, $V_{DD} =$ 5.0 V leads to $P_S = 0.225$ mW. If it is assumed that T1 spends half the time on then the average power dissipation falls to 0.113 mW.

Transient response

Inside an integrated circuit the load driven by the inverter will be capacitive, as shown in Fig. 6.16(*a*). Here C_L includes p-n junction depletion capacitances, input-gate capacitances and parasitic capacitance due to interconnections, for example. Since some of these are functions of voltage C_L will change with the output voltage.

Suppose that T1 is cut off initially and C_L is charged up to V_{DD}. When V_1 is suddenly increased T1 will conduct and discharge C_L (Fig. 6.19(*a*)). During this process T2 will be conducting and will oppose the fall in V_0 by current flow into C_L. However, the inverter ratio of 4 : 1 or more ensures that the relatively low resistance of T1 predominates and pulls V_0 down to the low level.

When V_1 is suddenly reduced T1 is cut off and V_0 is pulled up to the high level by current flow through T2 causing C_L to be charged up. Due to the relatively high resistance of T2 this is a slower process than the discharge of C_L which is evident from the SPICE simulation results shown in Fig. 6.20. These were obtained with C_L represented by the input to a second, identical, inverter which provides simulation of voltage dependent capacitances (Fig. A.3.1).

The switching time may be expressed as the average of two propagation delays, t_{pd-}, corresponding to the falling output voltage and t_{pd+} corresponding to the rising output voltage (Fig. 6.20). These are measured between the points where input and output voltages have changed by 50%. The average propagation delay is then

$$t_{pd} = \frac{t_{pd-} + t_{pd+}}{2} \tag{6.54}$$

From Fig. 6.20 $t_{pd-} = 0.55$ ns and $t_{pd+} = 2.3$ ns, so $t_{pd} = 1.4$ ns. It may be noted that $t_{pd+}/t_{pd-} \simeq 4$ which is the same as the inverter ratio of 4.0 and this is a common feature of nMOS circuits.

When n-channel enhancement-mode MOSTs first became available they were used for the load as well as the driver of an inverter but faster switching is now obtained with a depletion load. The previous p-channel devices were replaced because they were slow due to their relatively large size (minimum dimension 10–20 μm) and the lower mobility of holes compared with electrons. The negative supply voltages were also incompatible with the standard 5 V TTL supply.

nMOS gates are still widely used in integrated circuits due to their small size

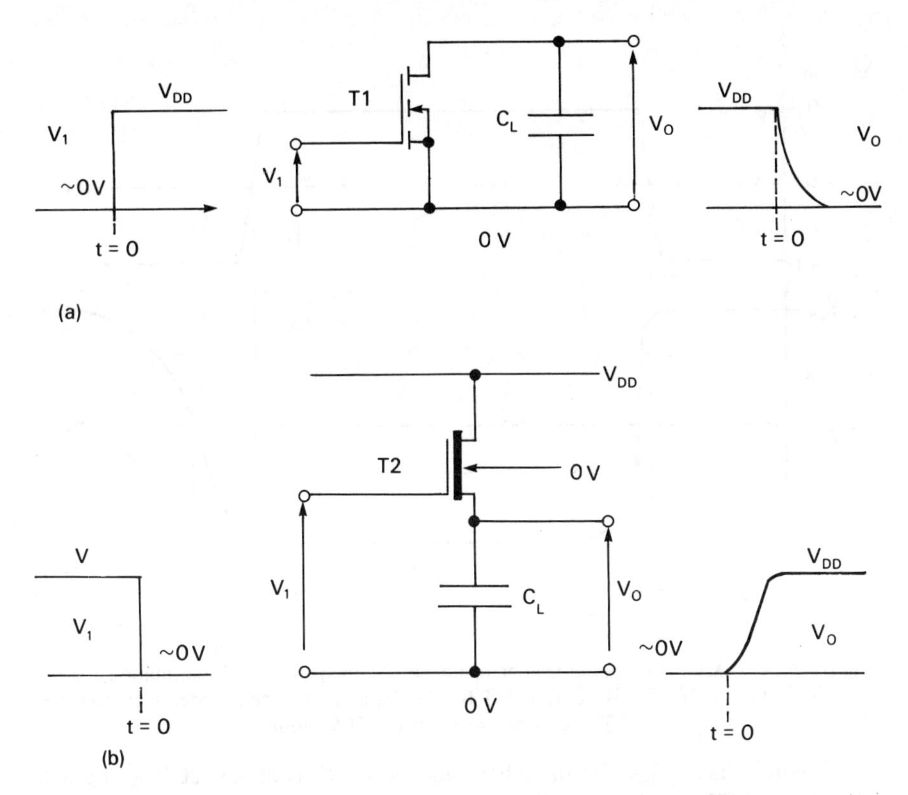

Figure 6.19 nMOS inverter. (*a*) T1 acting as a pull-down transistor; (*b*) T2 acting as a pull-up transistor.

and speed but their static power dissipation is a disadvantage, particularly in large electronic systems and battery-operated equipment. This has led to the development of the CMOS circuits described below.

The complementary MOST (CMOS) inverter

The second form of MOS inverter is shown in Fig. 6.21(*a*) where T1 is an n-channel and T2 is a p-channel device. Both transistors normally have similar dimensions, as shown in the layout of Fig. 6.22, and, unlike the previous MOS inverter, are operated with their drains connected to provide a common output terminal. In each case the substrate is connected to source, which is possible since the n-MOST is contained within an isolating island (*see* Fig. 7.18(*a*)). The gates are also connected to provide a common input to the two devices. Either MOST will conduct if its $V_{GS} > V_T$ since V_{DS} is positive for the n-channel and negative for the p-channel device.

Suppose now that the input voltage V_1 is close to 0 V. T1 will be cut off since $V_{GS1} < V_{T1}$ but T2 will conduct since $V_{GS2} \sim V_{DD}$. This is much greater than V_{T2},

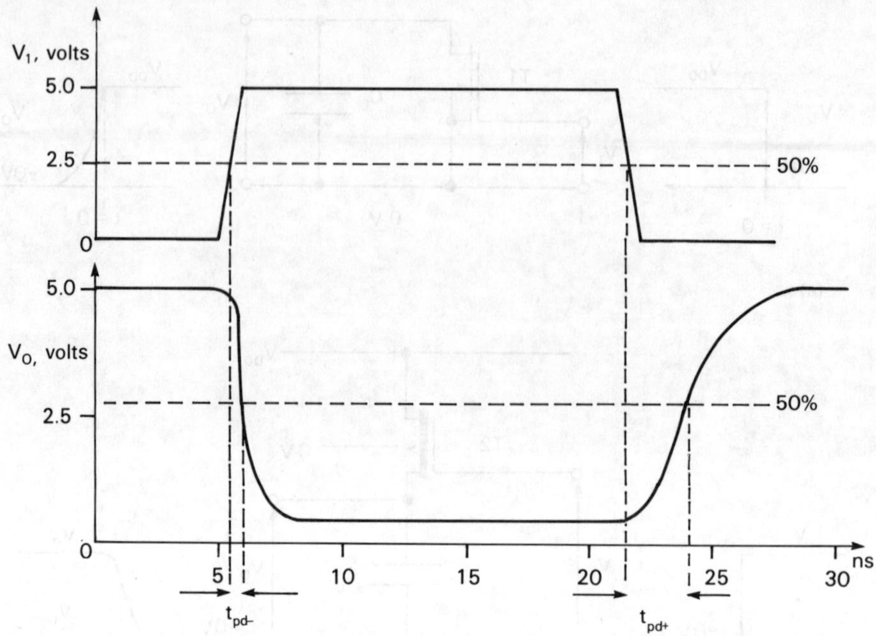

Figure 6.20 nMOS inverter transient response obtained by SPICE simulation for the circuit of Fig. 6.16. C_L is represented by the input to a second, identical inverter. $t_{pd-} = 0.55$ ns, $t_{pd+} = 2.30$ ns as measured at the 50% points.

so V_0 will be near V_{DD}. On the other hand, when V_1 is almost at V_{DD}, T1 will conduct and T2 will be cut off, so V_0 is nearly 0 V. The output voltages are within approximately 1 μV of V_{DD} and 0 V respectively because the drain current of the conducting MOST equals the leakage current of the cut off MOST. The conducting device is thus working close to the origin of its $I_D - V_{DS}$ characteristic. These features allow CMOS logic circuits to work with TTL circuits. With $V_{DD} = +5$ V, the same as the TTL supply voltage, a logical 0 and 1 are within a few mV of 0 V and 5 V respectively. These are well within the TTL limits of 0.2 V maximum for logical 0 and 2.8 V minimum for logical 1.

Power dissipation

Since the leakage current could be around 1 nA, the power dissipation under static conditions when either T1 or T2 is conducting is extremely low. The input current at the gates is also very low so the total dissipation in a static condition is around 1 nW. However, while the output is changing from one level to the other there are two effects which lead to a higher dissipation under dynamic conditions.

The first effect concerns the energy stored in the load capacitance C_L which is due mainly to the input capacitance of a following gate. When $V_0 = V_{DD}$ the energy stored is $1/2\ C_L V_{DD}^2$ and this is dissipated mainly within T1 when V_0 falls

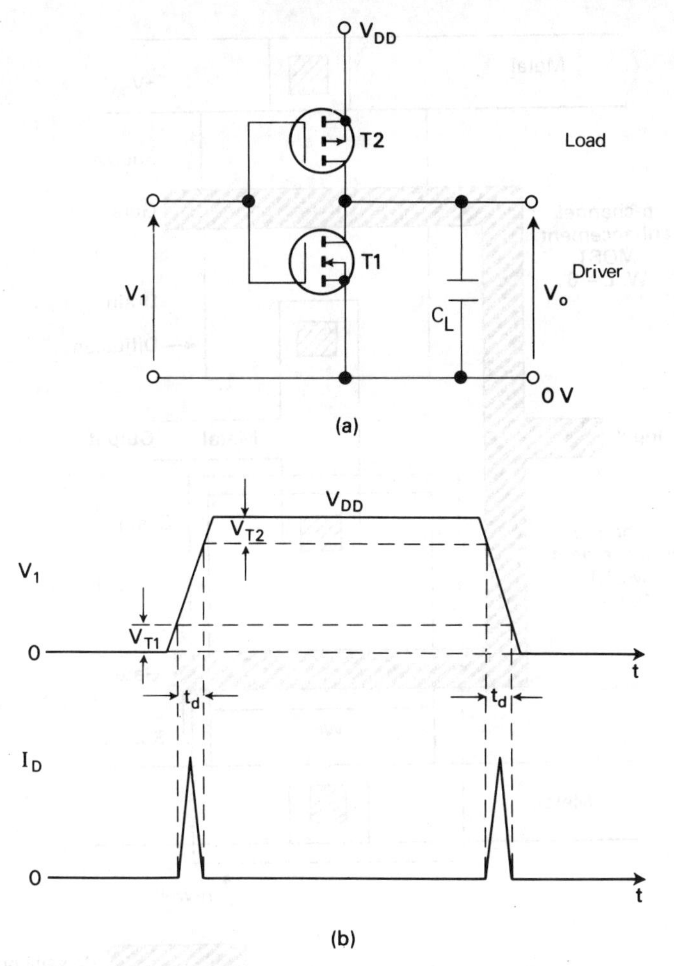

Figure 6.21 CMOS inverter. (*a*) Circuit; T1 is an n-channel enhancement driver or pull-down transistor, T2 is a p-channel enhancement load or pull-up transistor; (*b*) effect of gate voltage rise and fall times on drain current (transistion currents).

to 0 V. If the frequency at which V_0 changes is f then the power dissipated in T1 is 1/2 $C_L V_{DD}^2 f$. A similar power dissipation will occur in T2 when C_L is being charged, so the total dissipation is given approximately by

$$P = C_L V_{DD}^2 f \tag{6.55}$$

Worked example

A CMOS inverter operates from a 5 V power supply into a load of 1 pF at a switching frequency of 1 MHz. Obtain the dynamic power dissipation.

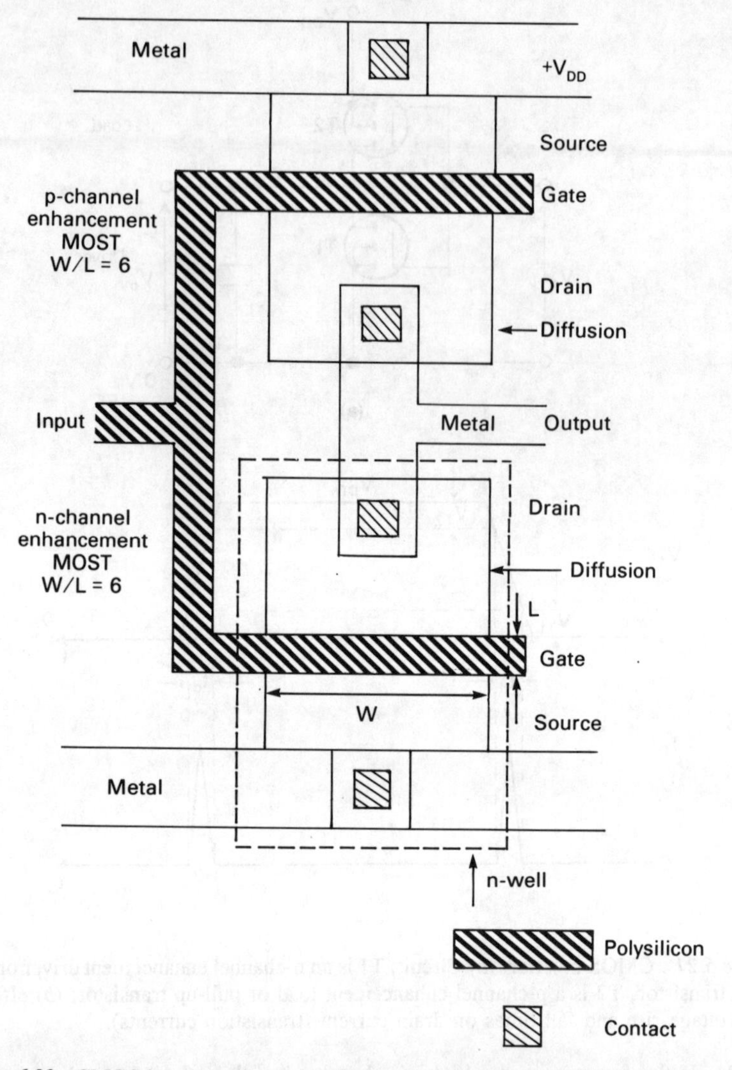

Figure 6.22 CMOS inverter, plan view of a possible layout. The formation of the diffusion and polysilicon layers is explained in Chapter 7.

Solution

From eq. (6.55) the dynamic power dissipation is given by

$$P = C_L V^2 f$$
$$= 10^{-12} \times 25 \times 10^6$$
$$= 25\,\mu\text{W}$$

The dynamic power dissipation is proportional to the clock frequency and so is often expressed in μW/MHz. In this example it would be 25 μW/MHz which allows values at other clock frequencies to be easily obtained. This concept is extended further in Problem 7.6 in the next chapter, where units of μW/pF/MHz are used to include the effect of the load capacitance.

The second effect involves the switching time between one output level and the other, which is controlled by the rate at which V_1 changes (Fig. 6.21(*b*)). There will be a short time t_d when V_1 has exceeded the threshold of T1 but not yet reached the threshold of T2 during switch on. During the time t_d both MOSTs are conducting together and a similar situation arises during switch-off. The shorter the rise or fall time of V_1 the shorter t_d will be and the lower the resulting dissipation. Power dissipation well above the static level will occur during time t_d and the dissipation level will be directly related to the rise or fall time of V_1. When these times are fixed the average dissipation will depend on the number of times per second that switching occurs. Thus power dissipation due to the effect of finite switching speed will be a function of frequency, as it was for the capacitive effect.

Of the three sources of power dissipation the charge and discharge of the load capacitance accounts for more than 90% of the total dissipation. The transition currents occurring due to the effect of finite input switching speed account for less than 10% of the total and the static power dissipation is almost negligible.

Transfer characteristic

In each steady state one transistor is conducting and the other is cut off, so the individual transistor resistances are less important than in nMOS and the inverter is therefore *ratioless*. The design is often simplified by making the aspect ratio of the two transistors equal, as in Fig. 6.22 which leads to the resistance of the p-channel device being greater than that of the n-channel device. This affects both the transfer characteristic and the propagation delay.

For the whole inverter

$$I_{D1} = I_{D2}$$

$$V_1 = V_{GS1} = V_{DD} + V_{GS2}$$

$$V_0 = V_{DS1} = V_{DD} + V_{DS2} \qquad (6.56)$$

where V_{GS2} and V_{DS2} are both negative. The two steady states occur when V1 is nearly 0 V (low) and nearly V_{DD} (high) and these are illustrated through the transistor characteristics in Fig. 6.23. In each case the operating point is given by the intersection of the T1 and T2 characteristics, since I_{D1} and I_{D2} must be equal, at the very low leakage current value.

For a low input T2 operates in the ohmic region, so from eq. (6.31), putting $V_{DS2} = V_0 - V_{DD}$ and $V_{GS2} = V_1 - V_{DD}$

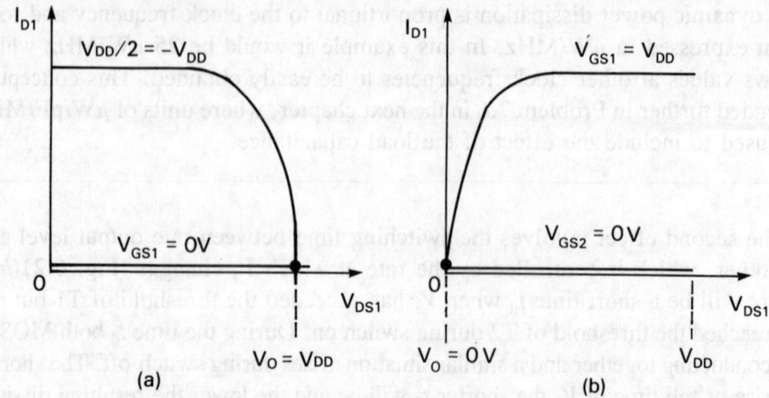

Figure 6.23 CMOS inverter. (*a*) Combined I_D/V_{DS} characteristic, $V_{GS1} = 0$ V; (*b*) combined I_D/V_{DS} characteristic, $V_{GS1} = V_{DD}$.

$$I_{D2} = K_2 \left[(V_O - V_{DD})\ (V_1 - V_{DD} - V_{T2}) - \frac{(V_O - V_{DD})^2}{2} \right] \tag{6.57}$$

and I_{D2} equals the leakage current of T1. Then V_O can be obtained from eq. (6.57) and will be almost equal to V_{DD}.

For a high input T1 operates in the ohmic region, so from eq. (6.31)

$$I_{D1} = K_1 \left[V_O\ (V_1 - V_{T1}) - \frac{V_O{}^2}{2} \right] \tag{6.58}$$

and I_{D1} equals the leakage current of T2. Again, V_O can be obtained from eq. 6.56 and will be almost 0 V. The circuit behaves as an inverter since when the input is low the output is high and vice versa.

The transfer characteristic of Fig. 6.24 shows how the output voltage changes between the static levels. This has been obtained by SPICE simulation using the aspect ratio of Fig. 6.22 and the typical values given in Table 6.3 for KP, CGSO, CGDO, CJ and TOX. VTO = +1 V for T1 and –1 V for T2 and AS = AD = 100 μm^2, L = 4 μm and W = 24 μm for each transistor. A gate length of 4 μm was used and the complete program is given in Appendix 3.

As V_1 is increased from 0 V, V_O remains constant until V_1 is just above V_{T1}. V_O then falls almost linearly as T1 and T2 conduct together and reaches 0 V just before T2 cuts off, where $V_1 = V_{DD} + V_{T2}$ with V_{T2} being negative. The output becomes $V_{DD/2}$ for an input below $V_{DD/2}$ because T1 requires a lower voltage to achieve the same current as T2. This is due to K_1 being about $3K_2$ in this design, since $(W/L)_{T1} = (W/L)_{T2}$ and the electron mobility is about three times the hole mobility. For a design with $(W/L)_{T2} = 3(W/L)_{T1}$ the transfer characteristic would be symmetrical, with $V_O = V_{DD/2}$ for $V_1 = V_{DD/2}$. When V_1 is increased to 5 V V_O has fallen to the very low value of 0.014 μV.

In either case, the linearity of the falling part of the characteristic is good enough

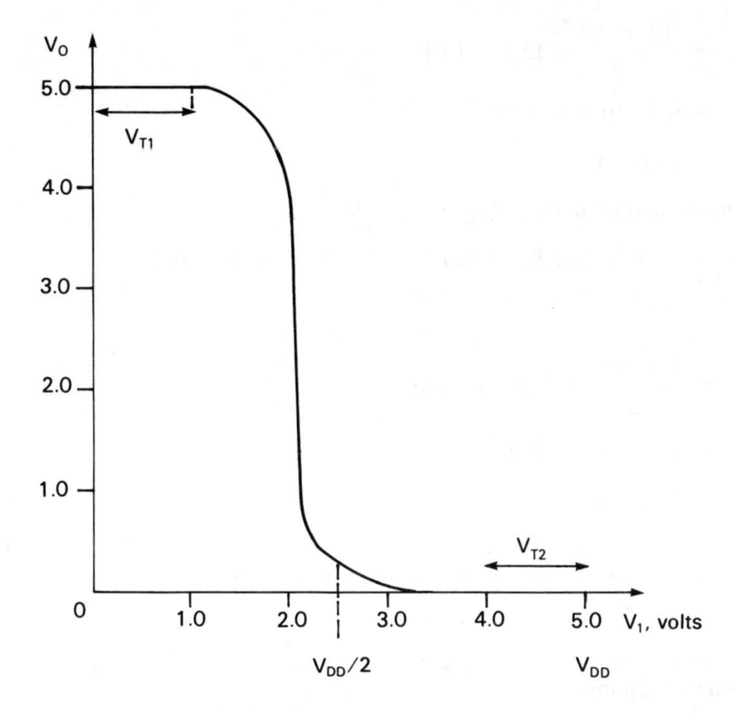

Figure 6.24 CMOS inverter transfer characteristic obtained by SPICE simulation for the circuit of Fig. 6.21(*a*). $V_0 = 0.014\ \mu V$ for $V_1 = 5$ V, $V_0 = 2.5$ V for $V_1 = 2.05$ V.

to allow the circuit to be used as an amplifier and CMOS integrated circuit amplifiers are available commercially, as well as CMOS logic circuits and digital systems.

Worked example

Estimate the maximum value of the supply current for the CMOS inverter with the transfer characteristic of Fig. 6.24.

Solution

The maximum supply current occurs during the transition between high and low output levels when both transistors are conducting equally. From Fig. 6.24 this will occur for an estimated input voltage of 2.1 V and both transistors will be in the saturated region so that eq. (6.35) is applicable.

For the n-channel device $V_{GS1} = 2.1$ V, $V_{T1} = 1.0$ V and $K_1 = 2 \times 10^{-5} \times 6 = 12 \times 10^{-5}$ A/V^2. Then

$$I_{D1} = \frac{12 \times 10^{-5}}{2} [2.1 - 1.0]^2$$
$$= 6 \times 10^{-5} \times 1.21$$
$$= 72.6 \ \mu A$$

For the p-channel device $V_{GS2} = -2.9$ V

$$V_{T2} = -1.0 \text{ V and } K_2 = 6.67 \times 10^{-6} = 4 \times 10^{-5} \text{ A/V}^2$$

Then

$$I_{D2} = \frac{4 \times 10^{-5}}{2} [-2.9 - (-1.0)]^2$$
$$= 2 \times 10^{-5} \times 3.61$$
$$= 72.2 \ \mu A$$

The maximum supply current is therefore about 72 μA, the difference between I_{D1} and I_{D2} occurring because the required input voltage was not exactly 2.1 V.

Transient response

For $V_1 = 0$ V T1 is cut off and C_L has been charged up to V_{DD} through T2 (Fig. 6.25). When V_1 is suddenly increased T2 is cut off and C_L is discharged through

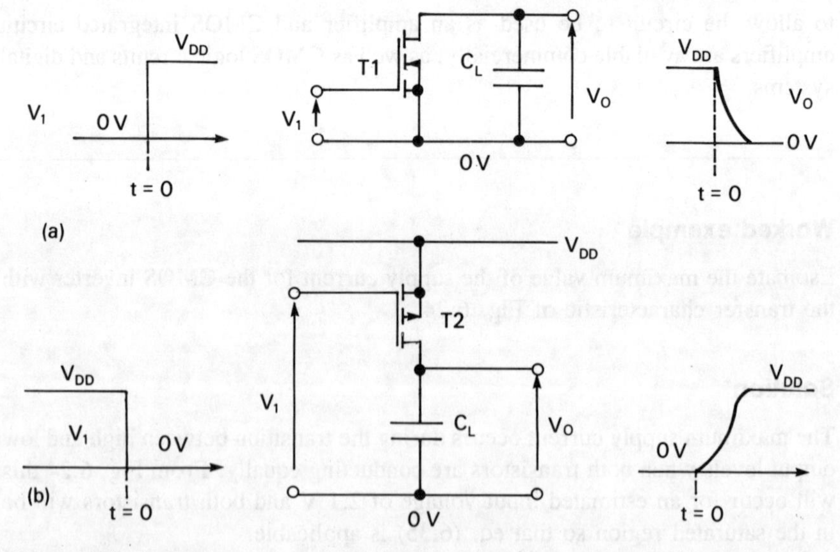

Figure 6.25 CMOS inverter. *(a)* T1 acting as a pull-down transistor; *(b)* T2 acting as a pull-up transistor.

T1. When V_1 is suddenly reduced T1 is cut off and C_L is charged up to V_{DD} through T2 again. The propagation delays between 50% points will depend on the resistances of T1 and T2, which means that t_{pd-} due to T1 is less than t_{pd+} due to T2 (Fig. 6.26). The values obtained from the SPICE simulation are 0.7 ns for t_{pd-} and 1.6 ns for t_{pd+} again with a similar inverter used as a load to provide a voltage dependent C_1 (Fig. A.3.2). The propagation delay t_{pd} taken as an average is 1.2 ns, less than the value for the nMOS inverter mainly due to the high aspect ratios and low resistances of the CMOS devices.

Scaling of nMOS and CMOS inverters

A basic approach to reducing the size of an MOST is to avoid short-channel effects by selective scaling, which means shallower junctions, thinner oxide and heavier substrate doping. The minimum channel length for which long-channel behaviour is maintained has been shown (Ref. [5]) empirically to be

$$L_{min} = 0.4\ [x_j\ d(w_s + w_d)^2]^{1/3} \qquad (6.59)$$

Here x_j is the junction depth in μm, d the gate oxide thickness in Å and $(w_s + w_d)$ the sum of the source and drain depletion layer widths in μm. w_s and w_d are obtained from eq. (3.74) with $N_j = N_a$ for n-channel devices, since the source

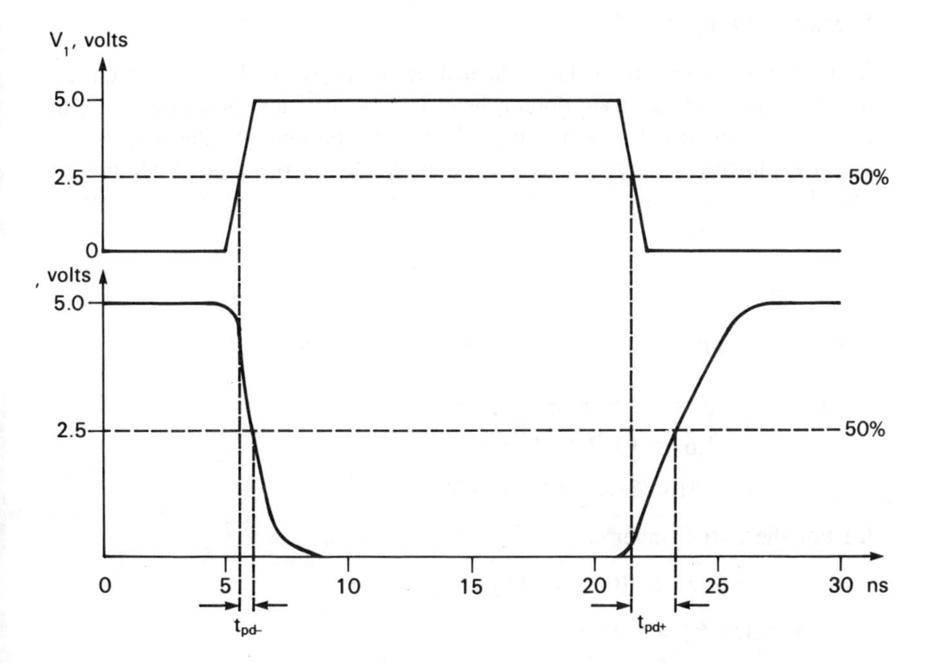

Figure 6.26 CMOS inverter transient response obtained by SPICE simulation for the circuit of Fig. 6.21. C_L is represented by the input to a second, identical inverter. t_{pd-} = 0.7 ns, t_{pd+} = 1.6 ns as measured at the 50% points.

and drain doping levels N_d are much greater than the substrate doping level N_a. The voltage V in eq. (3.74) becomes V_{BS} for the source and $V_{DS} + V_{BS}$ for the drain so that

$$w_s = \left[\frac{2\epsilon_r \epsilon_0}{eN_a}(\psi + V_{BS})\right]^{1/2} \tag{6.60}$$

and

$$w_d = \left[\frac{2\epsilon_r \epsilon_0}{eN_a}(\psi + V_{DS} + V_{BS})\right]^{1/2} \tag{6.61}$$

since the substrate is reverse biased with respect to source and drain.

As a starting point eq. (6.59) will be applied to the MOSTs in the nMOS and CMOS inverters described above. For nMOS, L_{min} applies only to the driver transistor due to the relative aspect ratios, but for CMOS L_{min} applies to both transistors. The SPICE values used to start the scaling simulation are $x_j = 1$ μm, $d = 1000$ Å $= 0.1$ μm, $N_a = 4 \times 10^{21}$ m^{-3} and $\psi = 0.8$ V with a channel length of 4 μm.

Worked example

In a MOS inverter circuit the n-channel driver transistor has $x_j = 1$ μm and $d = 0.1$ μm. The substrate doping level is 4×10^{21} acceptor atoms/m^3, the diffusion potential is 0.8 V and the substrate is connected to the source.

Determine the minimum channel length of the driver for (*a*) an nMOS inverter with $V_{DS} = 0.3$ V and (*b*) a CMOS inverter with $V_{DS} = 0$ V, both voltages representing saturation levels.

Solution

For both circuits $V_{BS} = 0$ V and from eqs (6.60) and (6.61)

$$\frac{2\epsilon_r \epsilon_0}{eN_a} = \frac{2 \times 12 \times 8.85 \times 10^{-12}}{1.6 \times 10^{-19} \times 4 \times 10^{21}}$$
$$= 3.318\ 75 \times 10^{-13}\ \text{m}^2\ \text{V}^{-1}$$

(*a*) For the nMOS inverter

$$w_s = (3.318\ 75 \times 10^{-13} \times 0.8)^{1/2}$$
$$= 5.152\ 67 \times 10^{-7}\ \text{m}$$

$$w_d = (3.318\ 75 \times 10^{-13} \times 1.1)^{1/2}, \text{ since } V_{DS} = 0.3\ \text{V}$$
$$= 6.042\ 0 \times 10^{-7}\ \text{m}$$

From eq. (6.59)

$$(w_s + w_d)^2 = (1.194\,7 \times 10^{-6})^2$$
$$= 1.253\,2 \times 10^{-12}\ \text{m}^2$$
$$L_{min} = 0.4(10^{-6} \times 10^3 \times 1.253\,2 \times 10^{-12})^{1/3}$$
$$= 0.4 \times 1.078\,1 \times 10^{-5}$$
$$= 4.3\ \mu\text{m}$$

(*b*) For the CMOS inverter

$$w_s = w_d = 5.152\,67 \times 10^{-7}\ \text{m}$$
$$(w_s + w_d)^2 = (1.030\,5 \times 10^{-6})^2$$
$$= 1.062\,0 \times 10^{-12}\ \text{m}^2$$
$$L_{min} = 0.4(10^{-6} \times 10^3 \times 1.062\,0 \times 10^{-12})^{1/3}$$
$$= 0.4 \times 1.020\,2 \times 10^{-5}$$
$$= 4.1\ \mu\text{m}$$

In each case, L_{min} is within 8% of the channel length of 4 μm, so long-channel (level 1) simulations are appropriate as a first approximation.

In carrying out the scaling simulations the input data are divided into 'electrical' parameters and the 'processing' parameters which are used to derive the electrical parameters. Thus VTO and LAMBDA are necessary electrical parameters, specifying the MOST type and the output conductance respectively. Although KP is an electrical parameter it is not specified but derived from the processing parameters UO and TOX and, similarly, GAMMA and PHI are derived from NSUB using eqs (6.43) and (6.44). CJ is derived from NSUB, PB and MJ together with source and drain areas As and AD. XJ is included to conform with eq. (6.59) and CGSO and CGDO define the overlap capacitances as before.

Table 6.5 shows the effect of applying scaling factors to the MOSTs in the nMOS and CMOS inverters previously discussed, which used a minimum dimension of 4 μm. This has been scaled down to 3 and 2 μm with some input parameters remaining constant, others changing according to the scaling factors of 1.33 and 2.0 and the remaining parameters derived within the simulation program. The effects of output conductance and substrate bias which were previously omitted have now been included through LAMBDA, GAMMA and PHI. The main improvement in performance is the reduction is propagation delay which is due to smaller internal capacitance as a result of area reductions.

As expected KP rises due to the reduction in TOX as the devices are scaled

Table 6.5 Scaling effects simulated in nMOS and CMOS inverters using SPICE, with constant 5 V power supply

Input parameters independent of scaling

VTO = –3.0 V for n-channel depletion MOST
VTO = +1.0 V for n-channel enhancement MOST
VTO = –1.0 V for p-channel enhancement MOST
UO = 600 cm^2/V s for n-channel MOST
UO = 200 cm^2/V s for p-channel MOST
LAMBDA = 0.02 V^{-1}
PB = 0.8 V
MJ = 0.5
CGSO = 3×10^{-10} F/M of channel width
CGDO = 3×10^{-10} F/M of channel width

Input parameters dependent on scaling

nMOS	Load	W	4	3	2	(μm)
		L	8	6	4	(μm)
	Driver	W	8	6	4	(μm)
		L	4	3	2	(μm)
		AS,AD	60	40	20	(μm^2)
CMOS		W	24	18	12	(μm)
		L	4	3	2	(μm)
		AS,AD	100	60	30	(μm^2)
		XJ	1.0	0.75	0.5	(μm)
		TOX	0.1	0.075	0.05	(μm)
		NSUB	4	6	8	($\times 10^{15}$ cm^{-3})

Input parameters derived by SPICE

	L	4	3	2	μm
nMOST	KP	2.07	2.76	4.14	($\times 10^{-5}$ A/V^2)
pMOST	KP	0.691	0.921	1.38	($\times 10^{-5}$ A/V^2)
	GAMMA	1.055	0.969	0.746	($V^{1/2}$)
	PHI	0.648	0.669	0.684	(V)
	CJ	2.04	2.49	2.88	($\times 10^{-4}$ F m^{-2})

Results obtained from simulations

	L	4	3	2	(μm)
nMOS	t_{pd-}	0.50	0.40	0.20	(ns)
	t_{pd+}	3.10	1.65	0.65	(ns)
	t_{pd}	1.80	1.03	0.43	(ns)
	Power dissipation	2.27	3.06	4.70	($\times 10^{-4}$ W)
CMOS	t_{pd-}	0.65	0.45	0.25	(ns)
	t_{pd+}	1.40	0.90	0.55	(ns)
	t_{pd}	1.03	0.68	0.40	(ns)
	Power dissipation	3.46	3.46	3.46	($\times 10^{-11}$ W)

Scaling factor $K = 1.33$ from 4 to 3 μm and 2.0 from 4 to 2 μm.

down and since V_{DD} has not been scaled but kept constant at 5 V, the nMOS static power dissipation has increased. This could be maintained constant by scaling V_{T2} with the dimensions, thus reducing I_{D2} (eq. (6.52)) and P_s. CJ also increases with scaling due to the increase in NSUB, as derived from eq. (3.78) with $N_j = N_a =$ NSUB. However, CJ is multiplied by the areas AS and AD, which are reduced, so the source- and drain-to-substrate capacitances fall. This also applies to the overlap capacitances derived from CGSO and CGDO and so the propagation delay t_{pd} falls with scaling in both nMOS and CMOS. The CMOS static power dissipations remain constant and very low as they are due to leakage currents which are assumed to be constant. The transfer characteristics for both nMOS and CMOS inverters remained substantially unchanged with scaling, except that the drain voltage at which the nMOS driver transistor entered saturation fell to about 1.3 V, due to the inclusion of extra parameters.

These level 1 simulations provide an introduction to the capabilities of the SPICE program and illustrate how preliminary first-order results can be produced when an integrated circuit design is being developed or modified. The parameter values are chosen to be typical here but in practice the values appropriate to the manufacturing process would be used. Level 1 would be followed by level 2 simulations to improve accuracy by including second-order effects, as described in Ref. [4]. This is especially necessary as dimensions are reduced to 2 μm and below and, ultimately, the level 3 simulation might be used.

Charge transfer devices

An important recent development of MOS technology is the *charge transfer device*, which is essentially an analogue shift register. In one common form a number of gate electrodes are deposited in sequence and a small packet of charge held under each gate. These packets are transferred from one gate to the next when clocking waveforms are applied to the gates. The transfer rate depends on the clock frequency and the time delay between input and output is then proportional to the number of gates. The charge packets can represent the instantaneous level of an analogue signal, or presence and absence of charge can represent the two levels of a digital signal. These properties lead to applications in the three main areas of analogue delay lines, digital shift registers and optical imaging devices.

Charge transfer devices (CTD) can be divided into surface channel devices (SCCD), buried channel devices (BCCD) and bucket brigade devices (BBD) depending on the mechanism of charge transfer and each is considered in turn.

Surface channel charge-coupled device, SCCD

A single CCD structure is shown in Fig. 6.27 for a p-type substrate, which is similar to the gate structure of an n-channel enhancement mode MOST. As V_G is increased from zero, holes are repelled from the substrate below the gate where a depletion layer is formed. When V_G exceeds a threshold voltage V_T electrons can form a thin inversion layer at the surface of the substrate, as in the MOST. In practice these electrons are either drawn from an n-type input region or generated through the absorption of light energy by the substrate. As the amount of charge stored is increased, the width of the depletion region decreases in order to preserve charge neutrality. The charge density can lie between zero and a maximum value obtained from eq. (6.23) with $V_x = 0$, so that

$$Q_{max} = \epsilon_{ins}\epsilon_0 \frac{V_G - V_T}{d} \text{ C/m}^2 \tag{6.62}$$

The maximum charge then depends on gate area, as well as gate voltage and oxide thickness d. Q_{max} consists of depletion layer charge as well as signal charge from the minority carriers. Thus the maximum signal charge is kQ_{max} where $k \sim 0.5$

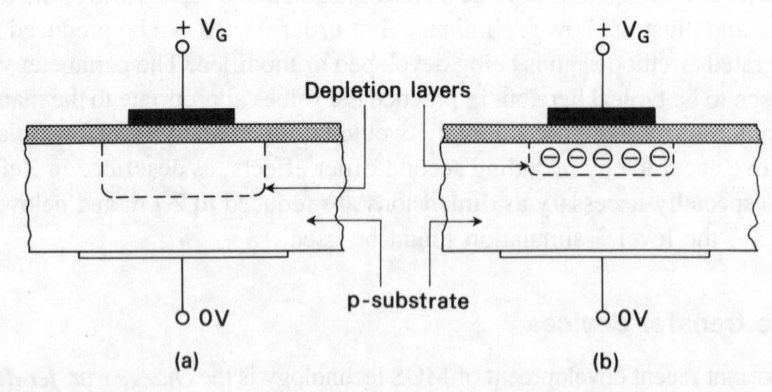

Figure 6.27 Single gate structure with positive bias above threshold. (*a*) Without charge packet; (*b*) with charge packet.

for a typical structure. Taking typical values of $\epsilon_{ins} = 3.9$, $V_G = 10$ V, $V_T = 2$ V, $d = 0.1$ μm and a gate area of 300 μm^2 leads to a maximum signal charge of about 0.4 pC. It should be noted that an alternative approach considers the electrons to be held in a *potential well* formed by the potential minimum in the silicon which constrains electrons to remain under the gate.

Charge transfer

When a charge packet has been formed it is transferred from one gate to the next by applying three clocking waveforms ϕ_1, ϕ_2 and ϕ_3 to gates 1, 2 and 3. Suppose that at time t_0 (Fig. 6.28(*a*)) ϕ_2 is at a high voltage V_{CC} and ϕ_1 and ϕ_3 are at a low voltage V_{SS}, which positions the charge packet under gate 2. Then at time t_1 (Fig. 6.28(*b*)) ϕ_3 is at V_{CC} and ϕ_2 is falling slowly to V_{SS}. Under gate 2 the depletion layer thickness decreases while under gate 3 it has increased and coupled with the previous depletion layer due to the close spacing of the gates. Finally, at time t_2 (Fig. 6.28(*c*)) ϕ_2 has reached V_{SS} and the charge packet is now under gate 3. Gate 1 has been held at V_{SS} throughout to prevent charge moving backwards and ϕ_2 has been reduced slowly to allow a finite time for the electrons to diffuse across the gate.

Thus three gates are required to store and transfer one charge packet and together they form one element of the charge-coupled device. Since the gates are arranged in triplets each gate in a triplet has the same clocking waveform as the corresponding gate in another triplet. Continuation of the clocking sequence, as shown in Fig. 28(*d*), causes the charge packet to move from gate 3 to the next gate 1 and so on to the output. A charge packet only occurs under every third electrode and these packets all move at the same time.

Since the surface channel must be maintained at all times V_{SS} exceeds V_T by a small amount. V_{SS} could be applied as a bias to the substrate, which would then allow the clocking waveforms to start at zero and lead to a simpler generating circuit.

During each transfer of a charge packet power is dissipated because the carriers fall through a potential difference nearly equal to the clock pulse amplitude. Thus for a 3-phase clock the maximum power dissipation is given approximately by

$$P_{max} = 3fVkQ_{max} \tag{6.63}$$

where f is the clock frequency and $V = V_{CC} - V_{SS}$.

Worked example

In a charge-coupled device the oxide has a relative permittivity of 3.9 and a thickness of 0.1 μm. A bias voltage of 2 V is applied to the substrate and a maximum voltage of 10 V is applied to the gates at a frequency of 1 MHz. The

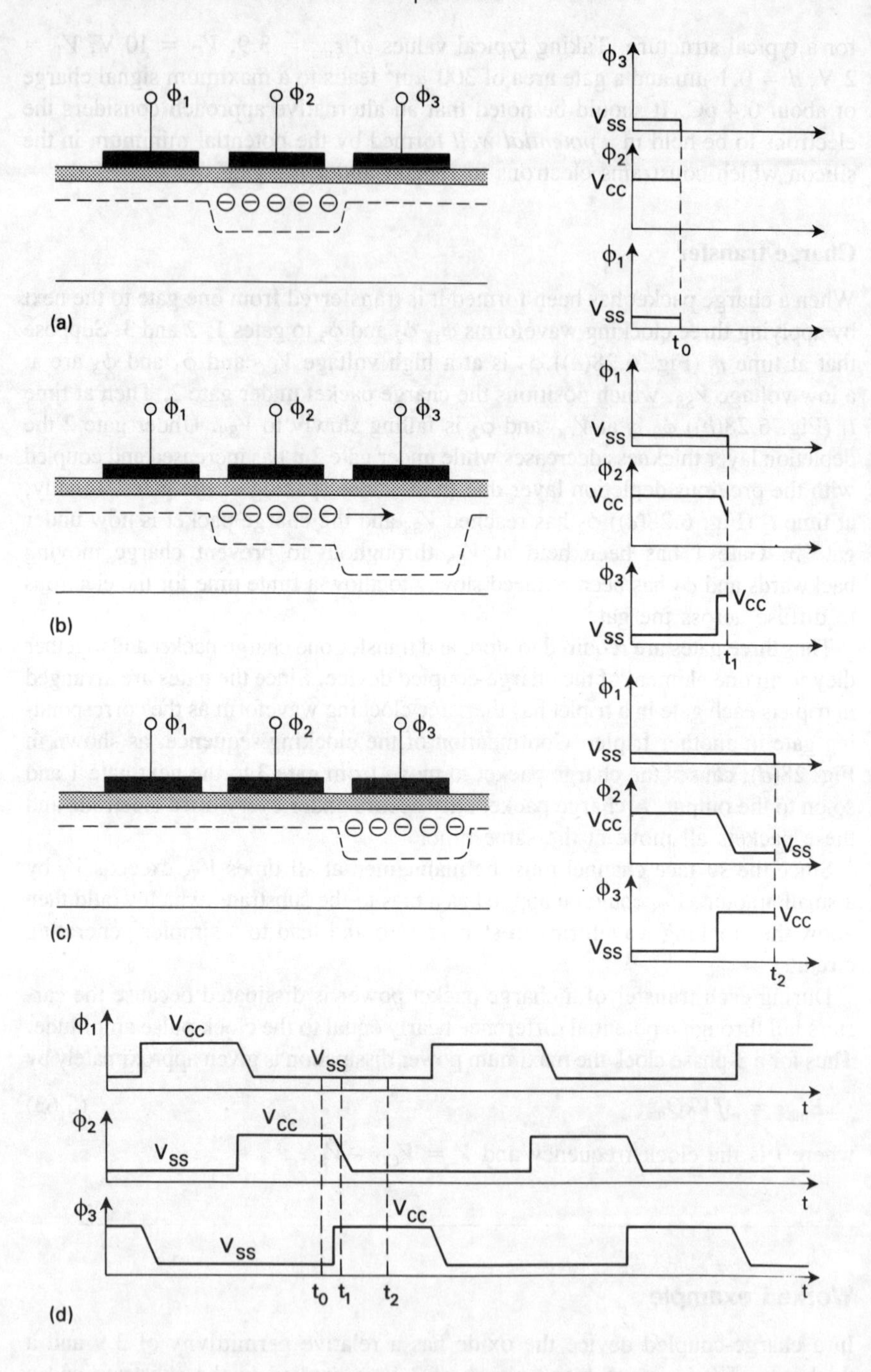

Figure 6.28 Clocking sequences for charge transfer in 3-phase charge-coupled elements.

maximum signal charge is 50% of the charge stored under each gate which has an area of 300 μm^2.

Determine the power dissipation for each element of the device.

Solution

From eq. (6.56)

$$Q_{max} = \frac{3.9 \times 8.85 \times 10^{-12} \times (10 - 2) \times 3 \times 10^{-10}}{10^{-7}}$$

$$= 8.28 \times 10^{-13}\ \text{C}$$

From eq. (6.57)

$$P_{max} = 3 \times 10^{6} \times 8 \times 0.5 \times 8.28 \times 10^{-13}$$

$$= 9.94\ \mu\text{W per element.}$$

Input and output of charge

Inputs can be either electrical or optical. For electrical inputs an n^+ source region is diffused into the substrate (Fig. 6.29(*a*)) and slightly reverse biased so that the depletion layer couples with that under the first gate 1. The amount of charge injected into the surface channel is controlled by a positive voltage on the input gate, with the depletion layer under the first gate 1 acting as a virtual drain.

When used as an imaging device the CCD converts incident light into charge packets. The light quanta generate electron-hole pairs when they enter the substrate, with electrons and holes being separated at the depletion layer boundary. The quanta may enter between the electrodes, through transparent electrodes or through the back face of a thin substrate. The charge packets collect as a pattern under the array of electrodes to produce an analogue replica of the variation of light intensity. Since the CCD is continuous, although it may be folded to cover a comparatively large area, the charge pattern is extracted sequentially as a series of current levels corresponding to the intensity levels of the image.

Output is normally taken through a reverse-biased p-n junction (Fig. 6.29(*a*)) so that its depletion layer couples with that under the last gate. The output gate, which has a fixed bias, minimizes the capacitive voltage pick-up between the output diode and the clock pulses on the last gate 3. The average output current is given by $i = dQ/dt$ and for a 0.1 pC charge packet clocked at 1 MHz this is around 0.1 μA. The signal could be measured as a voltage fluctuation across a series resistor between the output diode and a power supply. However, since the current is so small a low capacitance amplifier is often included on the chip and the simplest form is an MOS transistor (Fig. 6.29(*b*)). The reset gate is connected

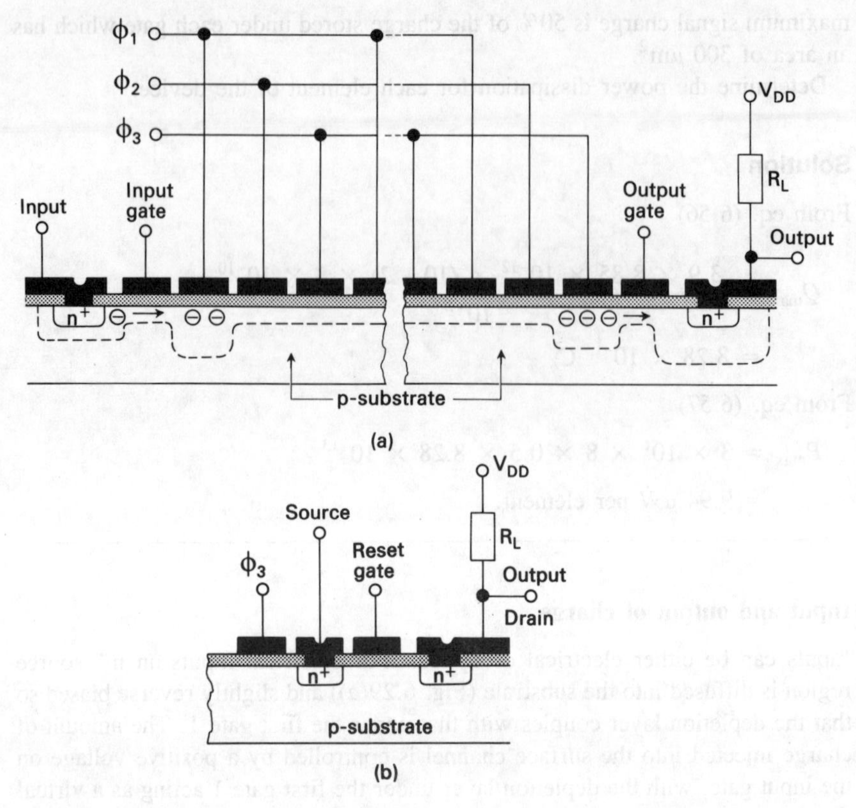

Figure 6.29 Surface channel charge-coupled device. (*a*) Diode output stage; (*b*) transistor output stage.

to the source and pulsed momentarily after the arrival of a charge packet. This clears the source and the charge packet moves rapidly towards the drain, which is held at the most positive potential. Thus charge is extracted more quickly from under the last gate, 3, and the output current is larger than if the output had been taken directly from the source.

Frequency limitations

The charge-coupled device cannot retain its charge packets indefinitely since, if the clock voltages were held constant, the depletion layers would gradually fill with minority carriers. These are due to the thermal generation of electron-hole pairs, providing a *dark current*, and the charge packets carrying information would eventually be masked. Thus the CCD is a dynamic device with a minimum clock frequency, typically of about 2 kHz.

The maximum clock frequency depends on the finite time for charge packets to transfer from under one gate to the next, which is reduced by closely spacing the gates. The clock edge must also fall slowly to prevent a rapid collapse of the depletion layer and consequent recombination of holes with electrons in the charge packet. These effects limit the maximum frequency typically to around 10 MHz. If the maximum frequency is exceeded some charge is left behind at each transfer and again information may be lost, this time due to inefficient transfer. Charge may also be left behind due to trapping at surface states, as explained below.

Transfer inefficiency ϵ

The *transfer efficiency* η is the fraction of a charge packet correctly transferred at each transfer and is typically in the range 99.90 to 99.99%. The *transfer inefficiency* ϵ is more convenient and is given by $\epsilon = 1-\eta$. This is cumulative since if Q_n is the charge *remaining* under the *n*th electrode after *n* transfers and Q_0 is the initial charge

$$Q_n \sim Q_0\eta^n \sim Q_0(1-n\epsilon) \tag{6.64}$$

Apart from a high clock frequency the other important cause of transfer inefficiency is the effect of carrier trapping states, which occur mainly at the interface between the silicon and the silicon dioxide. As each charge packet arrives under a gate a small part is immediately trapped and as the packet is moved on the states empty again more slowly. This leads to small residual charges trailing behind the main packet, with a consequent rise in transport inefficiency and fall in frequency response.

The effect can be reduced by ensuring that the interface states remain permanently filled and this is partially achieved by passing a constant charge through the array. This is called a *fat zero* and in practice the charges need to be about 10 to 25% of Q_{max}. The main disadvantage, however, is that less signal charge can be used so that the dynamic range of the device is reduced below 100 : 1 (40 dB).

Practical devices use silicon gates which overlap to reduce the effective gate spacing to the oxide thickness which is significantly less than the 1 μm that can be achieved with normal metallizing. 4-phase and 2-phase clocks are also used (Ref. [6]).

Buried channel charge-coupled device BCCD

In the surface channel CCD the charge packets travel along the $Si-SiO_2$ interface and so are subject to surface trapping. The problem can be overcome by constraining the charges to travel along a channel below the interface and this is achieved in the buried channel CCD (Fig. 6.30(*a*)). An n-type layer is formed on the p-substrate and reverse biased typically at 30 to 40 V. This reinforces the depletion layer due to the gate voltage and the resulting potential distribution is

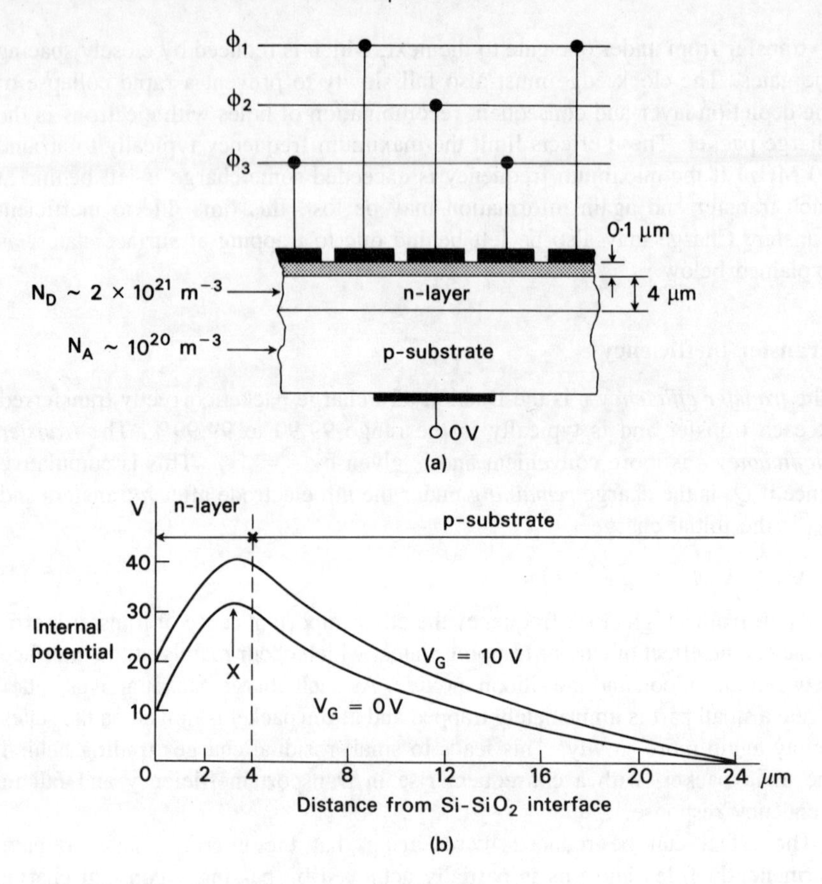

Figure 6.30 Buried channel charge-coupled device (not to scale). (*a*) 3-phase element; (*b*) potential distribution through device.

shown in Fig. 6.30(*b*) for two values of gate voltage. The potential reaches a maximum X at 2 to 3 μm away from the interface and so the charge packets will travel along a channel in this position. The charge packet will move from under a gate at low potential towards the next gate if it is at a higher potential (Ref. [7]).

Similar clocking waveforms are applied to the gates as in the surface channel device, although the voltage levels may be somewhat different. Similar gate structures, input and output arrangements can also be used, with the reverse channel bias applied at the output.

Since surface trapping is virtually eliminated the transfer inefficiency is less than that of the SCCD. The operating frequency is increased since the mobility of electrons is greater in the n-type layer than along the interface. Typical values are transfer inefficiencies of 5×10^{-4} up to 20 MHz for buried channel devices, which indicates an order of magnitude improvement in each parameter.

The bucket-brigade device BBD

The bucket-brigade device was introduced in 1969 before the charge-coupled device (Ref. [8]) and the basic structure is shown in Fig. 6.31(*a*) with an integrated circuit form in Fig. 6.31(*b*). It consists of a series of NMOS devices, each with a capacitor connected between drain and gate. Charge is transferred between the capacitors when alternate transistors conduct and so only a two-phase clock is

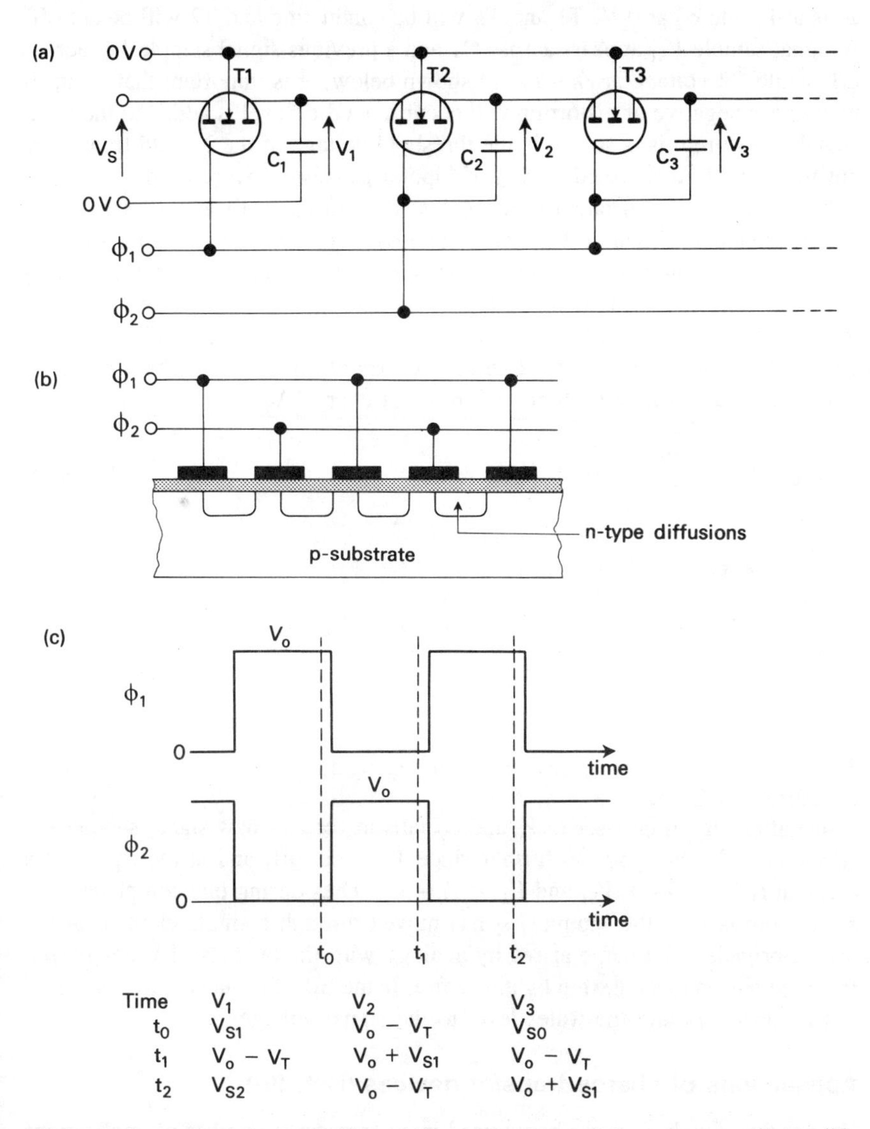

Time	V_1	V_2	V_3
t_0	V_{S1}	$V_o - V_T$	V_{SO}
t_1	$V_o - V_T$	$V_o + V_{S1}$	$V_o - V_T$
t_2	V_{S2}	$V_o - V_T$	$V_o + V_{S1}$

Figure 6.31 Bucket-brigade device. (*a*) Input sampling and first element; (*b*) possible integrated circuit form; (*c*) 2-phase clocking waveforms and voltage levels.

required using complementary square waveforms. A version can be built using discrete components and either bipolar or field-effect transistors, which is not possible with the charge-coupled devices described above.

The input signal V_S is introduced through the sampling transistor T1 on to capacitor C1 which is grounded. The first BBD element consists of the pair of transistors T2 and T3 with capacitors C2 and C3 and these transistor pairs are repeated along the length of the device. Then at time t_0 (Fig. 6.31(*c*)), where ϕ_1 is at V_0 and ϕ_2 at 0 V, T1 and T3 will be conducting and T2 will be cut off. A signal sample V_{S1} appears across C1 and a previous signal sample V_{S0} across C3, while C2 voltage is $V_0 - V_P$, as shown below. It is important that V_S itself never goes negative or distortion will occur, so a d.c. level is added to the input signal to ensure this is so. V_T is the threshold voltage of the NMOS transistors but this would be replaced by V_{BE} if bipolar transistors were used.

Just after time t_0, ϕ_1 falls to zero and ϕ_2 rises to V_0 so T1 and T3 are cut off and T2 conducts. Then at this rising edge ϕ_2 is added to the existing value of V_2, since the voltage across a capacitor cannot change instantaneously, and V_2 becomes $2V_0 - V_T$. This is much higher than V_1 so charge Δq flows from C2 into C1 and at time t_1 V_2 has fallen to $V_0 + V_{S1}$. The decay time of V_2 depends on the effective resistance of T2 and the values of C1 and C2, which normally have equal capacitance *C*. Since C2 has *lost* charge Δq

$$2V_0 - V_T - \frac{\Delta q}{C} = V_0 + V_{S1} \tag{6.65}$$

so

$$\Delta q = C(V_0 - V_{S1} - V_T) \tag{6.66}$$

C1 has *gained* charge Δq and V_1 has risen by Δq/C1, so if C1 = C2 = C

$$\begin{aligned} V_1 &= V_{S1} + \Delta q/C \\ &= V_0 - V_T \end{aligned} \tag{6.67}$$

Thus at time t_1 C2 has gained the signal sample V_{S1} from C1 due to charge transfer from C2 to C1.

Just after time t_1 ϕ_1 rises to V_0 and ϕ_2 falls to zero. A new signal sample V_{S2} appears on C1, but goes no further since T2 is cut off, and at the end of this phase at t_2 $V_3 = V_0 + V_{S1}$ and $V_2 = V_0 - V_T$. Thus during one complete clock cycle from t_0 to t_2 the sample V_{S1} has moved through a single element of the bucket-brigade. The name arises by analogy with the buckets of water passed from one person to the next in fighting a fire. In the BBD the buckets are analogous to the capacitors and the water level to the signal voltage.

Applications of charge transfer devices (Ref. [9])

Charge transfer devices are being used in an increasing number of applications which at present fall into two main areas:

(i) delay and processing of analogue signals, and
(ii) imaging of visible radiation.

(i) Where there are N elements and the clock frequency is f_c the delay introduced is N/f_c.

In practice, each element may be followed by an extra MOST with a fixed gate voltage and no capacitor. This reduces the transfer inefficiency but means that two such stages are needed to introduce a delay of one clock period. The delay is then $n/2f_c$, where n is the number of stages and for a 512 stage delay line with f_c between 5 and 300 kHz the delay will range between 51.2 and 0.853 ms. f_c must be at least three times the input signal frequency to avoid sideband interference and a dynamic range of 83 dB (14 000 : 1) can be achieved. Such a delay line has many applications in acoustics, for example introducing vibrato into electronic organs and simulating reverberation in recorded music.

The bucket-brigade device is considered to be the most suitable for sampling (clock) rates below about 5 MHz and for delays not exceeding 3500 elements, above which transfer inefficiency rises too much. It has the advantage of requiring only a simple two-phase clock and also the signal can be easily tapped at specified points along it. This feature is particularly useful in signal processing by methods such as filtering and correlation.

Charge-coupled devices with their low transfer inefficiency are required for longer delays and where sampling rates are above about 5 MHz. The packing density is greater for CCDs which also favours the larger number of elements.

(ii) The buried channel CCD is particularly suitable as an image sensor in which light from a particular scene is focused on to an array of charge-coupled elements. One example uses a 576 × 384 element array with a diagonal of 11 mm on a standard TV aspect ratio of 4:3. The buried channel design allows a wide dynamic range of 1500:1, since the fat zero of the surface channel device is not required.

Transfer inefficiency is low, which is very important where 221 184 elements are being used. An advantage of BCCD image sensors over a Vidicon TV camera tube is that they can operate at lower light levels and, due to their small size and 15 V power supply miniature, battery-powered TV cameras are now available.

Three possible organizations of an image-sensing device are shown in Fig. 6.32, where 'line' and 'frame' refer to these features in the television scanning system. In each case data is transferred into the output register in parallel and read out in serial form, as indicated by the arrows. Storage registers, not exposed to the light source, are used in series in the frame transfer organization and in parallel in the interline transfer organization, while direct transfer into the output register occurs in the line transfer organization.

Points to remember

* A field-effect transistor or FET is characterized by a very high input

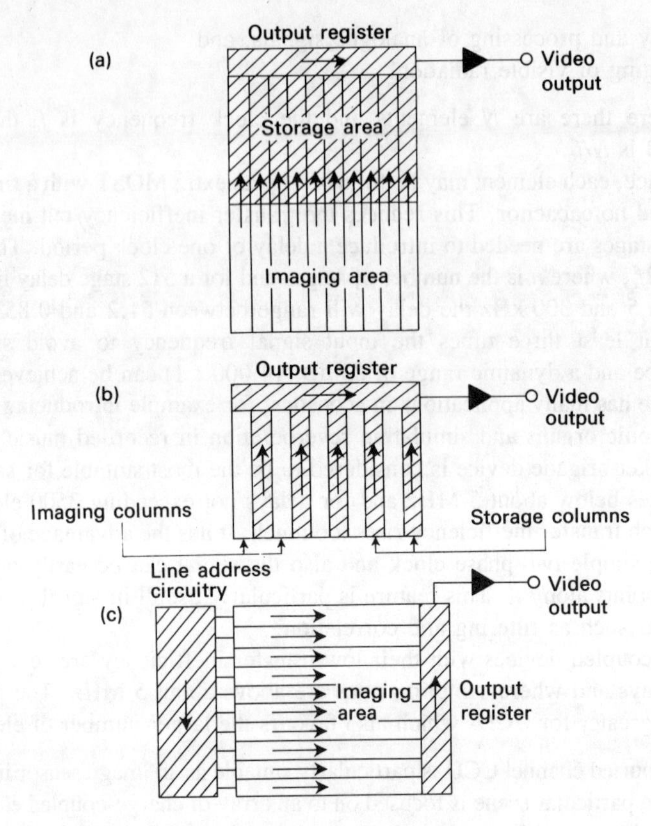

Figure 6.32 Basic types of organization for a charge-coupled area image sensor. (*a*) Frame transfer; (*b*) interline transfer; (*c*) line transfer.

resistance, with majority current flow controlled by the field due to a gate electrode.

* In a junction FET or JFET the gate is a p-n diode which is reverse-biased and formed at the side of the conducting channel. In a high-frequency FET the gate is normally a Schottky diode and the device becomes a metal-semiconductor FET or MESFET.

* In a metal-oxide-semiconductor FET (a MOSFET or MOST) the gate is one plate of a capacitor with the other plate being at the conducting channel.

* Depletion-mode MOSTs have a built-in channel, while in enhancement-mode MOSTs the channel is only formed when a sufficiently large gate voltage is applied.

* The current-voltage characteristics of a MOS transistor can be defined

by a simple formula common to all types of MOST, which also forms the basis for computer simulation.

* High-current versions of the MOSFET are the VMOST, the DMOST and the HEXFET. The MOS transistor is also widely used in read-write and read-only memories of various types.
* An nMOS logic inverter uses two n-channel MOSTs, a depletion-mode device for the load and an enhancement-mode device for the driver.
* A complementary MOS or CMOS logic inverter uses two enhancement-mode MOSTs, the load being a p-channel device and the driver an n-channel device.
* A charge-coupled device or CCD is a development of the MOSFET with many gate electrodes and applied mainly as a solid state imaging device.

References

[1] Middlebrook, R.D. and Richer, I. 'Limits on the power-law exponent for field-effect transistor transfer characteristics', *Solid State Electronics*, 1963, vol. 6, p. 542.
[2] Maloney, T.J. and Frey, J. 'Frequency limits of GaAs and InP field-effect transistors at 300 K and 77 K with typical active layer doping' *IEEE Trans. Electron Devices*, 1976, ED-23, p. 519.
[3] Sze, S.M. *Physics of semiconductor devices*, 2nd ed., (Wiley), 1981.
[4] Vladimirescu, A. and Liu, S. 'The simulation of MOS integrated circuits using SPICE 2'. *Electronics Research Laboratory, Berkeley, California, Memorandum UCB/ERL M80/7*, February 1980.
[5] Brews, J.R., Fichtner, W., Nicollian, E.H. and Sze, S.M. 'Generalised guide for MOSFET miniaturisation' *IEEE Electron Devices Lett.*, 1980, EDL 1–2.
[6] Burt, D.J. 'Basic operation of the charge coupled device', *International Conference on Technology and Applications of Charge-Coupled Devices, CCD* 74, Edinburgh, 1974, pp. 1–12.
[7] Walden, R.H. *et al.* 'The buried-channel charge coupled device', *Bell System Technical Journal*, 1972, vol. 51, pp. 1635–40.
[8] Sangster, F.L.J. and Teer, K. 'Bucket brigade electronics — new possibilities for delay, time-axis conversion and scanning', *IEEE J. Solid-State Circuits, SC*-4, 1969, pp. 131–6.
[9] Hobson, G.S. *Charge Transfer Devices*, (Arnold), 1978.

Problems

6.1 Show that the depth of the depletion layer in the channel of a JFET is given by

$$[(2\epsilon_r\epsilon_0 V)/(eN_a)]^{1/2}$$

where V is the gate-source voltage. The silicon channel of a JFET has a donor density of $6 \times 10^{21}/\mathrm{m}^3$ and a depth of 1 μm. If the relative permittivity of silicon is 12, calculate the pinch-off voltage.
[4.52 V]

6.2 The drain current of an n-channel junction field-effect transistor (JFET) below saturation is given by

$$I_D = K[V_{DS} - \tfrac{2}{3}(V_{DS} - V_{GS})^{3/2}/V_P^{1/2} + \tfrac{2}{3}(-V_{GS})^{3/2}/V_P^{1/2}]$$

Show that when $V_{DS} \ll V_{GS}$ the I_D/V_{DS} characteristics are ohmic. Calculate the drain resistance for $K = 6 \times 10^{-3}$ A/V, $V_P = 5.0$ V and $V_{GS} = -2.0$ V.

Define *pinch-off* in terms of V_{DS}, V_{GS} and V_P and obtain an expression for I_D in the saturation region. Show that the saturation current at $V_{GS} = 0$ is given by $I_{DSS} = KV_P/3$. Sketch the transfer characteristics of the JFET showing significant values.

[453 Ω]

6.3 The donor density in the channel of a silicon junction FET is $10^{22}/\text{m}^3$ and its dimensions are $a = 1.0\ \mu$m, $W = 100\ \mu$m and $L = 25\ \mu$m (Fig. 6.1(*a*)). Obtain the open-channel conductance. If $V_P = 5.0$ V calculate I_{DSS} and the mutual conductance at $V_{GS} = -2.0$ V.

[12.8 mS; 3.2 mA; 0.77 mS]

6.4 A basic MOST device has an input capacitance C_{gs} of 3 pF and the length of the channel is 10 μm. For an enhancement-mode device with $V_T = +1.5$ V, calculate I_{Dsat} and g_m when $V_{GS} = +3$ V. For a depletion-mode device with $V_T = -2$ V calculate I_{Dsat} and g_m for $V_{GS} = 0$ V.

[4.5 mA; 6.1 mS; 8.1 mA; 8.1 mS]

6.5 An FET has the following small-signal parameters: $C_{gs} = 4$ pF, $C_{gd} = 1$ pF and $g_m = 3$ mS. It is used as an amplifier between a 600 Ω source and a drain load of 2 kΩ. Calculate (i) the gain at low frequencies; (ii) the high frequency at which the gain has fallen 3 dB below the value in (i) (*see* the analysis of the hydrid-π circuit for the BJT in Chapter 4).

[–6; 9.2 MHz]

6.6 Calculate the value of r_d at $V_{DS} = 0$ for an n-channel MOST with $\mu = 0.06\ \text{m}^2/\text{V s}$, $C_0 = 3.5 \times 10^{-4}\ \text{F/m}^2$, $W/L = 2$, $V_{GS} = 5.0$ V, $V_T = 0.8$ V.

How could a p-channel device, with $\mu = 0.02\ \text{m}^2/\text{V s}$, be designed to give the same r_d with the same voltage magnitudes?

[5668 Ω, make $W/L = 6$]

6.7 Determine the gate capacitance of an MOST of length 5.0 μm and having values of K and carrier mobility of $2 \times 10^{-5}\ \text{A/V}^2$ and $0.06\ \text{m}^2/\text{V s}$ respectively.

[8.33×10^{-15} F]

6.8 Using the values in the text for the nMOS inverter of Fig. 6.16 show by calculation that $V_0 = 0.29$ V for $V_1 = 5$ V.

6.9 Using the values in the text for the CMOS inverter of Fig. 6.22 show by calculation that $V_0 = 0.014\ \mu$V for $V_1 = 5$ V.

6.10 Determine the effective channel length for a device with $L = 2\ \mu$m, given the following data for three different MOSTs operating at the same value of $(V_{GS} - V_T)$ and fabricated by the same process:

g_d mA/V	L μm
1.0	3
1.7	2
4.2	1

This may be done graphically by plotting g_d^{-1} against L.

[0.5 μm]

6.11 A charge transfer device has 512 elements each with a transfer inefficiency of 10^{-4} and clocked at a frequency of 100 kHz. Determine (i) the delay time between input and output; (ii) the percentage of input charge appearing at the output; (iii) the bandwidth.

[5.12 ms; 94.88%; 50 kHz]

7 Integrated circuits

In this chapter the principles of fabricating integrated circuits are discussed and applied to common digital and linear circuits.

The construction of integrated circuits follows from the invention of the planar process for transistor manufacture in 1959, in which the device is formed only on one side of a silicon wafer. This principle was soon extended to producing complete circuits containing transistors, resistors, capacitors and diodes on a single 'chip' of silicon, typically 1 mm square and 0.1 mm thick. Thus a circuit sub-system is made available, such as a complete amplifier or logic circuit and more recently the microprocessor, which is the essential part of a small computer. The first integrated circuits were based on bipolar junction transistors, which are still widely used, but with the development of large and very large scale integration (LS1 and VLS1) the majority of integrated circuits now use MOS transistors.

Compared with a design using separate (discrete) components and wired interconnections integrated circuits offer several significant advantages. The relative cost is low since the technology favours mass production, and high reliability is achieved since all similar components, including the interconnections, are fabricated at the same time. Performance is improved since highly complex circuitry becomes more easily available and the size of electronic equipment is thereby much reduced. Indeed, the electronic requirements of spacecraft and micro-computers could only have been met by integrated circuits.

Fabrication of monolithic integrated circuits

The majority of integrated circuits are of the monolithic type, as described above, and are normally formed on a substrate of p-type silicon. This is obtained from a cylindrical single crystal of doped silicon, which is produced by cooling slowly from a high temperature. A controlled amount of acceptor impurity, such as boron, is added to the melt and the finished material has acceptor densities around $10^{21}/m^3$, giving a resistivity in the range 0.01 to 0.5 Ω m.

A typical crystal might be about 25 cm long with a diamenter up to 15 cm. A substrate slice or wafer of typical thickness 0.1 mm is cut from the crystal with a diamond-tipped saw and the surface of the wafer is then chemically etched and mechanically polished until it is perfectly smooth. After fabrication each 15 cm wafer could hold around 16 000 individual circuits if each circuit occupied a 1 mm square chip (Fig. 7.1(*a*)). In turn each chip could contain about 100 logic gates, when one common technology for bipolar devices is used and about 1000 or more gates when a MOS technology is used. Other chip sizes extend up to about 10 mm square.

n-p-n bipolar processes

The fabrication of a bipolar integrated circuit will be illustrated by considering a single n-p-n transistor with a resistor in series with its base. A number of prepared wafers are mounted in a diffusion furnace as shown in Fig. 7.1(*b*) in which they can be heated in a gaseous atmosphere. For example, phosphorus atoms are released from PH_3, phosphine, for n-type doping and boron atoms are released from BCl_3, boron trichloride, for p-type doping.

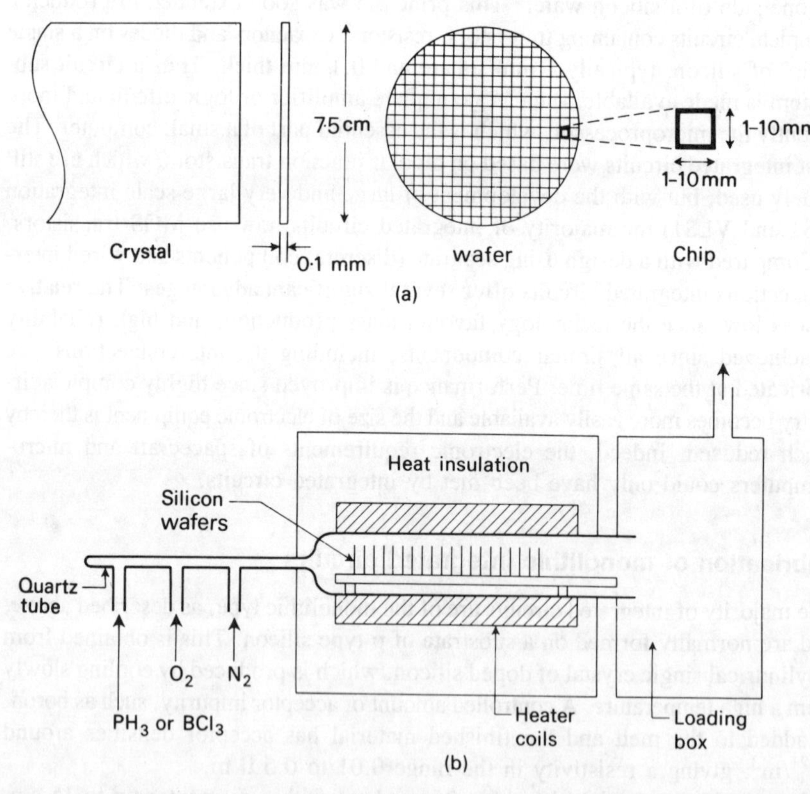

Figure 7.1 (*a*) Formation of chips for integrated circuits from a doped silicon crystal; (*b*) diffusion furnace for integrated circuit production.

Oxidation

At first a layer of silicon dioxide is formed (Fig. 7.3(*b*)) by heating in oxygen. If water vapour is added the growth rate at 1100 °C is increased, giving a final thickness around 0.1 μm in about 10 minutes. The oxide layer is used to mask the silicon wafer from any gas introduced during a following diffusion, the mask being formed by closely defined openings, or *windows*, etched into the oxide layer (Fig. 7.3(*c*)) through which impurities can diffuse into the wafer. The impurities such as boron, phosphorus, arsenic, and antimony do not penetrate the oxide layer.

Window formation

Each window is defined by a photographic technique which involves first coating the oxide with a photoresist emulsion. The window is represented by an opaque region on a transparent plate or *mask*, which is then exposed to ultraviolet light (Fig. 7.2(*a*)). After this the photoresist can be removed from under the opaque region by developing and etching (Fig. 7.2(*b*)). Outside the opaque region the photoresist is hardened and acts as another mask through which the oxide layer

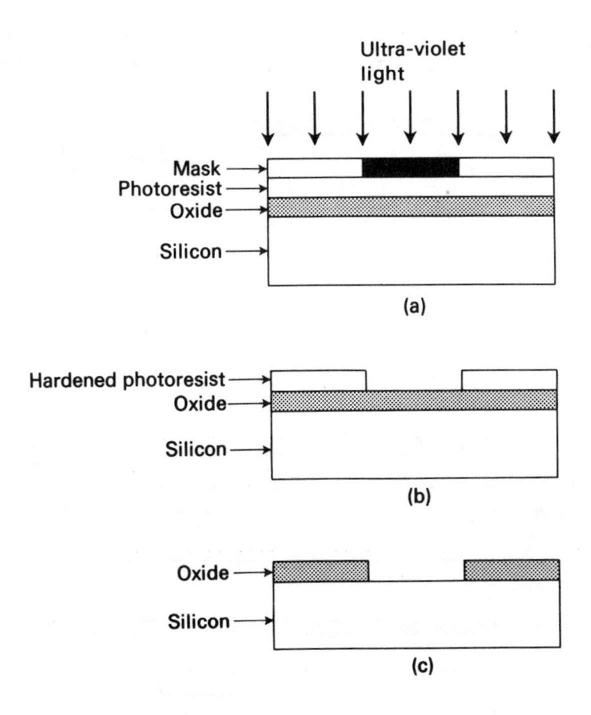

Figure 7.2 Window formation. (*a*) Exposure of photoresist to ultra-violet light through mask; (*b*) after exposure, development and etching of photoresist; (*c*) after etching of oxide and removal of photoresist by acid and washing.

can be etched away, exposing the required area of semiconductor underneath. The remaining photoresist is finally removed by acid treatment and washing, leaving the window in the oxide (Fig. 7.2(*c*)).

After a diffusion the oxide is removed and then completely restored before the next diffusion so that new windows can be introduced. Thus the whole window formation process is repeated before each of the diffusions.

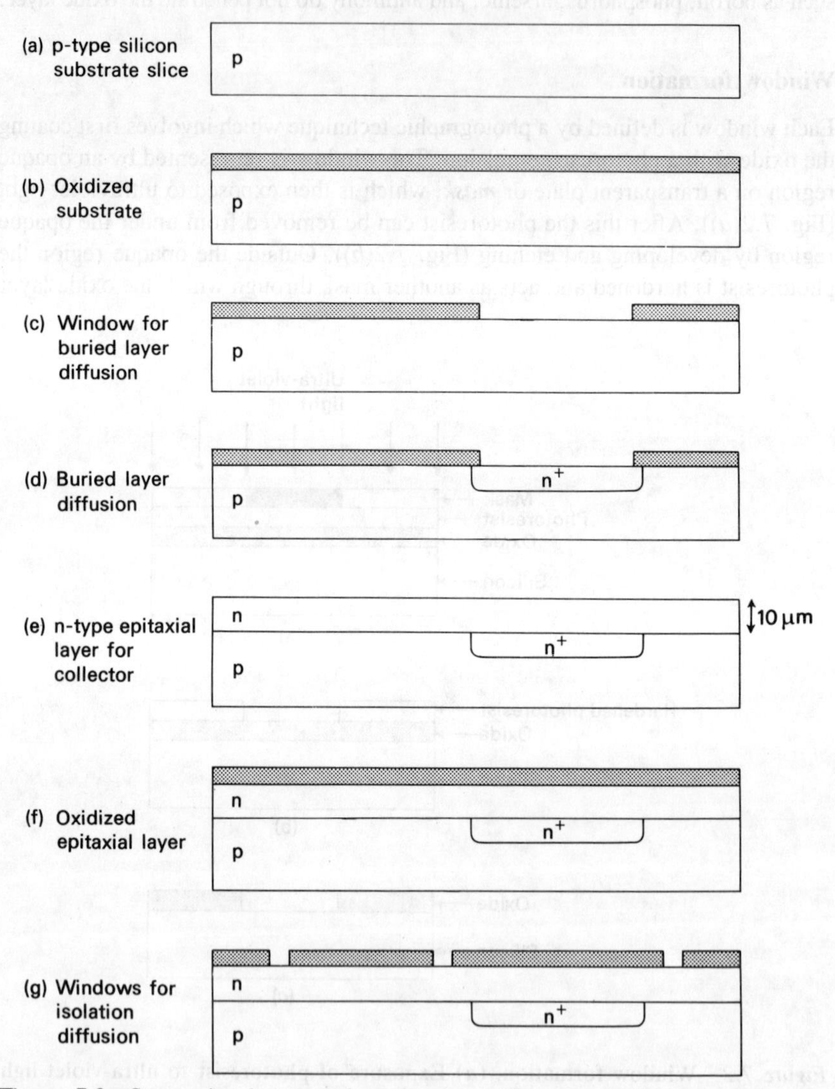

Figure 7.3 Stages in the manufacture of an integrated circuit containing a transistor with a resistor in series with its base (not to scale).

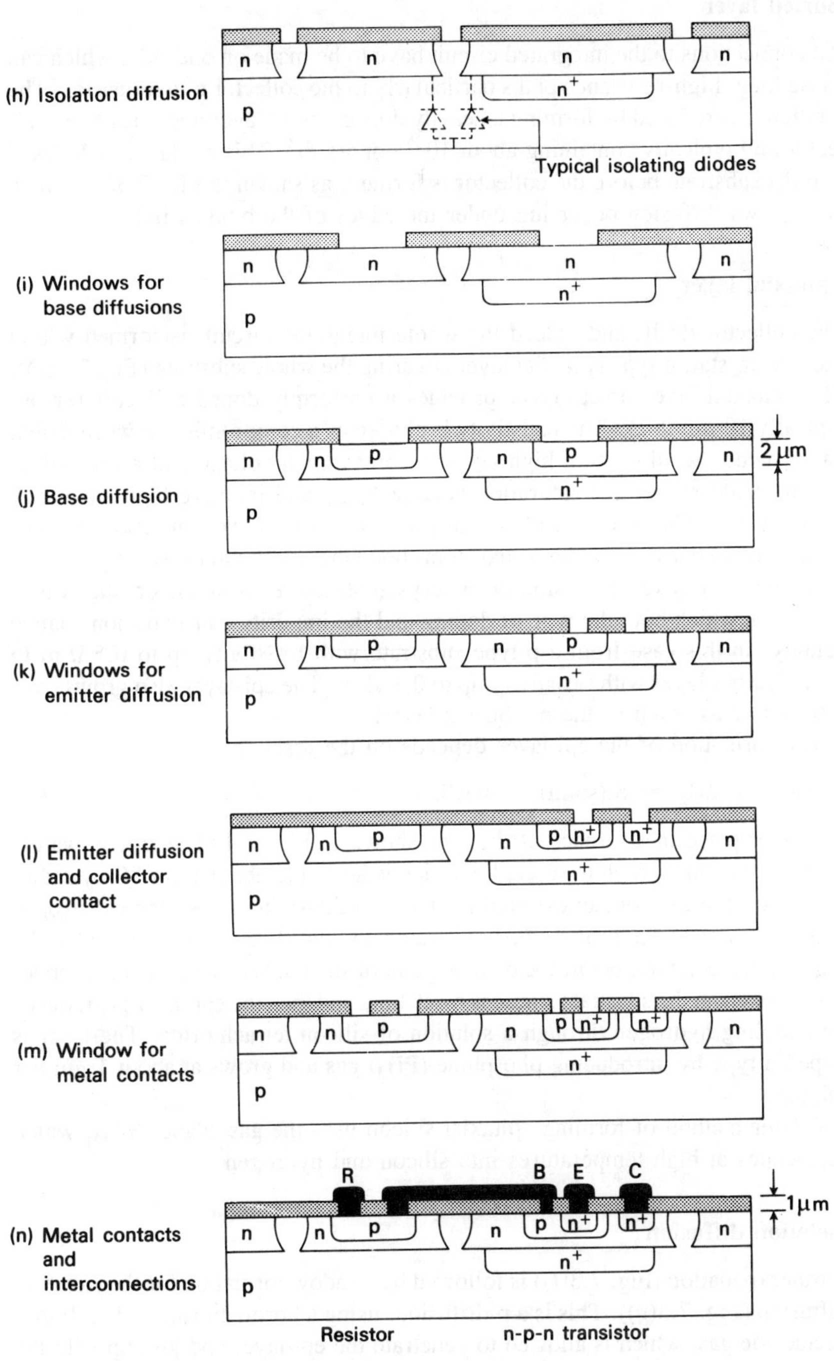

Figure 7.3 (*Continued*)

Buried layer

All connections to the integrated circuit have to be made on one side, which can cause long, high-resistance paths particularly to the collector of a transistor. The resistance is reduced by forming a heavily doped buried layer underneath the collector and typically containing about 10^{25} donors/m^3. This n^+ layer is diffused into the substrate before the collector is formed, as shown in Fig. 7.3(*d*), which also shows diffusion occurring under the edges of the oxide window.

Epitaxial layer

The collector itself, and indeed the whole integrated circuit, is formed within a single crystal, n-type epitaxial layer covering the whole substrate (Fig. 7.3(*e*)). The epitaxial layer or epi-layer provides a uniformly doped collector region, typically 10 μm thick. The resistivity is chosen as a compromise between a high value, which would give a high collector breakdown voltage and a low value, which would reduce the saturation voltage V_{CEsat} and improve high frequency performance. The layer is called *epitaxial* (from the Greek 'epi'-upon, 'taxis'-arrangement) because the deposited atoms bond themselves to the substrate atoms to form an unbroken extension of the crystal structure. A single crystal is thus formed in which both the type of doping and the impurity concentration change abuptly, in this case from a p-type substrate with resistivity up to 0.5 Ω m to an n-type epi-layer with resistivity up to 0.1 Ω m. The epi-layer also grows over diffused areas, such as the n^+ buried layer.

The formation of the epi-layer depends on the reaction

$$SiCl_4 + 2H_2 \rightleftharpoons Si(solid) + 4HCl \tag{7.1}$$

and takes place in a furnace of the type shown in Fig. 7.4. The silicon wafers are mounted on a graphite susceptor which is heated to about 1200 °C by radio-frequency energy from an external induction coil. An inert gas such as nitrogen is passed through the furnace during loading and the slices are then raised to the required temperature, below the melting point of silicon. Hydrochloric acid vapour is used to clean the substrate surfaces by etching and then the epi-layer is produced by bubbling hydrogen through a solution of silicon tetrachloride. The layer is doped n-type by introducing phosphine (PH_3) gas and grows at about 1 μm per minute.

A later method of forming epitaxial silicon uses the gas silane, SiH_4, which dissociates at high temperatures into silicon and hydrogen.

Isolation diffusion

Further oxidation (Fig. 7.3(*f*)) is followed by window formation for the isolation diffusion (Fig. 7.3(*g*)). This is a p-diffusion, using a boron dopant such as boron trichloride gas, which is allowed to penetrate the epi-layer and join up with the substrate (Fig. 7.3(*h*)). In this way islands are created in the n-type epi-layer in

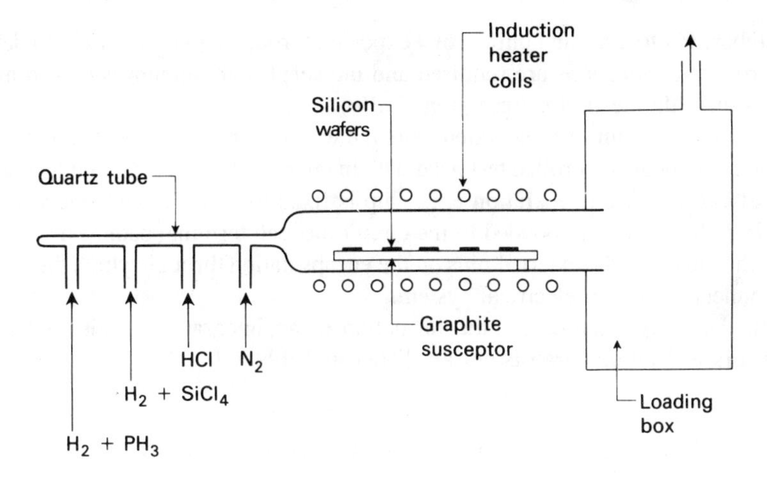

Figure 7.4 Induction furnace for epitaxial layer formation.

which all the devices and components are fabricated, thus achieving physical isolation. A p-n junction occurs between each island and the substrate which is held at the most negative potential during operation. Thus the diodes are reverse biased to provide electrical isolation, but voltage dependent junction capacitances are introduced at the same time resulting in a complex equivalent circuit.

Another method of providing isolation between components is to surround each island by a thick dielectric insulating layer, normally silicon dioxide. This reduces stray capacitance between components and substrate, the leakage current between components is small and no isolation biasing is required. Dielectric isolation is not often used since it requires additional processing steps which increase costs and may reduce yield.

Base and emitter diffusions

The areas for bases and monolithic resistors are now defined by further oxidation and window formation (Fig. 7.3(*i*)). A second p-type diffusion follows, of lower concentration than before and penetrating only about 2 μm (Fig. 7.3(*j*)). At depths less than 2 μm the epitaxial layer has been turned into a p-region where diffusion has occurred, due to the excess of p-type impurities. Finally, the emitters and the collector contacts are formed in an n^+-diffusion using phosphorus dopant such as phosphine (Fig. 7.3(*k*) and (*l*)). Here the p-region has been turned into an n^+-region where diffusion has occurred.

Metallization

After further oxidation and window formation (Fig. 7.3(*m*)) interconnections are normally made with aluminium evaporated over the whole slice in a vacuum chamber. Photoresist and ultra-violet exposure through a mask are used to define where interconnections are required and the surplus aluminium is etched away to produce the desired pattern (Fig. 7.3(*n*)).

Before the circuits are separated from the wafer each one is electrically tested, using a computer-controlled test head with many needle probes, and faulty wafers are discarded. After separation each chip is fixed to a header and fine wires of gold or aluminium are bonded to the circuit through terminal pads. These wires link the circuit to the external pins on the encapsulation through which the circuit is connected into an electrical system.

All the steps required for the formation of an integrated circuit containing resistors and bipolar transistors are listed in Table 7.1.

Table 7.1 Steps in the production of the integrated circuit illustrated in Fig. 7.3

1	p-type single crystal obtained
2	p-type substrate cut and polished (a)
3	oxidation (b)
4	window formation (c)
5	n^+-buried layer diffusion (d)
6	oxide removal
7	n-type epitaxial layer growth (e)
8	oxidation (after removal of previous oxide) (f)
9	window formation (g)
10	p-type isolation diffusion (h)
11	oxidation
12	window formation (i)
13	p-type base and resistor diffusion (j)
14	oxidation
15	window formation (k)
16	n^+-emitter and collector contact diffusion (1)
17	oxidation
18	window formation (m)
19	metallization (n)
20	electrical testing
21	chip separation
22	bonding to header
23	wire bonding
24	encapsulation
25	final test

The whole process as described uses six masks and four diffusions.

Integrated bipolar transistors

Schottky transistor

The normal n-p-n transistor, as shown in Fig. 7.3, has a propagation delay of about 10 ns in a switching circuit, which is due mainly to the transistor being driven into saturation when the collector-base junction becomes forward biased. This can be prevented by connecting a Schottky diode between collector and base as shown in Fig. 7.5(*b*). The forward bias voltage of the diode is only about 0.3 V compared with about 0.7 V for the collector-base junction. Thus as the collector voltage falls during switch-on, the Schottky diode conducts before the collector-base diode. The load current is thereby diverted from the collector which never saturates, and the saturation charge is zero. The propagation delay is reduced to about 3 ns and the Schottky diode itself has a low storage time so that the collector voltage rises very rapidly, typically in about 100 picoseconds, during switch-off.

To form the Schottky diode aluminium metallization is deposited directly on to the n-type collector to give a rectifying contact, as shown in Fig. 7.5(*a*), and extended to make an ohmic contact with the base. No extra processing steps are needed and the collector connection is made ohmic through an n^+ region as before. The circuit symbol for this *Schottky transistor* is shown in Fig. 7.5(*c*).

Lateral p-n-p transistor

The n-p-n structure is normally preferred to the p-n-p structure since electrons have a higher mobility than holes, giving a device with better frequency response and lower propagation delay. However, p-n-p transistors are required in linear amplifiers, for example, and the technology does not allow simultaneous, economical production of the two types. Instead a *lateral p-n-p transistor* is formed using the n epi-layer as a base and the p-diffusion for both emitter and collector

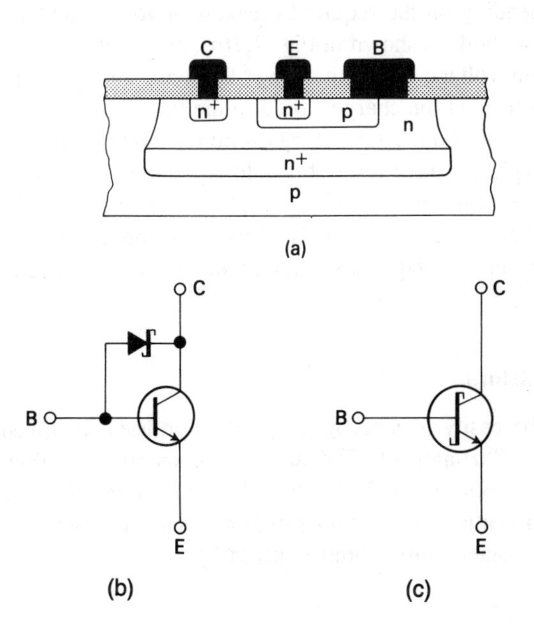

Figure 7.5 Schottky transistor. (*a*) Section; (*b*) circuit; (*c*) symbol.

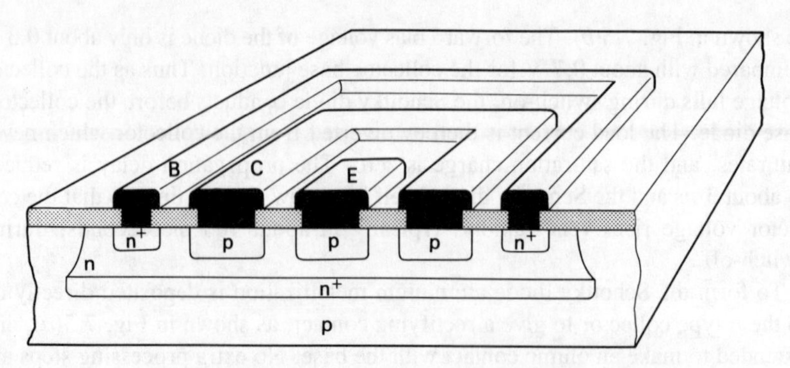

Figure 7.6 Lateral p-n-p transistor.

(Fig. 7.6). The collector and base contacts can be in the shape of a ring as shown or each electrode may be contacted at several points. The design gives a transistor with a wide effective base region having a low current gain, typically 0.5–5.0 instead of 50–200 for the n-p-n device, so that p-n-p transistors must be used in circuits not requiring current gain.

Integrated diodes

Either of the two junctions produced in a transistor can be used as diodes, with the choice depending on the required breakdown voltage and switching speed. A collector-base diode is shown in Fig. 7.7(*a*), and is likely to have a relatively high breakdown voltage of about 50 V. The base and emitter are connected together to ensure that the charge stored in the base is removed rapidly to give a switching time of about 70 ns. A base-emitter diode is shown in Fig. 7.7(*b*) and this will typically have a low breakdown voltage of about 5 V due to the high doping level of the emitter. However, the switching time with collector and base connected can be made as low as 20 ns. Thus the first diode would be used where a high reverse voltage was required and the second where fast switching was necessary.

Integrated resistors

Resistors can be easily formed by the p-diffusion used for transistor bases, as shown in Fig. 7.8(*a*) and (*b*). The current is constrained to flow only through the p-region between terminals X and Y by maintaining the n-type island at a more positive potential. Suppose the p-region has length l, width w and thickness t, so that its resistance throughout is given by

$$R = \frac{\rho l}{wt} \tag{7.2}$$

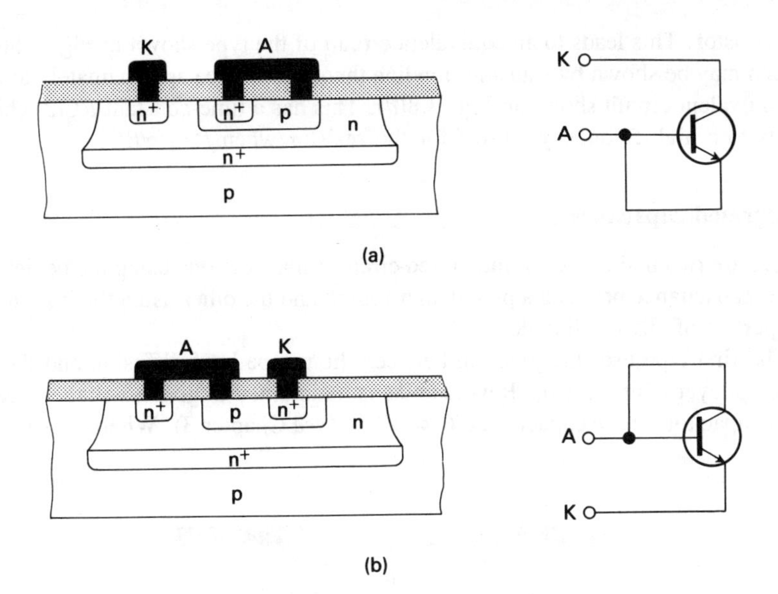

Figure 7.7 Integrated diodes. (*a*) Collector-base; (*b*) base-emitter.

If $l = w$ the plan view is a square and the resistance becomes

$$R_s = \frac{\rho}{t} = \rho_s \tag{7.3}$$

ρ_s is the *resistance per square* of the material and may be used instead of resistivity so that in general

$$R = \rho_s \frac{l}{w} \tag{7.4}$$

This equation is useful in comparing the resistances of regions of the same thickness, such as occur in integrated circuits, and shows the dependence of such resistances on their length-to-width ratio, l/w. Typically ρ_s lies between about 50 and 300 Ω per square and l/w ratio up to about 50 : l can be achieved where w is around 25 μm. Hence diffused resistors can be formed in the range 100 to 10 000 Ω, with high value resistors often folded to conserve chip area (Fig. 7.11). Their temperature coefficient is rather large at about $+3 \times 10^{-3}$ per degree Celsius above room temperature. The tolerance of single resistors is about ±10%, which is quite acceptable for a switching circuit, but it is possible to make a pair of resistors whose resistance ratio may be held to about ±1% by diffusing them at the same time and close together on the chip.

A feature of a diffused resistor is that it is purely resistive only at low frequencies, which is due to the capacitance of the reverse-biased island distributed along the resistor. This leads to an equivalent circuit of the type shown in Fig. 7.8(*c*), which may be shown by transmission line theory to reduce approximately to the T-equivalent circuit shown in Fig. 7.8(*d*). This has a time-constant $RC/2$ which leads to a high-frequency cut-off for the resistor when $f \geq 1/\pi RC$.

Integrated capacitors

There are two main types of integrated circuit capacitor, one using the depletion layer capacitance between a p- and an n-region and the other using the insulating properties of silicon dioxide.

The first type uses the junction between the p-type base diffusion and the n-type epilayer (Fig. 7.9(*a*)). Reverse bias is maintained to give a voltage dependent capacitor where capacitance $C \propto V^{-1/3}$ (*see* Chapter 3). Where V is in the

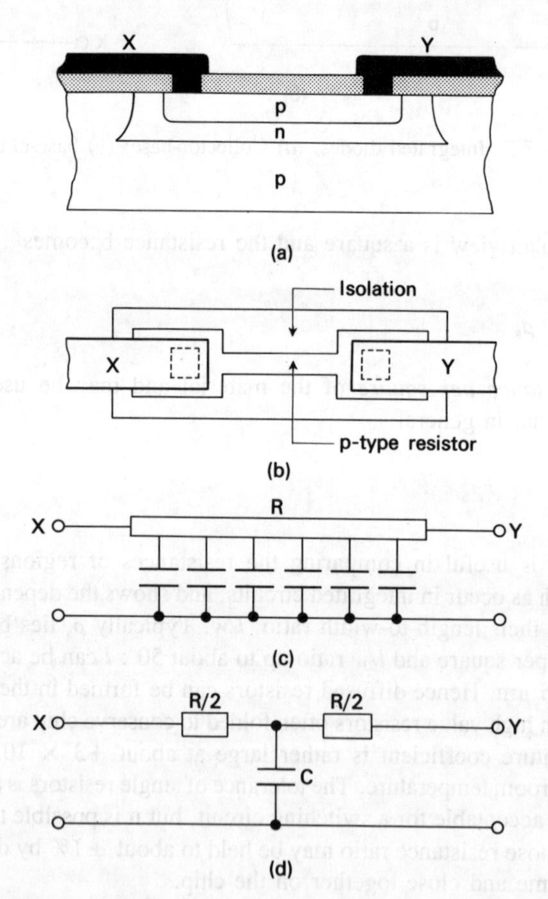

Figure 7.8 **Diffused resistor. (*a*) Section; (*b*) plan; (*c*) distributed equivalent circuit; (*d*) T-equivalent circuit.**

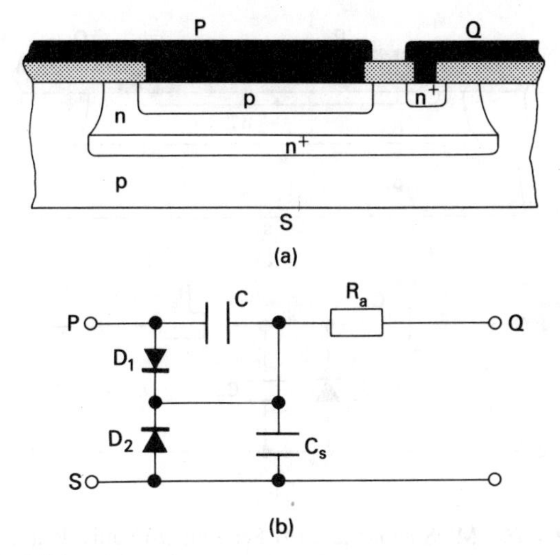

Figure 7.9 Diffused junction capacitor. (*a*) Section; (*b*) equivalent circuit.

order of 1 V, C is only about 300 pF/mm^2, so that capacitors of value above a few hundred picofarads would use excessive chip area. The breakdown voltage is typically in the range 5 to 20 V.

An equivalent circuit is shown in Fig. 7.9(*b*) where terminal Q is held positive with respect to terminal P, leading to a polarized dielectric but ensuring that diode D_1 is reverse biased. D_2 is also reverse biased, since the substrate S is at the most negative potential, but the diode introduces stray capacitance C_s. The access resistance R_a is the effective resistance of the n-region and is typically 10–50 Ω.

The second type uses metal-oxide-silicon (MOS) technology to form a parallel plate capacitor (Fig. 7.10(*a*)). The lower electrode uses an n$^+$ emitter diffusion which also ensures a low access resistance, the upper electrode is the aluminium metallization P and the dielectric is normally silicon dioxide with a relative permittivity of about 3.8. The thinnest oxide layer obtainable at present is about 0.1 μm which leads to a maximum capacitance of about 350 pF/mm^2. The breakdown voltage is typically in the range of 50 to 200 V.

An equivalent circuit is shown in Fig. 7.10(*b*) where terminal Q is always held positive with respect to the substrate S to ensure that the isolation diode D_s is reverse biased. Typically R_a is 5 to 10 Ω due to the low resistance of the n$^+$ region.

Where a higher value capacitor is required the area is increased by fabrication over the top of an existing circuit, with two aluminium electrodes. The lower electrode is insulated from the other components by a second oxide layer.

High-performance bipolar technologies

The chip area required for the n-p-n transistor in the above process is about

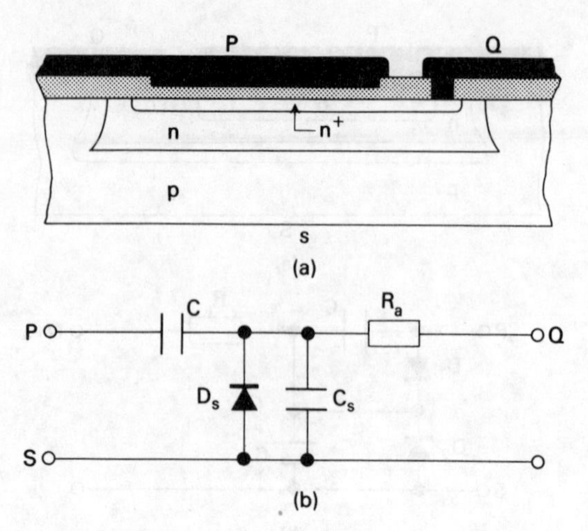

Figure 7.10 MOS capacitor. (*a*) Section; (*b*) equivalent circuit.

3000 μm^2, as illustrated in Fig. 7.11, which would restrict the number of devices in a large-scale integrated circuit. Also the reverse-biased isolating diodes shown in Fig. 7.3(*h*) introduce depletion layer capacitances, which increase switching time.

The performance of a bipolar transistor is improved by replacing diffusion isolation with *oxide isolation* (Fig. 7.12). This eliminates the diodes by means of thermally grown silicon dioxide which penetrates the epitaxial layer. Since the oxide is an insulator it can be used to define the edges of the emitter and collector regions

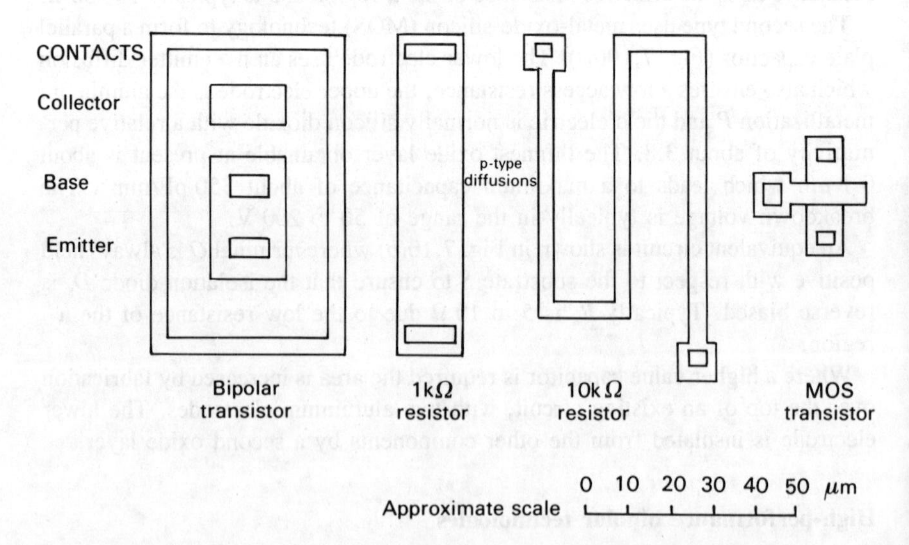

Figure 7.11 Relative sizes of a selection of integrated circuit components.

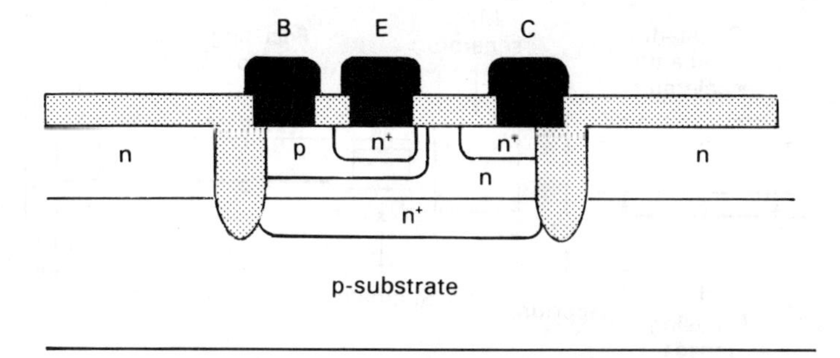

Figure 7.12 Section through an oxide-isolated (isoplanar) n-p-n bipolar transistor.

so that the process is *self-aligned* and leads to a reduction in area down to about 200 μm^2 (Ref. [1]). Oxide-isolation processes have now been developed to give a base width less than 1 μm and a propagation delay of less than 70 ns.

Where large-scale bipolar integrated circuits are concerned it is normal to use transistors without resistors, as in integrated injection logic described later in this chapter. This is due to the large areas required by diffused resistors in integrated circuits (Fig. 7.11). Although compatible with the bipolar transistor the area of a resistor is much greater than that of a MOS transistor. This leads to the different approach in MOS integrated circuits, described in Chapter 6, where the resistor is replaced by another MOS transistor.

nMOS silicon-gate processes

Ion implantation

The fabrication of a silicon-gate nMOS integrated circuit, as exemplified by the inverter of Fig. 6.16, uses the window formations and diffusions as described above for the bipolar process but an additional process, *ion implantation*, is also employed (Fig. 7.13). The ion implantation system is complex (Ref. [2]) and only the essential elements are illustrated, with positively charged ions being produced from a gas containing the required impurity by, for example, exposing the gas to a radioactive source. The ions are then focused and accelerated through a constant potential gradient to voltages between about 50 and 200 kV. Separation of the required ions from others produced is achieved by passing the beam through a strong magnetic field, as in the mass spectrometer (Chapter 8). The ion beam is concentrated in the stigmator and focused, after which waveforms applied to the deflector plates cause the beam to scan the wafer.

The high energy ions penetrate the silicon surface to a depth of a μm or so, controlled by the accelerating field. A high temperature phase drives in the impurity to achieve the specified penetration depth and repairs any damage caused by the bombardment. The areas where the impurity is needed are defined by masks which can be thick oxide, photoresist or metal. A thin oxide is penetrated by

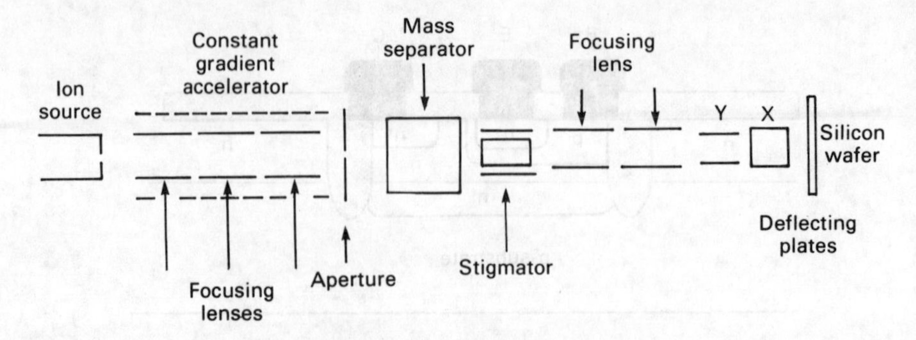

Figure 7.13 Sketch of an ion implantation system.

the ions to create the n-channel of a depletion MOST or to adjust the threshold voltage of an enhancement MOST.

Boron implantation

The nMOS inverter is fabricated on a p-substrate which is lightly doped to ensure an adequate drain breakdown voltage of about 25 V. However, spurious n-channels may occur between circuits, due to a low threshold voltage, so they are prevented from forming by implanting heavily doped p^+ regions, known as *channel stoppers*. These increase the threshold voltage, so that a supply line, for example acting as a gate cannot cause a channel to be formed. The channel stoppers are situated in the *field region* between circuits. The circuit is contained in the *active region* which is defined by a composite mask of photoresist, thin oxide, silicon nitride and more thin oxide. The nitride prevents oxidation of the active region during the deposition of field oxide described below. Boron implantation is now carried out and the channel stoppers produced (Fig. 7.14(*a*)). *Field oxide* 1 μm thick is now grown on top of the implanted p^+ regions, which further improves the electrical isolation of circuits (Fig. 7.14(*b*)).

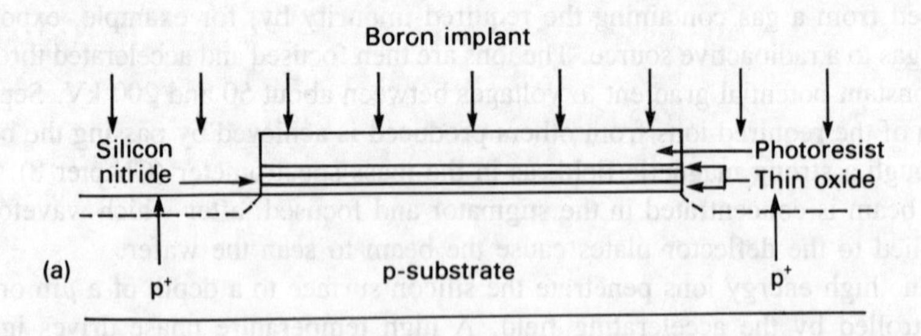

Figure 7.14 nMOS silicon-gate process. (*a*) Boron implantation for channel stoppers; (*b*) definition of active regions; (*c*) depletion implant; (*d*) buried contact and polysilicon diffusion; (*e*) source and drain diffusions; (*f*) contacts and metallization.

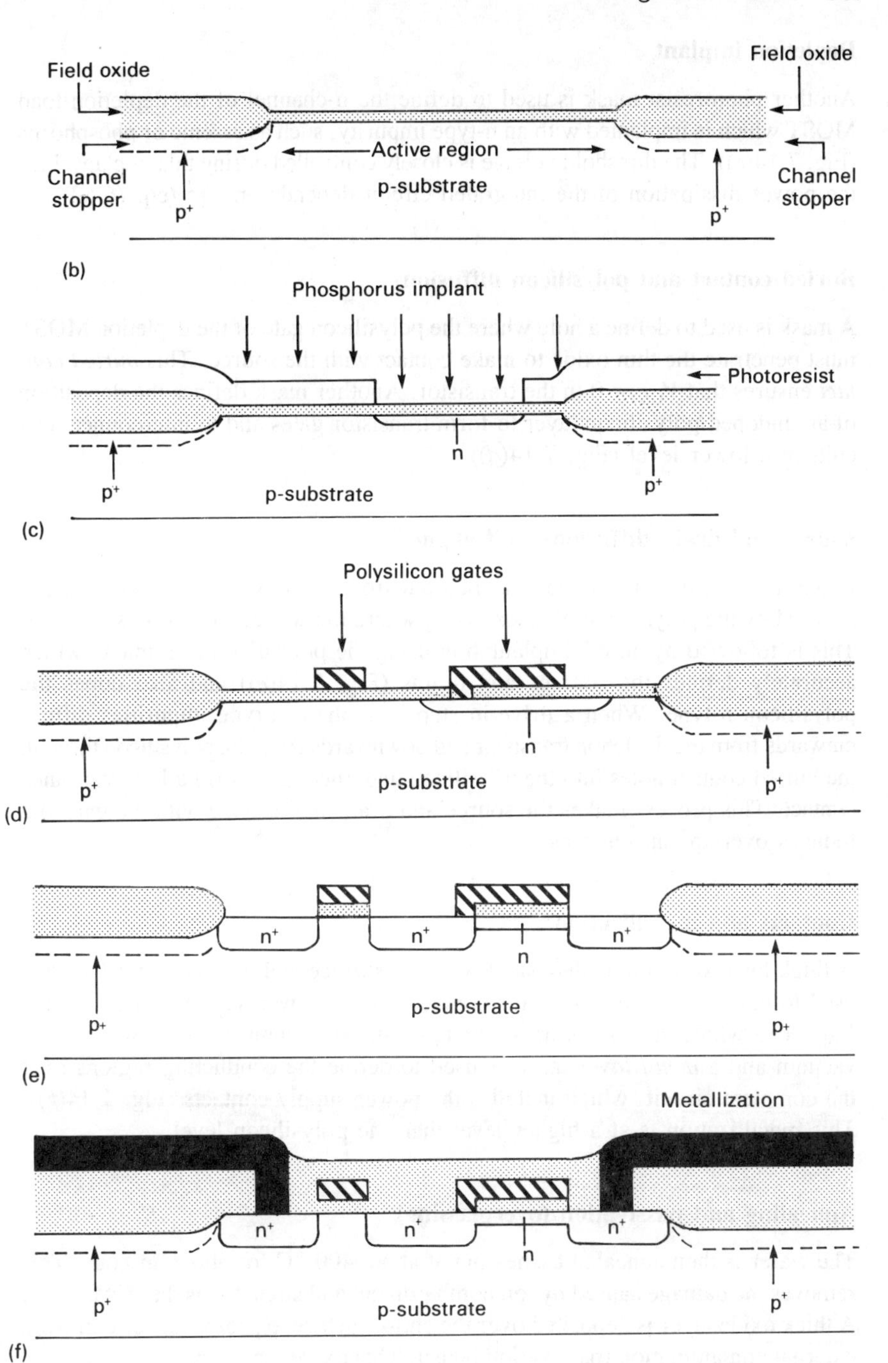
Field oxide
Field oxide
Active region
Channel stopper
p-substrate
Channel stopper
p+
p+
(b)
Phosphorus implant
Photoresist
n
p+
p-substrate
p+
(c)
Polysilicon gates
n
p+
p-substrate
p+
(d)
n+
n+
n+
n
p+
p-substrate
p+
(e)
Metallization
n+
n+
n+
n
p+
p-substrate
p+
(f)

Depletion implant

Another photoresist mask is used to define the n-channel of the depletion-load MOST which is implanted with an n-type impurity, such as arsenic or phosphorus (Fig. 7.14(*c*)). The threshold voltage is closely controlled during this implant since the power dissipation of the integrated circuit depends on V_T^2 (eq. (6.53)).

Buried contact and polysilicon diffusion

A mask is used to define a hole where the polysilicon gate of the depletion MOST must penetrate the thin oxide to make contact with the source. This *buried contact* ensures that $V_{GS} = 0$ in the transistor. Another mask defines the deposition of an undoped polysilicon layer to form transistor gates and to interconnect circuits at a lower level (Fig. 7.14(*d*)).

Source and drain diffusions (self-aligned)

Using a mask, the thin oxide is etched out from regions in the active area not covered by the polysilicon, thus exposing where the sources and drains must be. This is followed by an n^+-implantation using the polysilicon as a mask, which accurately defines the sources and drains (Fig. 7.14(*e*)) and also dopes the polysilicon n-type. When a drive-in step is applied, n-type impurities diffuse outwards from the depletion transistor and downwards from the polysilicon through the buried contact holes into the n^+ silicon underneath, to form a low-resistance contact. This process makes the source and drain *self-aligned* with the gate and reduces overlap capacitances.

Contacts and metallization

A thick layer of oxide is deposited over the surface and a *contact mask* is then used to cut holes in the oxide for contacts to the power supply rails, 0 V and V_{DD}. The whole surface is then covered with aluminium by evaporation in a vacuum and a *metal layer* mask is used to define the conducting regions over the complete circuit, which includes the power supply contacts (Fig. 7.14(*f*)). This metallization is at a higher level than the polysilicon level.

Annealing and passivation (overglassing)

The wafer is then annealed by heating at about 400 °C for about an hour. This removes the damage caused by ion bombardment and strengthens the metal layer. A thick oxide layer is deposited over the entire surface to protect the circuit from external contamination (passivation layer). This oxide may contain phosphorus, when it becomes a glass, and an *overglass mask* is used to open windows in the oxide corresponding to the bonding pads which are connected by gold wire to the external pins of the integrated circuit. Finally, the circuit (or chip) is encapsulated so that it can be handled without damage.

CMOS silicon-gate processes

Since a CMOS circuit requires both p- and n-channel enhancement MOST, only one of them will have a different type of doping to the substrate. A p-substrate allows an n-channel MOST to be formed directly on it, so the p-channel MOST must be contained in an n-well. This has the advantage that it is compatible with nMOS, so that processing steps can be shared in a mass production environment. On the other hand, an n-substrate supports the p-channel MOST directly and requires a p-well for the n-channel MOST. This tends to produce a slower n-channel device and so reduces the difference in speed between the two types of transistor. This approach has been used from the beginning of CMOS and has proven reliability, since it is easier to produce a p-well in an n-substrate than an n-well in a p-substrate.

At present the p-well process is mainly used but in the future it may be replaced by the n-well process. A further variation is the *twin-well* process in which two separate wells are formed for the p- and n-channel transistors on an n-type substrate, thus giving the advantages of p- and n-well designs. Much experimental work is being carried out at present to achieve a standard CMOS process, as has been attained with nMOS.

p-well process (isoplanar, oxide isolated)

Boron implantation

The p-well is formed by implantation of boron ions through thin oxide using a photoresist mask, the ions then being driven in (Fig. 7.15(*a*)). The photoresist is removed, field oxide is grown and the active regions are defined as for the nMOS process but in this case in both the p-well and the n-substrate.

Transistor definition

Active areas are defined in both the p-well and n-substrate regions by similar methods to that already used in nMOS (Fig. 7.14(*a*) and (*b*)). After that, transistor regions are defined by masking to produce thin oxide (Fig. 7.15(*b*)) and then a polysilicon layer is deposited for the gates. Two further masks define the openings for the source and drain of each type of transistor (Fig. 7.15(*c*)), with one mask being the negative of the other.

Source and drain implantations

Two separate implantations are now being used instead of diffusions to form the sources and drains. Figure 7.15(*c*) shows the p-well region covered with photoresist and the source and drain for the p-channel MOST being formed by

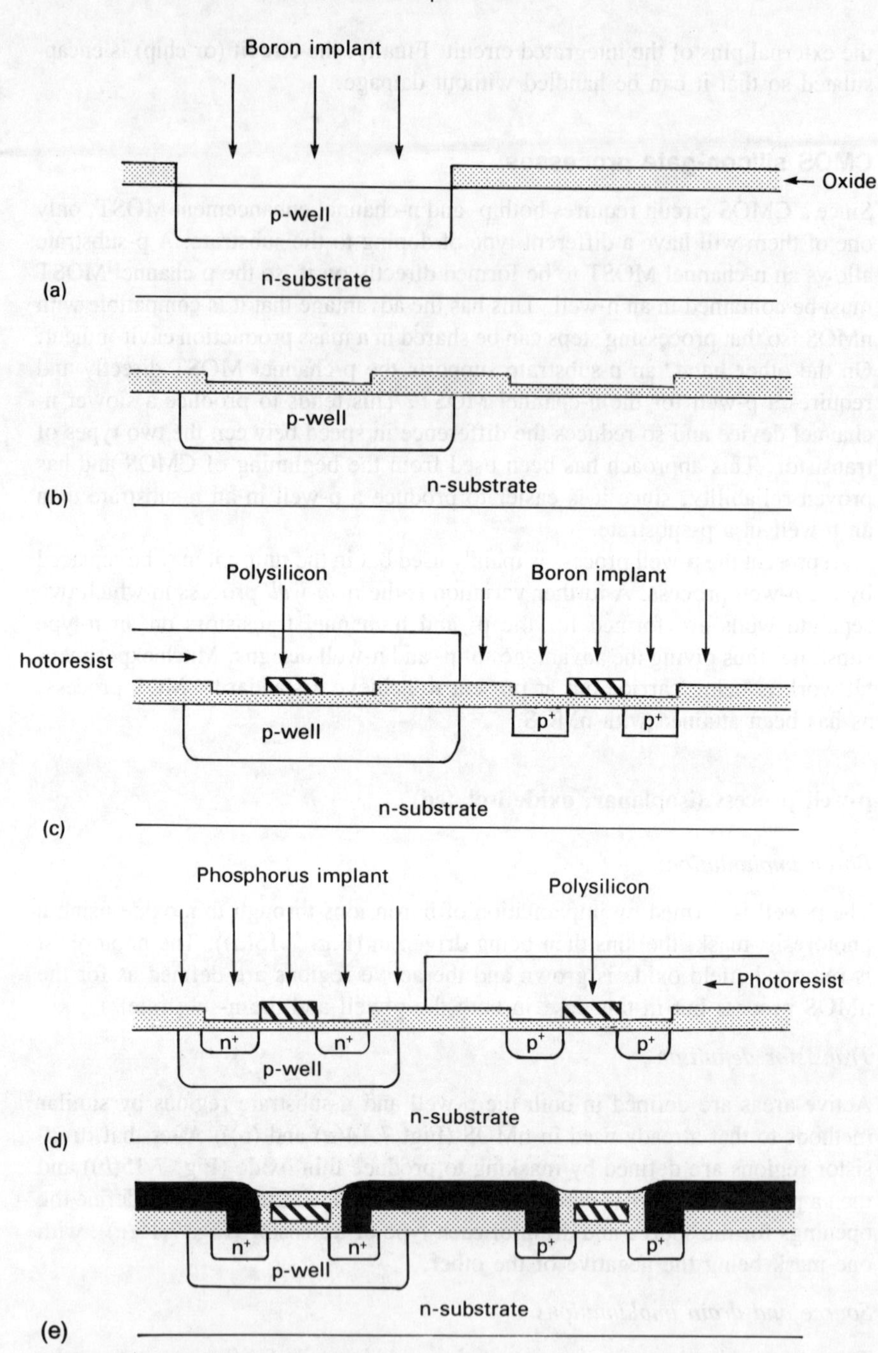

Figure 7.15 Isoplanar CMOS process. (*a*) Boron implantation; (*b*) definition of transistors; (*c*) boron implantation for p-channel device; (*d*) phosphorus implantation for n-channel device; (*e*) contacts and metallization; (*f*) oxide isolation.

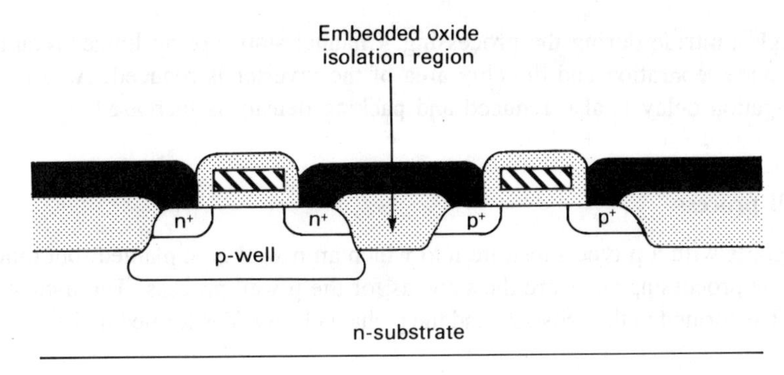

a boron implantation, which also dopes the polysilicon gate. In Fig. 7.15(*d*) the p-channel MOST is protected by photoresist and a phosphorus or arsenic implantation is forming the source and drain of the n-channel MOST and also doping the polysilicon gate. In this way each transistor is accurately self-aligned and overlap capacitance is reduced. This has led to the process being termed *isoplanar*.

Contacts, metallization, annealing and passivation

These steps are similar to those carried out for the nMOS process and the section of the final circuit is shown in Fig. 7.15(*e*). However, the p-well has to be connected to ground for correct operation, which can be achieved by inserting a p^+-region (or plug) into the well and connecting the plug to ground by extra metallization. Similarly, the n-substrate would be connected to V_{DD} through an n^+-plug and metallization (Fig. 7.17(*a*)).

The steps in nMOS and CMOS processing are summarized in Table 7.2.

A further development of the process is shown in Fig. 7.15(*f*). Here oxide isolation regions have been embedded in the substrate by oxidation of the silicon

Table 7.2 Steps in the production of MOS integrated circuits

nMOS	CMOS
Active region definition	Active region definition
Depletion implant	p-well implant
Buried contact	Polysilicon
Polysilicon	p-channel source and drain
Sources and drains	n-channel source and drain
Contacts	Contacts
Metallization	Metallization
Passivation	Passivation

Eight masks are required in each case

through a nitride during the processing. Channel stops are no longer required for device separation and the chip area of the inverter is reduced. As a result propagation delay is also reduced and packing density is increased.

n-well process

This starts with a p-type substrate into which an n-well is implanted, but otherwise the processing steps are the same as for the p-well process. The n-channel MOST is formed in the substrate and the p-channel MOST is formed in the n-well.

Twin-well process

In order to reduce chip area per gate and increase speed it is necessary to scale CMOS devices and this affects the performance of the p- and n-channel MOSTs differently. In the twin-well process an n^+-substrate is used and the doping levels can be independently optimized in the p- and n-wells, which tends to reduce second-order effects and to compensate for differences between the devices. Further advantages are protection against high voltage and minimization of latch-up (see below). The section is similar to Fig. 7.15(*e*) but with an n^+-substrate and an n-well round the p-channel device.

Silicon-on-sapphire process (SOS CMOS)

The doped silicon substrate used so far leads to the introduction of parasitic capacitance between the devices and the substrate. Both a.c. and d.c. isolation of devices are improved if an insulating substrate, such as sapphire (Al_2O_3) is used.

A layer of silicon is grown on the sapphire, which is possible since the crystal lattices of the two materials are compatible. A section through a SOS CMOS circuit is given in Fig. 7.16 which shows that the junctions are vertical and of small area. Thus only source-channel and drain-channel capacitances can occur and no isolation junctions are required. Since the substrate is insulating there are no parasitic MOSTs so the propagation delay is about half that of bulk devices and latch-up is eliminated. SOS is expensive but is particularly useful in military systems since it is little affected by radiation and this is true also for other silicon-on-insulator (SOI) technologies.

Latch-up in CMOS

The CMOS structure has parasitic p-n-p and n-p-n transistors built into it, particularly when the n^+- and p^+-plugs are included, as shown in Fig. 7.17(*a*). These parasitic transistors are effectively connected as a thyristor and correspond

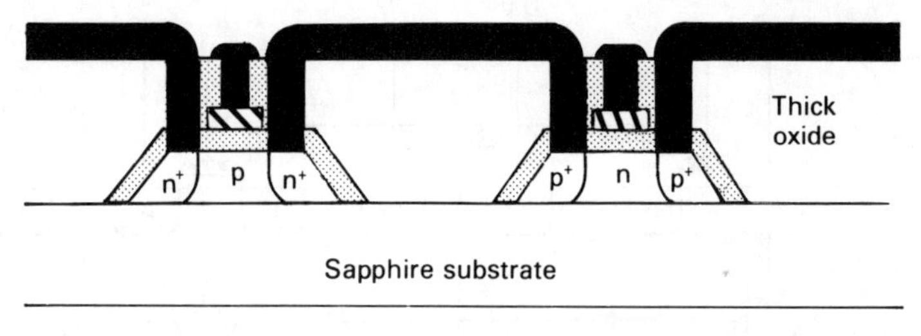

Figure 7.16 Section through silicon-on-sapphire CMOS (SOS CMOS).

to the two-transistor model of that device (Fig. 7.17(*b*)). If the two transistors have gains α_1 and α_2 then when $\alpha_1+\alpha_2>1$ positive feedback action turns on the thyristor, which then presents a very low resistance between anode A and cathode K. In the CMOS circuit A is connected to V_{DD} and K to 0 V, so that when the parasitic thyristor is turned on there is virtually a short-circuit across the power supply. This phenomenon is known as *latch-up* and can lead to destruction of the CMOS circuit (Ref. [3]).

If there is a positive surge above V_{DD} current i_s will flow through the substrate resistance R_s which will turn on the p-n-p transistor when its V_{BE} falls below about -0.6 V. This causes current i_p to flow through the p-well resistance R_p which can also turn on the n-p-n transistor, leading to positive feedback. Similarly, if there is a negative surge below 0 V the n-p-n transistor will turn on when its V_{BE} exceeds about 0.6 V, due to the voltage drop across R_p. This causes current i_s to flow through R_s, turning on the p-n-p transistor and again causing positive feedback. Typically, latch-up can occur for voltage excursions above +0.5 V for V_{DD} and below -0.5 V for the 0 V supply. This is particularly important in the larger input buffer and output driver circuits that occur round the periphery of a chip, such as the gate array described below.

One way of minimizing latch-up is to introduce *guard-rings* between the n- and p-channel MOSTs (Fig. 7.17(*a*)) with the n^+ guard ring taken to V_{DD} and the p^+ guard ring taken to 0 V. In this way resistances are connected in parallel with R_s and R_p, respectively, and the overall values are reduced, together with the base emitter voltages of the parasitic transistors. In practice, these extra n^+ and p^+ regions can form a closed ring round the p- and n-channel MOSTs.

Another method for controlling latch-up is to separate the two MOSTs and thereby reduce the current gain of the p-n-p transistor. For example, if the p-channel MOST is placed further away from the p-well then α_1 is reduced due to the increased base width. This physical separation can be reinforced electrically by inserting a deep groove between the p-well and the p-channel MOST. Either method increases the gate area, which is a disadvantage in large-scale integrated circuits. However, in the twin-well process device size is not increased at the expense of electrical isolation.

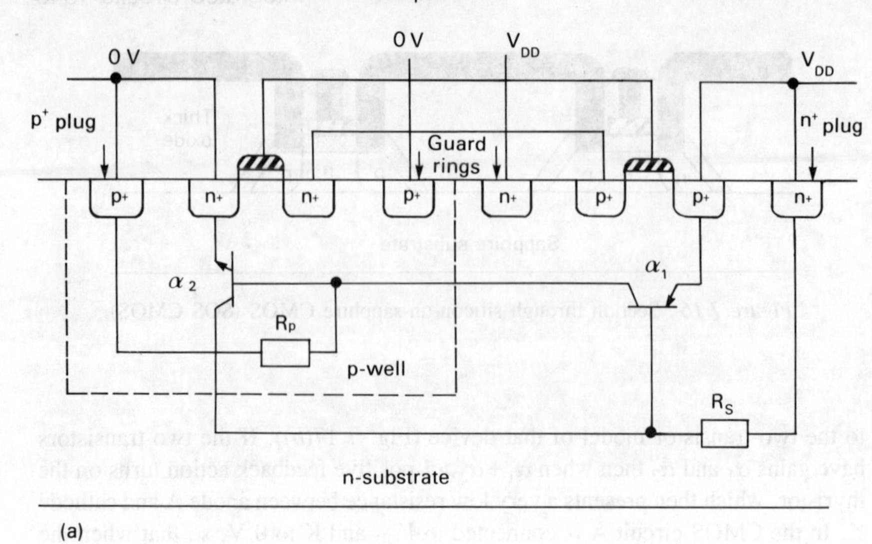

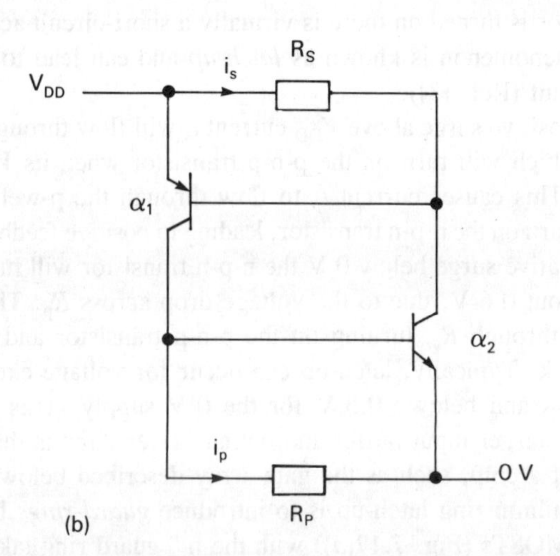

Figure 7.17 Latchup in CMOS. (*a*) Parasitic thyristor: the p-well represents the base of the n-p-n transistor and the collector of the p-n-p transistor; (*b*) equivalent circuit of parasitic thyristor.

Further details of manufacturing processes

The manufacturing processes of bipolar, nMOS and CMOS integrated circuits have been described in general but before times and temperatures can be specified more detailed analysis is required. Before this is done the important process of pattern generation is described.

Computer-aided pattern specification

It may be noted from the inverter layouts (Figs 6.16(*b*) and 6.22) and the comparison of components (Fig. 7.11) that each layer on an integrated circuit consists of rectangular shapes joined together. Many variations are possible in the shapes and depend both on the designer's skill and the design rules for the process. A common approach (Refs [4], [5]) is to build up each shape from elementary squares of side λ. The value of λ depends on the process and is continually being scaled down as technology improves. Gate length is typically 2λ, power rail width 4λ and contacts 2λ square, while source and drain areas are adjusted to give the required aspect ratios. A conservative value of λ would lie between 1.5 and 2.5 μm, giving a gate length between 3 and 5 μm with an average value of 4 μm.

λ also defines the maximum expected deviation in any linear dimension for a given process, such that the width of a shape will differ from its specified size by less than λ. The maximum misalignment of one shape relative to another will also be λ and both of these constraints on λ are examples of *design rules*. The design remains constant for each value of λ used, which is of great assistance in scaling down.

The mask geometry for an LSI design of, say, 3000 transistors might easily comprise 50 000 rectangles, which requires a computer-aided design tool or package to avoid errors due to the large amount of data. A geometric description language defines the coordinates of each shape and allows the production of plots in APPLICON or CALMA formats from the description, as well as checking the design rules. Such a CAD tool would support polygons, rather than simple rectangles and allow groups of shapes to be called and recalled. Thus the layouts of complete circuits, such as inverters, gates and registers, can be stored in subroutines and used as necessary.

Coordinates may be entered either as numbers from a keyboard or directly from the drawing by means of an electronic 'mouse', joystick, light-pen or digitizer. Each mask shape can be displayed as differently coloured areas on a graphics terminal and multi-colour plotters can then produce hard copy of the artwork for checking. Examples of such packages are GAELIC (Graphics Assisted Engineering Layout of Integrated Circuits) and PRINCESS. The output from the packages would be used as input to a mask making machine or pattern generator.

Pattern generation

Each shape is synthesized from a rectangular aperture which is illuminated and projected on to a photographic plate (Fig. 7.18). The coordinates of each aperture are held in a memory as pattern generator files, so that the aperture size and position can be adjusted through a control processor. The aperture position is defined by the *X* and *Y* coordinates of a stage which carries the photographic plate.

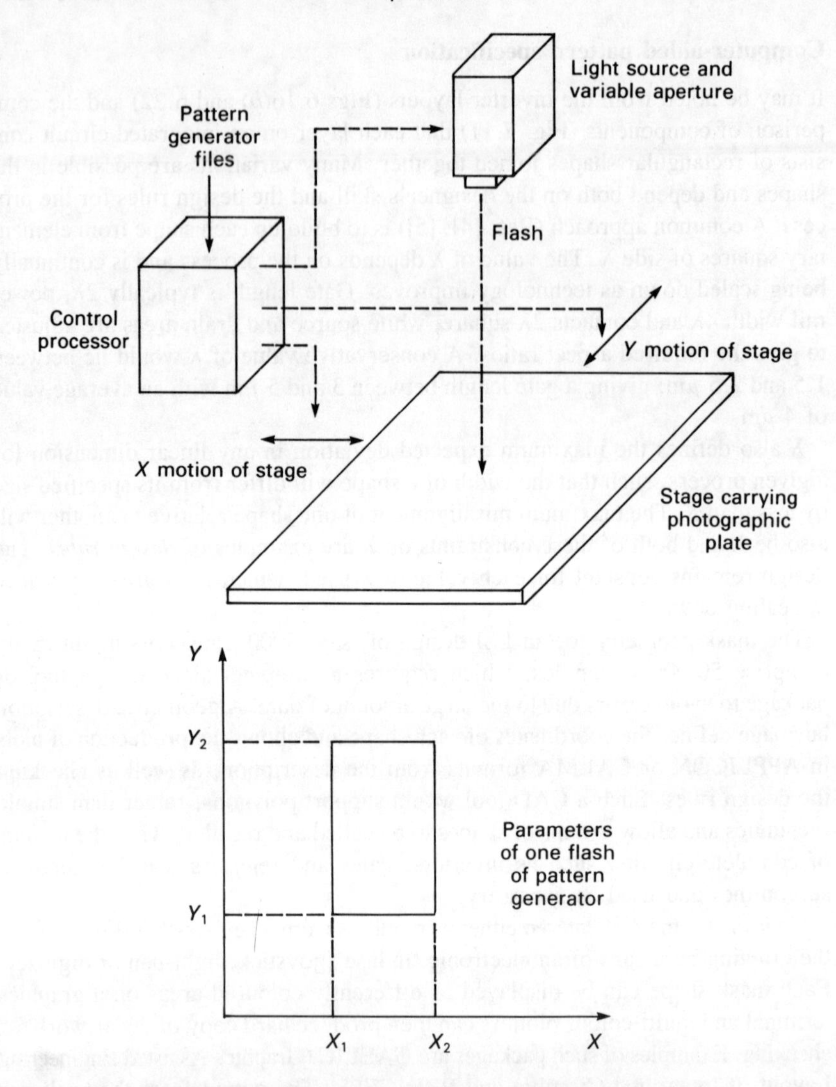

Figure 7.18 Pattern generation for integrated circuits.

Each aperture is exposed through a flash and the exposures accumulate and merge on the plate to form the required shape. Digital control allows precise position and dimensions for each shape.

The photographic plate carries an image of one complete shape at 10 × full size, which is called a *reticle*. The final-size pattern is photographically reduced and corresponds to one chip. It is then replicated by a *step-and-repeat* process to provide a pattern for all the chips on the silicon wafer. This pattern is photographically reproduced on a glass plate to form the *mask*, which is used to define windows in an oxide surface covering, as previously described. The mask image can either be projected directly on to the wafer without contact or copies of the mask can be made so that mask and wafer are brought together for a contact print.

The method is known as *photolithography* since ultra-violet light is used with a wavelength of about 0.4 μm which can define patterns down to about 2 μm. For smaller dimensions *electron-beam lithography* is used and the shorter wavelength allows pattern definition down to about 0.5 μm. The deflection coils of the electron beam generator are driven from the computer files, so no reticle is required and the mask is produced directly. A special resist, sensitive to electrons, is necessary.

A further technique is known as *direct-step-on-wafer* in which the reticle is demagnified and projected directly on to the wafer. This eliminates the need for step-and-repeat masks and the wafer itself is moved between exposures until it has been completely covered. This is much slower than exposing the wafer with all the chips in parallel but it improves placement accuracy down to 1–2 μm for VLSI devices, such as 64K RAMs. The method is applicable to both photo- and electron-beam lithography.

A more rapid technique employs electronic movement of an electron beam rather than mechanical movement of the wafer. Here the electron beam is used to write the pattern directly on to the wafer without using any masks, and the production time is greatly reduced. Although equipment costs are very high, relative to other methods, it is likely that this method will be used more and more as smaller devices are mass-produced.

Oxidation

Silicon dioxide is used extensively in the manufacture of integrated circuits, for example as a diffusion mask, in MOST gate insulation and in passivation. For a diffusion mask the most important properties are the relative diffusion rates of dopants through it. For this, electrical properties are unimportant, but in gate insulation they are extremely important and the highest oxide purity and quality are required. In passivation the oxide must be capable of resisting contamination, so that the circuit is protected from the atmosphere surrounding it.

During thermal oxidation an SiO_2 layer is grown on the Si wafer surface by holding it at high temperature in oxygen or water vapour, leading to the reactions

$$Si + O_2 \rightarrow SiO_2$$

$$\text{or} \quad Si + 2H_2O \rightarrow SiO_2 + 2H_2 \qquad (7.5)$$

Alternatively, the temperature may be reduced to 250 °C or above when silane gas is used with the oxygen, which gives

$$SiH_2 + O_2 \rightarrow SiO_2 + 2H_2 \qquad (7.5)$$

O_2 or H_2O molecules diffuse from the vapour, through the slowly forming oxide layer to combine with silicon atoms on the wafer surface. Thus when an oxide layer of thickness d is formed a silicon layer of thickness $0.45d$ is consumed.

At the beginning of the process the oxide grows linearly with time but as the oxidation proceeds the growth rate becomes parabolic, so that

$$d^2 = \text{const.} \times t \tag{7.6}$$

The constant increases with temperature, depends on the crystal orientation at the surface and is 5 to 10 times higher for steam than for dry oxygen (Ref. [1]). For example, after one hour at 1050 °C the oxide thickness after dry oxidation is about 0.08 μm, while in steam it is about 0.6 μm. After 10 hours at this temperature the thicknesses rise to about 0.35 μm and 2.0 μm, respectively.

Diffusion of impurities

The diffusion of impurities into the silicon chip is clearly a very important process in the fabrication of integrated circuits, although it is gradually being replaced by ion implantation. The process is comparable to the diffusion of minority carriers in a crystal described in Chapter 2, since the impurity atoms are constrained to move from a high concentration in the gas to a low concentration in the chip. Since there is no recombination the process is described by an equation of the form

$$D \frac{\partial^2 N}{\partial x^2} = \frac{\partial N}{\partial t} \tag{7.7}$$

where $N(x, t)$ represents the density of impurity atoms at a distance x from the crystal surface after a diffusion time t. The diffusion coefficient D rises with temperature, since the diffusing atoms thereby acquire more energy, and in practice the process is carried out at about 1100 °C, as explained above. Two main diffusion methods are used, the first mainly for emitters, diodes, sources and drains, and the second mainly for bases, resistors and isolation.

Constant source diffusion (erfc diffusion)

The concentration of impurity atoms at the surface of the slice, N_0, remains constant throughout so that $N(0, t) = N_0$ (Fig. 7.19). It may be shown that the solution of eq. (7.7) in these circumstances is given by

$$N(x, t) = N_0 \operatorname{erfc} \frac{x}{2\sqrt{(Dt)}} \tag{7.8}$$

where erfc is the complementary error function, which is tabulated or may be read from a graph such as Fig. 7.20. The use of this expression in junction formation is explained below. The variation of N with distance below the surface is sketched on a linear scale in Fig. 7.21(a), which shows that the depth at which a particular value of N occurs increases with the diffusion time t. In the limit,

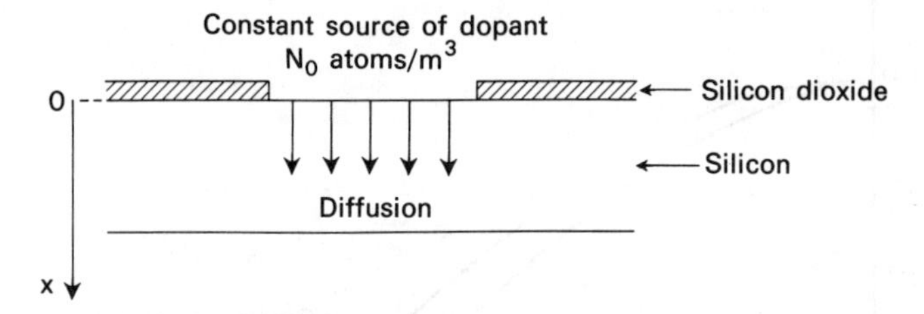

Figure 7.19 Constant source diffusion.

where t is very long, the whole slice would acquire a dopant concentration of N_0. The surface concentration N_0 depends on the upper limit of concentration of the impurity which can be absorbed. This is the *solid solubility*, typically in the order of 10^{27} atoms/m^3 for relevant materials, or about 1 in 100 silicon atoms replaced. Since this diffusion method provides a high surface concentration it is used for example, where an n^+ region is required in the emitter of an n-p-n transistor (*see* Chapter 4).

Limited source diffusion (Gaussian diffusion)

An initial *deposition cycle*, using erfc diffusion, provides a finite quantity of dopant, Q atoms/m^2, in a thin surface layer to form the limited source (Fig. 7.22). A *drive cycle* follows when the source is turned off but the temperature is maintained at around 1000 °C. The surface concentration falls with increasing time as the dopant penetrates the slice and leads to a more uniform distribution than before. The limiting conditions are $\int_0^\infty N(x)\mathrm{d}x = Q$ and $N(x) = 0$ at $t = 0$ which leads to a solution of eq. (7.7) given by

$$N(x,\ t) = \frac{Q}{\sqrt{(\pi Dt)}} \exp\left(-\frac{x^2}{4DT}\right) \tag{7.9}$$

This is a Gaussian distribution shown in Fig. 7.20 and the variation of N with x is sketched on a linear scale in Fig. 7.21(*b*) which shows the effects of increasing diffusion time. The method is used, for example, in forming the base of a transistor.

Both eqs (7.8) and (7.9) involve the diffusion coefficient D, which rises with temperature as shown in Fig. 7.23 for the common dopants in silicon. D can double itself for a few degrees rise in temperature, so that the diffusion furnace temperature must be held constant in the range 1000 to 1300 °C to within ± 0.5 °C. Phosphorus (n-type) and boron (p-type) have the same values of D at all relevant temperatures. An important consideration in choosing them is that their diffusion coefficients in silicon dioxide are respectively 1×10^{-4} and 3×10^{-3} of their diffusion coefficients in silicon. This means that they penetrate the slice far more rapidly than the oxide, which can thus act as an effective mask.

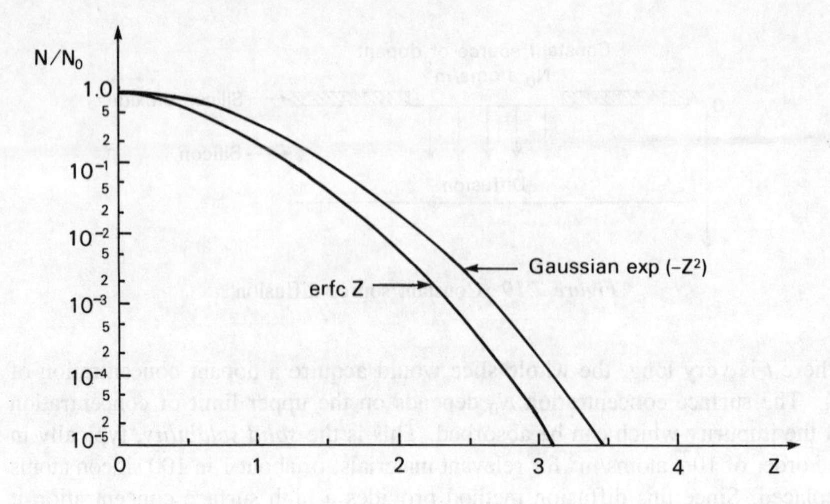

Figure 7.20 Complementary error function, erfc, and Gaussian function of $Z = x/2\sqrt{(Dt)}$

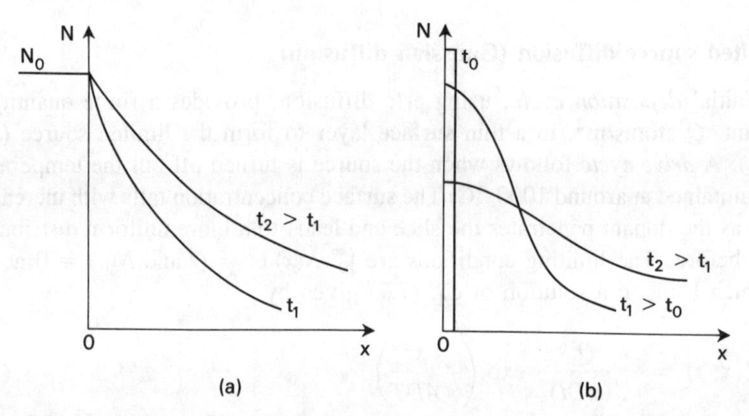

Figure 7.21 Impurity profiles on a linear scale as a function of time. (*a*) Constant source diffusion; (*b*) limited source diffusion.

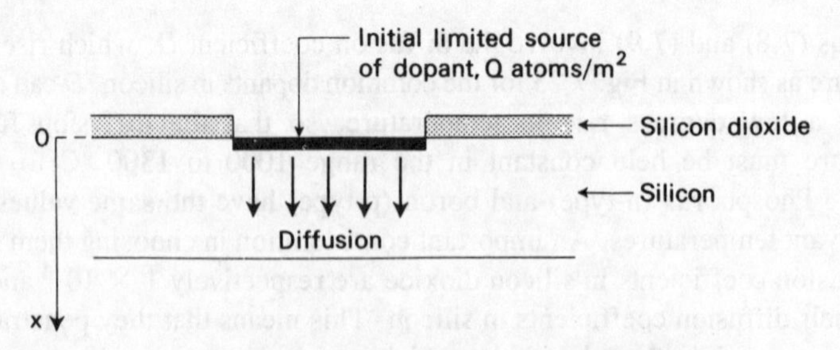

Figure 7.22 Limited source diffusion.

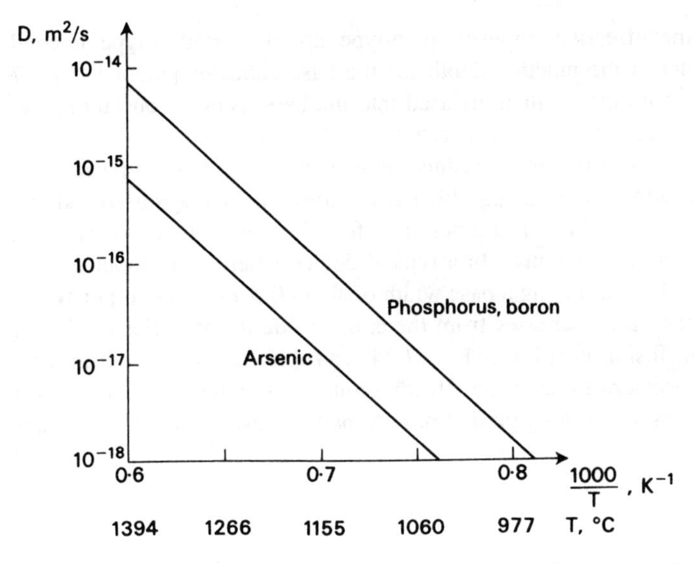

Figure 7.23 Variation of diffusion coefficient with temperature for common dopants in silicon.

Arsenic may be used as the dopant of the epitaxial layer, since it has a much lower diffusion coefficient than boron or phosphorus and so is only slightly disturbed during subsequent base and emitter diffusions.

Both diffusion methods depend on the product of D and t, so a change in D or t has the same effect on N. All diffusions proceed simultaneously; so the buried layer diffusion is affected by the isolation diffusion, both are affected by the base diffusion and all three by the emitter diffusion. The effective Dt product is given by

$$(Dt)_{\text{eff}} = D_1t_1 + D_2t_2 + \ldots \tag{7.10}$$

with the values of D being controlled by temperature, different values being chosen in succeeding stages as explained below.

Junction formation

After the epitaxial layer has been deposited, and the isolation diffusion has defined the collector region, a p-type diffusion is required for the base region. This is normally a limited source diffusion to provide a relatively deep junction and at a certain depth x_1 the p and n concentrations are exactly equal. Thus where N_D is the impurity concentration of the epi-layer

$$N(x_j, t) = \frac{Q}{\surd(\pi Dt)} \exp\left(-\frac{x_1^2}{4Dt}\right) = N_D \tag{7.11}$$

Since the effective impurity is p-type up to x_1 and n-type beyond x_1 this parameter is the junction depth for the base-collector junction (Fig. 7.24).

The n^+ emitter is then diffused into the base using a constant source since a high surface concentration is required. This takes place at a lower temperature than the base diffusion to reduce movement of the base impurities. However, the diffusion time is comparable since a constant source is used and the junction is relatively shallow. The junction is formed at a depth x_2 where the n^+ and p concentrations are equal. In a typical device where x_1 is about 2 μm x_2 would be about 1.5 μm giving a base width of about 0.5 μm. The impurity distribution across the base decreases from the emitter side to the collector side, as shown in the diffusion profiles of Fig. 7.24, and this leads to a built-in electric field which accelerates electrons from emitter to collector. The operation and characteristics of such graded-base transistors are discussed in Chapter 4.

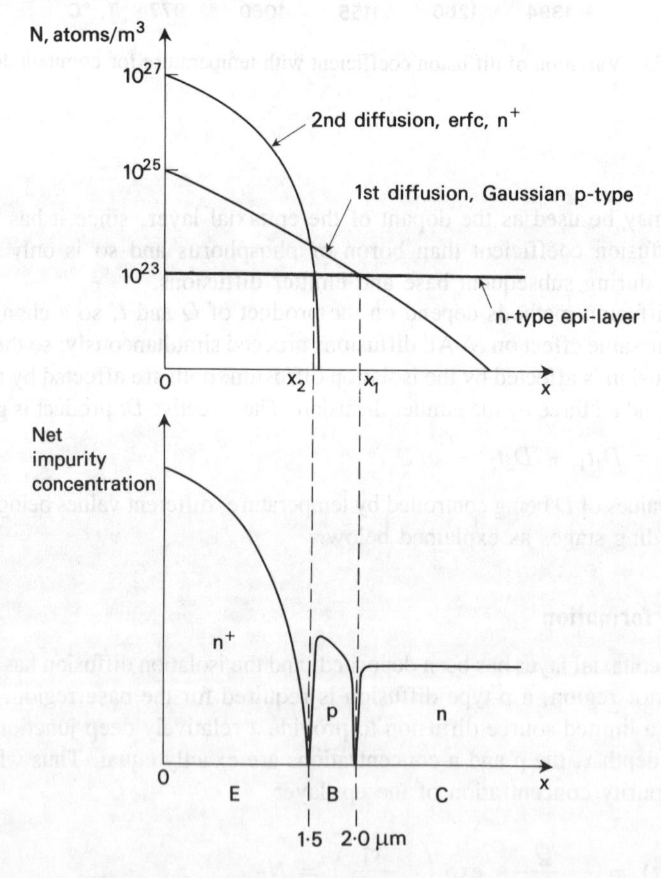

Figure 7.24 Impurity profiles on a log scale for a planar n-p-n transistor.

Ion implantation

This technique, illustrated in Fig. 7.12, produces the same results as diffusion but with the advantages that it is a low temperature process and that there is no sideways diffusion. Doping levels can be controlled between 10^{17} and 10^{22} ions/m^2 but the depth of doping is limited to about 1 μm (Ref. [6]). Thus it is used both for MOST fabrication and for bipolar transistors and it is much employed in controlling the threshold voltages of silicon-gate MOSTs.

Where the ion beam current I flows for a time t an integrated charge Q is implanted in the silicon and $Q = \int I dt$. The implanted dose is defined as N_s =

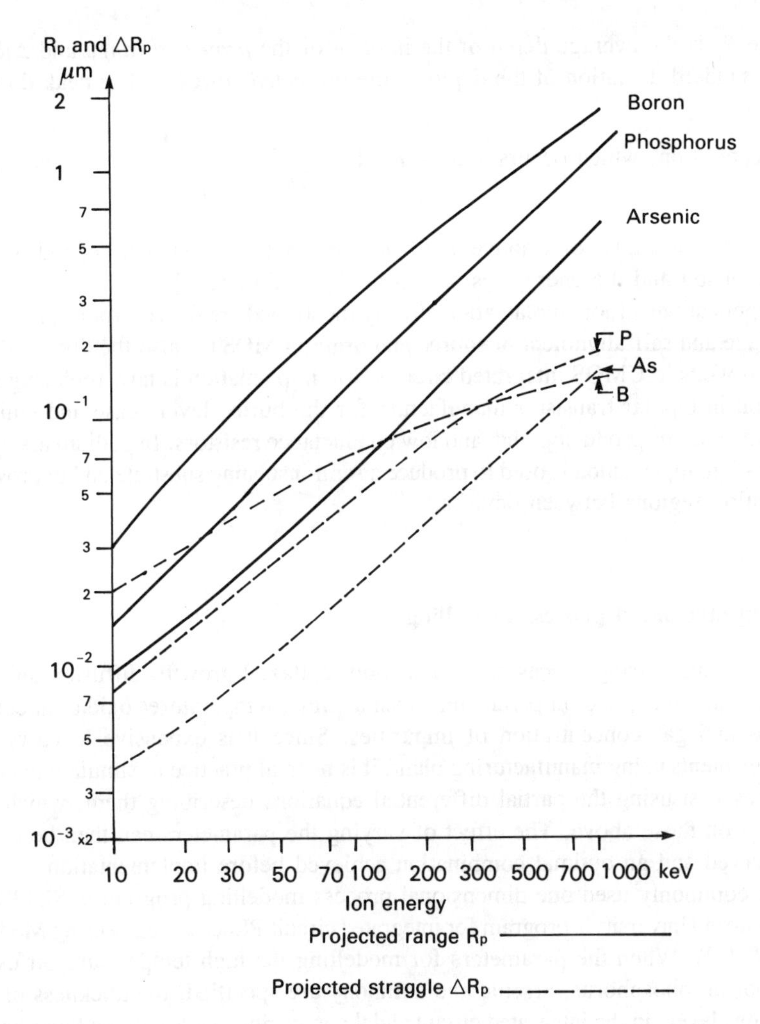

Figure 7.25 Variation of projected range R_p and projected straggle ΔR_p with energy in ion implantation.

Q/qA ions/m^2, where q is the charge on each ion and A is the area of the aperture through which the beam is swept. For example, using ions with $q = e$, a beam current of 1 mA swept over an area of 0.01 m^2 for 100 seconds gives a dose of 6.25×10^{19} ions/m^2.

The impurity profile normal to the surface at $x = 0$ after implantation is approximately Gaussian and given by

$$N(x) = \frac{N_s}{(2\pi)^{1/2}\,\Delta R_p} \exp\left[\frac{-(x - R_p)^2}{2(\Delta R_p)^2}\right] \tag{7.12}$$

Here R_p is the average depth of the implant or the *projected range* and ΔR_p is the standard deviation of the depth or the *projected straggle*. The peak doping concentration, which occurs at $x = R_p$, is $\dfrac{N_s}{(2\pi)^{1/2}\,\Delta R_p}$ and N_x falls exponentially for values of x above and below R_p. Both R_p and ΔR_p depend on the type of ion and the energy, as shown in Fig. 7.25 (Ref. [7]).

Applications of ion implantation already mentioned are the control of threshold voltage and self-alignment of source and drain in MOSTs, also the doping of n- and p-wells in CMOS integrated circuits. Ion implantation is now replacing diffusion in bipolar transistor manufacture for the buried layer, base and emitter doping and for producing high and low conductance resistors. In gallium arsenide FETs ion implantation is used to produce a semi-insulating substrate and to provide isolation regions between devices.

Computer-aided process modelling

The manufacturing processes of oxidation, epitaxial growth, diffusion and ion implantation depend on parameters such as time, temperature, oxidation conditions and gas concentration of impurities. Since it is expensive to carry out experiments using manufacturing plant, it is normal practice to simulate the processes first using the partial differential equations describing them, which are based on those above. The effect of varying the parameters can then be easily observed and an optimal combination achieved before implementation.

A commonly used one-dimensional process modelling program is SUPREM (Stanford University's program for integrated circuit PRocess Engineering Models) (Ref. [8]). When the parameters for modelling the high temperature diffusion of boron, phosphorus, arsenic and antimony are specified, the thickness of the various layers in the integrated circuit and the impurity profiles within those layers are calculated. Similarly, when the ion dose and energy are specified, the distribution of impurities following an ion implantation process is obtained.

Digital integrated circuits

A digital or logic circuit has to respond only to one of two conditions at its input terminals, normally the absence or presence of a voltage level designated logical 0 or 1, which represent binary digits or *bits*. Such a circuit can operate satisfactorily with a wide tolerance of component values and complete digital systems tend to require large numbers of just a few basic circuits. For example a memory may contain 4096 identical cells each capable of holding a 0 or 1 level. Again the essential elements of a computer are accommodated on a single chip, known as a *microprocessor*, which operates on bits in groups of 18, 16 or 32 bits.

Integrated circuit technology is thus ideal for digital applications and a number of different types have been developed. Each of these optimizes one or more of the following criteria:

(1) the number of logic gates per unit area of the chip or the *packing density*;
(2) the product of propagation delay and input power or the *delay-power product*;
(3) the size of spurious signal which will be ignored or the *noise margin*.

Generally, the main choice lies between bipolar and field-effect devices, with bipolar devices having a higher speed and field-effect devices have a higher packing density as well as a high speed.

The main methods of implementing a digital system are indicated in Fig. 7.26. The four broad levels of integration can be defined in terms of the number of four input logic gates, or their equivalent, provided on a single chip, as follows.

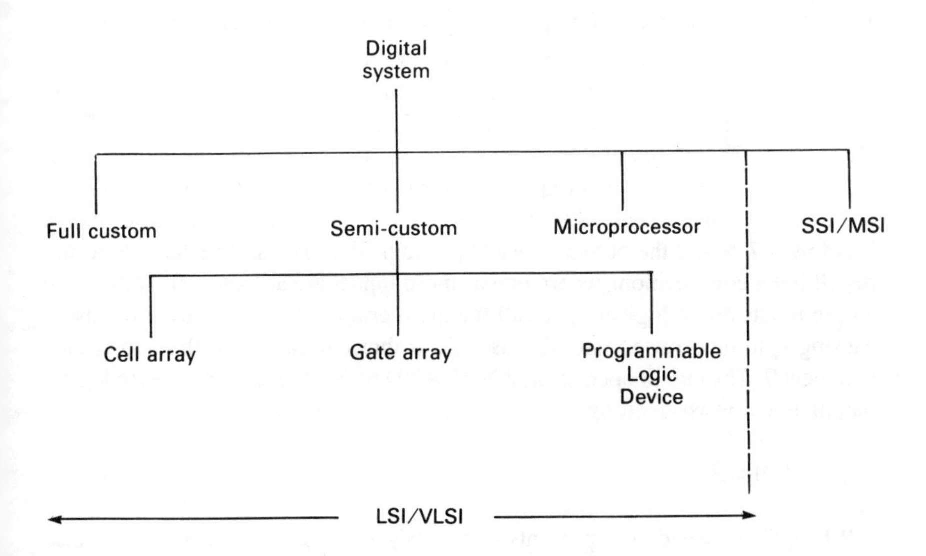

Figure 7.26 Methods of implementing a digital system.

Small-scale integration, SSI
1–20 gates of all types, either separate as NAND gates, for example or combined in flip-flops of varius kinds.

Medium-scale integration, MSI
20–100 gates combined to form, for example, registers, adders, encoders. Many linear integrated circuits, such as operational amplifiers (opamps) would also come in this category.

Large-scale integration, LSI
100–7000 gates in, for example small memories, 8-bit microprocessors, digital watch chips, semi-custom designed chips such as gate arrays.

Very large-scale integration, VLSI
Over 7000 gates in, for example, large memories, 16- and 32-bit microprocessors, custom-designed chips.

As technologies are scaled down so the boundary between LSI and VLSI is likely to increase above 7000 gates/chip. The terms *custom* and *semi-custom* are explained later in this chapter. In general, CMOS is used at all levels of integration, nMOS is used in LSI and VLSI only and bipolar technologies are used mainly for SSI and MSI but also for some LSI chips.

Transistor-transistor logic circuit, TTL (T^2L)

The circuit for the simplest form of 4-input TTL gate is shown in Fig. 7.27(*a*), where T1 is a multi-emitter transistor shown in section in Fig. 7.27(*b*). Suppose all four input emitters are taken to a low voltage near 0 V, logical 0. Current I_1 then flows through the four emitters of T1 rather than into the base of T2. This is because the forward bias of the four parallel emitter diodes is only about 0.7 V, while the forward bias required for conduction through the two series diodes at T1 collector and T2 base is about 1.4 V. T2 is then cut off since its base voltage is below 0.7 V and the output is near V_{CC}, logical 1. The same situation occurs for all input combinations where one or more inputs are at logical 0. Only when all the inputs are at logical 1 will all the input emitter diodes be reverse-biased, causing I_1 to be diverted into T2 base. T2 is then saturated and the output falls to logical 0. The circuit operates as a NOT AND or NAND gate, where the logical output function is given by

$$f = \overline{A.B.C.D} \qquad (7.13)$$

When T2 is turned on it presents a low resistance across the load capacitance C_L, which is quickly discharged. When T2 is turned off C_L is charged up to V_{CC} more slowly through R_2, which has a much higher resistance than T2. In practice, the number of inputs can lie between two and eight and corresponds to the number of separate emitters.

TTL was the first digital integrated circuit technology to become established in SSI and MSI and it has defined levels, such as the 5 V supply, for later technologies like CMOS. It has been continuously developed since 1969, with the operation of all forms of NAND gate, for example, based on the standard circuit which is numbered 7400 in the 74 series of gates.

Standard TTL gate (74 series)

In the two-input standard TTL NAND gate shown in Fig. 7.27(*c*) the output fall and rise times are made more equal by employing a *totem-pole* output stage. Here C_L is discharged through T4 and charged up through T3, which reduces the total propagation delay since the transistor resistances are similar. Diodes D1 and D2 are open-circuit for normal operation and only conduct if the inputs accidentally go negative. In this way I_1 is limited to a safe value.

When one or both inputs are low (logical 0) I_1 flows *out* of the input, V_b is about 0.7 V, T2 is cut off, T3 conducts and the output is high. When both inputs are high (logical 1) I_1 flows into T2 base, V_b is about 2.1 V (i.e. three forward diode voltages), T2 and T4 conduct, T3 is cut off and the output is low. Diode D3 ensures that the base-emitter voltage of T3 is below 0.7 V when T2 and T4 conduct. All conducting transistors are saturated, leading to stored base charge, but the output resistance is always low, giving a propagation delay of about 10 ns. The 130 Ω resistor limits the current pulse or 'spike' that occurs during the short time that T3 and T4 conduct together, when the level is changing as in CMOS. It may be noted that if an input is disconnected and allowed to 'float' the input diode is effectively cut off, so the effect is the same as if the input had been taken as high.

Transfer characteristic

As explained in Chapter 6, the transfer characteristic shows the relationship between the input and output voltages. It is given in Fig. 7.28 for the standard TTL gate of Fig. 7.27(*c*) and has been obtained by SPICE simulation with TTL load. The complete program with parameter values is given in Appendix 3 but all transistors have IS = 10^{-16} A and BF = 50 with V_{CC} 5 V. Input A is at 3.5 V, which represents a high output from a previous gate and input B is changed between 0 and 5 V.

With input voltage V_{in} at 0 V the output voltage V_{out} is at about 3.6 V, which is less than 5 V due to the drop across T3 and D3 acting as a current source. V_{out} is typically 4 V and depends on the current flowing into a following input, which can have a maximum value of 40 μA.

As V_{in} is increased from 0 V, V_{out} remains constant at the 1 level until the input emitter-base diodes begin to cut off. This occurs at the voltage V_{IL}, which is about 0.6 V in the simulation. V_{out} then falls with increasing V_{in} at T2 and T4 begin to conduct until it reaches the level V_{OH} which is about 2.5 V here. Just

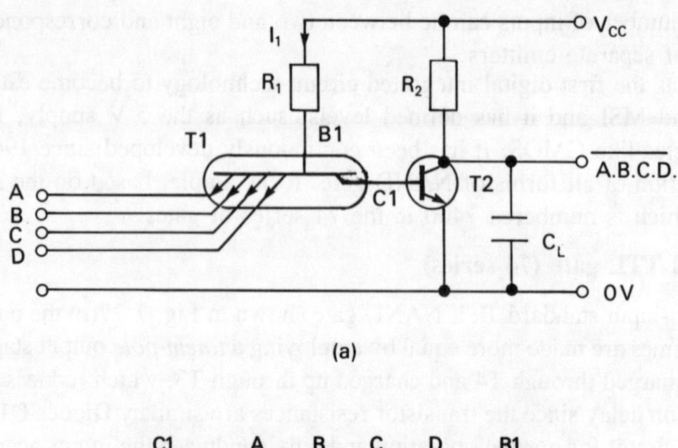

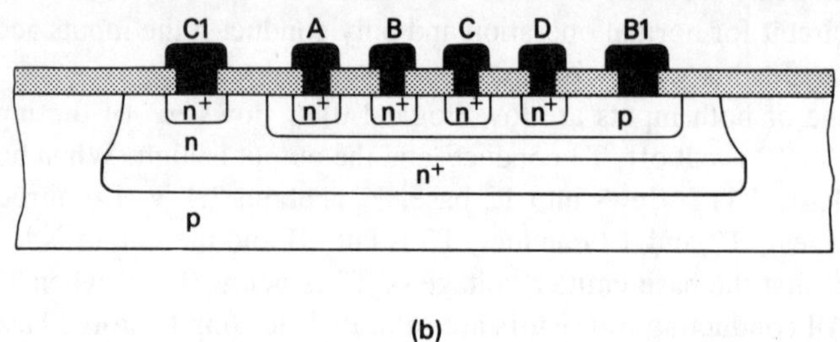

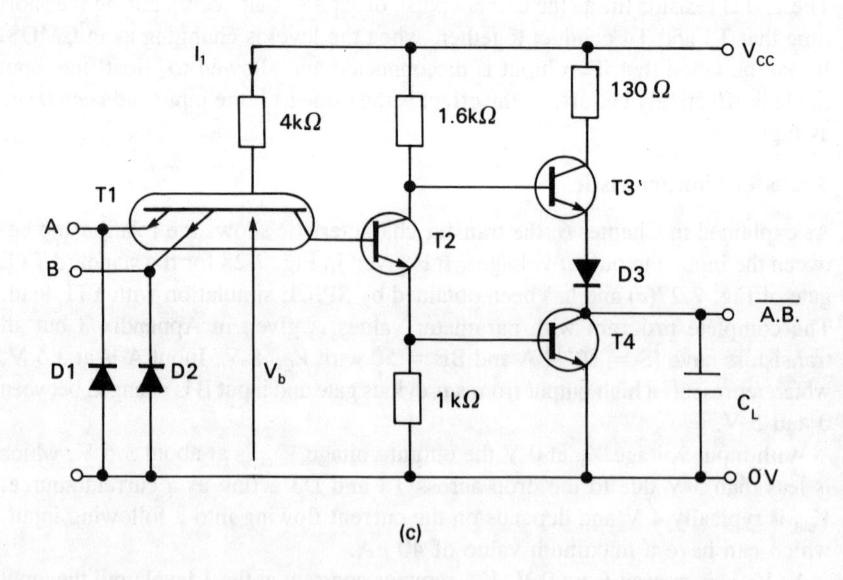

Figure 7.27 **TTL NAND gates. (*a*) Basic gate circuit; (*b*) transistor section; (*c*) standard TTL 2-input NAND gate circuit.**

below V_{OH} all transistors are operating as amplifiers in the active region and a current spike flows through T3 and T4, until V_{out} levels out when $V_{in} = 1.7$ V. This allows V_b to be high enough for each of the three diodes (T1 collector, T2 base and T4 base) to begin to conduct.

Finally V_{out} remains constant at V_{OL} as V_{in} is increased towards 5 V. When V_{in} exceeds V_{IH} at about 2.1 V the gate input has reached the 1 level beyond which the output remains at the 0 level. VOL is 0.06 V here and depends on the current flowing *out* of a following input. The maximum value of this current is 1.6 mA for each input, which flows into the output transistor T4 acting as a current sink.

Noise margins

In a practical digital system the 0 and 1 levels are by no means constant with time but subject to random variation about a constant value. This may be caused by interference from external sources, such as electrical machinery or generated internally, for example through the power supply when current spikes occur. The random variation or noise can be minimized by connecting a 0.01 μF capacitor, say, across the power supply for every two or three chips. However, safety margins are still required, called *noise margins* and defined from the transfer characteristic, separately for the 0 and 1 levels. The voltages V_{IL}, V_{OL}, V_{IH} and V_{OH} are used, which can be obtained for all technologies but are not always so clearly defined as for TTL in Fig. 7.28. A gate is designed to satisfy the following criteria.

A voltage is recognized as a 0 up to V_{IL} but a 0 gate output will not rise above V_{OL}. The 0 level noise margin is given by

$$NM_0 = V_{IL} - V_{OL} \tag{7.14}$$

A voltage is recognized as a 1 down to V_{IH} but a 1 gate output will not fall below V_{OH}. The 1 level noise margin is given by

$$NM_1 = V_{OH} - V_{IH} \tag{7.15}$$

Typical and worst case values are quoted for each technology and for TTL these are 1.0 V and 0.4 V, respectively. The worst case voltages for the 0 and 1 levels are obtained from

$$V_{IL} = 0.8 \text{ V}, \ V_{OL} = 0.4 \text{ V}, \ NM_0 = 0.4 \text{ V}$$

$$V_{OH} = 2.4 \text{ V}, \ V_{IH} = 2.0 \text{ V}, \ NM_1 = 0.4 \text{ V}$$

The values obtained from Fig. 7.28 are

$$V_{IL} = 0.6 \text{ V}, \ V_{OL} = 0.06 \text{ V}, \ NM_0 = 0.54 \text{ V}$$

$$V_{OH} = 2.5 \text{ V}, \ V_{IH} = 2.1 \text{ V}, \ NM_1 = 0.40 \text{ V}$$

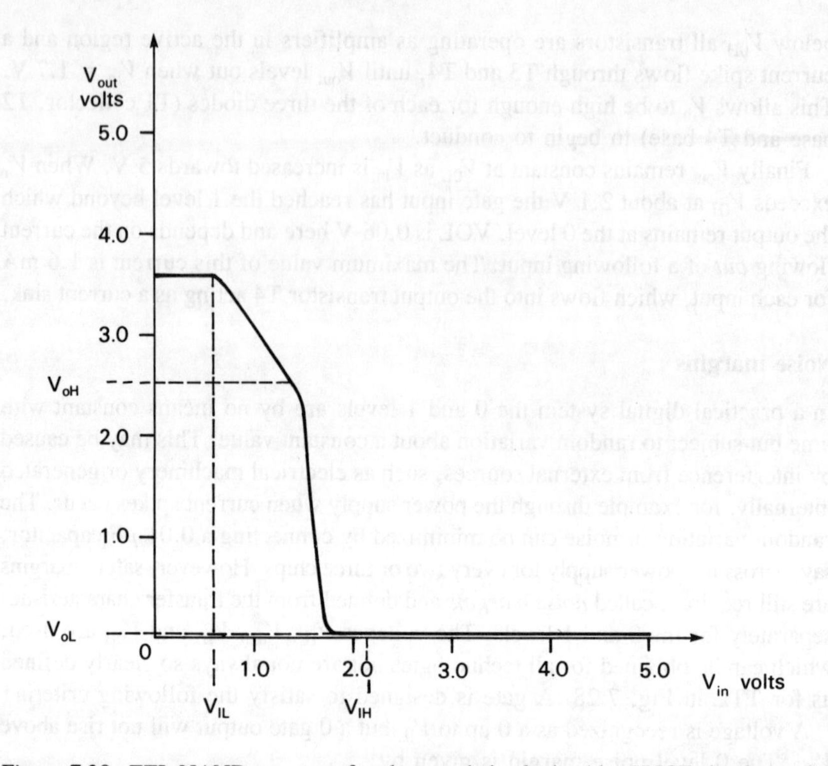

Figure 7.28 TTL NAND gate transfer characteristic obtained by SPICE simulation for the circuit of Fig. 7.27 with TTL load (Fig. A.3.3).

Fanout

Each input or *logic load* connected to a gate output will tend to reduce noise margins and increase propagation delay. The maximum number of logic loads that will allow the gate to operate within its specification is known as the *fanout*.

For TTL a logic load is 40 μA for a high output and 1.6 mA for a low output, with specified fanouts of 20 and 10, respectively. This implies that the gate must supply a maximum 1 level current of 800 μA and sink a maximum 0 level current of 16 mA.

Transient response

The transient response shown in Fig. 7.29 has been obtained by SPICE simulation with input A at 3.5 V and input B switched between 0 and 3.5 V, well above V_{IH} in each case. When T4 conducts it pulls down the output voltage, leading to t_{pd-}, and when T3 conducts it pulls up the output voltage leading to t_{pd+}. During part of each transient both transistors conduct together and since T4 tends

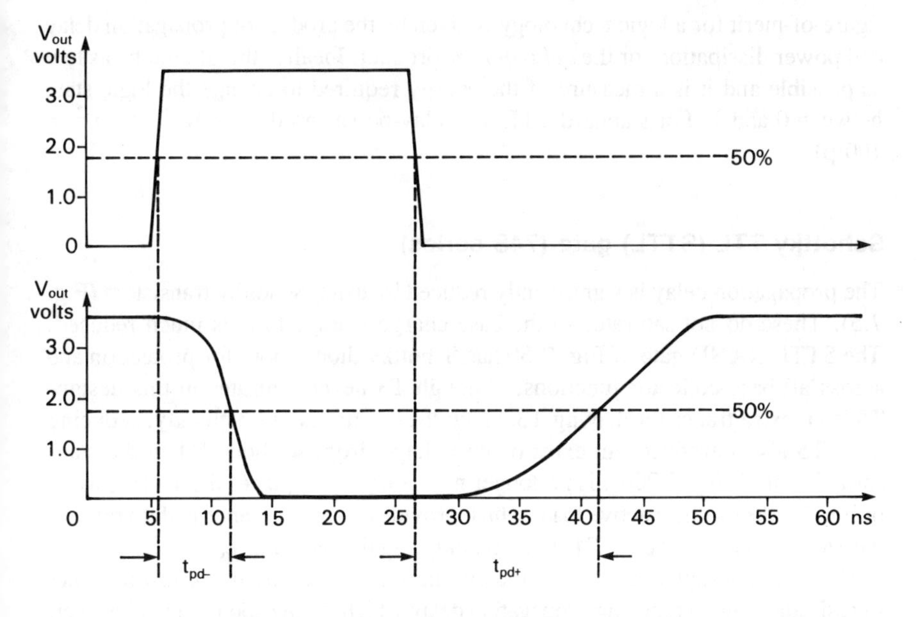

Figure 7.29 TTL transient response obtained by SPICE simulation for the circuit of Fig. 7.20 with TTL load (Fig. A.3.3). t_{pd-} = 6.0 ns, t_{pd+} = 13 ns as measured at the 50% points.

to be more heavily saturated than T3 it takes longer to switch off, so that t_{pd+} is longer than t_{pd-}. In the standard TTL simulation t_{pd-} is about 6.0 ns and t_{pd+} is about 13 ns, measured between 50% points, so the mean propagation delay t_{pd-} is 9.5 ns with t_{pd+} being the larger component.

Power dissipation

The d.c. power required by the gate is obtained by considering the supply current for high and low inputs and taking the mean value, which assumes a square wave input voltage waveform. With V_{CC} = 5 V and with 5 V on all inputs the specified standard TTL supply current is 3 mA, which flows mainly through T2 and into T4. With 0 V on all inputs the current falls to 1 mA, flowing mainly through the input diodes. The mean supply current is then 2 mA and the mean d.c. power is 10 mW.

Delay-power product

Two very important parameters of a digital system are the clock frequency and the supply current. A high clock frequency implies a low propagation delay and a low supply current implies low power dissipation. Consequently, a common figure-of-merit for a logic technology is given by the product of propagation delay and power dissipation, or the *delay-power* product. Ideally, this should be as low as possible and it is a measure of the energy required to change the logic state between 0 and 1. For standard TTL the delay-power product is $10^{-8} \times 10^{-2} =$ 100 pJ.

Schottky TTL (STTL) gate (74S series)

The propagation delay is significantly reduced by using Schottky transistors (Fig. 7.5). These do not saturate, so the base charge storage time is much reduced. The STTL NAND gate of Fig. 7.30 has Schottky diodes both for protection and across all base-collector junctions, although T3 never saturates in this design. T5 is an extra transistor driving T3 which thus switches more quickly, reducing t_{pd+}. T5 also introduces an extra diode voltage drop, so diode D1 in the standard circuit of Fig. 7.20(c) is no longer necessary. T6 replaces the 1 kΩ resistor of Fig. 7.20(c) with an active load, which provides a low resistance path to remove the stored base charge of T4 more quickly, again reducing t_{pd+}.

The time constants throughout the circuit are reduced by using lower values of resistance, again reducing propagation delay which is now about 3 ns. However, the supply currents are increased thereby and the power dissipation rises to about 20 mW, giving a delay-power product of about 60 pJ.

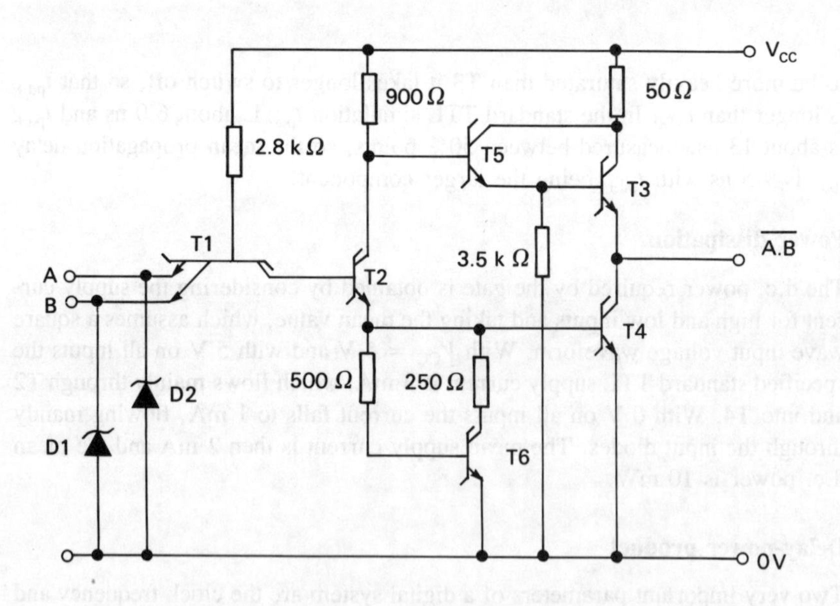

Figure 7.30 Schottky TTL 2-input NAND gate circuit.

Low-power Schottky TTL (LSTTL) gate (74LS series)

The STTL circuit is used where very high speed is required and the high power dissipation can be tolerated. However, for many applications the speed of standard TTL is adequate and a power reduction would be an advantage. This is achieved in the LSTTL gate of Fig. 7.31, which is similar to the STTL gate of Fig. 7.30 but with resistor values higher even than those of standard TTL.

Changes from the STTL gate are that T1 has been replaced by two input diodes D3 and D4 which eliminates collector capacitance. T2 has a 12 kΩ base-emitter resistor and T3 has a 4 kΩ base-emitter resistor, which in each case helps to remove charge more quickly from the base. With the 4 kΩ emitter load of T5 now connected to the output D5 protects the base-emitter diode of T5 against breakdown and D6 adjusts the output voltage level.

The propagation delay has now returned to 10 ns, due to the larger resistor values but power dissipation is much reduced at 2 mW, giving a delay-power product of 20 pJ. Low-power Schottky TTL gates have replaced standard TTL gates as a first choice in many applications, as speed is comparable and power dissipation is much lower. However, the circuit has been further developed as described in the next section.

Worked example

Estimate the power dissipation of the LSTTL circuit of Fig. 7.31 with a 5 V supply voltage:

(*a*) for one input at 0.2 V and the other at 4.0 V
(*b*) for both inputs at 4.0 V.

Assume that the forward voltages are 0.7 V for a silicon diode and 0.3 V for a Schottky diode and that the saturated collector voltage of a Schottky transistor is 0.4 V.

Solution

(*a*) In this condition one of the input diodes is conducting with a forward voltage of 0.3 V and the other is cut off with a reverse voltage of 3.7 V. There will be no current through the 8 kΩ and 120 Ω resistors since T2 and T4 are cut off.

The current through the 20 kΩ resistor is

$$\frac{5 - 0.3 - 0.2}{20 \times 10^3} = 0.23 \text{ mA}$$

so the power dissipation is

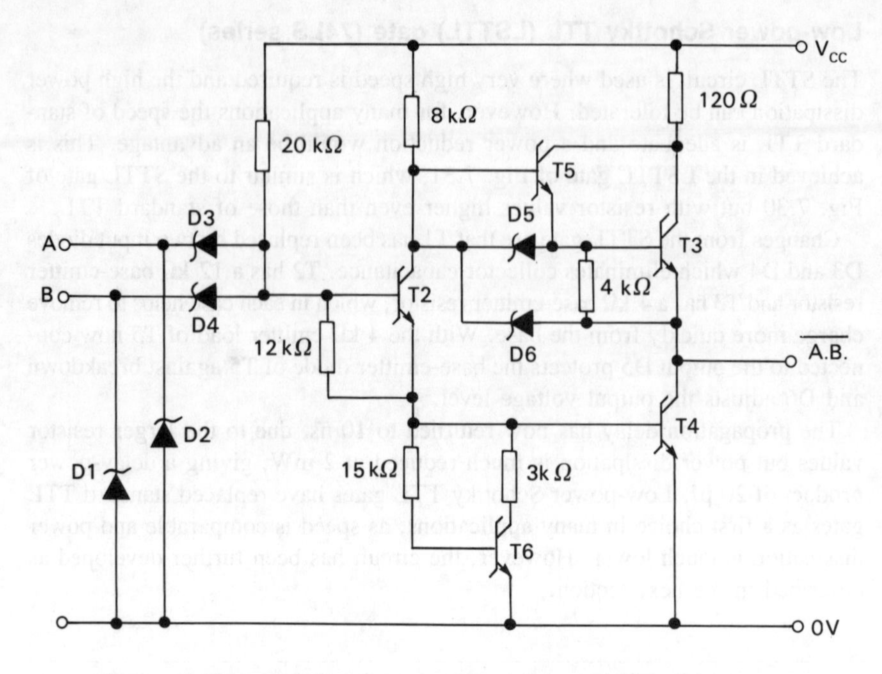

Figure 7.31 Low-power Schottky TTL 2-input NAND gate circuit.

$5 \times 0.23 = 1.15$ mW.

(*b*) Here both input diodes are cut off and transistors T2 and T4 are conducting, the reverse bias across D3 and D4 being $4 - 0.7 - 0.7 = 2.6$ V.

Ignoring any current through T6 but including the 12 kΩ resistor the current through the 20 kΩ resistor is

$$\frac{5 - 1.4}{20 \times 10^3} + \frac{0.7}{12 \times 10^3} = 0.24 \text{ mA}$$

The current through the 8 kΩ resistor is

$$\frac{5 - 0.4 - 0.7}{8 \times 10^3} = 0.49 \text{ mA}$$

The current through the 120 Ω resistor is zero since T3 is cut off.

The total current is $0.24 + 0.49 = 0.73$ mA so the power dissipation is $5 \times 0.73 = 3.65$ mW.

The mean power dissipation is 2.4 mW and this estimate compares with the quoted value of 2 mW.

Advanced and FAST TTL gates (74AS, 74ALS and 74F series)

Advanced Schottky and advanced low-power Schottky gates use transistors with small geometries and an advanced oxide isolation construction, similar to Fig. 7.12. In each case they provide approximately half the propagation delay of the corresponding Schottky and low-power Schottky gates, as illustrated in Table 7.3.

FAST gates have an advanced isoplanar construction, obtained from a further development of the oxide isolation process. Size is even further reduced and the stray capacitances are so small that f_T exceeds 5 GHz. They are intended to replace Schottky TTL gates, being only slightly faster but with about one-third of the power dissipation.

Integrated-injection logic circuit*, IIL (I^2L)

The elements of an I^2L gate are shown in Fig. 7.32 where the transistor T1 with multiple collectors is supplied from a constant current source *I*. The inputs are connected together to the base of T1 and isolation occurs at the output through the separate collectors, each of which is loaded by a following gate (Ref [9]). When current *I* flows into T1 base the collector currents flow and the output falls to V_{CEsat}, about 50 mV, corresponding to logical 0. When current *I* is diverted from T1 base to one or more of the inputs the collectors present a virtual open circuit and rise to the 0.7 V threshold level of the following gates, corresponding to a logical 1.

Current switching depends on the inputs A, B and C which are connected to the collectors of preceding gates with their load being the constant current source. Suppose all three inputs are at logical 0. T1 base voltage is then well below the 0.7 V threshold level for conduction so the outputs are at logical 1. This occurs

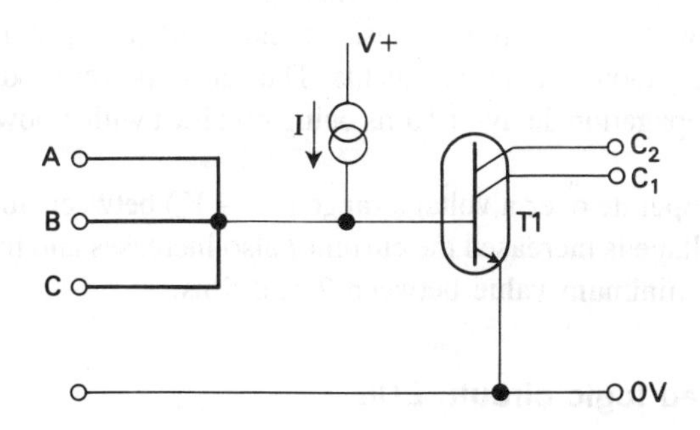

Figure 7.32 Basic I^2L NAND gate.

* or current-injection logic.

Table 7.3 Comparison of properties of TTL logic families

	Standard 74	Schottky 74S	Low-power Schottky 74LS	Advanced Schottky 74AS	Advanced low-power Schottky 74ALS	FAST 74F
Propagation delay t_{pd}, ns	10	3	10	1.5	5	3
Power dissipation, mW	10	20	2	20	1	6
Delay-power product, pJ	100	60	20	30	5	18
Maximum clock frequency, MHz	30	110	30	220	80	110

The maximum clock rate assumes three gate delays, as might occur in a flip-flop, and is given by $\frac{1}{3\,t_{pd}}$.

for all input combinations where one or more is at logical 0 and it is only when all the inputs are at logical 1 that current I can flow into T1 base, causing outputs to become logical 0. This corresponds to the NAND function and it is clear that such a gate would not normally be used alone but in combination with other gates to form an LS1 chip, such as a programmable sound generator.

The practical form of a 4-output I^2L NAND gate is shown in Fig. 7.33(*a*) and requires no resistors in contrast to other bipolar transistor gates. The injector p-n-p transistor T1, supplied from the voltages V_+ and V_-, operates both as a current source and as an active load. Its collector is common with the base of the multiple collector n-p-n transistor T2, which leads to the very compact layout of Fig. 7.33(*b*), with a lateral p-n-p transistor and a vertical n-p-n transistor. Versions with Schottky diodes are also available. The delay–power product is typically 0.5 pJ with a propagation delay of 10 ns being obtained with a power dissipation of 50 μW.

I^2L gates can operate over a voltage range ($V_+ - V_-$) between about 0.8 V and 15 V. As the voltage is increased the current I also increases and the propagation delay falls to a minimum value between 2 and 5 ns.

Emitter-coupled logic circuit, ECL

The TTL and I^2L circuits previously described require that the bipolar transistors are driven well into saturation, which increases the propagation delay. In Schottky TTL diodes are added to prevent this, but in emitter-coupled logic only small

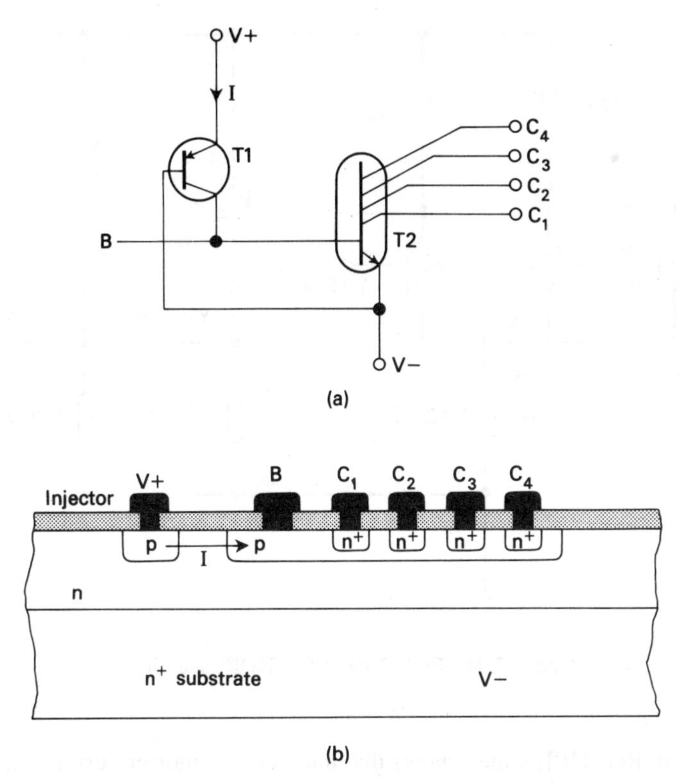

Figure 7.33 I^2L NAND gate. (*a*) Circuit; (*b*) section.

changes of current are used. This means that saturation never occurs and the propagation delay is reduced to about 2 ns.

A typical ECL OR/NOR gate is shown in Fig. 7.34 where T1 and T2 are the input transistors. They operate with T3 to form a differential amplifier in which the current through R_E remains almost constant at about 3 mA. T4 and T5 are emitter followers and ensure that there is no d.c. level shift between input and output of the gate.

The base of T3 is connected to a fixed reference voltage, normally −1.15 V. If the two input voltages are low enough then T1 and T2 are cut off, all the current I_E flows through T3 and the output Y is also low. This occurs for an input of −1.55 V, logical 0. If the two input voltages are high enough then all the current I_E flows through T1 and T2, T3 is cut off and the output Y is high. This occurs for an input of −0.75 V, logical 1.

If one or both inputs is high then T1 or T2 conduct and the Y output is also high, which corresponds to the OR function. Under these conditions the $\bar{Y}$ output is low, corresponding to the NOR function. A complete analysis of the circuit

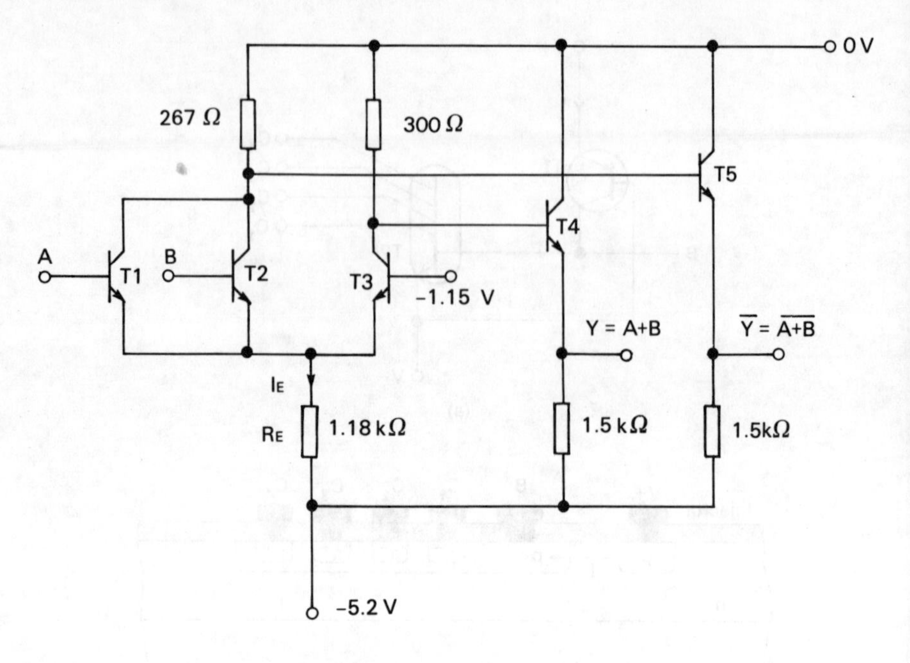

Figure 7.34 ECL 2-input OR/NOR gate circuit.

is given in Ref. [10], which shows that the emitter follower currents add up to about 5.4 mA. When combined with an average I_E or 3 mA this leads to a power dissipation of 43.7 mW and a delay–power product of 87 pJ. This is a high value so that ECL, and its development with a propagation delay of 0.8 ns, would only be used in systems where SSI gates with the highest speed are necessary, having clock rates in the 1 GHz region.

nMOS logic gates

An nMOS two-input NAND gate is shown in Fig. 7.35(*a*). It is much simpler than the TTL circuit of Fig. 7.27(*c*) since it requires only three transistors and no resistors and so it is well suited to LSI and VLSI applications. It is a development of the nMOS inverter of Fig. 6.16(*a*) with two driver transistors in *series*. It can be extended with extra inputs, up to a total of about four, provided by extra series driver transistors. A low input lies below the threshold voltage V_T, which is typically about 1 V, so that the corresponding transistor is cut off. A high input is well above V_T and is near V_{DD}, typically 5 V.

The output is high when either input A or input B or both are low, since either T2 or T3 or both are cut off. T1 is always conducting which pulls the output to V_{DD}. The output is low only when both A and B are high, since T2 and T3 are then conducting together. Their combined resistance is lower than that of T1,

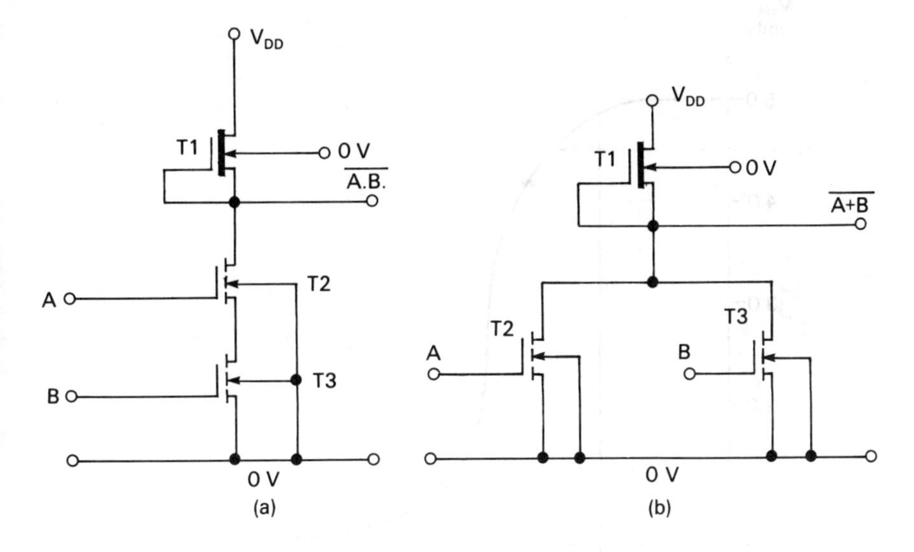

Figure 7.35 nMOS logic gates. (*a*) 2-input NAND gate; (*b*) 2-input NOR gate.

so the output is pulled down to just above 0 V. It may be noted that the aspect ratios of T2 and T3 should be twice that of T2 in the inverter in order to maintain the same low logic level, typically 8 : 1 instead of 4 : 1.

In the nMOS two-input NOR gate of Fig. 7.35(*b*) the two driver transistors are in *parallel* and again further parallel driver transistors can be added to extend the gate. The output is low when either A or B or both are high, since T2 or T3 or both are conducting. The output is high only when both A and B are low, since both T2 and T3 are cut off and T1 pulls the output high.

Transfer characteristic

The transfer characteristic of the NAND gate is shown in Fig. 7.36 and was obtained by SPICE simulation with input A at 5 V and input B changed between 0 and 5 V. The complete program is given in Appendix 3.

The main features are the same as in the nMOS inverter transfer characteristic of Fig. 6.18 but since the inverter aspect ratios of Fig. 6.16(*b*) were maintained in the simulation the output voltage has risen from 0.29 V to 0.61 V for a 5 V input. Noise margins can be obtained from this characteristic but the required voltages are less well defined than for TTL. Approximate simulated values are

$$V_{IL} = 1.3 \text{ V}, \; V_{OL} = 0.6 \text{ V}, \; NM_0 = 0.7 \text{ V}$$

$$V_{OH} = 4.5 \text{ V}, \; V_{IH} = 4.0 \text{ V}, \; NM_1 = 0.5 \text{ V}$$

The noise margins are thus similar to the simulated TTL values. The fanout is,

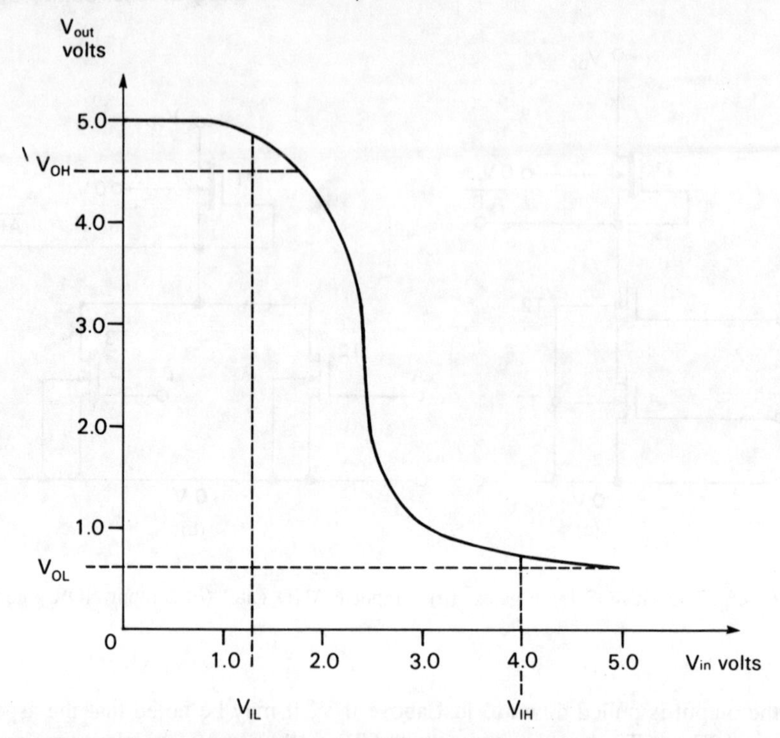

Figure 7.36 nMOS NAND gate transfer characteristic obtained by SPICE simulation for the B input of Fig. 7.35(*a*) with A at 5 V.

however, much larger since the gate input current is negligible, but a limitation is imposed by capacitive load increasing the propagation delay.

Transient response

The transient response of the NAND gate of Fig. 7.35(*a*) was obtained by SPICE simulation as for the transfer characteristic, with the gate output loaded with an inverter having the same aspect ratios (Fig. 7.37). This allowed the transient response to be obtained for both the NAND and the AND outputs, with the waveforms being similar to that of Fig. 6.20 for the inverter alone. Furthermore, the AND response was obtained when the output was loaded with a 0.1 pF capacitance, representing approximately 1 mm of metallized track on an integrated circuit.

The propagation delays are given in Table 7.4 and compared with those from the inverter or NOT gate, obtained in Chapter 6. It may be noted that t_{pd-}, which depends on the resistance of the driver transistors, has doubled from the NOT to the NAND gate. This is because there are twice as many driver transistors in the NAND gate. Also t_{pd+}, which depends on the resistance of the load transistor, rises more slowly than t_{pd-} from NOT to NAND to AND gates. There

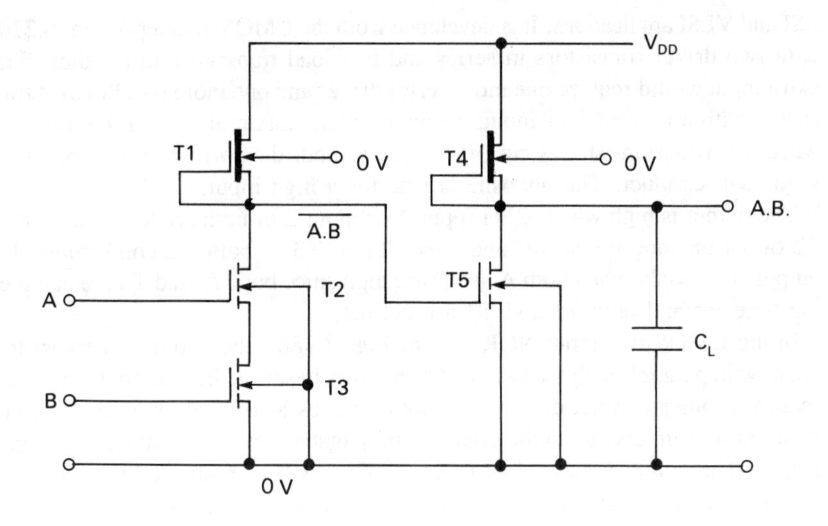

Figure 7.37 nMOS 2 input AND gate with capacitive load.

Table 7.4 Propagation delays of simulated nMOS and CMOS gates

		t_{pd-}	t_{pd+}	t_{pd}
nMOS	NOT	0.55	2.3	1.4 (ns)
	NAND	1.1	3.6	2.4 (ns)
	AND	2.6	4.7	3.3 (ns)
	AND + 1 mm track	5.9	7.7	6.8 (ns)
CMOS	NOT	0.7	1.6	1.2 (ns)
	NAND	1.4	2.3	1.9 (ns)
	AND	2.5	2.9	2.7 (ns)
	AND + 1 mm track	3.6	3.7	3.7 (ns)

is, however, a large increase in propagation delay when the load capacitance is included. This suggests that the design should be modified to have higher aspect ratios and therefore lower transistor resistances when the track capacitance is high. Output capacitance will also impose a limit on the fanout to avoid too large an increase in propagation delay. The power dissipation is similar to that of the inverter but will also depend on the load capacitance due to dynamic power dissipation, as in CMOS.

CMOS logic gates

A CMOS two-input NAND gate is shown in Fig. 7.38(*a*), which requires an extra load transistor compared with the nMOS NAND gate but is still well suited to

LSI and VLSI applications. It is developed from the CMOS inverter of Fig. 6.21(*a*) with two driver transistors in series and two load transistors in parallel. Each extra input would require one more series driver and one more parallel load transistor, with a total of four inputs being a typical maximum. As in the inverter, when a driver transistor is cut off by a low input the corresponding load transistor will conduct. The opposite is true for a high input.

The output is high when either input A or input B or both are low, since either T2 or T4 or both are cut off and either T1 or T3 or both are conducting. The output is low only when both A and B are high since both T2 and T4 are conducting together and both T1 and T3 are cut off.

In the CMOS two-input NOR gate of Fig. 7.38(*b*) the two driver transistors are now in parallel, with the two load transistors in series. Each extra input would require another parallel driver and another series load, again with a maximum of about four inputs due to the effect on propagation delay, discussed under gate arrays. The output is low when input A or input B or both are high, since T2 or T4 or both are conducting and T1 or T3 or both are cut off. The output is high only when both A and B are low, since T2 and T4 are both cut off and T1 and T3 are both conducting, pulling the output to V_{DD}.

Transfer characteristic

The transfer characteristic of the NAND gate is shown in Fig. 7.39 and was obtained by SPICE simulation with input A at 5 V and input B changed between 0 and 5 V. The complete program is given in Appendix 3. The main features are the same as in the CMOS inverter transfer characteristic of Fig. 6.24. The

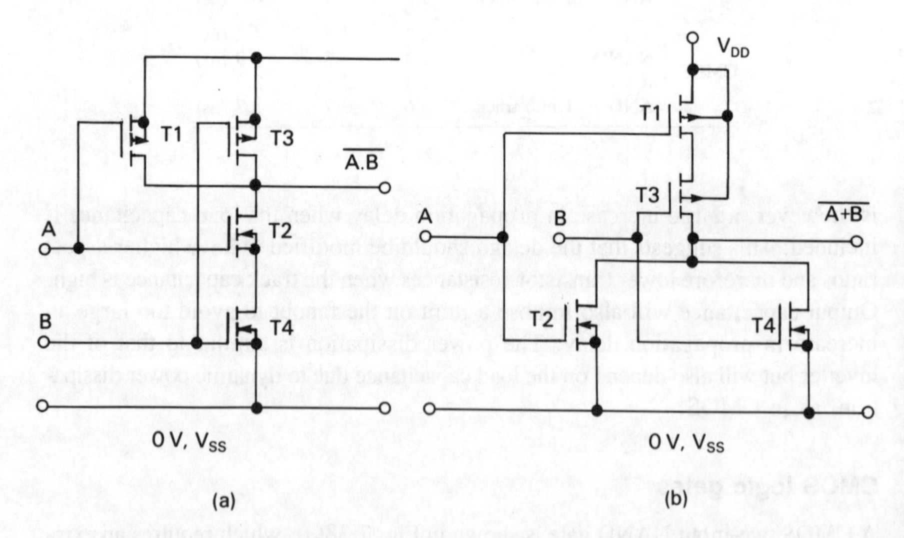

Figure 7.38 CMOS logic gates. (*a*) 2-input NAND gate; (*b*) 2-input NOR gate.

output voltage for a 5 V input has risen from 0.014 μV to 0.058 μV due to the increased potential drop across the two driver transistors but is still extremely low. In obtaining the noise margins the required voltages are less well defined than for TTL but better defined than for nMOS. Approximate simulated values are

$$V_{IL} = 1.6\text{ V},\ V_{OL} = 0\text{ V},\ NM_0 = 1.6\text{ V}$$

$$V_{OH} = 5.0\text{ V},\ V_{IH} = 3.0\text{ V},\ NM_1 = 2.0\text{ V}$$

These are somewhat better than the simulated TTL values. Again the fanout is much larger due to negligible gate input current but it is limited by capacitive load, which increases the propagation delay.

Transient response

The transient response of the NAND gate of Fig. 7.38(*a*) was obtained by SPICE simulation with an inverter load (Fig. 7.40). Again, as for nMOS, both NAND and AND outputs were considered, with the waveforms obtained for the inverter

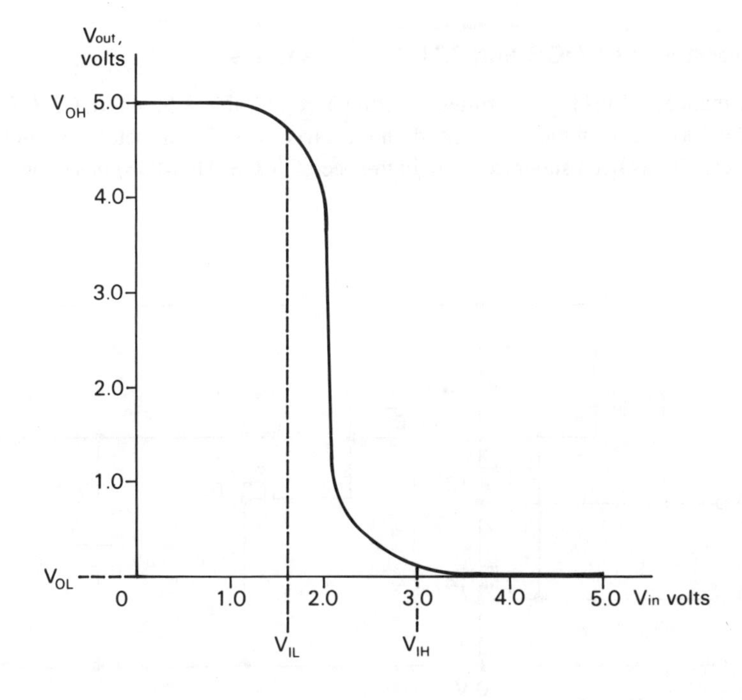

Figure 7.39 CMOS NAND gate transfer characteristic obtained by SPICE simulation for the B input of Fig. 7.38(*a*) with A at 5 V.

above being similar to Fig. 6.26. Track capacitance was again simulated with a 0.1 pF load on the AND gate to represent about 1 mm of track on an integrated circuit.

The CMOS propagation delays are given in Table 7.4 for comparison with the nMOS simulation and again t_{pd-} has doubled from the NOT to the NAND gate, due to the two series driver transistors. t_{pd+} again rises more slowly than t_{pd-} from NOT to NAND to AND, being nearly equal to t_{pd-} for the AND gate, which has two load transistors in parallel, thus reducing the overall pull-up resistance. Load capacitance again leads to an increase in propagation delay, which is less marked than for the nMOS simulation due to the higher aspect ratios of the transistors but still imposes a fanout limitation. Power dissipation will be almost entirely due to the load capacitance as described below.

The above discussion applies mainly to silicon-gate LSI and VLSI circuits but it may be noted that when metal-gate CMOS is used in discrete logic circuits (SSI/MSI) the supply voltage can lie between 3 V and 15 V. This allows metal-gate CMOS (4000 series) to be used directly with linear integrated circuits, which typically have a 15 V power supply. The transfer characteristic is still approximately symmetrical but noise margins increase from about 1 V at 5 V to about 2.5 V at 15 V.

Properties of CMOS and TTL Logic families

The main SSI/MSI logic families are compared in Table 7.5, where CMOS refers to the 4000 series which was based on the early 10–15 μm metal gate p-channel process. It has been superseded by high-speed CMOS (HCMOS) based on a 3 μm

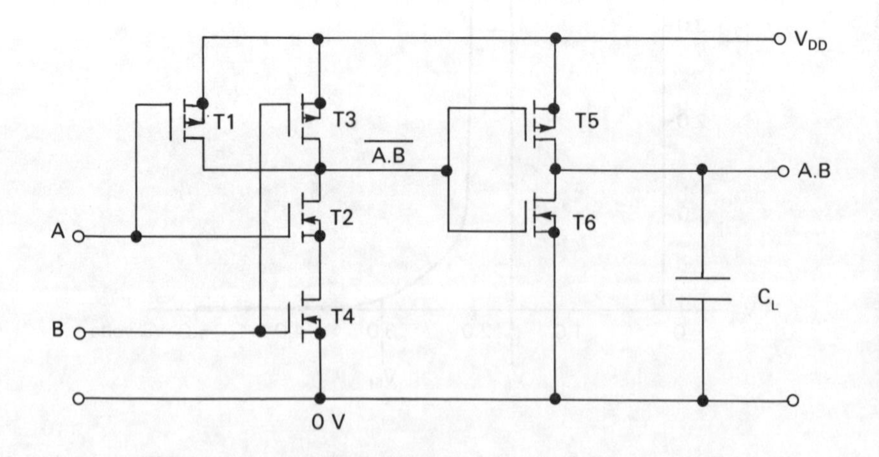

Figure 7.40 CMOS 2-input AND gate with capacitive load.

silicon gate oxide-isolation process. The 74 HC series has CMOS compatible inputs and the 74 HCT series, with a similar performance, has inputs directly compatible with TTL. HCMOS has a performance compatible with LSTTL but with a far lower static power dissipation.

Table 7.5 Comparison of properties of discrete logic families

	CMOS 4000	HCMOS 74HC	LSTTL 74LS
Supply voltage, V	3 to 15	2 to 6	5
Propagation delay, C_L = 50 pF, ns	125	10	12
Power dissipation at 1 MHz, C_L = 50 pF, mW	2.5	1.4	2.8
Delay–power product, pJ	313	14	34
Maximum clock frequency, MHz	2.7	33	27

The total power dissipation P is obtained by combining static and dynamic components from eqs (6.53) and (6.55), respectively, to give

$$P = VI + C_L V^2 f \tag{7.16}$$

V and I are the power supply voltage and current, C_L is the load capacitance and f is the operating frequency. In Table 7.5, V is 5 V, C_L is 50 pF and f is 1 MHz. C_L is chosen to represent the higher capacitance of printed-circuit board tracks or wiring and the choice of f can be explained from Fig. 7.41, which shows the variation of power dissipation with frequency for LSTTL and HCMOS gates. Below 1 MHz, static power predominates for LSTTL, while it is negligible throughout for HCMOS. At about 5 MHz the dynamic power dissipations for the two technologies are about equal, and above this frequency the LSTTL dissipation rises more slowly with frequency than does the CMOS dissipation. HCMOS is available with the same pin connections and in many of the LSTTL functions, so choice between them may depend on the importance of static power dissipation and maximum clock frequency.

Application-specific integrated circuits (ASICs)

When a new electronic system is required a typical design approach is to produce it with SSI/MSI integrated circuits mounted on one or more printed circuit boards. A decision is then taken whether to leave the circuit as it is or to have

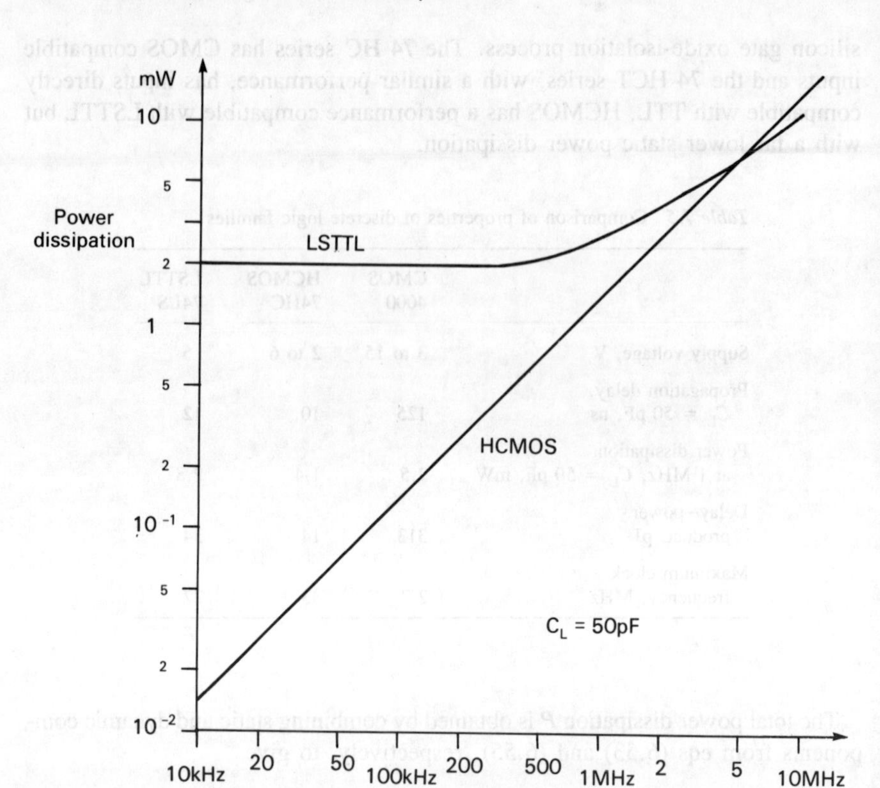

Figure 7.41 Power dissipation and frequency for LSTTL and HCMOS gates.

it integrated on to one or more chips, using semi-custom or full-custom techniques. The advantages of an LSI custom chip (designed to meet a customer's specification) are as follows.

(1) Reduced manufacturing costs, due to a lower number of components, simplified assembly and a high packing density.
(2) Improved system performance, due to less dissipation per gate and higher switching speeds.
(3) Improved system reliability, due to a higher level of integration.
(4) Greater design security than either a circuit board or a read-only memory.

Full-custom implies that the chip is designed from the beginning and requires four or more masks. Very few circuits use this method, due to the high development costs involved in more than 20 man years for a design. The method would be used where large numbers of chips would be sold, such as for a microprocessor.

Semi-custom implies that all the components on the chip are already available and that only interconnections need be specified, requiring one to three masks. The semi-custom options include programmable logic devices, gate arrays and standard cell arrays, in ascending order of development cost and time.

Programmable logic devices

There are three forms of programmable logic device all of which have a matrix of AND gates feeding into a matrix of OR gates (Fig. 7.42). This allows AND–OR combinational logic to be implemented, and in some devices flip-flops and registers can also be provided by interconnecting gates.

The required connections can be either pre-programmed by masking during manufacture (fixed) or programmed by the user (field programmable). The three forms are the PROM (Programmable Read-Only Memory), the PLA (Programmable Logic Array) and the PAL (Programmable Array Logic). They are distinguished by the way in which the AND and OR arrays are programmed, as shown in Fig. 7.42. User-programming is under software control and may lead to fuses being blown where connections are not required in a bipolar device, or to a memory cell being electrically programmed in a MOS device. PALASM is one software package used for this purpose.

A typical PLA could have 16 inputs, 48 AND gates and 16 outputs and can replace about 15 LSTTL chips with a cost saving of over 50%. The LSTTL chips would require a current of about 40 mA, compared with about 120 mA for a bipolar device, and less than 1 mA for a MOS device.

A 32K PROM would have 15 inputs (addresses) and 8 outputs (one byte), with other address ranges being available but with the byte output being very common.

Where up to about 1000 systems are required the programmable logic device is the most economical semi-custom approach but it becomes too expensive above this number due to production costs.

Gate arrays

A gate array or Uncommitted Logic Array* chip contains a regular matrix of devices which is processed up to the metallization stage. Thus about 90% of the processing steps have been completed before the chip is committed to the function required by a customer. A variety of technologies are used, both bipolar (particularly ECL) and MOS, with CMOS being very widespread (Ref. [6]).

A general layout for a gate array chip is shown in Fig. 7.43 where rows of array elements are separated by *highways*, along which interconnections will pass. Round the edge of the chip are peripheral cells that form an interface between the internal logic and outside circuits. For example, if the array elements are in CMOS the peripheral cells would provide TTL inputs and outputs. The connection between elements normally occurs in two layers, along the highways and

* Uncommitted Logic Array (ULA) is a trademark of the Ferranti Company.

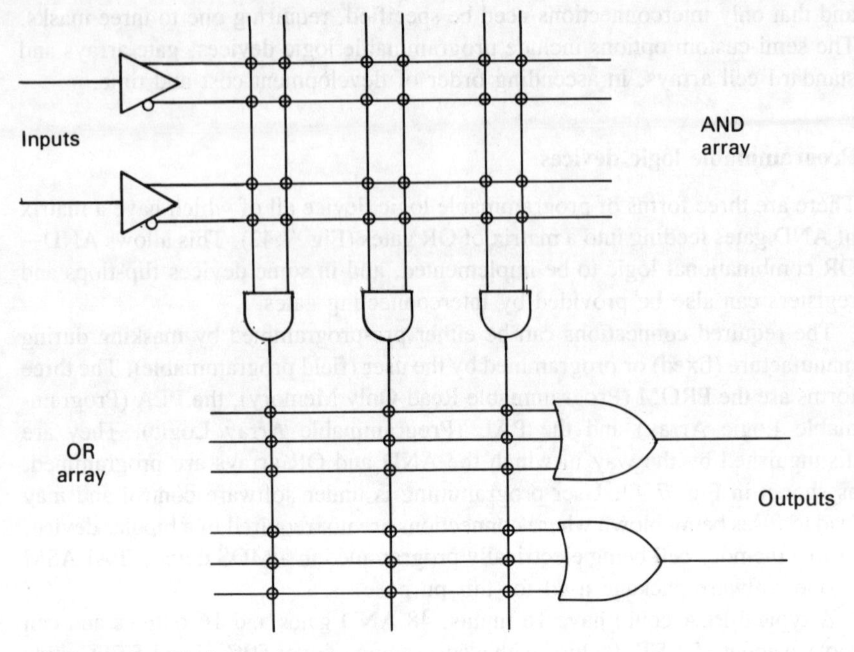

o Programmable interconnection

	AND array	OR array
PROM	Fixed	Programmable
PAL	Programmable	Fixed
PLA	Programmable	Programmable

Figure 7.42 Programmable logic device showing AND and OR arrays, with programming for PROM, PLA and PAL.

across the elements at right angles, the two layers being separated by insulating material. This orthogonal arrangement lends itself to straightforward definition of the interconnections by computer, or auto-routing. GARDS is a typical gate array layout package, which can handle up to 10 000 gates in any technology.

In most cases the array elements consist of a small group of transistors and other components arranged so that they can easily be interconnected to form logic gates. For example, the plan view of a CMOS element shown in Fig. 7.44(*a*), like many others, contains two p-channel and two n-channel MOS transistors. Its equivalent circuit is shown in Fig. 7.44(*b*) and the dotted lines indicate the internal connections required to turn the element into a NAND gate. The program specifying these internal connections would be provided by the manufacturer, as part of a computer library of gates, flip-flops and registers.

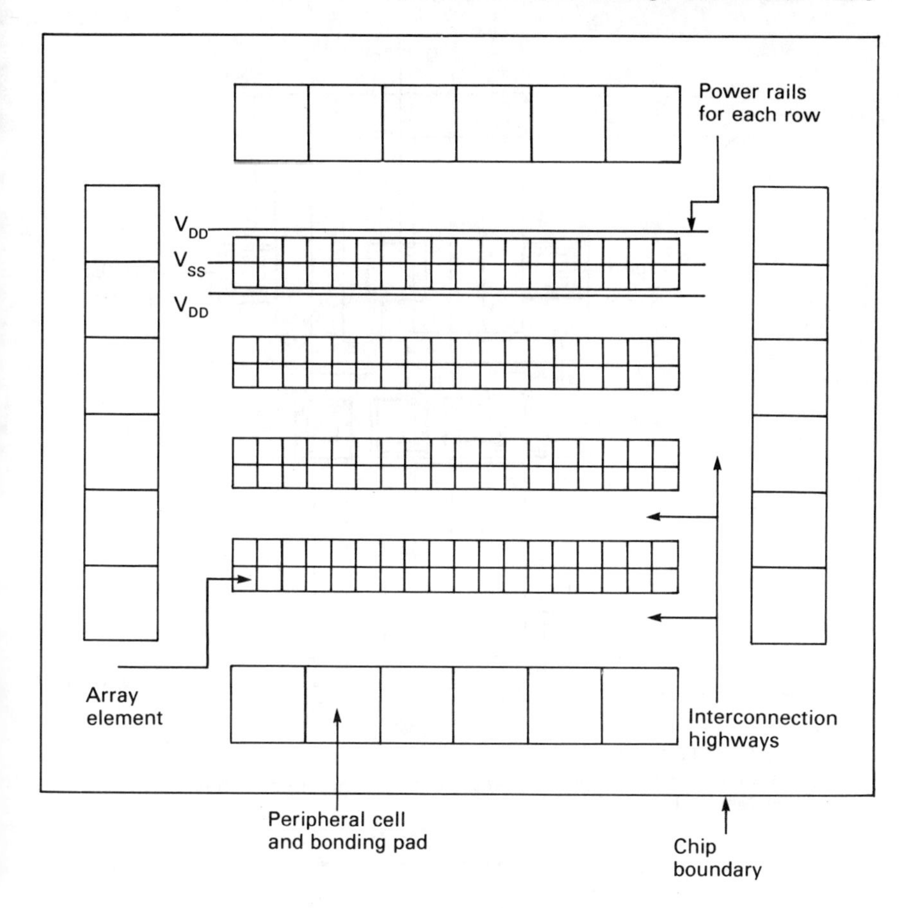

Figure 7.43 Gate array layout of the row-cell type.

The number of array elements on a chip and the propagation delay of a gate depend on the device dimensions, as illustrated in Table 7.6. Normally, a maximum of about 80% of the elements are used to ensure that automatic routing can be carried out successfully.

Typically, a 2000-gate array can replace about 100 LSTTL chips, with a power reduction from about 10 W to about 50 mW for a CMOS gate array. Gate arrays provide an economical semi-custom approach when between about 10^3 and 10^4 systems are required, although these figures are very approximate.

Standard cell arrays

The cells correspond to standard logic elements such as gates, flip-flops and registers, which have been custom-designed to have a constant height but varying width according to complexity. The mask specifications for each cell are

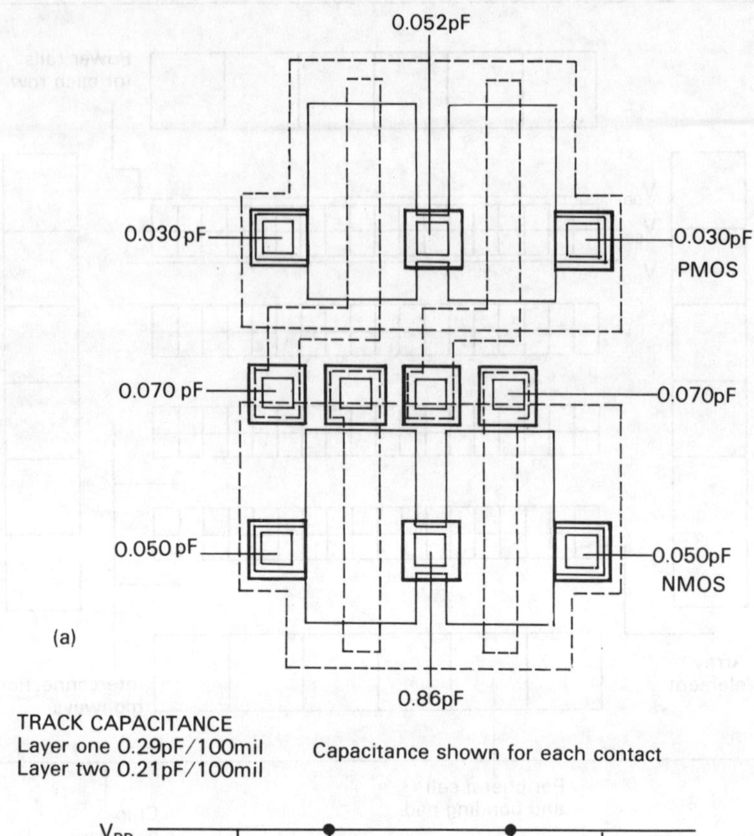

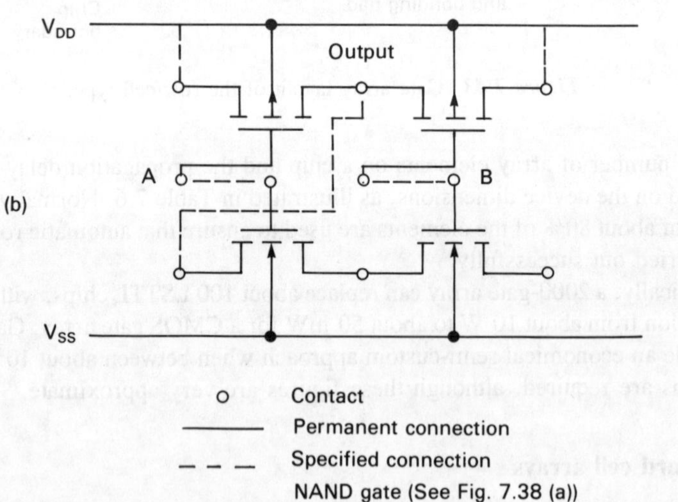

Figure 7.44 CMOS gate array element. (*a*) Chip layout [1 mil = 0.001 in = 25 μm] (reproduced by permission of Plessey Semiconductor); (*b*) equivalent circuit with NAND gate connections.

Table 7.6 Gate array performance and scaling

Number of gates	840 to 2400	840 to 6000	640 to 10 044
Minimum dimension, μm	5	3	2
Propagation delay, ns	3.5	2.5	1.2
With 2 mm track, ns	7	5	2.5
Maximum clock frequency, MHz	14	20	40

required for all levels and are held in a computer library, which may correspond to a particular logic family. This may be the TTL family, but the cells themselves are most likely to be in CMOS. The development time for a library is quite long, for example seven man-years for three libraries each of 150 cells. Once the cells have been developed, the masks can be used again and again.

For a given system design the selected cells are placed in rows across the chip, with highways between them for interconnections (Fig. 7.45). Again this layout is ideal for automatic routing and since there are no unused gates the utilization can be up to 100%. However, the silicon area is used inefficiently, with a packing density perhaps half that of a full custom chip.

When between about 10^4 and 10^5 systems are required, cell arrays become more economical than other semi-custom techniques. Above about 10^5 systems the full-custom approach takes over and, again, these figures are very approximate. It seems likely that in the future gate arrays and cell arrays, even including microprocressors, will be available on the same chip, rather than gate arrays with more than 10 000 elements for a VLSI semi-custom chip being produced. Full-custom techniques are less likely to be used, except where the ultimate in speed and packing density is required. An overview of the available computer-aided design tools is given in Ref. [11].

CMOS design with gate array elements

Instead of building a prototype system using LSTTL and then converting it into a CMOS gate array chip, the logical design can be simulated directly. This simplifies the development, and there are a number of simulators available such as BIMOS, HELIX, HILO and TEGAS. A logic simulator such as CLASSIC (Custom Logic Analysis and Simulation System for Integrated Circuits) operates in terms of cells which range in complexity from inverters to master-slave flip-flops. These cells are formed from one or more array elements interconnected to carry out the required logic function. Programs within CLASSIC aid the placement of cells on the locations of the gate array, and routing data for the cell-to-cell interconnection is then automatically generated. The end result is a set of

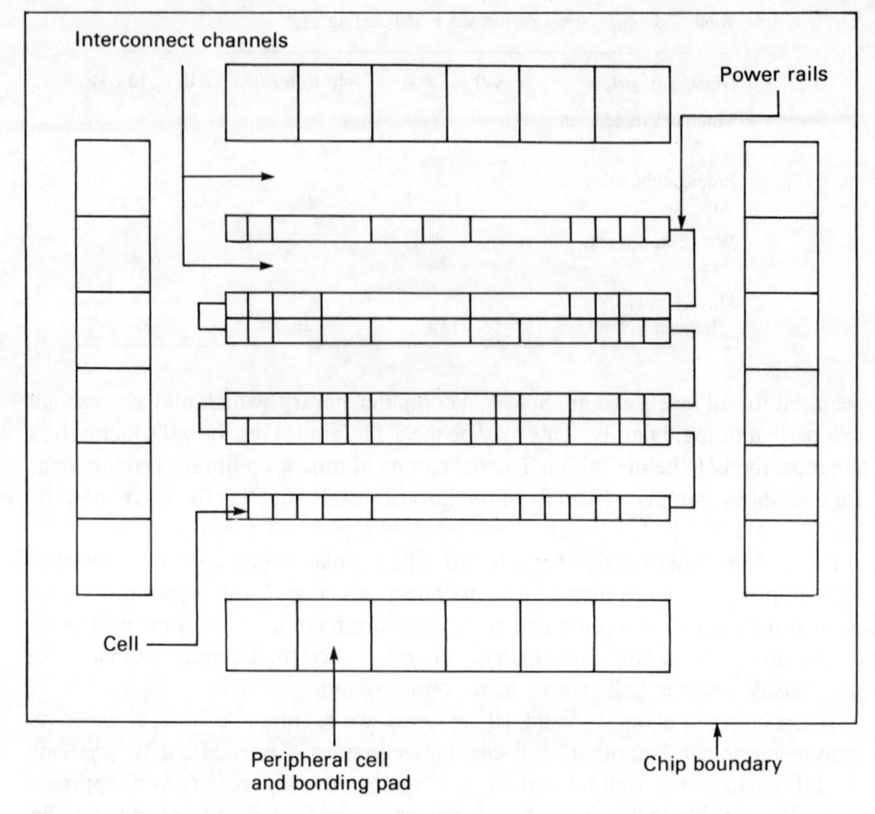

Figure 7.45 Standard cell array layout. Each cell has a constant height but the width increases with complexity from gates to flip-flops, to registers and so on.

metal patterns, in two layers insulated from each other, that customizes the gate array into the desired logic system.

CMOS lends itself to economical cell design, and some basic configurations are now discussed.

NAND and NOR gates

These gates are shown in Fig. 7.38 for CMOS and each would require one array element. The p-channel loads, which control the t_{pd+} part of the propagation delay, normally have higher resistance than the n-channel drivers. Since the loads are in parallel in the NAND gate, t_{pd+} is lower than for the NOR gate, where the loads are in series. Consequently, NAND gates are generally preferred to NOR gates in designs with a high clock frequency.

Transmission gate

This consists of p- and n-channel enhancement transistors in parallel, driven by antiphase clock waveforms ϕ and $\bar{\phi}$ (Fig. 7.46(*a*)). V_1 and V_0 are either at 0 V or V_{DD} and ϕ and $\bar{\phi}$ also switch between these levels. When $\phi = V_{DD}$, T1 is on when $V_1 = 0$ V, and T2 is on when $V_1 = V_{DD}$, so V_0 follows V_1. When $\phi = 0$ V both transistors are off whether $V_1 = 0$ V or V_{DD}, so V_0 is unaffected by V_1 (Fig. 7.46(*b*)). The overall switch resistance is due to the two parallel

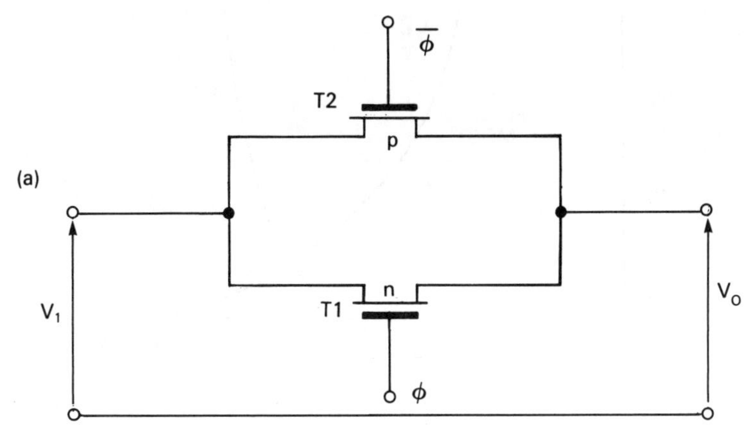

Transmission gate symbols

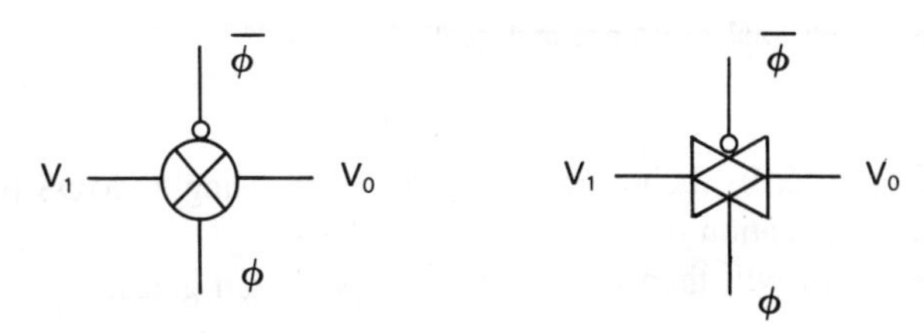

Transistor symbols are alternative forms for enhancement devices.
Sources are at the input end.

(b)

$\phi = V_{DD}$ $\quad \bar{\phi} = 0$ V

V_1	V_{GS1}	T1	V_{GS2}	T2	V_0
V_{DD}	0 V	off	$<V_{T2}$	on	V_{DD}
0 V	$>V_{T1}$	on	0 V	off	0 V

$\phi = 0$ V, $\quad \bar{\phi} = V_{DD}$

V_1	V_{GS1}	T1	V_{GS2}	T2	V_0
V_{DD}	$<V_{T1}$	off	0 V	off	undefined
0 V	0 V	off	$>V_{T2}$	off	undefined

Figure 7.46 CMOS transmission gate. (*a*) Circuit; (*b*) table of levels and operation;

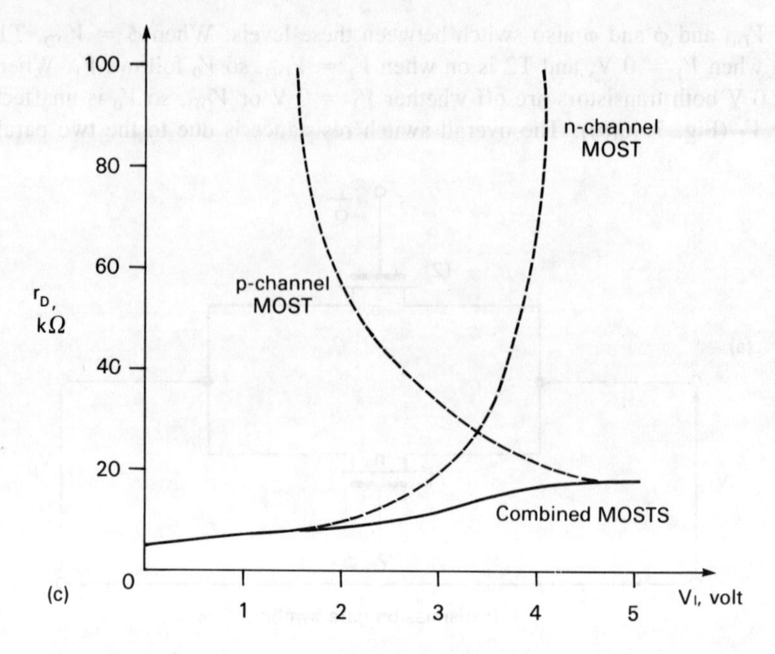

Fig. 7.46 (cont.) (*c*) total resistance and applied voltage.

resistances (Fig. 7.46(*c*)), being much less than for a single nMOS transistor, which reduces propagation delay.

One array element will then provide two transmission gates.

Exclusive-OR gate

The exclusive-OR gate or modulo-2 adder is used extensively in arithmetic circuits. The function is expressed by the symbol ⊕, so for two inputs

$$A \oplus B = \bar{A}B + A\bar{B} \tag{7.17}$$

This requires two inverters and three NAND gates when implemented conventionally, as shown in Fig. 7.47(*a*), and would need four array elements.

The CMOS version uses one transmission gate and two inverters as shown in Fig. 7.47(*b*). Inverter 1 has input B and operates normally from the supply rails V_{DD} and 0 V. The transmission gate operates with $\bar{\phi}$ = B and ϕ = $\bar{B}$. Inverter 2 has input A and its supply rails are replaced by B and $\bar{B}$. The circuit relies on the zero static power consumption of CMOS, so that the input supplies no static current to inverter 2 when B is high at V_{DD}.

The XOR function is implemented because when B is low the transmission gate is on and inverter 2 is isolated. The output is then the same as A. When B is high the transmission gate is off and inverter 2 operates normally. The output is then the inverse of A.

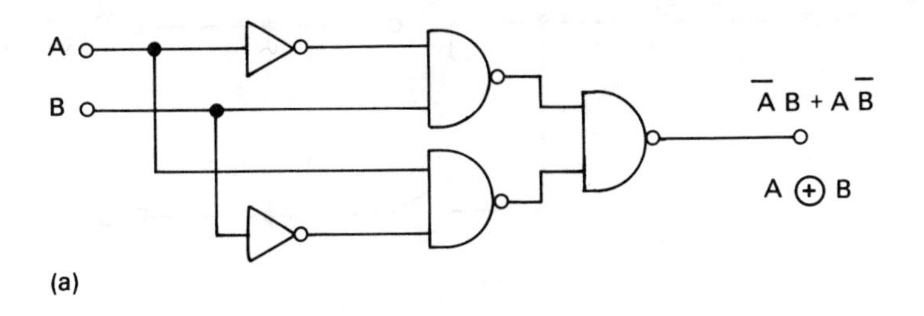

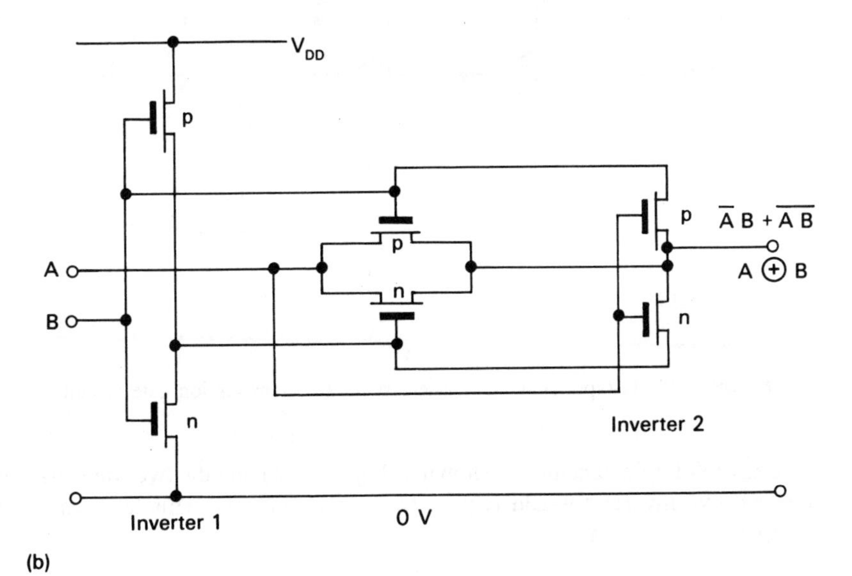

Figure 7.47 Exclusive-OR gate. (*a*) NAND/NOT circuit; (*b*) CMOS circuit.

The circuit of Fig. 7.47(*b*) requires $1^{1/2}$ array elements, which is a big saving over the conventional approach. It could be specified with an inverter for two complete array elements.

D-type latch (flip-flop)

The basic configuration of a D-type latch is shown in Fig. 7.48(*a*). It is effectively a one-bit memory cell, with the logic level at the D input being retained at the Q output.

When switch C is closed and switch $\bar{C}$ is open $\bar{Q} = \bar{D}$ and $Q = D$ in the write mode, where Q follows any change in D. When C is open and $\bar{C}$ is closed the input is isolated and the levels on Q and $\bar{Q}$ are maintained in the latch mode. If C represents a high clock pulse and $\bar{C}$ a low clock pulse then the circuit corresponds to a D-type latch, with *two* clock inputs.

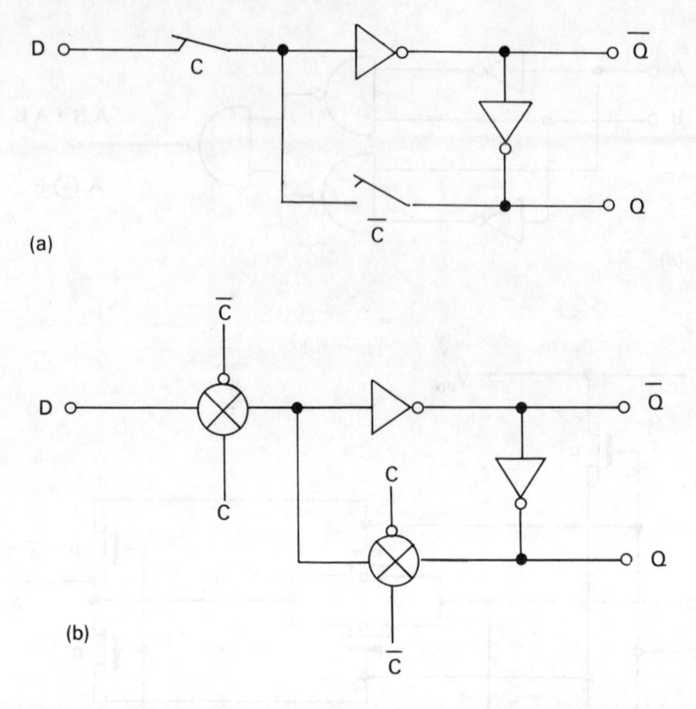

Figure 7.48 D-type latch. (*a*) Basic circuit; (*b*) transmission gate circuit.

The CMOS implementation is shown in Fig. 7.48(*b*) and the two transmission gates and two inverters would require two array elements. This latch can also be implemented in nMOS.

Linear integrated circuits

A linear circuit responds to continuously variable signals and for design purposes a complete amplifier is required. The operational amplifier, or *opamp*, is the integrated circuit version and is available with many different specifications. In general it has a differential input with a wide voltage range, high voltage gain at low frequencies, high input impedance and low output impedance, and a high frequency at which the gain has become unity. Typical values of these parameters for an opamp of the widely used 741 type are ± 13 V, 10^5, 2 MΩ, 100 Ω and 1 MHz. The supply voltages are normally ± 15 V.

The design of a bipolar integrated circuit amplifier differs considerably from that of an amplifier using discrete components, due to the constraints imposed by the available types and values of integrated components. A basic layout for an opamp of the 741 type is shown in Fig. 7.49(*a*) and consists of input, driver and output stages, although the details of each stage are considerably modified in a practical circuit. The input stage consists of two coupled transistors forming

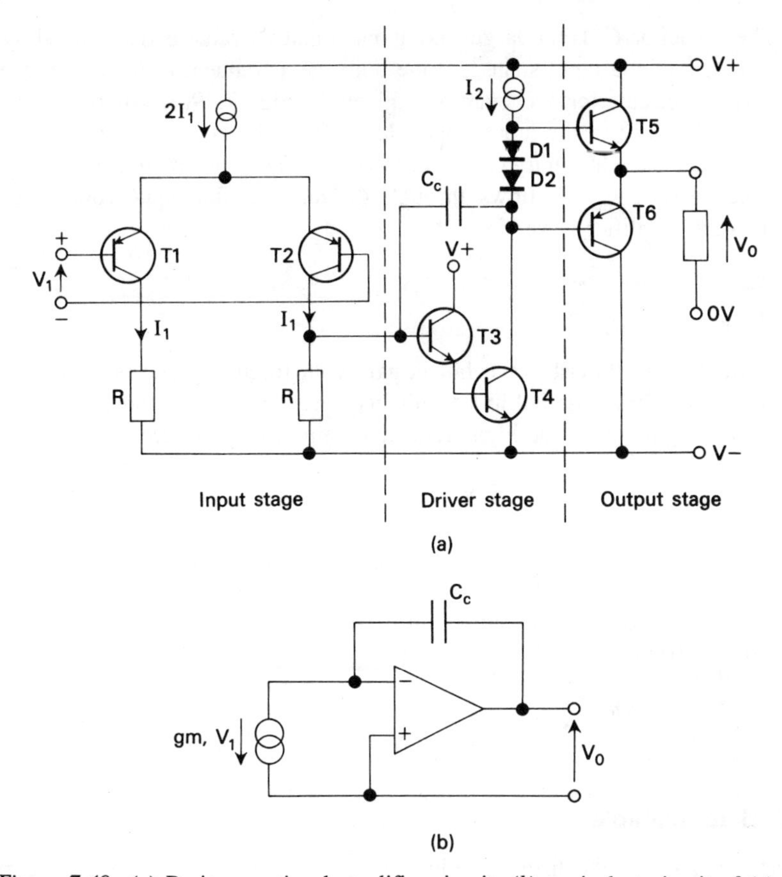

Figure 7.49 (*a*) Basic operational amplifier circuit; (*b*) equivalent circuit of (*a*).

a differential amplifier and operating from a constant current source $2I_1$ with differential input V_1. If V_1 increases then T1 current rises and T2 current falls since the total current is constant. The output from T2 is passed to the driver amplifier consisting of the two transistors T3 and T4 connected as a Darlington pair. The emitter current of T3 supplies the base current of T4, so that the two transistors behave as a single transistor with a very high current gain, where $(I_{C3}+I_{C4})/I_{B3} = h_{fe3} \times h_{fe4}$. The stage is supplied by constant current source I_2 and includes the frequency compensation capacitor C_c which limits the response of the whole amplifier to prevent the occurrence of oscillations. T5 and T6 form a complementary emitter follower output stage, where the n-p-n and p-n-p transistors increase the load current on the positive and negative parts of the signal respectively and the two diodes ensure that the correct d.c. bias is applied. The whole amplifier is direct coupled throughout rather than a.c. coupled, since this would need capacitors, which require a large chip area.

The capacitor C_c is the largest component and its value is determined by the parameters of the input stage and the unity gain frequency. A high-frequency equivalent circuit for the opamp is given in Fig. 7.49(*b*) where g_{m1} is the transconductance of the input stage (Ref. [12]). The effect of the output stage is ignored since it has unity gain over the whole frequency range. Assuming that all the current $g_{m1}\ V_1$ flows through C_c and that the input voltage of the equivalent amplifier is very small

$$g_{m1}\ V_1 = j\omega C_c\ V_O \qquad (7.18)$$

$$V_O/V_1 = g_{m1}/j\omega C_c$$

Equation (7.18) describes the fall in gain with frequency and is similar to the expression defining the fall in h_{fe} with frequency for a single transistor (Chapter 4). The magnitude of the high frequency gain is unity where

$$\omega = \omega_h = g_{m1}/C_c \quad \text{or} \quad f_h = \frac{g_{m1}}{2\pi C_c}$$

From Chapter 4 $g_{m1} = eI_1/kT$ for a single transistor, such as T1 or T2, while for the whole input stage $g_{m1} = g_m/2$. Since I_1 is about 10 μA in this case g_{m1} is about 0.2 mS at room temperature. If $f_h = 1$ MHz this allows C_c to be calculated from

$$C_c = g_{m1}/2\pi f_h \qquad (7.19)$$

Worked example

Each transistor of the differential input stage of an opamp has a collector current of 10 μA and the high frequency gain is unity at 1 MHz.

Determine the value of the frequency compensating capacitor at room temperature.

Solution

In the equivalent circuit of the basic opamp $g_{m1} = eI_1/2kT$, where I_1 is the collector current of one input transistor.

$$g_{m1} = \frac{10^{-5}}{2 \times 0.025} = 2 \times 10^{-4}\ \text{S}$$

The frequency compensating capacitor is given by

$$C_c = \frac{g_{m1}}{2\pi f_h} = \frac{2 \times 10^{-4}}{2 \times \pi \times 10^6} = 3.18 \times 10^{-11}$$

$$= 32\ \text{pF}$$

In practice C_c is about 30 pF and this requires a chip area of about 0.1 mm^2 for an MOS capacitor.

Such an area, while acceptable for a package containing one opamp, must be reduced for a package containing four opamps, corresponding to large-scale integration of linear circuits. From eq. (7.19), C_c is proportional to g_{m1} if f_h is maintained constant, and g_m is proportional to I_1. If T1 and T2 are made with multiple collectors (Fig. 7.50(*a*)) the current divides in the ratio of the collector areas, $n : 1$. The larger area of collector is connected to V_- so that the effective collector current is reduced to $I_1/(1+n)$. typically $n = 5$ so C_c is reduced by a factor of 6 and the required capacitor area is reduced by the same amount, allowing the four opamps to be contained in the same package.

A possible layout for the complete opamp is given in Fig. 7.51, still showing three main stages but each one being more complicated than in the basic circuit of Fig. 7.49(*a*). One extra feature is the *current mirror* circuit of Fig. 7.50(*b*), consisting of two transistors with equal base-emitter voltages and hence equal collector currents. Provided the characteristics of the two transistors are identical, the current of one mirrors the current of the other, and this is achieved by forming the transistors close to each other on the chip.

On the circuit of Fig. 7.51 T9 and T10 form the first current mirror and T11 and T12 the second one, sharing the same current. D1 is an emitter-base diode

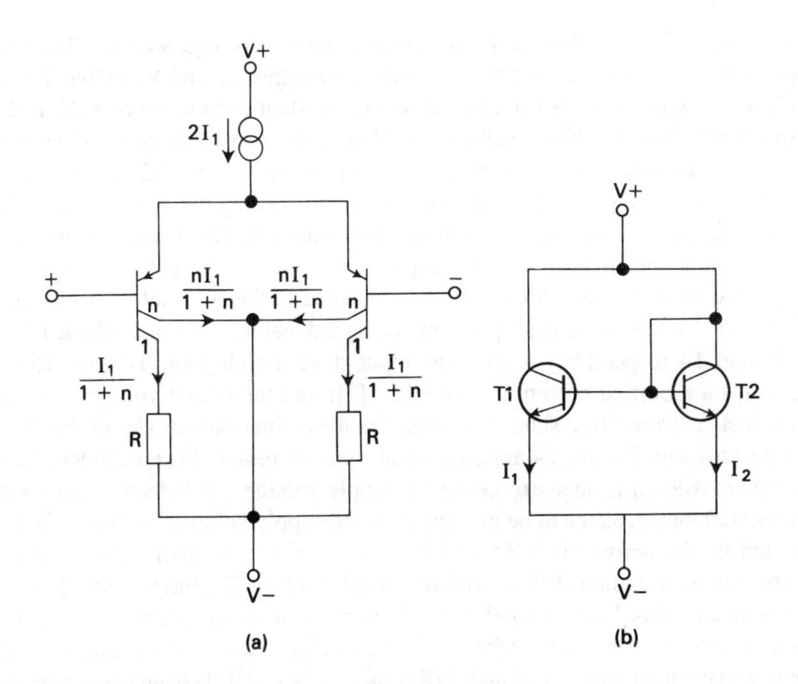

Figure 7.50 (*a*) g_m reduction by multiple collectors; (*b*) current mirror circuit, $I_1 = I_2$.

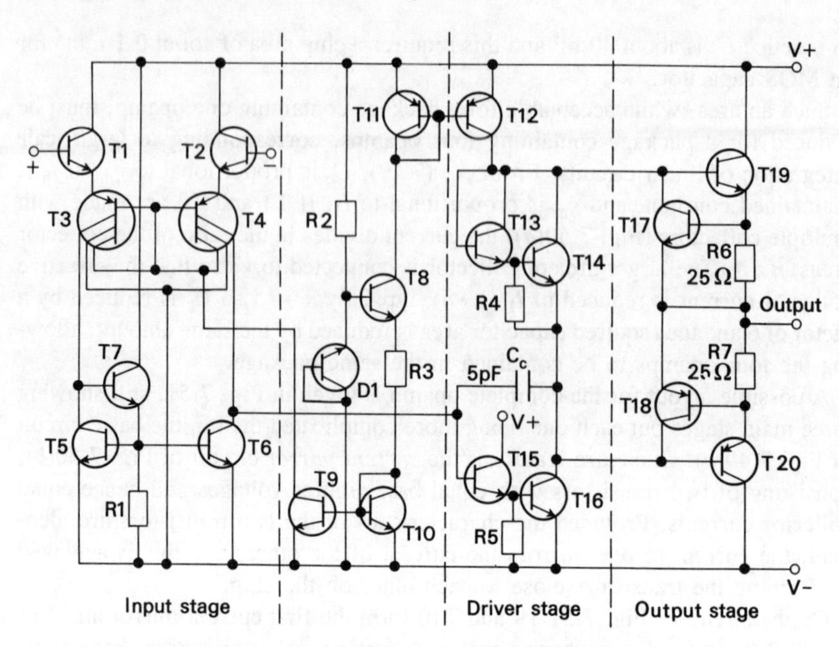

Figure 7.51 Integrated operational amplifier (opamp) circuit.

which breaks down at about 5 V and acts as a constant voltage source. This holds V_{BE8} constant against variations in the supply voltages V_+ and V_- so that T9 and T12, with high output resistances, act as the constant current sources $2I_1$ and I_2 respectively. Thus the bias conditions and hence the gain are largely independent of supply variations. The matched input transistors T1 and T2 act as emitter followers presenting a high input impedance and feeding the common base differential amplifier using matched transistors T3 and T4. The + and – signs imply non-inverting and inverting inputs respectively. Two of the collectors are used for g_m reduction and the third is diode-connected to the constant current source T9. T5 and T6 are active loads presenting a high dynamic resistance (about 1 MΩ) to T3 and T4 respectively to give the input stage a high gain. T7 with T5 and T6 forms a modified current mirror (Ref. [7]) and the output voltage from the stage is taken from T6. It should be noted that the emitter-base diodes of the lateral p-n-p transistors T3 and T4 are equivalent to n-p-n base-collector diodes and so have a breakdown voltage exceeding the supply voltages. This feature allows the differential input voltage to be just less than the supply voltages, which is further assisted by the active loads T5 and T6 requiring only a small voltage drop.

The common-emitter driver amplifier consists of the Darlington pair T15 and T16 with an active load T13 and T14, all supplied by the constant current source T12. Differential outputs are taken to the bases of T19 and T20 forming the complementary output stage. T17 and T18 protect T19 and T20 from excessive current, caused perhaps by a short circuit. If the current through R6, for example, exceeds the maximum allowable T17 turns on and diverts current from T19 base, thus reducing the excess current. T18 acts in a similar way to protect T20. It should be noted that throughout the design the number of resistors is kept to a minimum by using the dynamic resistance of transistors as active loads wherever possible.

Computer simulation of an opamp

Before a circuit containing one or more opamps is constructed it is very desirable to be able to simulate it. This allows optimum values of resistance and capacitance to be obtained beforehand, for example in an active filter. The circuit of Fig. 7.51 could certainly be simulated in SPICE, but due to the number of transistors much computer time would be required. The basic circuit of Fig. 7.49 would require less computer time on its own but again more than one such circuit might be employed in a complex active filter design.

The circuit symbol for an opamp is shown in Fig. 7.52(*a*) and a simple equivalent circuit, suitable for SPICE simulation is given in Fig. 7.52(*b*). Here the input impedance is 2 MΩ and the output impedance 75 Ω. The two internal voltage-controlled current sources do not correspond with any particular devices but serve to define the low-frequency voltage gain and the break frequency, from which

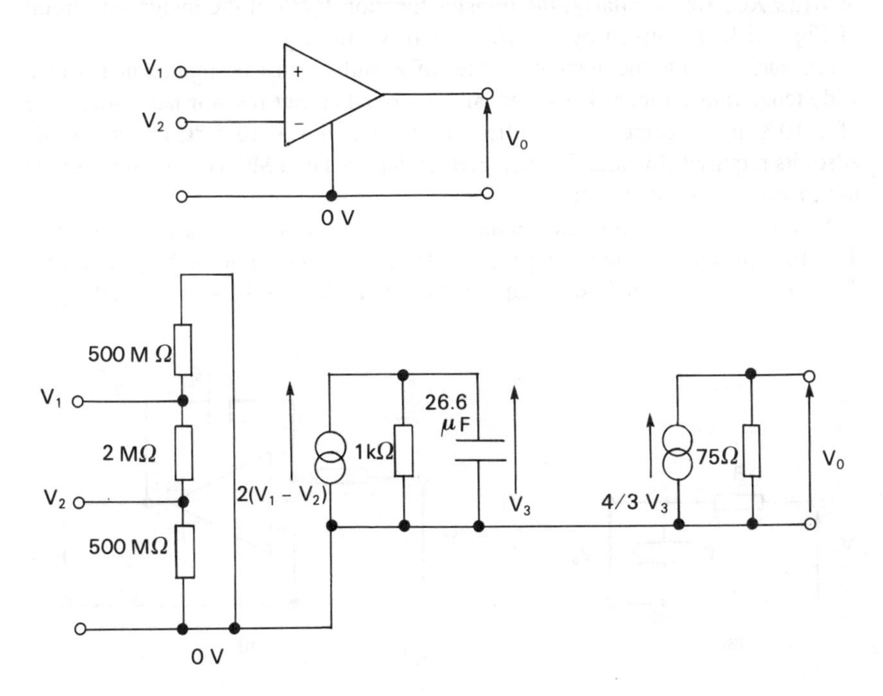

Figure 7.52 Opamp representation. (*a*) Circuit symbol; (*b*) equivalent circuit of 741-type for computer simulation.

the unity-gain frequency can be obtained. The low-frequency voltage gain is given by $2 \times 10^3 \times \frac{4}{3} \times 75 = 200\ 000$ or 106 dB. The break frequency is given by $1/(2\pi \times 10^3 \times 26.6 \times 10^{-6}) = 5.98$ Hz and the unity–gain frequency is given by $200\ 000 \times 5.98 = 1.2$ MHz.

This circuit lacks the refinements provided by the complete circuit but does provide the main features of a 741-type opamp and does not require excessive computer time, even when several are simulated in the same circuit.

An opamp with a FET input stage provides a very high input impedance of about $10^{12}\ \Omega$. CMOS opamps are also available, using the linear part of the transfer characteristic, which have similar input impedance and very low power consumption. Provided there is only one break frequency they can also be simulated using the circuit of Fig. 7.52(*b*).

Switched-capacitor techniques

A passive frequency-selective circuit can be formed from resistors, capacitors and inductors. Circuits using inductors are only used at microwave frequencies, as coils are very difficult to manufacture in an integrated circuit, so a practical filter circuit relies on accurate values of *RC* time constants. For instance, a simple low-pass filter (Fig. 7.53(*a*)) will pass signals from d.c. to a break frequency of $1/(2\pi RC)$ Hz. Similarly, the transfer function V_0/V_1 of the integrator circuit of Fig. 7.53(*b*) is given by $-1/sRC$, where s can be $j\omega$.

For such circuits the absolute values of R and C must be maintained over a wide temperature range. However, an integrated circuit resistor has a tolerance of ±10% and a temperature coefficient of about $+3 \times 10^{-3}$/°C above 293 K. Also, its required chip area is much greater than that of a MOST, so it is desirable not to use the resistor if at all possible.

A MOS capacitor, with a silicon dioxide dielectric, offers a capacitance of about 4×10^{-4} pF/mm², so that a 1 pF capacitor may occupy a 50 × 50 μm square. The tolerance on an individual capacitor is again about ±10% but the tolerance

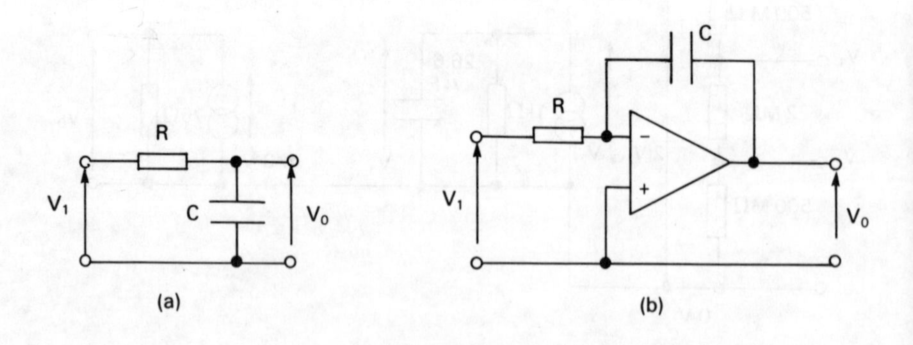

Figure 7.53 Low-pass circuits. (*a*) Simple filter; (*b*) integrator.

on capacitance *ratios* can be made ±1% or less. this is because individual areas can be accumulated as unit squares by means of a pattern generator (Fig. 7.54(*a*)).

Resistor emulation

The problems of high chip area for R and high tolerance for R and C can be solved by using switched-capacitor techniques, which are based on ratios of capacitance. The a.c. characteristics of a resistor are emulated by switching a capacitor between two terminals using a high-frequency clock signal (Fig. 7.54(*b*)). This is achieved in practice by means of two nMOSTs with antiphase clock signals applied to their gates, which may be non-overlapping, so as to ensure that capacitor charge is retained during the changeover from ϕ_1 to ϕ_2 (Fig. 7.54(*c*)).

With the switch initially to the left (ϕ_1 high, ϕ_2 low), capacitor C is charged to V_2. If the switch is then closed to the right (ϕ_1 low, ϕ_2 high) the capacitor is charged to V_2. In general, V_1 and V_2 are unequal so the amount of charge transferred from left to right is

$$Q = C(V_1 - V_2) \tag{7.20}$$

If the switch is operated at frequency $f_c = \frac{1}{T}$ then the continuous flow of charge packets appears as current I, given by

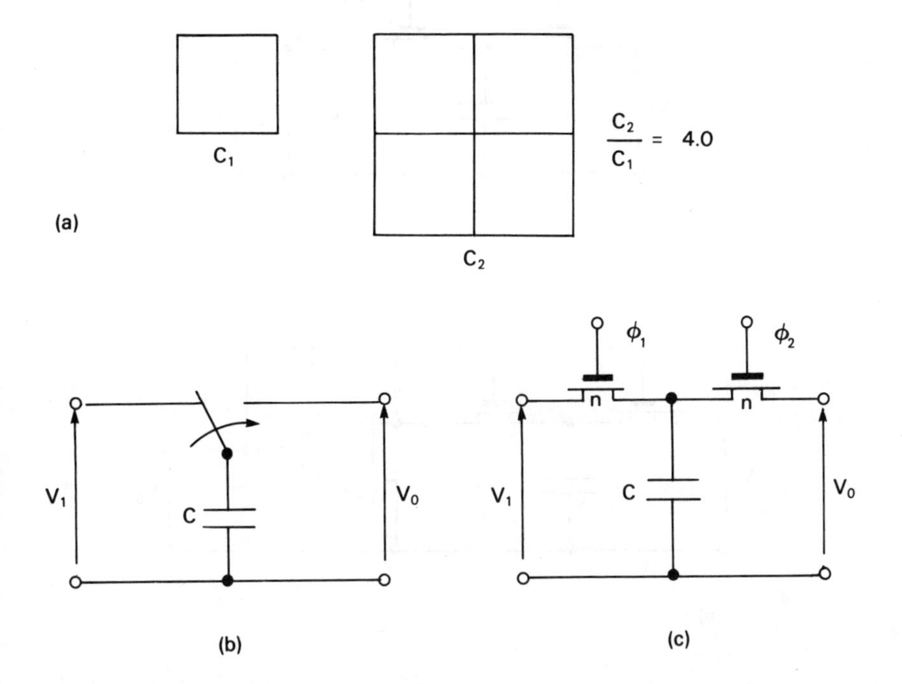

Figure 7.54 (*a*) Capacitor from accumulated squares. (*b*) Switched-capacitor circuit. (*c*) Practical form of (*b*).

$$I = \frac{Q}{T} = Qf_c = C(V_1 - V_2) f_c \tag{7.21}$$

The voltage/current ratio is represented by a resistor R where

$$\frac{V_1 - V_2}{I} = \frac{1}{Cf_c} = R \tag{7.22}$$

and R is the apparent a.c. resistance emulated by the switching circuit. It is particularly useful that R is proportional to C^{-1}, since the larger the value of R the smaller the value of C and the less chip area is required.

Switched-capacitor circuits

The circuit of a simple switched-capacitor filter is shown in Fig. 7.55(a), with $RC = \frac{C_2}{C_1 f_c}$ and break frequency $\frac{f_c C_1}{2\pi C_2}$. Thus the frequency response now

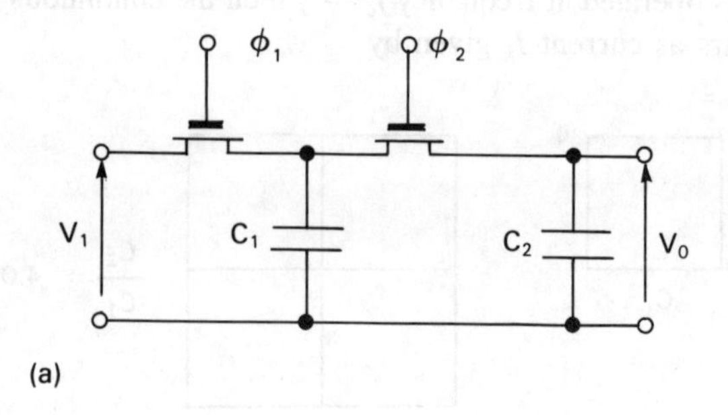

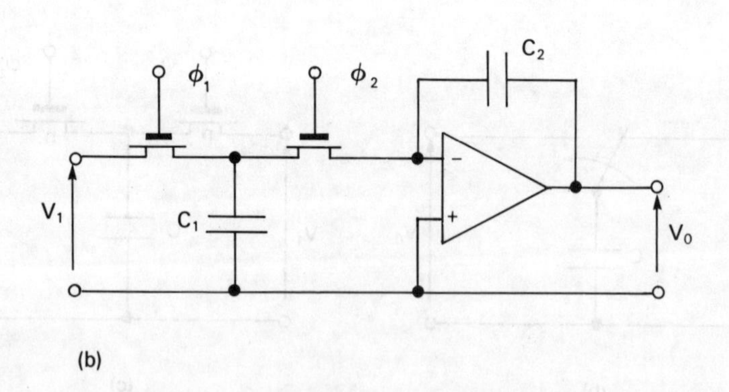

Figure 7.55 (a) Simple low-pass switched-capacitor filter. (b) Switched-capacitor integrator.

depends on the ratio of capacitance and on the clock frequency, both of which can be held to better than 1% tolerance. Similarly, in the switched-capacitor integrator of Fig. 7.55(*b*) *RC* is again $\frac{C_2}{C_1 f_c}$ and the transfer characteristic $\frac{V_0}{V_1} = -\frac{f_c C_1}{s C_2}$.

Since the input signal is effectively sampled the simple theory holds only when the clock frequency is much higher than the input frequency, typically by a factor of 50 to 100 times, which means that the technique is best suited to the audio range. A practical switched-capacitor filter would have several active stages employing opamps (Ref. [5]) and facilities for implementing low-pass, high-pass and band-pass filters. Break or centre frequencies can be controlled up to 20 kHz by varying the clock frequency, providing a form of electronic tuning. The design of such a filter on an integrated circuit is greatly assisted by a switched-capacitor network simulator, such as SWAP.

Gallium arsenide integrated circuits

In the quest for faster integrated circuits the main approach has been to reduce device size and to minimize stray capacitance by improved manufacture using silicon. A different approach is to change the semiconductor from silicon to gallium arsenide which has superior electrical properties, although it is much rarer and more expensive. However, specialized integrated circuits are now being produced in this material and it is likely that they will represent a significant part of the market in the future.

As discussed in Chapter 2, the effective mass of electrons in gallium arsenide is very low, being about 7% of the mass in free space, while in silicon the effective mass is almost equal to that in free space. This leads to a mobility of 0.85 m^2/V s in gallium arsenide, compared with 0.15 m^2/V s in silicon, and an electron velocity about five times higher than in silicon. This provides a low propagation delay, about 0.06 ns (60 ps), and also means that MESFETs can operate at lower voltages for the same current (eq. (6.31)). Power dissipation per gate is thus about 0.5 mW, which gives the very low delay–power product of 0.03 pJ.

Semi-insulating (SI) gallium arsenide (Chapter 2) is used for the substrate, which leads to low parasitic capacitances, so that the low propagation delay can be exploited in an integrated circuit. The large energy gap also allows a wide operating temperature range of –200 to +200 °C to be achieved. Since it is a direct-gap material the carrier lifetime is low and so it has high radiation hardness.

Gallium arsenide FETs are manufactured by ion implantation with n-channels and aluminium gates and are available in both depletion- and enhancement-mode forms, D-MESFET and E-MESFET (Ref. [13]). In each case a depletion region occurs at zero bias under the gate which forms a Schottky-barrier diode with the n-layer. The D-MESFET has the thicker layer, and a negative gate voltage is required to pinch off the n-channel between source and drain. The E-MESFET has the thinner layer and the n-channel is already pinched off at zero bias, so a small positive threshold voltage must be exceeded before drain current can flow.

The characteristics of a D-MESFET (Fig. 7.56(*a*)) are similar to those of an n-channel JFET with a pinch-off voltage of about 1 V and it is used in both digital and linear circuits. The transconductance is very high, due to the high mobility, and the input capacitance is very low, which leads to a gain–bandwidth product of 15 to 20 GHz. Normally, two power supplies are required to ensure pinch off and level-shifting circuits are built into the gates (Fig. 7.57(*a*)).

The characteristics of an E-MESFET (Fig. 7.56(*b*)) are more like those of an n-channel enhancement MOST, with a small threshold voltage of about 0.1 V. The logic swing is limited to about 0.5 V, beyond which the Schottky gate diode is turned on. The device is used in logic circuits (Fig. 7.57(*b*)), where only one

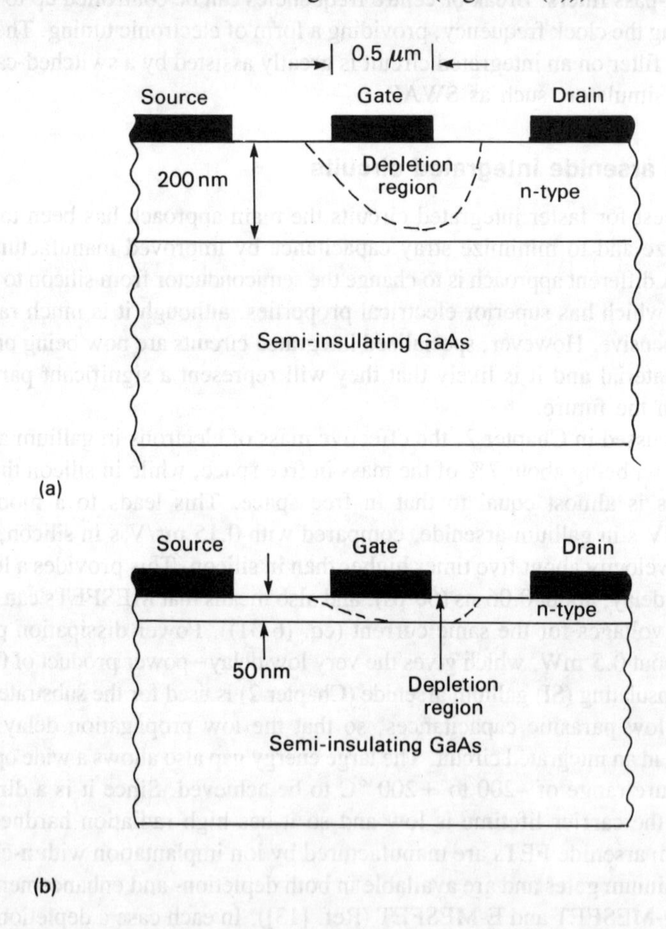

Figure 7.56 Gallium arsenide metal-silicon FETs. (*a*) n-channel depletion mode FET (D-MESFET); (*b*) n-channel enhancement mode FET (E-MESFET).

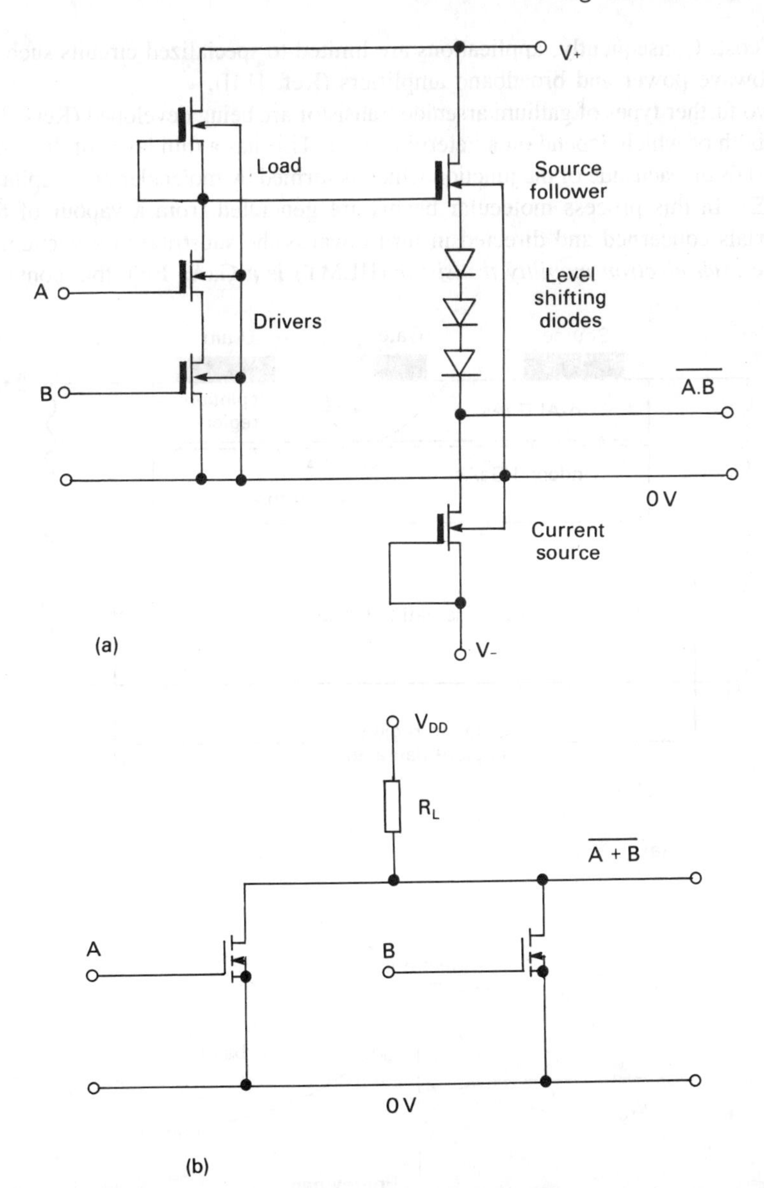

Figure 7.57 (*a*) GaAs FET logic gates. (*a*) D-MESFET NAND gate with level shifter. Typically, logic 1 is at 0.7 V and logic 0 at $-V_p$; (*b*) E-MESFET NOR gate. Typically, logic 1 is at 0.6 to 0.4 V and logic 0 at 0 V.

power supply is required but the pinch-off voltage must be controlled within about 25 mV, which is very hard to achieve in production. In addition, at present, the fabrication of gallium arsenide integrated circuits is generally difficult due to a high incidence of crystal imperfections and dislocations, leading to low yield and high cost. Consequently, applications are limited to specialized circuits such as microwave power and broadband amplifiers (Ref. [14]).

Two further types of gallium arsenide transistor are being developed (Refs [15, 16]) both of which depend on a heterojunction. This has a thin layer of different materials on each side of the junction which is formed by molecular beam epitaxy (MBE). In this process molecular beams are generated from a vapour of the materials concerned and directed in turn towards the substrate in a vacuum.

The *high electron mobility transistor* (HEMT) is a GaAs FET that contains

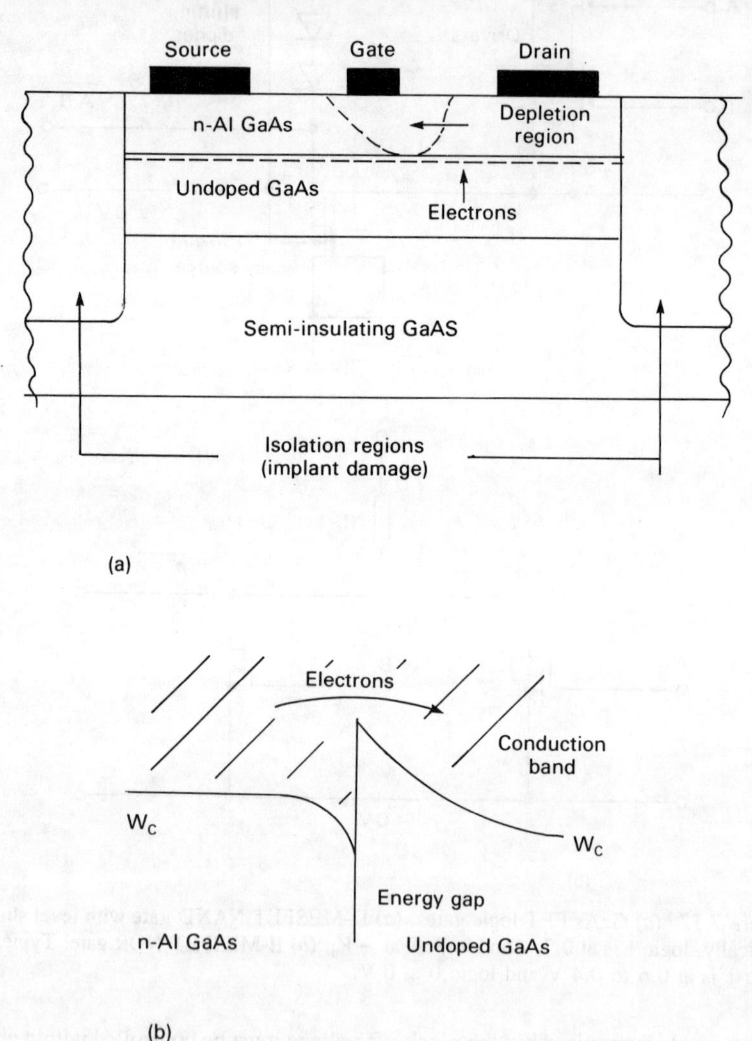

Figure 7.58 High electron mobility transistor (HEMT). (*a*) Simplified section; (*b*) energy barrier between n-type AlGaAs and undoped GaAs layers.

an n-type AlGaAs layer next to an undoped GaAs channel layer. This is shown in the simplified section of Fig. 7.58(*a*) which omits other layers. The device is sometimes called a modulation doped FET.

The energy gap of AlGaAs is greater than that of GaAs, which leads to bending of the energy bands and a barrier between the layers shown for the conduction band in Fig. 7.58(*b*). Electrons then diffuse from the n-type AlGaAs into the GaAs layer where they are retained by the energy barrier. The mobility of electrons in the n-type material is limited by scattering due to the impurity atoms (Chapter 2), so when the electrons are separated from their donors their mobility is much higher in the undoped material.

In the HEMT the channel mobility can rise to 4.9 m^2/V s which provides improved switching speed and high-frequency performance. The operation of the HEMT is similar to that of the E-MESFET, being cut off at $V_{GS} = 0$ V and with a threshold voltage around 0.7 V.

The *heterojunction bipolar transistor* (HJBT) has an n-type AlGaAs emitter and a p^+-type GaAs base (Fig. 7.59(*a*)). Thus the relative doping levels of emitter and base are the reverse of the bipolar transistor with homojunctions.

Normally, the base doping is only a fraction of the emitter doping level, which limits the hole current in the emitter and makes the emitter efficiency γ just less than 1. However, the low base doping increases the resistance $r_{bb'}$ between the active region and the base contact, which reduces the high-frequency performance (Chapter 4).

This problem is overcome in the HJBT by choosing an emitter material with an energy gap about 0.25 eV higher than that of the base. As shown in the transistor energy band diagram of Fig. 7.59(*b*) the barrier for electrons diffusing into the base is less than the barrier for holes diffusing into the emitter. Thus the hole current in the emitter is negligible and $\gamma = 1$ for any base doping, leading to very high values of current gain h_{fe}. The light emitter doping reduces the base-emitter capacitance and the heavy base doping reduces $r_{bb'}$, both of which lead to improved high-frequency performance.

Both the HEMT and the HJBT are likely to become commercially available in the future. As yet their respective manufacturing technologies have not been fully developed but the HEMT is likely to be available first.

Points to remember

* Fundamental techniques in integrated circuit (IC) manufacture such as oxidation, window formation, diffusion and expitaxy were first introduced in bipolar ICs.
* These techniques allow transistors, resistors and capacitors to be formed on a single chip of silicon and then interconnected to produce a complete circuit.

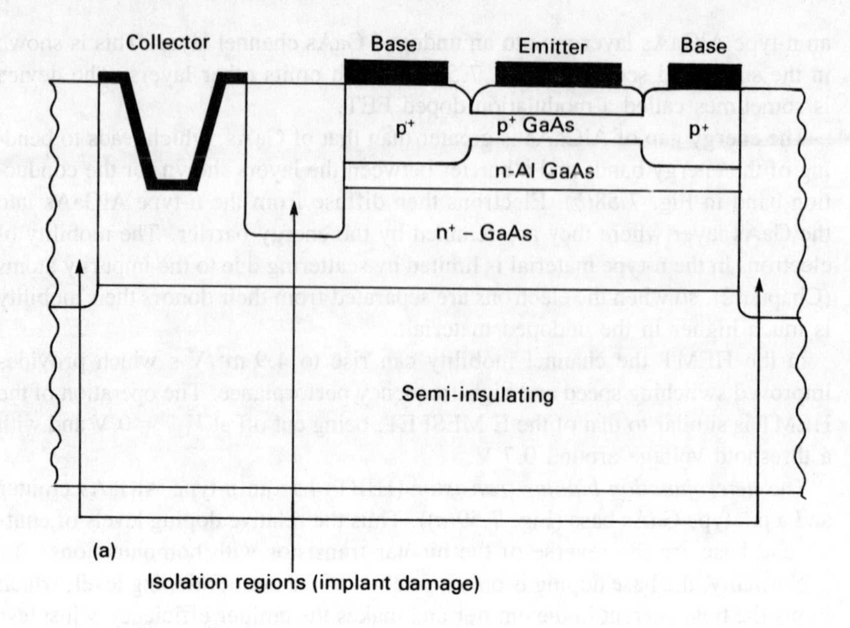

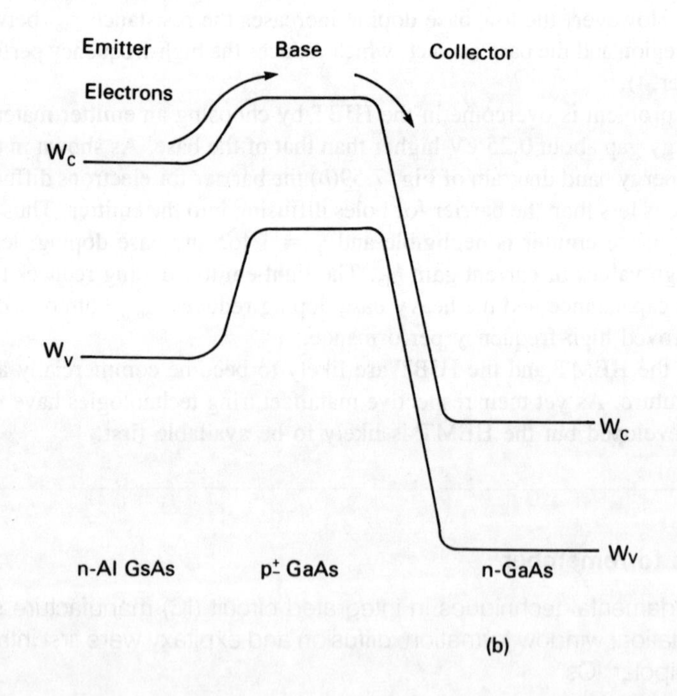

Figure 7.59 **Heterojunction bipolar transistor (HBJT). (*a*) Section; (*b*) energy band diagram for positive values of V_{BE} and V_{CE}. The energy gap $W_c - W_v$ is greater for the emitter than for the base or the collector.**

* Integrated circuits are now mainly formed from MOS transistors with silicon gates and no longer use resistors. Both nMOS and CMOS techniques are used, with CMOS becoming increasingly important.
* The early small- or medium-scale ICs with just a few logic gates on a chip are now accompanied by large- and very large-scale ICs. These contain scaled-down MOS transistors and provide complex electronic systems.
* Transistor-transistor logic (TTL) was the first major IC family and so has set standards that are followed by later families. Schottky, low-power Schottky, advanced and fast versions are now available and all use bipolar transistors.
* Other bipolar logic families are ECL, which is very fast, and I^2L, which uses no resistors and so is suitable for large-scale ICs.
* The HCMOS logic family provides the same logic functions and similar propagation delay to LSTTL but with much less power dissipation.
* Application-specific ICs allow a logic system developed on one or more printed circuit boards to be integrated on to a single chip, with CMOS gate arrays being commonly used.
* The opamp is a linear IC based on the differential amplifier. Other linear ICs use the switched-capacitor technique to emulate a resistor and provide filter circuits.
* ICs are beginning to employ gallium arsenide FETs which are faster than silicon devices. New structures are being developed, based on heterojunctions, and both digital and linear circuits are being produced.

References

[1] Sze, S.M. *Physics of semiconductor devices*, 2nd ed., (Wiley), 1981.
[2] Wilson, R.G. and Brewer, G.R. *Ion beams*, (Wiley), 1973.
[3] Mukherjee, A. *Introduction to nMOS and CMOS VLSI systems design*, (Prentice Hall), 1986.
[4] Mead, C. and Conway, L. *Introduction to VLSI system*, (Addison-Wesley), 1980.
[5] Mavor, J., Jack, M.A. and Denyer, P.B. *Introduction to MOS LSI design*, (Addison-Wesley), 1983.
[6] Hicks, P.J. *Semi-custom IC design and VLSI*, (Peter Peregrinus), 1983.
[7] Sze, S.M. *VLSI Technology*, (McGraw-Hill), 1983.
[8] Antoniadis, D.A., Hansen, S.E., Dutton, R.W., Gonzalez, A.G. and Rondoni, M. *SUPREM* 1 *–A program for process engineering models, Stanford Electronics Laboratory, Stanford, California, Report SU SEZ-77-006*, May 1977.
[9] Herman, J.M. and Evans, S.A. Special issue on I^2L, IEEE J. *Solid State Circuits*, SC-12(2), 1977.
[10] Millman, J. and Grabel, A. *Microelectronics*, 2nd ed., (McGraw-Hill), 1987.

[11] Kinniment, D.J., Chester, E.G. and McLauchlan, M.R. *CAD for VLSI*, Van Nostrand Reinhold), 1985.
[12] Solomon, J.E. 'The monolithic opamp: a tutorial study', *IEEE Journal of Solid State Circuits SC-9*, 1974, pp. 314–32.
[13] Eden, R.C., Livingston, A.R. and Welch, B.M. 'Integrated circuits, the case for gallium arsenide', *IEEE Spectrum vol. 20.*, 1983, pp. 30–7.
[14] Bierman, H. 'Move over, silicon! Here comes GaAs', *Electronics*, vol. 58, 1985, pp. 39–46.
[15] Eden, R.C. 'Comparison of GaAs device approaches for ultra high-speed VLSI', *Proc. IEEE*, vol. 70, no. 1, 1982, pp. 5–12.
[16] Morkoc, H. Solomon, P.M. 'The HEMT: a superfast transistor', *IEEE Spectrum*, vol. 21, Feb. 1984, pp. 28–35.

Problems

7.1 A silicon substrate has an initial n-type impurity concentration of 10^{22} atoms/m^3. Prior to a limited source diffusion 5×10^{21} boron atoms/m^2 are deposited on the surface of the substrate. If the diffusion proceeds for 45 minutes at 1100 °C determine the junction depth. (Diffusion coefficient of boron at 1100 °C is 3×10^{-17} m^2/s.)
[2.1 μm]

7.2 Determine the peak doping concentration and its depth for a 100 keV boron implantation, when the incident flux is 2×10^{19} ions/m^2.
[1.14×10^{26} impurities/m^3, 0.3 μm]

7.3 Explain why the parameter ρ_s is used for a diffused resistor rather than resistivity. In practice, what factors affect the value of ρ_s?

Discuss why diffused resistors are not generally employed in large-scale integrated circuits. Use as an example a 1 kΩ resistor, 10 μm wide with ρ_s = 100 Ω per square. What replaces a resistor in a large-scale integrated circuit?

7.4 In the circuit of Fig. 7.27(*a*) R_1 = 4 kΩ, R_2 = 1 kΩ, V_{CC} = 5 V, logical 0 is at 0.1 V and logical 1 is at 5 V. The emitter and collector diodes of T1 may be assumed to be independent and all diodes have an infinite resistance for all voltages below a forward bias of 0.6 V and a resistance of 200 Ω for all voltages above 0.6 V.

Calculate I_1 when (i) one or more of the inputs is at logical 0; (ii) when all the inputs are at logical 1. What is the minimum value of h_{FE} for T2 to be saturated?
[1.02 mA; 0.86 mA; 5.8]

7.5 Write a SPICE program to give the transfer characteristic and transient response of the ECL gate of Fig. 7.34, for input changes between −1.55 V and −0.75 V.

7.6 An integrated circuit design, based on a 2400 element CMOS gate array uses:

(i) 1920 array elements with power/element of 15 μW/MHz;
(ii) 30 input elements each consuming 45 μW/MHz;
(iii) 32 output elements, each of which has a 40 pF load and consumes 100 μW/MHz plus 25 μW/pF/MHz. If 15% of all elements are clocked at 10 MHz estimate the power consumed by the integrated circuit.

[98 mW]

7.7 Obtain the number of CMOS gate arrangements (cells) for the following logic functions

(i) 3 input NAND + inverter [2]
(ii) 4 input NAND [2]
(iii) Exclusive OR + inverter + 2*i*/*p* NAND [3]
(iv) Half-adder (Σ and $\bar{\Sigma}$) + inverter [4]

7.8 A switched-capacitor element operates at a clock frequency of 2 MHz. Calculate the capacitance value to produce an equivalent resistance of (i) 50 kΩ, (ii) 500 kΩ. What are the relative sizes of chip area required?
[10 pF, 1 pF, 10 : 1]

7.9 A simple low-pass switched-capacitor filter is to have an upper 3 dB frequency which is variable between 2 and 20 kHz. Determine the range of clock frequencies assuming a C_1/C_2 ratio of 0.1.
[125 kHz to 1.25 MHz]

8 Vacuum and gas-filled devices

Although devices based on the emission of electrons from a surface have, in many cases, been superseded by semi-conductor devices, an important group still remains. At present the cathode-ray tube is unrivalled for most display purposes and the electron microscope also performs a unique function. These are both vacuum devices operating at very low pressure, equivalent to about 10^{-6} torr or less. The effects of gas pressure are discussed later in this chapter but it is assumed initially that the pressure is so low that any residual gas offers no obstruction to electron emission or motion.

Emission of electrons

Inside a solid an electron moves between the atoms in a random manner owing to thermal agitation. Near the surface it may have a component of velocity which will take it away from the atoms, but as soon as it escapes from the solid it leaves behind a positive charge. This exerts a force trying to prevent its escape, and this force may be obtained by the method of electrical images. If the electron, charge $-e$, is distance x outside the surface, the force F acting on it may be represented by a charge of $+e$ at a distance x below the surface (Fig. 8.1). Then, by Coulomb's law,

$$F = \frac{e^2}{4\pi\epsilon_0(2x)^2} = \frac{e^2}{16\pi\epsilon_0 x^2} \tag{8.1}$$

When x is large, $F \to 0$ and the electron has then escaped from the solid. However, when $x = 0$ the electron is just on the surface and eq. (8.1) suggests that F then becomes infinite. In fact F must tend to zero, since within the solid the electron is surrounded by atoms and the net force on it is zero. Thus the image theory does not apply close to the solid, and in practice the image value of F is reached at about 1 nm from the surface (Fig. 8.2(a)). From this distance the row of atoms at the surface appears to be almost continuous, since their spacing is about 0.1 nm.

The work done in moving the electron a distance x from the surface is then

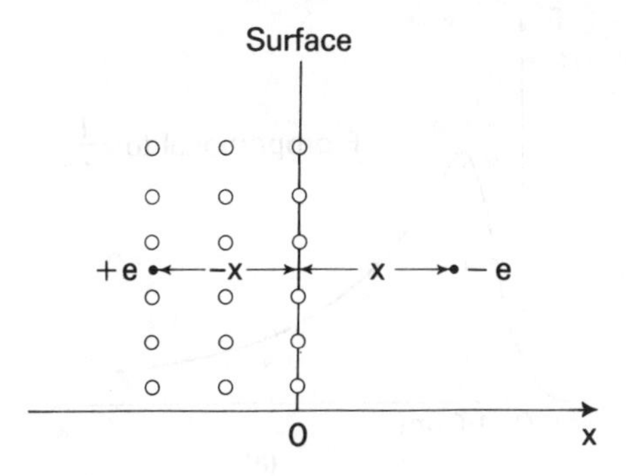

Figure 8.1 Electron image near a surface.

$$W = \int_0^x F \, dx \tag{8.2}$$

which is the area under the F/x curve. The variation of W with x is shown in Fig. 8.2(*b*), where it will be seen that it reaches a constant value, W_1, asymptotically, in practice at about 10 nm from the surface. In order to escape, the electron must have potential energy greater than the energy barrier W_1. If its kinetic energy of emission is less than W_1 the electron will travel beyond the surface until it comes to rest and then returns to the solid. On the energy level diagram W_1 may be taken as the zero energy an electron possesses when it leaves the solid, so that its energy within the solid is negative with respect to W_1. The maximum energy it can possess at the absolute zero of temperature is W_F, the Fermi level. The difference between these two energies is given by

$$W_1 - W_F = \phi \tag{8.3}$$

where ϕ is the *work function*, usually measured in electronvolts, which corresponds to the minimum extra energy that must be given to the solid to allow an electron to escape. The values of ϕ range from 1 to 6 eV, depending on the atomic structure of the material, and at room temperature very few electrons have enough energy to escape. This energy can be supplied in four ways:

(1) radiation, causing photoemission;
(2) heat, causing thermionic emission;
(3) bombardment, causing secondary emission;
(4) external field, causing field emission.

Electron motion in a uniform electric field

Before discussing in detail the different types of electron emission and the devices based on them, let us consider what happens to an electron after it has escaped. The solid from which it has been emitted is called the *cathode*, which may be

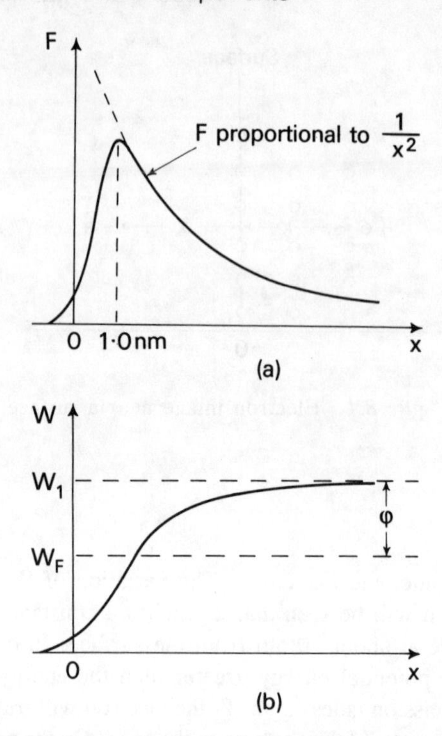

Figure 8.2 Work function of a surface. (*a*) Force acting on an electron; (*b*) energy barrier.

in the form of a flat plate, and it can be collected by a second plate called the *anode*. The plates are known as *electrodes* and many other forms are possible. They are mounted in a vacuum so that the motion of the electron is not interrupted by collision with air particles.

Suppose the spacing between the plates is d and the anode is held at a potential V_A which is positive with respect to the cathode (Fig. 8.3). An electric field E_x is set up which is uniform away from the edges of the electrodes, leading to electron motion in the x-direction. It is assumed that there is negligible motion in the y- and z-directions. The force on the electron is given by

$$F_x = -eE_x = e\frac{dV}{dx} \tag{8.4}$$

and is in the direction of increasing potential so that work will be done on the electron in moving it towards the anode. In moving between the points 1 and 2 the work done is

$$W_{12} = \int_1^2 F_x\,dx = e\int_1^2 dV \tag{8.5}$$

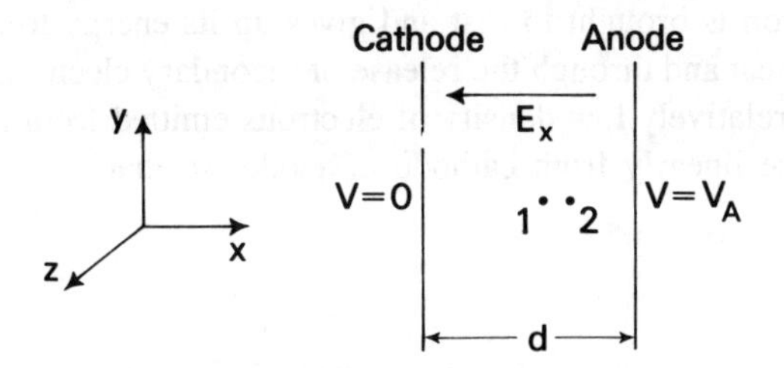

Figure 8.3 Electron in a uniform electric field.

$$= e(V_2 - V_1) \tag{8.6}$$

where V_1 and V_2 are the potentials at 1 and 2 respectively. The force is also given by the rate of change of momentum, so that, if u is the velocity in the x-direction,

$$F_x = \frac{\mathrm{d}}{\mathrm{d}t} mu = m \frac{\mathrm{d}u}{\mathrm{d}t} + u \frac{\mathrm{d}m}{\mathrm{d}t} \tag{8.7}$$

When relativistic effects are negligible, $\mathrm{d}m/\mathrm{d}t = 0$ and

$$W_{12} = \int_1^2 F_x \,\mathrm{d}x = \int_1^2 m \frac{\mathrm{d}u}{\mathrm{d}t} \,\mathrm{d}x = \int_1^2 mu \,\mathrm{d}u \tag{8.8}$$

$$= \tfrac{1}{2} m(u_2^2 - u_1^2) \tag{8.9}$$

Equating (8.6) and (8.9),

$$e(V_2 - V_1) = \tfrac{1}{2} m(u_2^2 - u_1^2) \tag{8.10}$$

where u_1 and u_2 are the velocities at 1 and 2 respectively. Thus the kinetic energy of the electron has been increased because it has gained energy from the electric field, and this is of fundamental importance in vacuum devices.

In the common case of an electron starting from rest at the cathode and moving through a potential difference V, both V_1 and u_1 are zero, so that the velocity u is obtained from the expression

$$eV = \tfrac{1}{2} mu^2 \tag{8.11}$$

whence

$$u = \sqrt{\frac{2eV}{m}} \tag{8.12}$$

The velocity will reach its greatest value at the anode, where it is given by

$$u_A = \sqrt{\frac{2eV_A}{m}} \tag{8.13}$$

Here the electron is brought to rest and gives up its energy to the anode, both in the form of heat and through the release of secondary electrons. For one electron and for a relatively low density of electrons emitted from the cathode, the voltage will rise linearly from cathode to anode, so that

$$E_x = -\frac{V_A}{d} \tag{8.14}$$

everywhere between anode and cathode. The acceleration of an electron from rest may be obtained by combining eqs (8.4), (8.7) and (8.14) to give

$$\frac{du}{dt} = \frac{eV_A}{md} \tag{8.15}$$

and this is also uniform, so that u rises linearly with time as shown in Fig. 8.4. Then the average velocity is $\frac{1}{2}u_A$ and the transit time τ_A from cathode to anode is given by

$$\tau_A = \frac{2d}{u_A} = d\sqrt{\frac{2m}{eV_A}} \tag{8.16}$$

using eq. (8.13).

Owing to its very small mass it is quite easy to cause an electron to move at a velocity approaching that of light. Under these conditions its mass m is no longer constant but increases according to relativity theory. It may be shown that

$$m = \frac{m_0}{(1 - u^2/c^2)^{1/2}} \tag{8.17}$$

where m_0 is the rest mass, i.e. the mass for low velocities such that $u^2/c^2 \ll 1$. For electrons accelerated through a potential difference of 5 kV, the increase in

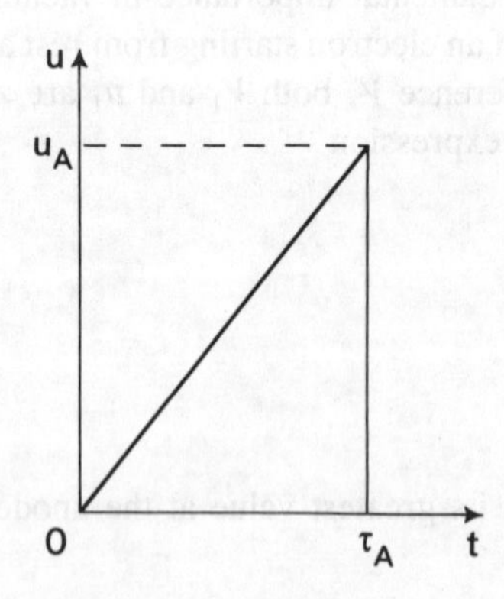

Figure 8.4 Velocity/time graph for an electron.

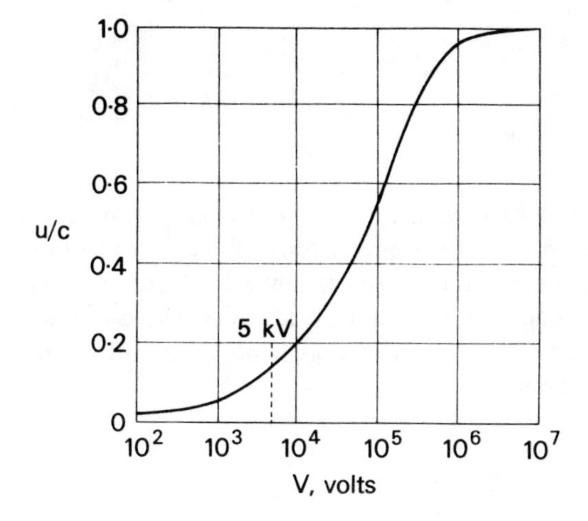

Figure 8.5 Relativistic variation of speed with energy.

mass is just less than 1% and so may be ignored. At voltages above 5 kV the curve relating u/c and V is given in Fig. 8.5. The mass increase is important in high-power microwave valves and in electron microscopes, the latter using electron guns with anode voltages at and above 100 kV.

Current in the external circuit

Although the emitted electrons are moving in the space between the electrodes a measuring instrument in the external circuit will indicate a flow of current. This is because each electron repels a negative charge from the anode, which must equal $-e$, in the time, τ_A, it takes the electron to travel from cathode to anode. Thus a current i flows in the external circuit for each electron and energy is supplied to it from the external power supply (Fig. 8.6(*a*)). For an electron moving from rest through a potential V this energy is given by

$$\frac{1}{2}mu^2 = Ve = V\int_0^t i\,dt \tag{8.18}$$

Differentiating,

$$mu\frac{du}{dt} = Vi$$

and, substituting from eqs (8.4) and (8.7),

$$mu\frac{du}{dt} = -eE_x u = e\frac{V}{d}u \tag{8.19}$$

so that

$$i = \frac{eu}{d} \tag{8.20}$$

Thus i increases linearly with time, since it is proportional to u (Fig. 8.6(*b*)). The total current for a number of electrons will be the sum of the individual currents, provided the electron density is low enough for interaction to be negligible. If the electrode area is S and the electron density is n, the effective volume is Sd and the total number of electrons is nSd. Thus the total current flowing in the external circuit is

$$I = nSdi \tag{8.21}$$

$$= neuS \tag{8.22}$$

from eq. (8.20). The current density between the electrodes is

$$J = neu \tag{8.23}$$

since ne is the charge density. Thus, since $J = I/S$, the same current flows in the external circuit as between the electrodes.

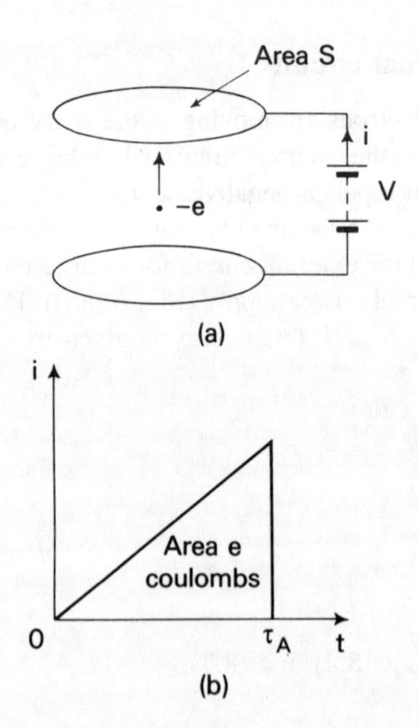

Figure 8.6 Induced current in a diode. (*a*) Production of induced current; (*b*) variation of induced current with time.

Photoemission

Let us now consider how electrons can be released from a cathode by the absorption of radiation. When radiation of frequency f falls on a solid, each energy packet or photon will release one electron if $hf = e\phi$. The electron will have received just enough energy to surmount the energy barrier and so will be released with zero initial velocity. If $hf > e\phi$ the electron will have received more than enough energy and so will have an initial velocity u_e. The value of u_e is obtained by equating energies to give Einstein's photoelectric equation:

$$hf = e\phi + \tfrac{1}{2} mu_e^2 \tag{8.24}$$

Threshold wavelengths λ_0

For a given surface the threshold wavelength λ_0 is the maximum wavelength that will allow an electron to be released with zero initial velocity. This is obtained from eq. (8.24) with $u_e = 0$ and corresponds to the minimum frequency f_0, where $hf_0 = e\phi$. By comparison with eq. (5.2) and with ϕ measured in electronvolts,

$$\lambda_0 = \frac{hc}{\phi} = \frac{1.24}{\phi}\ \mu\text{m} \tag{8.25}$$

Photoemission is not possible with light of wavelength longer than λ_0, and if visible light is taken to have an average wavelength of 0.6 μm, the work function of the cathode must be not greater than 2 eV for photoemission to occur (Fig. 5.1).

Maximum emission velocity u_e

u_e corresponds to the maximum velocity with which an electron can leave the surface. An electron can be released below the surface and be emitted with zero velocity due to collisions on its way through the material. Thus when $\lambda < \lambda_0$ electrons can be emitted with all values of velocity between 0 and u_e. If an anode is mounted near the cathode, photoelectrons will flow to it even when the anode-to-cathode voltage V_A is zero, owing to the initial energy with which they are emitted. If V_A is made increasingly negative the corresponding current will fall, since only those electrons with sufficient initial energy to overcome the potential barrier will be able to reach the anode. When the current is just zero only the most energetic electrons are reaching the anode, so that, if this occurs at an anode voltage V_0,

$$\tfrac{1}{2} mu_e^2 = eV_0 \tag{8.26}$$

and the maximum emission velocity u_e can be measured. Combining eqs (8.24) and (8.26),

$$V_0 = \frac{hf}{e} - \phi \tag{8.27}$$

so that if the cathode is illuminated with light of various frequencies and the corresponding values of V_0 are measured, a plot of V_0 against f will be a straight line (Fig. 8.7). The slope of this line will give the value of Planck's constant h, and the intercept on the V_0 axis will give the work function ϕ of the cathode in electronvolts. The line will cut the frequency axis at the threshold frequency f_0, so that an experiment of this type can be used to confirm the validity of the quantum theory.

The intensity of the light will not affect the initial velocity of emission, but the anode current is proportional to the intensity as may be seen by using an argument similar to that for eq. (5.4). The density of the photoelectrons is relatively low, so that the voltage is sufficient to draw all the photoelectrons to the anode. Thus a *vacuum photocell* has current/voltage characteristics as shown in Fig. 8.8 the current being nearly independent of voltage when $V_A > 15$ V.

Photocathode materials

In many applications photocells are illuminated by visible radiation and so are required to provide a response to wavelengths between about 0.4 and 0.7 μm. This means that the work function of the emitter must be less than 2 eV (eq. (8.25)) for the threshold wavelength to be high enough. Of the pure metals only caesium, with $\phi \approx 1.9$ eV satisfies this condition, so composite cathode materials with higher threshold wavelengths are used which incorporate a caesium coating. One of these is silver oxide caesium, also known as SI, which has a response similar to that of the human eye as shown in Fig. 8.9 and so is used in television camera tubes. Out of about 50 known cathode types only three are widely used which are Cs_3SbO (S11), Na_2KSbCs (S20) and Na_2KSb (bialkali). They all have a greater

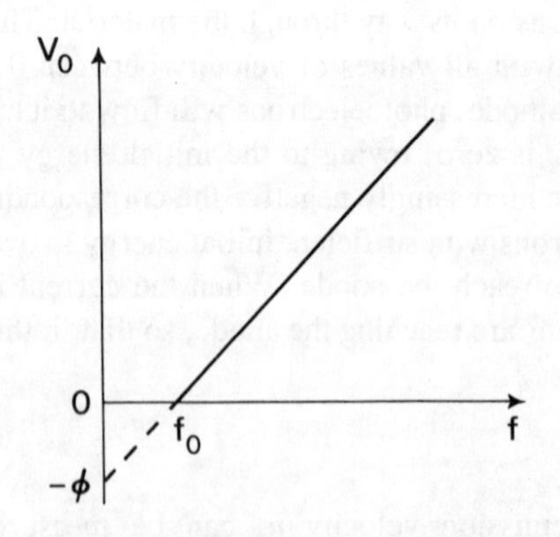

Figure 8.7 Verification of Einstein's equation (8.27).

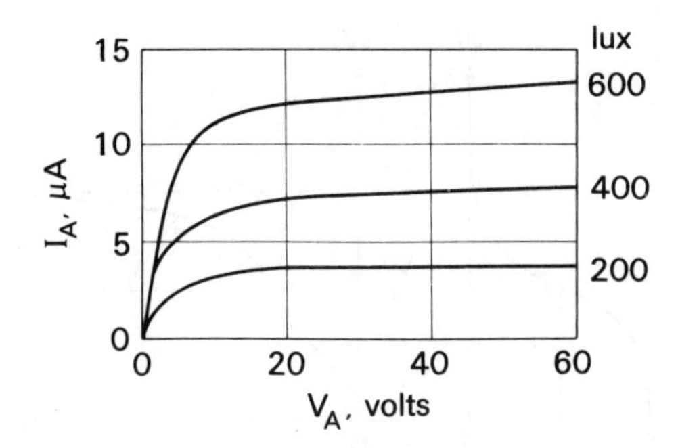

Figure 8.8 Characteristics of a vacuum photocell.

responsivity than AgOCs but a lower threshold wavelength (Fig. 8.9). They are essentially semiconductor materials with a low electron affinity due to the use of caesium, leading to a correspondingly low work function.

Thermionic emission

As the temperature of a material in a vacuum is raised above room temperature the probability of an electron escaping is increased, according to the Fermi-Dirac function (Fig. 8.10). The probability of an electron having energy W_1 is given by eq. (2.17):

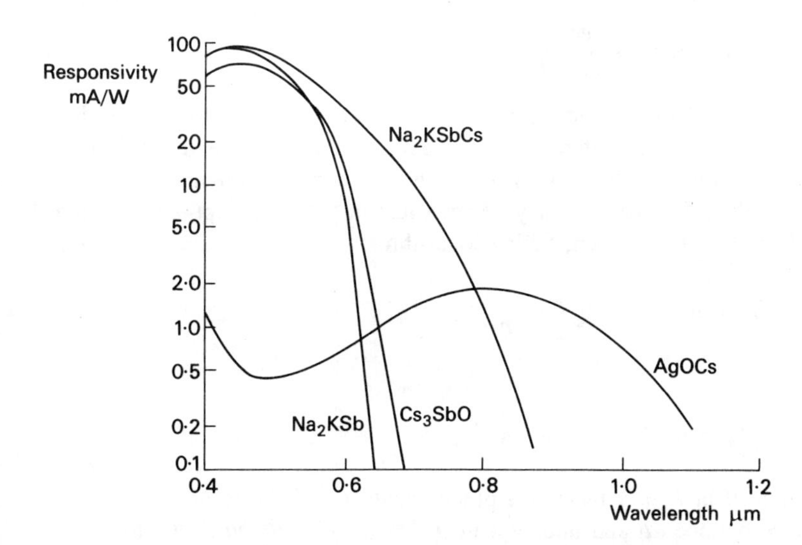

Figure 8.9 Responsivity of photocathode materials.

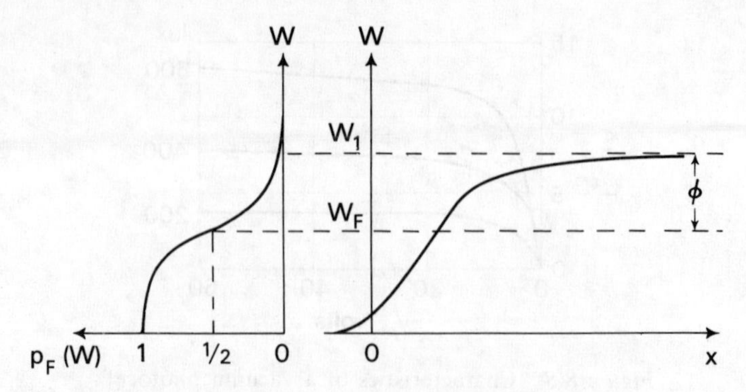

Figure 8.10 Probability of thermionic emission.

$$p_F(W_1) = \frac{1}{1 + \exp\left(\frac{W_1 - W_F}{kT}\right)}$$

$$\approx \frac{1}{\exp\left(\frac{e\phi}{kT}\right)} = \exp\left(-\frac{e\phi}{kT}\right) \quad (8.28)$$

The number of electrons available and hence J_s the maximum current per unit area of cathode depends on $p_F(W)$ but is almost independent of the anode voltage. J_s is known as the *saturated* or *temperature-limited* current density. The relationship is the Richardson–Dushman equation

$$J_s = AT^2 \exp\left(-\frac{e\phi}{kT}\right) \quad (8.29)$$

where A is a constant depending on the material of the cathode, varying from 60 A/cm^2 K^2 for tungsten to 0.01 for a mixture of barium and strontium oxides, the *oxide cathode* (Ref. [1]). The exponential term has a much greater effect on J_s than the AT^2 term and may be rewritten in the form $\exp(-b/T)$, where $b = e\phi/k = 11\,600\,\phi$. Then, taking logarithms,

$$\ln J_s = \ln A + 2 \ln T - \frac{b}{T}$$

or

$$\ln J_s + 2 \ln \frac{1}{T} = \ln A - b\,\frac{1}{T} \quad (8.30)$$

Then, if $\ln J_s + 2 \ln(1/T)$ is plotted against $1/T$, the result will be a straight line with slope $-b$ and intercept $\ln A$, known as a *Richardson plot*.

This is illustrated in Fig. 8.11 for the three main cathode materials, tungsten,

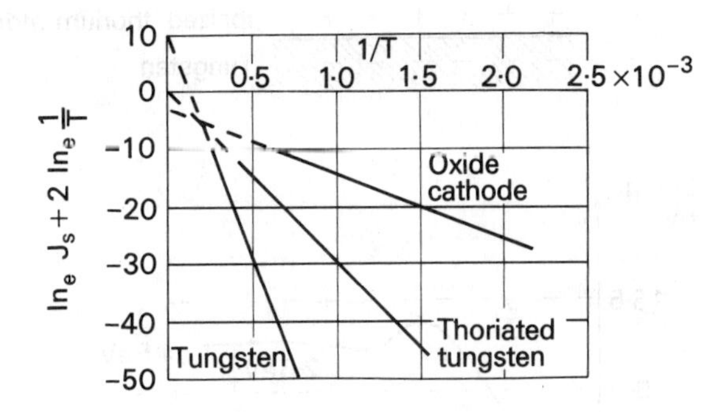

Figure 8.11 Characteristics of thermionic cathode materials.

thoriated tungsten and oxide. Their main properties are given in Table 8.1, where it can be seen that a low work function allows a low working temperature and a long life, which results mainly from reduced evaporation of cathode material. The oxide cathode is considered to be an n-type semiconductor with excess barium atoms providing the donor impurity. First introduced by Wehnelt in 1905, it gives the greatest current density of the three types and is used in the great majority of thermionic cathodes. However, where the anode voltage is greater than about 2 kV, bombardment of the cathode by positive ions quickly destroys the thin ($\approx 50\ \mu m$) coating of barium and strontium oxides. For anode voltages up to 15 kV, which occur in transmitting valves, thoriated tungsten is used, and where the anode voltage reaches 20 kV, as in high power transmitting valves and X-ray tubes a tungsten filament is used.

A thoriated-tungsten cathode consists of a tungsten filament coated with a layer of thorium about one atom thick (a monatomic layer). Since the work function of thorium is less than that of tungsten, valence electrons from the thorium mix with the free electrons in the tungsten leaving ionized thorium atoms on the surface, which is thus positively charged. Since the body of the cathode has acquired a negative charge a *dipole layer* or *electrical double layer* is formed on the surface (Fig. 8.12(*a*)). This layer creates an electric field at the surface which assists

Table 8.1 Thermionic cathode materials

Cathode	Work function	Working temperature	J_s	Life
	eV	K	A/cm^2	h
Tungsten	4.5	2500	0.25	3000
Thoriated tungsten	2.6	1900	1.5	10 000
BaO + SrO on nickel	1.0	1000–1100	1.0	20 000

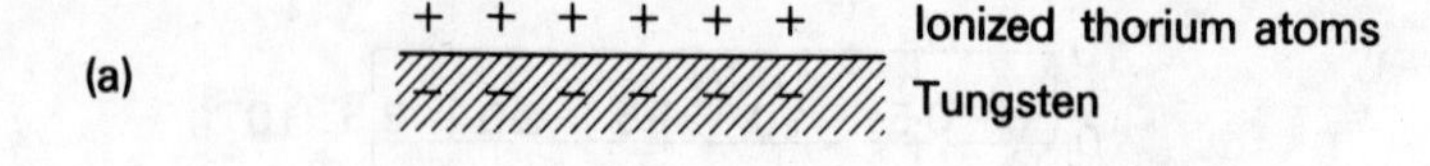

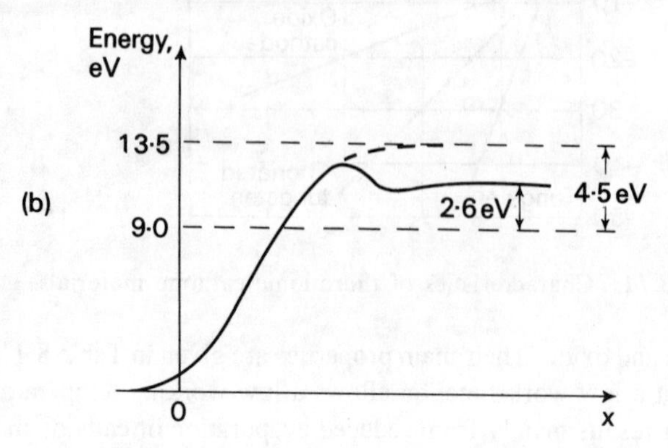

Figure 8.12 Thoriated-tungsten cathode. (*a*) Dipole layer; (*b*) energy barrier.

electrons to escape, so that the overall work function is less than that of tungsten. The surface energy barrier then has a peak (Fig. 8.12(*b*)), which is thin enough to allow a finite probability than an electron can 'tunnel' through it (Appendix 1).

The Schottky effect

When a sufficiently high voltage is applied to the anode all the electrons available for emission are collected so that the potential rises linearly from cathode to anode (Fig. 8.13(*a*)). The resulting electric field, $-V_A/d$, lowers the energy of the emitted electrons by an amount eV, where V is the potential at any point, so that the energy barrier is reduced by this amount at each point; the reduction may be significant for a high anode voltage but will occur to a lesser degree for lower voltages. A peak then occurs in the new barrier which is too wide for an electron to penetrate and the work function is reduced from ϕ to ϕ_e (Fig. 8.13(*b*)). This means that the saturated current increases slowly with voltage instead of being independent of voltage, which occurs for fields up to about 10^5 V/m.

The current/voltage characteristic of a thermionic diode is shown in Fig. 8.14. The saturated current, I_s, at the knee of each curve increases with cathode temperature. I_s would be independent of V_A but for the Schottky effect, which causes the current to rise slowly with voltage. The Schottky effect is more pronounced in the oxide cathode, which has a porous surface, than in the tungsten cathode, which has a relatively smooth surface. This is due to the penetration of the pores of the coating by the field due to the anode, which causes electrons to be extracted from the inside surfaces of the coating. For voltages less than the saturation value the current rises with voltage. in this region the current is limited by the space charge as discussed below.

Although the thermionic diode has by now been replaced in most circuit

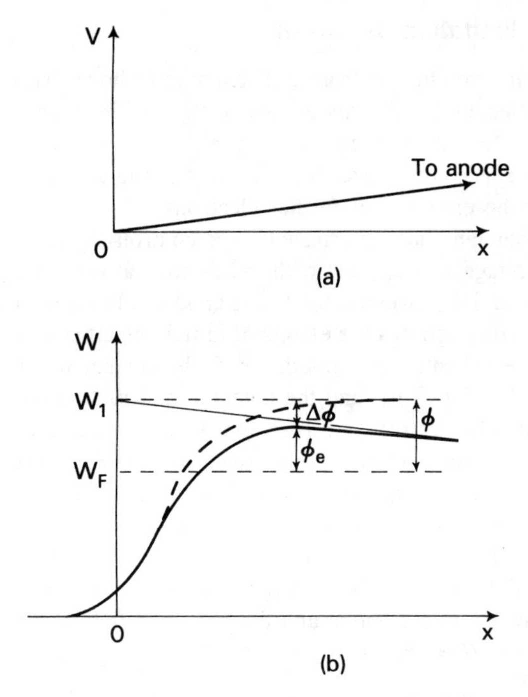

Figure 8.13 The Schottky effect. (*a*) Anode field; (*b*) reduction of work function.

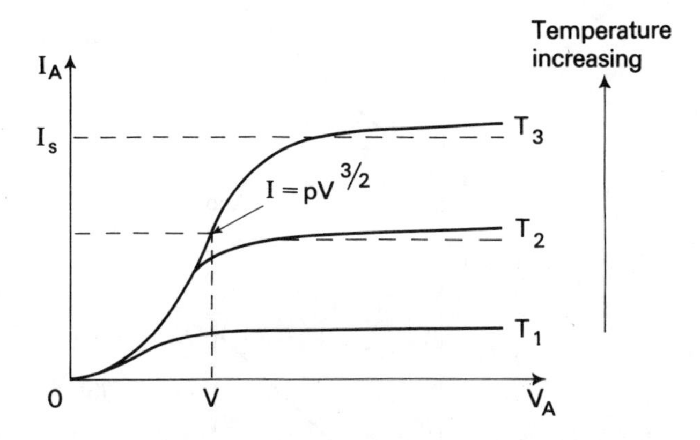

Figure 8.14 Characteristics of a thermionic diode.

applications by the semiconductor diode, it is still necessary to gain an understanding of its properties. This is because it is the basis of the electron gun which is used extensively in cathode-ray tubes, electron microscopes and microwave values. The diode is also fundamental to the operation of the triode, which again has been replaced almost entirely by the transistor but is still used in high fidelity (hi-fi) systems and in radio transmitters. These applications exploit the low distortion and high-power properties of thermionic valves.

Space-charge limitation of current

So many electrons are emitted from a thermionic cathode that only a fraction of them are collected by the anode when V_A has its working value. The remainder form a 'cloud' of negative charges, known as a *space charge*, near the cathode. The space charge has the effect of causing a retarding field, which tends to prevent the emission of further electrons.

In order to understand how the space charge controls the anode current, suppose an anode voltage V is applied with the cathode at room temperature. The distribution of potential between cathode and anode will then be linear, as shown in Fig. 8.15(*a*), which ignores the effects of initial velocities and contact potential. If the cathode is heated to temperature T_1 the current will be temperature-limited as shown in Fig. 8.14, and the potential distribution will still be linear since all the available electrons are reaching the anode and no space charge is formed. The field at the cathode is then given by $-\tan\theta$, where θ is the angle the voltage curve makes with the horizontal at the cathode ($x = 0$). When the cathode temperature is raised to its working value, T_3 in Fig. 8.14, say, the resulting current I is less than its saturated value I_s, so that a space charge will be formed. The field at the cathode is then due to the combined effects of the accelerating field due to the anode and the retarding field due to the electrons of the space charge (Fig. 8.15(*b*)).

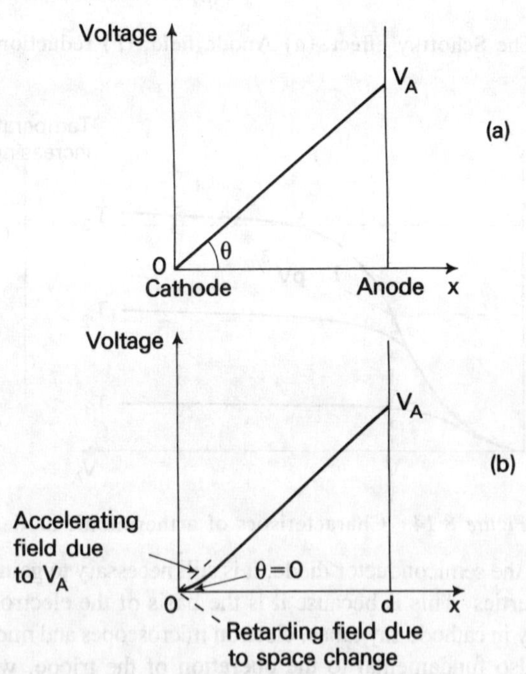

Figure 8.15 **Voltage distribution in a thermionic diode. (*a*) Temperature-limited current; (*b*) space-charge-limited current.**

An equilibrium will be set up in which the two fields are equal and opposite, so that $\tan\theta = \theta = 0$. Thus, when the space charge loses an electron to the anode, θ becomes positive and its place is taken by another electron drawn from the cathode. If the space charge gains a surplus electron, θ becomes negative and an electron is returned to the cathode. The equilibrium condition, $\theta = 0$, corresponds to the same number of electrons entering the space charge as are leaving it for the anode, so that the space charge is controlling the anode current.

The cathode at $x = 0$ is assumed to have an infinite supply of electrons and the anode at $x = d$ is at a potential V_A. Electrons are assumed to leave the cathode with zero initial velocity, and the electric field at the cathode is zero under space-charge-limited conditions. Thus the electron velocity at any point is related to the potential V at that point by eq. (8.12),

$$u = \sqrt{\frac{2eV}{m}}$$

The current density at any point is then

$$J = neu \tag{8.31}$$

where n is the density of electrons. The value of J is constant through the valve, so that the greatest density of space charge occurs near the cathode where the velocity is least.

It may be shown that (Ref. [1]) for a diode with plane parallel electrodes, that the current density at any distance x from the cathode is given by

$$J_x = \frac{2.33}{10^6}\frac{V^{3/2}}{x^2}\ \text{A/m}^2 \tag{8.32}$$

Where the electrode spacing is d the current density at the anode is given by

$$J_A = \frac{2.33}{10^6}\frac{V_A^{3/2}}{d^2} \tag{8.33}$$

which is known as the Langmuir–Child Law.

Properties of the thermionic diode

Although eq. (8.33) has been derived for a planar geometry the three-halves power relationship between current and voltage holds theoretically for *any* geometry. This is because the current density depends only on the density and velocity of the electrons (eq. (8.31)). Suppose the anode voltage of a diode of arbitrary geometry is changed from V_A to V'_A, so that

$$V'_A = aV_A \tag{8.34}$$

where a is any number greater than zero. Then, from eq. (8.12), the new electron velocity is $a^{1/2}u$, and from eq. (8.31) the new electron density is ane. The new current density is therefore

$$J'_A = a^{3/2} neu = a^{3/2} J_A$$

so

$$\frac{J'_A}{J_A} = a^{3/2} = \left(\frac{V'_A}{V_A}\right)^{3/2} \tag{8.35}$$

which can be written in the form

$$I_A = PV_A^{3/2}$$

where P is a constant depending on the geometry and electrode area and is known as the *perveance*. For a planar diode

$$P = \frac{2.33}{10^6}\frac{S}{d^2} \tag{8.36}$$

from eq. (8.33), and in general P varies inversely as the square of some characteristic length. This is important in the design of an electron gun, which is essentially a thermionic diode with a hole in the anode through which a beam of electrons emerges. When a gun of the desired geometry is available, all the dimensions cna be changed by the same factor without changing P, since the area S depends on the square of an electrode dimension. Thus the beam diameter can be controlled without changing the cathode current, provided that the current density does not exceed the saturated value.

The potential, the electron velocity and the density all vary with distance measured from the cathode. Thus, from eqs. (8.32) and (8.33).

$$V^{3/2} = V_A^{3/2}\left(\frac{x}{d}\right)^2 \quad \text{or} \quad V = V_A\left(\frac{x}{d}\right)^{4/3} \tag{8.37}$$

The velocity is proportional to $V^{1/2}$, so that

$$u = u_A\left(\frac{x}{d}\right)^{2/3} \tag{8.38}$$

where

$$u_A = \sqrt{(2eV_A/m)}$$

Putting ρ for the charge density ne,

$$J = \rho u$$

and at the anode

$$J_A = \rho_A u_A \tag{8.39}$$

Then

$$\rho_A = \rho \frac{u}{u_A} \tag{8.40}$$

so that

$$\rho = \rho_A \left(\frac{x}{d}\right)^{-2/3} \tag{8.41}$$

This equation cannot hold very close to the cathode since it implies that ρ becomes infinite at $x = 0$.

The final property to be considered is the transit time of electrons from cathode to anode in the presence of space charge. Since their velocity increases with distance we must write

$$\begin{aligned}\tau_A &= \int_0^d u \, dx \\ &= \frac{d^{2/3}}{u_A} \int_0^d x^{-2/3} \, dx\end{aligned}$$

using eq. (8.38). Hence

$$\begin{aligned}\tau_A &= \frac{d^{2/3}}{u_A} [3x^{1/3}]_0^d \\ &= \frac{3d}{u_A} \qquad (8.42)\end{aligned}$$

Thus the transit time has been increased by a factor of 3/2 compared with the value without space charge (eq. (8.16)).

Field emission

If a very high field is maintained between cathode and anode (up to 10^9 V/m), the effective potential barrier will be reduced further than in the Schottky effect (Fig. 8.16). ϕ_e will be smaller and the width of the barrier less than for the Schottky effect, with the result that there is a small but finite probability that electrons will be able to tunnel through the barrier (Appendix 1), as discussed by Fowler and Nordheim in 1928. Field emission can thus occur even at room temperature, and in practice the cathode is in the form of a point to increase the field intensity to the required value. Typically the effective cathode area is about 10^{-12} m^2 and the emission current density about 10^6 A/m^2, so that emission currents are of the order of microamperes. An application is in the *field emission microscope*, which provides such a large magnification that the positions of the atoms on the tip of the tungsten cathode are made visible by allowing the electrons to strike a fluorescent screen.

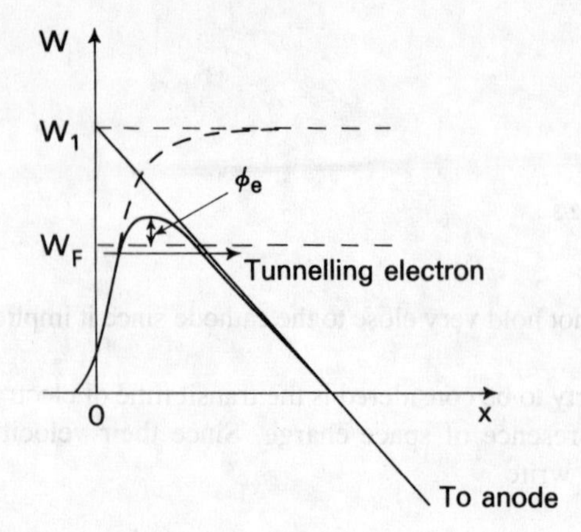

Figure 8.16 Energy barrier for field emission.

Secondary emission

When the surface of a solid is bombarded by a stream of electrons (or other particles), other electrons may be removed from the surface. These are called *secondary* electrons, the incident beam consisting of *primary* electrons. They transfer their energy to the surface electrons, which can then surmount the energy barrier $W_1 - W_F$. The number of secondary electrons removed per second in this way is directly proportional to the primary current, and the constant of proportionality is the *secondary emission coefficient* δ:

$$\delta = \frac{\text{Secondary electron current}}{\text{Primary electron current}} = \frac{I_S}{I_P} \tag{8.43}$$

δ depends on the nature of the surface and the energy of the primary beam, which must exceed a few electronvolts before secondary emission can occur.

The variation of δ with primary energy is illustrated in Fig. 8.17(*c*) for three materials. It will be seen that δ at first rises with energy; this is due to the removal of electrons from the surface, since these have the greatest probability of escape. As the primary energy is increased, the bombarding electrons travel faster and so spend less time at the surface before penetrating below it. Thus the surface electrons receive less energy and excitation of electrons *below* the surface occurs. For example, for primary energies of a few hundred electronvolts, the primary electrons striking a platinum target can penetrate up to 30 atomic layers below the surface before all their energy is expended. The primary electrons only come to rest when their energy has been reduced to that of the conduction electrons in the target. Many of the electrons excited below the surface lose their energy before reaching it due to collisions with the lattice atoms.

Hence, as the primary energy is increased, δ rises more slowly than at first, reaches a maximum value δ_{max} and then falls. For pure metals δ shows a similar variation to the curve for nickel in Fig. 8.17(*a*) and δ_{max} is normally less than 1.8. A larger value of δ_{max} is obtained with compound surfaces such as barium oxide, which is deposited on the surfaces of the electrodes of thermionic valves owing to evaporation from the cathode. Here δ_{max} is about 4.2, so considerable secondary emission occurs. Finally beryllium-copper has a value of δ_{max} around 8 and is used as a secondary source of electrons in the photomultiplier. A similar curve is obtained for insulators. Figure 8.17(*b*) shows a plot of δ against primary energy for mica. For $\delta < 1$ the specimen will become negatively charged and primary electrons will be repelled, so that δ cannot be determined in this region.

The phenomena associated with secondary emission may be studied by fixing the primary energy and measuring the distribution of the energy of the secondary electrons. A curve similar to that of Fig. 8.18 is obtained for pure metals for a primary energy of about 150 eV. The majority of the electrons have an energy less than 50 eV and these are considered to be the true secondary electrons, with the most probable energy for metals lying between 1.3 and 6 eV. The narrow peak occurring at the primary energy is due to primary electrons which have been reflected *elastically* from the surface, losing very little energy in the process. The energies between 50 and 150 eV are due to electrons which have penetrated the surface and have then been scattered out of it by the lattice vibrations. These

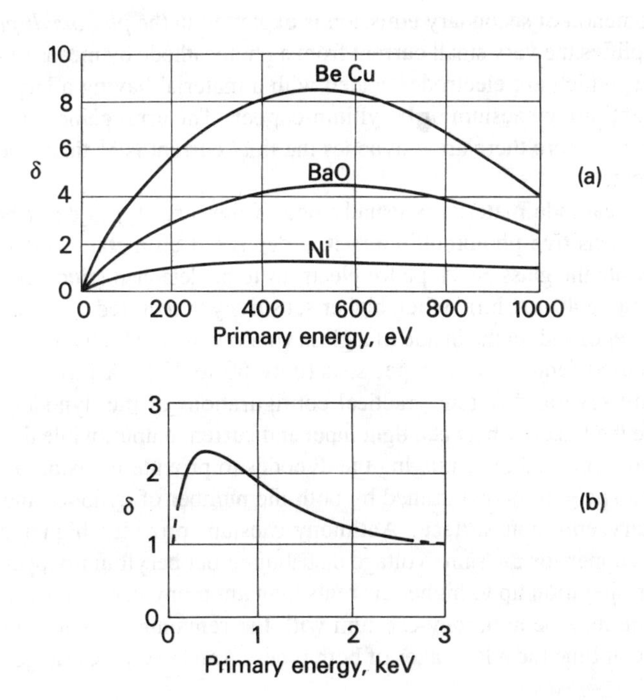

Figure 8.17 Secondary-emission coefficient and target voltage. (*a*) Conductors and semiconductors; (*b*) insulators.

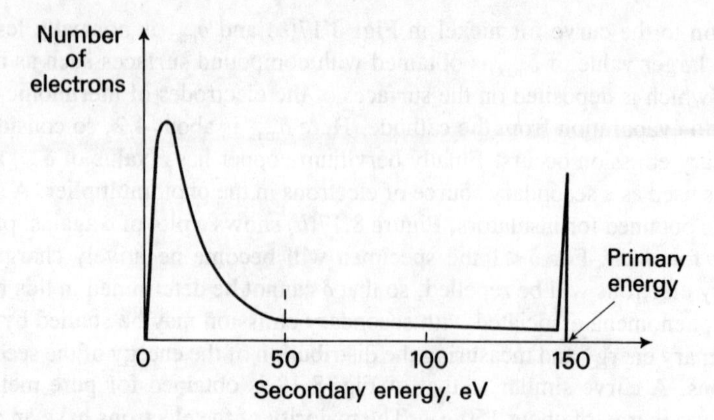

Figure 8.18 Distribution of energy of secondary electrons.

are known as *inelastically* reflected primary electrons, since they have given up energy to the lattice without causing electrons to be removed.

Photomultipliers

The phenomenon of secondary emission is exploited in the *photomultiplier*. This device amplifies the very small current from a photocathode by means of a number of *dynodes*, which are electrodes coated with a material having a large value of δ such as antimony-caesium or beryllium-copper. The arrangement is shown in Fig. 8.19 and where there are n dynodes the final current is δ^n times the original photocurrent.

The photocathode material is usually one of the three types described above and in less sensitive photomultipliers it is deposited on metal, so that photons must penetrate the glass envelope for electrons to be detached (side window type, sensitivity up to 40 μA/lm). When higher sensitivity is required a semitransparent cathode is deposited on the inside of a glass window on to which an optical image can be focused (end window type, sensitivity 50 to 130 μA/lm).

There are several different practical configurations of the dynodes designed to optimize the linearity between light input and current output, while the response time is also improved by arranging the dynodes to provide focusing of the electron beam. The gain is determined by both the number of dynodes and the type of secondary emission surface. Antimony-caesium provides higher gain than beryllium-copper for the same voltage distribution but beryllium-copper dynodes give linear operation up to higher currents than antimony-caesium dynodes. The first dynode may be antimony-caesium with the remaining dynodes beryllium-copper to combine the advantages of both types. Anode voltages range from 600 to 2000 V.

Photomultipliers are produced having from 6 to 15 dynodes and current gains between 10^4 and 10^8. They are applied in photometry and spectrophotometry,

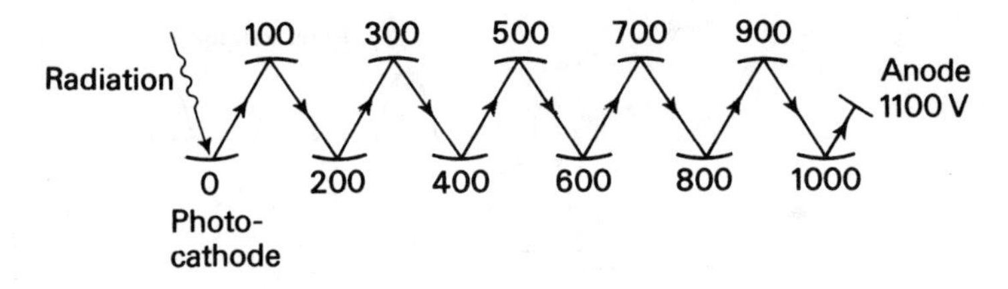

Figure 8.19 Principle of photomultiplier.

over the whole range of wavelengths from ultra-violet to infra-red, and also in scintillation counting.

Cathode-ray tubes (CRTs)

In the cathode-ray tube a beam of fast electrons from a thermionic cathode is focused on the centre of a fluorescent screen (Fig. 8.20). Where the electrons strike the screen their energy is used to produce light emission and secondary electrons are also produced. The end of the beam thus appears as a small, bright spot on the screen, which can be deflected horizontally and vertically using either electrostatic or magnetic fields. Normally the horizontal deflecting field increases linearly with time in order to move the beam across the screen at constant velocity (Fig. 8.20(*a*)), thus providing a *time base*. The vertical deflecting field is derived from the voltage or current waveform which is to be displayed, so that each position of the beam corresponds to the value of the waveform at a particular instant. When the frequencies of the time base and waveform are synchronized the waveform appears as a stationary trace on the screen.

The CRT is widely used for display purposes, notably in computer terminals, oscilloscopes, television sets and radar displays. In an oscilloscope the beam is often divided electronically into two parts. This allows two waveforms to be compared and input frequencies from d.c. to 100 MHz can commonly be accepted. Electrostatic deflection is normally used (Fig. 8.20(*b*)) which requires electrodes mounted inside the tube. This restricts the angle of deflection which can be obtained and so leads to a small screen, with a calibrated area typically 8 × 10 cm.

In a television set a tube with as wide a screen as possible is required, which means deflecting the beam through a wider angle than can be achieved with electrostatic deflection. This leads to the use of magnetic deflection in which coils are mounted outside the tube (Fig. 8.20(*c*)). A tube with a 26-inch diagonal and a viewing area about 53 × 39 cm is commonly available at present. Similar considerations apply to CRTs for radar displays which again use magnetic deflection.

In both types of tube the electrons are emitted from an electron gun. This consists of a flat oxide-coated cathode and a control electrode. The beam may then be focused to a small spot on the screen by means of two anodes and the action of such an electrostatic lens is described in a later section. Magnetic focusing

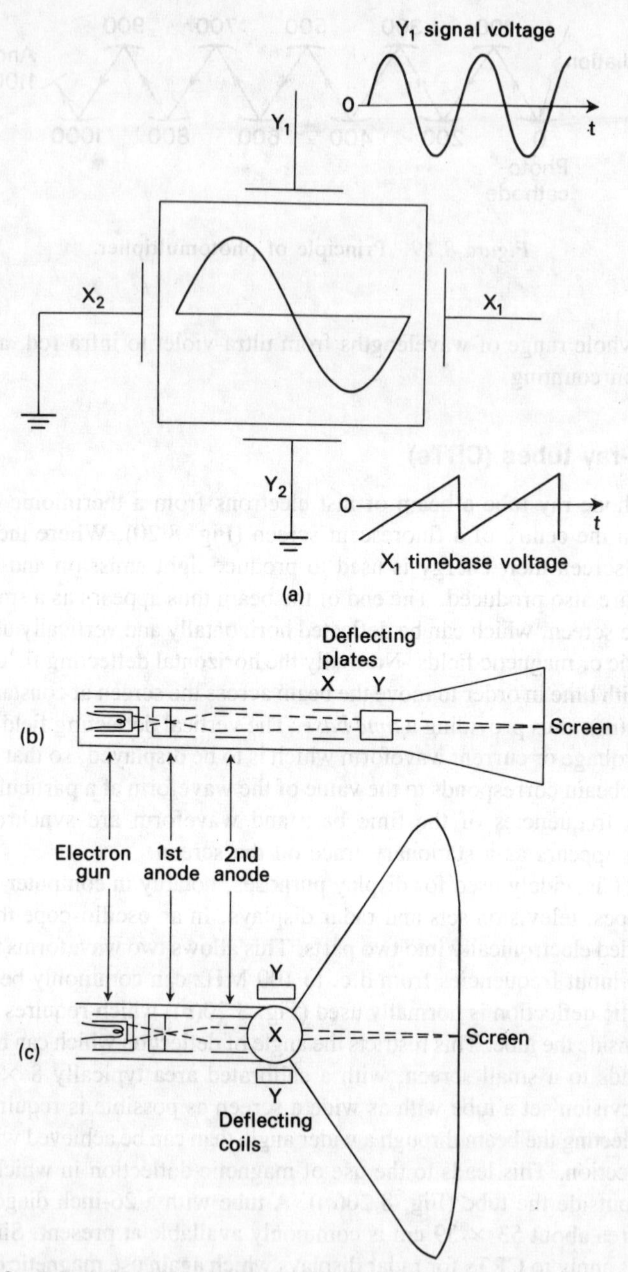

Figure 8.20 Principles of cathode-ray tubes (CRTs). (*a*) Signal and time base voltages for an electrostatic CRT. In practice the voltages may be divided between X_1 and X_2 and Y_1 and Y_2 to avoid distortion; (*b*) plan view of a CRT with electrostatic deflection; (*c*) plan view of a CRT with electromagnetic deflection.

by means of an external coil is also used, as in microwave valves and the electron microscope where the inside of the tube must be unrestricted. Since this requires a heavy current less power is consumed by electrostatic focusing which is used wherever possible.

The deflecting fields are perpendicular to the tube axis in each type of tube and consist either of two pairs of plates (Fig. 8.20(*a*)) or two pairs of coils (Fig. 8.20(*b*)). In each case deflection is provided in X and Y directions and again the actions of electrostatic and magnetic deflection are discussed in a later section.

The fluorescent screen is an insulator such as zinc orthosilicate, giving a green trace, or zinc sulphide, giving a white trace (Fig. 8.21). When the electron beam strikes the screen emission of both light and secondary electrons occurs. Provided that $\delta > 1$, a condition which depends on the final anode potential relative to the cathode, more electrons are emitted from the screen that arrive at it so that the screen takes up a potential slightly more positive than the final anode. The secondary electrons are attracted to a graphite coating on the inside walls of the glass bulb of the tube, which is held at final anode potential. If this process did not take place the screen would charge up negatively and serious defocusing of the primary beam would occur.

Electrostatic deflection

Consider a narrow beam of electrons which have been accelerated through a potential V_A. Their velocity in the x-direction (Fig. 8.22(*a*)) is given by

$$u_x = \sqrt{\frac{2eV_A}{m}} \tag{8.44}$$

Suppose the beam passes between two parallel deflecting plates, having separation s, and finally strikes a fluorescent screen as would occur in a cathode-ray oscilloscope. If there is no potential difference between the plates the beam is unaffected, but if a potential V exists between them, which sets up an electric

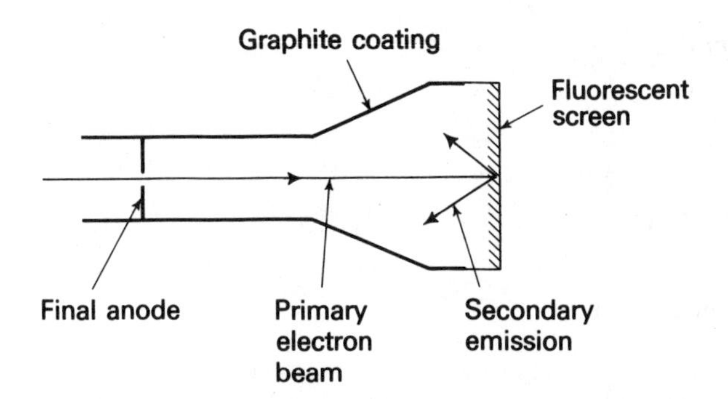

Figure 8.21 Secondary emission in a cathode-ray tube.

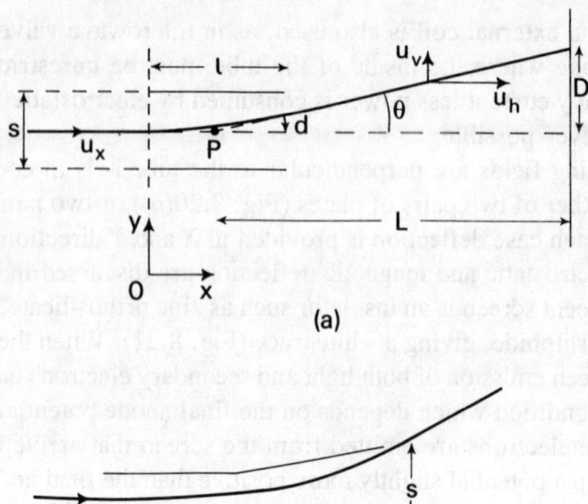

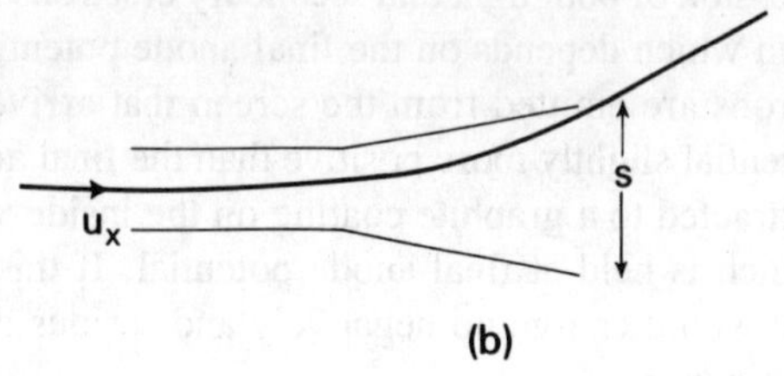

Figure 8.22 Electrostatic deflection. (*a*) Flat deflecting plates; (*b*) flared deflecting plates.

field assumed to be uniform over the length l of the plates, the electrons will be deflected and strike the screen at a distance D from the undeflected position.

The electrons will have acquired a vertical velocity u_v as they leave the plates if a potential V exists between them. If their horizontal velocity after leaving the plates is u_h then their total energy is unchanged after passing through the plates, so that

$$\tfrac{1}{2} m(u_v^2 + u_h^2) = \tfrac{1}{2} mu_x^2 \qquad (8.45)$$

Since u_v^2 and u_x^2 are proportional to V and V_A respectively, then provided that $V \ll V_A$, which is normally the case in a cathode-ray tube, $u_h \approx u_x$.

The deflecting force is given by

$$F_y = \frac{m \, du_y}{dt} = eE = e\frac{V}{s} \qquad (8.46)$$

Hence

$$u_y = \frac{e}{m}\frac{V}{s}t \qquad (8.47)$$

and, after integration,

$$y = \frac{1}{2}\frac{e}{m}\frac{V}{s}t^2 \qquad (8.48)$$

But $t = x/u_x$, so that

$$y = \frac{1}{2}\frac{e}{m}\frac{V}{s}\frac{x^2}{u_x^2} \tag{8.49}$$

Thus $y \propto x^2$ and the path of the electrons between the plates is a parabola, taking $x = 0$ at the left-hand edge of the plates.

The slope of the parabola at $x = l$ is

$$\frac{dy}{dx} = \frac{e}{m}\frac{V}{s}\frac{l}{u_x^2} = \tan\theta \tag{8.50}$$

where d is the deflection of the beam as it leaves the plates. The distance of point P from the end of the plates is $d/\tan\theta$. But

$$d = \frac{1}{2}\frac{e}{m}\frac{V}{s}\frac{l^2}{u_x^2} \tag{8.51}$$

so that

$$\frac{d}{\tan\theta} = \frac{l}{2} \tag{8.52}$$

Hence P is at the centre of the plates and distant L from the screen. Then the final deflection is given by

$$D = L\tan\theta = \frac{eVlL}{msu_x^2} \tag{8.53}$$

using eq. (8.50), so that

$$D = \frac{1}{2}\frac{V}{V_A}\frac{lL}{s} \tag{8.54}$$

using eq. (8.44).

Thus $D \propto V$, and if V is sinusoidal D will correspond to the instantaneous value. The upper frequency is limited by the time an electron spends between the plates.

Worked example

In a cathode-ray tube with electrostatic deflection the length of the deflecting plates is 2 cm and the anode voltage is 1000 V.

Calculate the time spent by an electron between the deflecting plates and estimate the maximum frequency for which the displayed voltage may be assumed equal to the instantaneous value.

Solution

From eq. (8.44) the electron velocity

$$u_x = (2 \times 1.76 \times 10^{11} \times 10^3)^{1/2}$$
$$= 1.876 \times 10^7 \text{ m/s}$$

An electron spends $\dfrac{2 \times 10^{-2}}{1.876 \times 10^7}$ seconds or 1.07×10^{-9} s between the plates. If the period of a signal is assumed to be 1000 times this value for an accurate display then the maximum frequency is approximately 1 MHz.

The deflection sensitivity is given by

$$\frac{D}{V} = \frac{1}{2}\frac{lL}{V_A s} \tag{8.55}$$

and is often quoted in milimetres per volt. It is proportional to the length of the plates and their distance from the screen, which are both limited by the required size of the tube. It is inversely proportional to V_A, but the electron velocity and brightness of the spot increase with V_A. The sensitivity is also increased by bringing the plates closer together, but the beam must not foul the edge of the plates. Consequently they are often flared to prevent this, the final separation being the same as for the parallel plates (Fig. 8.22(*b*)). The deflecting voltage V is normally applied to the plates through a phase-splitting amplifier, so that one plate is at a potential of $+V/2$ and the other at $-V/2$. This avoids distortion due to the position of zero voltage not being on the tube axis.

Motion in a uniform magnetic field

The force F on a conducting element length ds carrying a current I in a magnetic field of density B making an angle θ with the conductor (Fig. 8.23(*a*)) is given by

$$F = Bi \, ds \sin\theta \tag{8.56}$$

An electron moving with velocity u corresponds to a current moving in the opposite sense to a conventional current and also experiences a force in a magnetic field. From eq. (8.56)

$$F = B\frac{dq}{dt} ds \sin\theta$$
$$= B \, dq \frac{ds}{dt} \sin\theta \tag{8.57}$$
$$= Beu \sin\theta$$

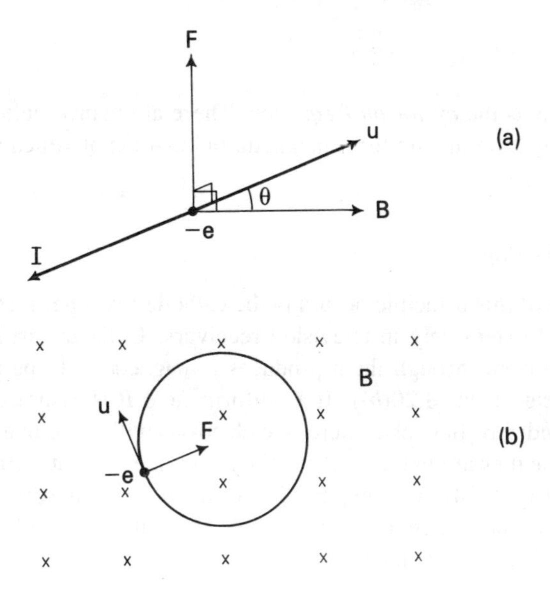

Figure 8.23 Electron in a uniform magnetic field. (*a*) Force acting on an electron; (*b*) circular motion of an electron.

The force acts at right angles to the directions of both B and u, and for electrons the relative directions are given by the right-hand rule. It should be noted that B can only alter the direction of u and not its magnitude, since the resulting force is perpendicular to the motion of the electron. The force on a stationary electron is zero and the maximum force occurs when $\theta = 90°$. Here $F = Beu$, and since it is always perpendicular to the direction of motion the electron moves in a circle (Fig. 8.23(*b*)). If r is the radius of the circle,

$$F = Beu = \frac{mu^2}{r} \tag{8.58}$$

and

$$r = \frac{mu}{Be} \tag{8.59}$$

The time for a complete orbit is $2\pi r/u$, so that the period of the motion is given by

$$T = \frac{2\pi m}{Be} \tag{8.60}$$

This depends only on the magnetic field B, since the radius increases linearly with the velocity. The angular frequency is given by

$$\omega_c = \frac{Be}{m} \tag{8.61}$$

which is known as the *cyclotron frequency*. There are many applications of the motion of an electron in a uniform magnetic field, some of which are discussed below.

Magnetic deflection

An application of this principle occurs in the cathode-ray tube, magnetic deflection being used extensively in television receivers. Coils are arranged outside the tube and current through them produces a magnetic field perpendicular to the electron beam (Fig. 8.20(*b*)). If a uniform field B is assumed over length l of the coils and zero field elsewhere, the electrons will move in a circular path of radius r while they are in the field and then continue without further deflection to the screen (Fig. 8.24). The length L is measured from the screen to the point of intersection of the tangent to the circle with the path of the undeflected beam. Then, for small angles of deflection θ,

$$\tan\theta = \frac{D}{L} = \frac{l}{r} \tag{8.62}$$

From eq. (8.59) $r = mu_x/Be$ so that

$$D = \frac{Ll}{r} = \frac{LlBe}{mu_x} \tag{8.63}$$

If V_A is the final anode voltage, $u_x = \sqrt{(2eV_A/m)}$, so that

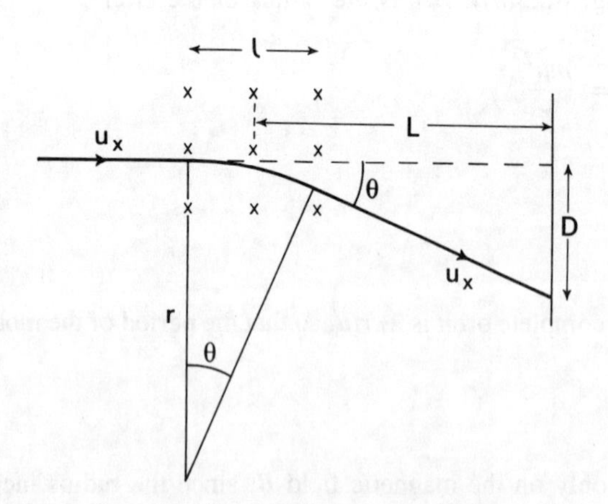

Figure 8.24 Magnetic deflection.

$$D = \frac{BLl}{\sqrt{(2V_A)}} \sqrt{\frac{e}{m}} \tag{8.64}$$

The value of B depends on the coil design and is proportional to the current, so that the deflection sensitivity is normally quoted in millimetres per milliampere of coil current. The sensitivity depends on $\sqrt{V_A}$ and so a larger accelerating potential may be used than in electrostatic deflection, where the sensitivity depends on V_A. This results in a greater spot brilliance on the screen.

Thomson's determination of *e/m*

J.J. Thomson used combined electric and magnetic fields to determine the value of e/m for cathode rays (electrons) (*see* Fig. 1.1). The magnitudes of the two fields were adjusted until the net deflection of the beam was zero, so that $Beu_x = eE$, or

$$u_x = \frac{E}{B} \tag{8.65}$$

The electric field was then removed and the deflection D due to the magnetic field alone was measured. The angle of deflection was given by

$$\tan\theta = \frac{Bel}{mu_x} = \frac{D}{L} \tag{8.66}$$

Hence

$$\frac{e}{m} = \frac{Du_x}{BlL} = \frac{DE}{B^2lL} \tag{8.67}$$

using eq. (8.65), or, since $E = V/s$,

$$\frac{e}{m} = \frac{DV}{B^2lLs} \tag{8.68}$$

The mass spectrometer

The mass spectrometer is a device which can be used to measure e/m for positive ions (Fig. 8.25). Ions are accelerated from the source to the slit S_1, beyond which a magnetic field B exists. This causes the ions to move in circular paths whose radii depend on their value of e/m (eq. 8.59). Where the path corresponds with the positions of slits S_2 and S_3 the ions can pass through but others are rejected. Those reaching the collector give rise to a current which can be amplified and measured, and the value of e/m is obtained by combining eqs (8.12) and (8.59) to give

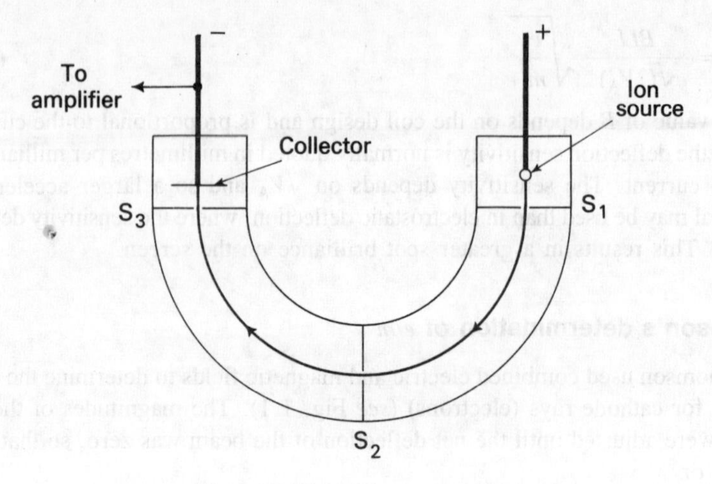

Figure 8.25 Mass spectrometer.

$$\frac{e}{m} = \frac{2V}{B^2r^2} \tag{8.69}$$

The accelerating voltage V is adjusted for maximum collector current, and B and r are also known. The mass spectrometer can also be used to detect the presence of an ion having a known value of e/m in a vacuum system.

Magnetic focusing

So far, the magnetic field B has acted at 90° to the direction of electron motion. Suppose it acts at angle θ, so that the electron velocity u may be resolved into two components, $u \cos \theta$ parallel to the field and $u \sin \theta$ perpendicular to the field (Fig. 8.26(*a*)). There can be no force on the electron due to the first component, since it has a zero angle with respect to B, but the second component will give circular motion of radius $mu \sin \theta/Be$ (Fig. 8.26(*b*)). At the same time the electron is moving parallel to the field with velocity $u \cos \theta$, so that it follows a helical path. Thus the electron will cross the x-axis after a complete revolution at points P_1, P_2, . . . (Fig. 8.26(*c*)) and where T is the period of the circular motion, and p is the pitch of the spiral, the distance between consecutive points such as P_1 and P_2

$$OP_1 = Tu \cos \theta \tag{8.70}$$

or

$$p = \frac{2\pi mu \cos \theta}{Be} \tag{8.71}$$

using eq. (8.60).

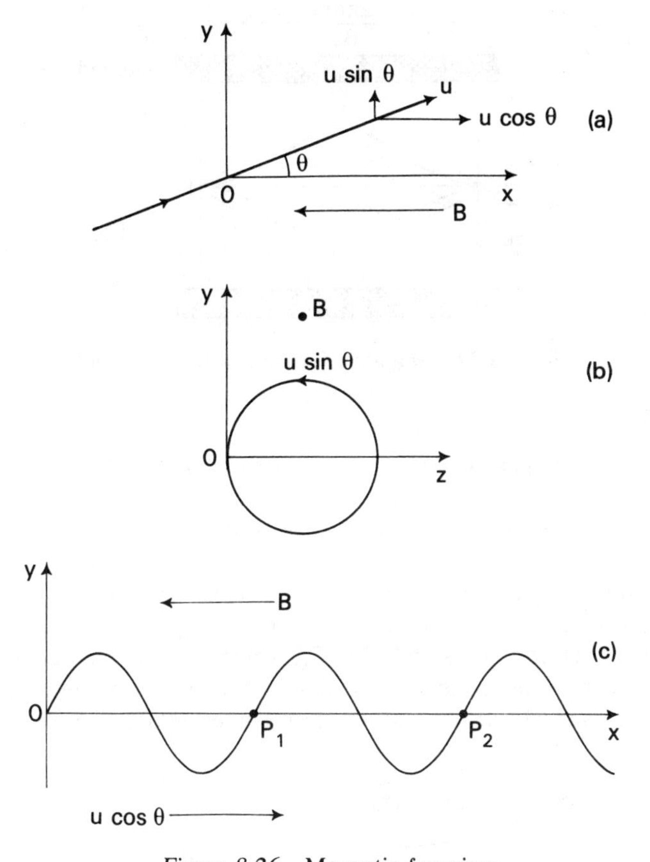

Figure 8.26 Magnetic focusing.

When θ is small, $\cos\theta \approx 1$, and for a stream of electrons with the same velocity u but diverging within a small angle, p will be approximately constant. For a circular aperture in the anode a solenoid may be used to focus the beam to a spot on a screen at P, provided the effective source is within the magnetic field of the solenoid (Fig. 8.27).

Electron optics

Electron optics is concerned with the production of electron beams which are focused to a narrow diameter by means of magnetic or electric fields. The use of *electron lenses* is analogous to the control of light beams by means of lenses and apertures. The focused electron beam may be used in a cathode-ray tube, in the electron microscope which has greater resolution than an optical microscope, or for welding and etching purposes.

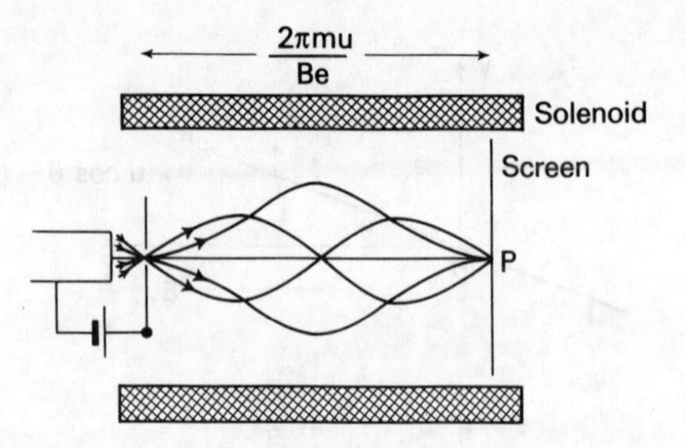

Figure 8.27 Magnetic focusing with a solenoid.

Magnetic lens

Consider a short coil carrying a current as shown in Fig. 8.28(*a*). The magnetic field due to the current can be resolved into a radial component B_r and an axial component B_x. B_r is directed towards the axis on the left-hand side, becomes zero in the plane of the coil and is directed away from the axis on the right-hand side. An electron moving parallel to the axis with velocity u_x will be unaffected by B_x, but B_r will cause it to move sideways out of the paper (Fig. 8.28(*b*)). This component of velocity, u_s, is normal to B_x, which will give the electron another component u_r towards the axis. As it passes through the centre of the coil B_r reverses and gradually reduces u_s, which becomes zero as the electron leaves

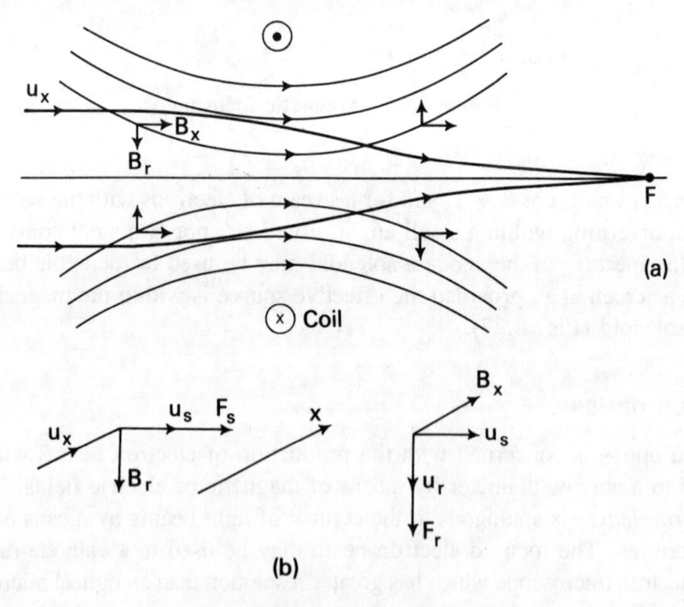

Figure 8.28 Magnetic lens.

the lens. The electron continues to travel in a straight line towards the axis intersecting it at F, the focal point of the lens. A similar argument shows that a diverging beam is first brought parallel to the axis and then converges towards it intersecting at a point beyond F. The effect of the coil on an electron beam is thus analogous to the effect of a convex lens on a beam of light. Magnetic lenses are used extensively in the electron microscope (*see* Fig. 8.32).

Electrostatic lens

Consider an electric field with two regions separated by a very small gap s in which the potential changes from a constant value V_1 to a higher value V_2 (Fig. 8.29). The two potential regions are separated by parallel equipotential surfaces between which there is a field E given by

$$E = \frac{V_2 - V_1}{s} \tag{8.72}$$

Suppose an electron moves in the first region in a direction making an angle i with the normal to the gap and with constant velocity $u_1 = \sqrt{(2eV_1/m)}$. The velocity may be resolved into two components $u_1 \sin i$ and $u_1 \cos i$. At the gap $u_1 \sin i$ will be unaffected, but $u_1 \cos i$ will be increased by the field E so that as the electron enters the second region its velocity will be $u_2 = \sqrt{(2eV_2/m)}$. If the direction of u_2 makes an angle r with the normal, the two components are $u_2 \sin r$, and $u_2 \cos r$, which is greater than $u_1 \cos i$. The components parallel to the gap are unaffected by the field so that

$$u_1 \sin i = u_2 \sin r \tag{8.73}$$

or

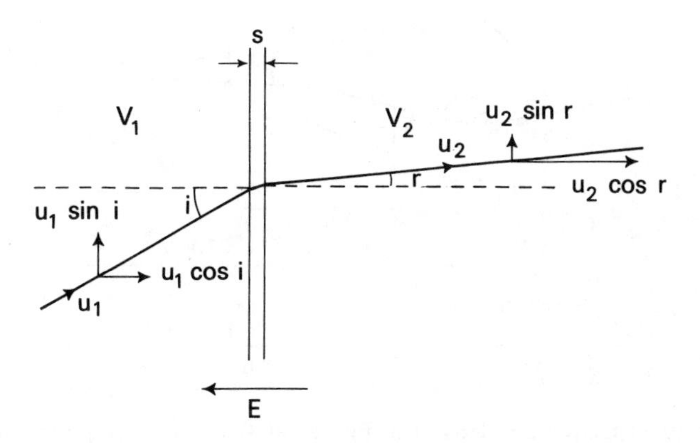

Figure 8.29 Refraction of an electron beam.

$$\frac{\sin i}{\sin r} = \frac{u_2}{u_1} = \sqrt{\frac{V_2}{V_1}} \tag{8.74}$$

A *refractive index* (n) for electrons may be defined where

$$\frac{\sin i}{\sin r} = \frac{n_2}{n_1} = n_{12} \tag{8.75}$$

or

$$n_{12} = \sqrt{\frac{V_2}{V_1}} \tag{8.76}$$

However, in practice a discontinuous change in V is not possible and V changes more gradually from one region to another. This is analogous to a changing optical refractive index.

A practical electrostatic lens consists of two coaxial cylinders, at different potentials V_1 and V_2, with a gap between them. The electric field is of the type shown

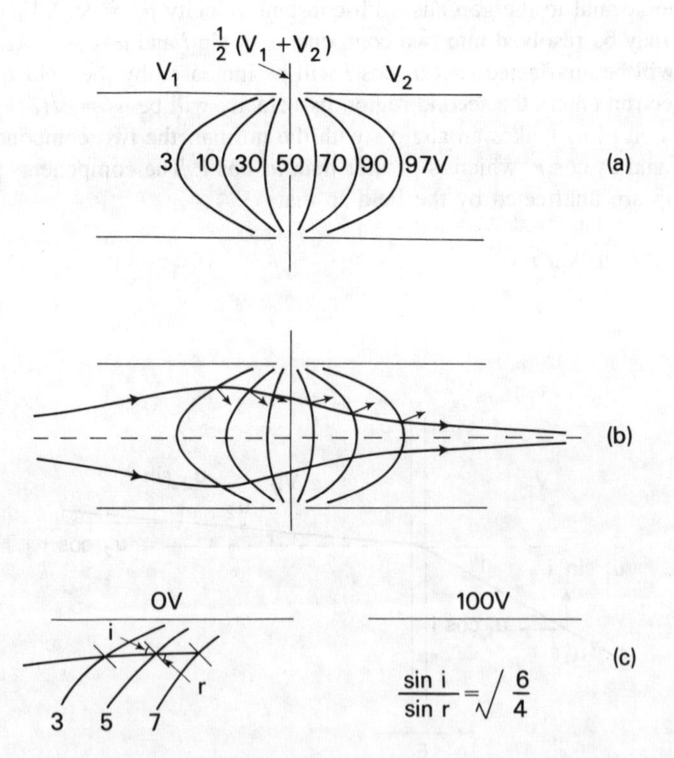

Figure 8.30 Electrostatic lens. The figures at (a) refer to equipotentials when $V_1 = 0$, $V_2 = 100$ V.

in Fig. 8.30(*a*), where the equipotentials correspond to $V_1 = 0$, $V_2 = 100$ V and may be obtained experimentally using an analogue of the lens (Ref. [2]). The electron enters the cylinder of lower potential first and the electrostatic force on it acts at right angles to the equipotential line. Thus an electron moving in a path divergent from the axis will experience forces on the low-potential side of the mid-plane which move it towards the axis. Beyond the mid-plane forces will tend to move it away from the axis, but since they are more nearly parallel to the axis than before the electron continues to move gradually towards it. Again the system acts like a converging lens whose focal length may be reduced by increasing $V_2 - V_1$.

The action of the lens may also be explained by considering the gradual increase in potential through the lens as a corresponding increase in refractive index (Fig. 8.30(*c*)). The angle through which the electron path is bent at each equipotential is given by eq. (8.74), where V_1 and V_2 refer to the mean potential between two equipotentials.

An aperture and an anode at a higher potential may also be used as a lens. An example is the electron gun of a cathode-ray tube, where the aperture is in the *control electrode* held at a negative potential with respect to the cathode (Fig. 8.31(*a*)). The equipotentials between cathode and anode ensure a focusing action on the beam of electrons leaving the cathode, and the potentials can be adjusted to bring the beam to a focus at the anode. The beam is finally focused at the screen by means of a second anode which with the first one forms a lens (Fig. 8.31(*b*)).

The electron microscope

The limit of usefulness of a microscope is set by the distance between two points on a specimen which can be seen as just separated at the eyepiece, which is known as the *resolving power*. Due to the finite wavelength of light the image of a point is spread into a pattern, so that even if the lens system is perfect the patterns of two points very close together will overlap to give a single blurred image. The size of the pattern increases with the wavelength and for a typical laboratory microscope the resolving power is about 500 nm with white light. This means that two points which have a separation less than about 500 nm will appear as a combined image.

This fundamental limitation can only be overcome by using radiation of a shorter wavelength, provided by a beam of electrons which is associated with a de Broglie wavelength (Chapter 1). From eq. (8.11) the electron velocity is given by

$$u = \sqrt{\frac{2eV}{m}}$$

where V is the anode voltage. From eq. (1.12) the electron wavelength is given by

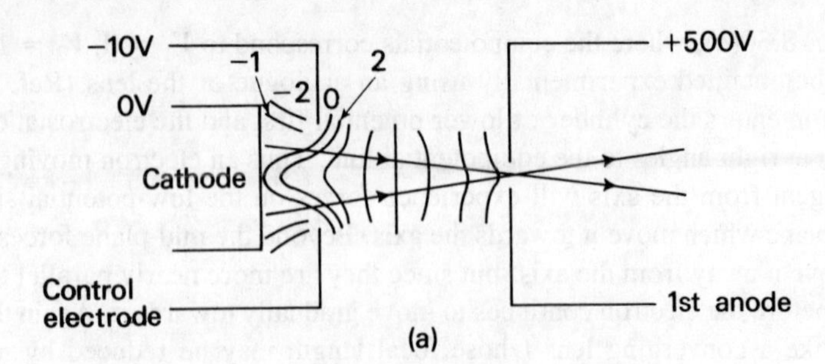

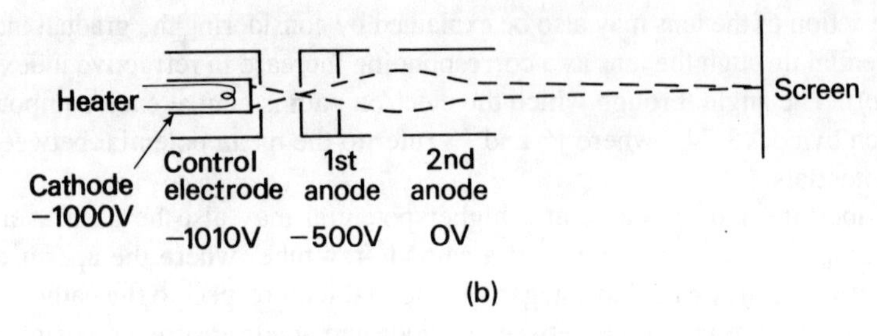

Figure 8.31 Electron gun.

$$\lambda = \frac{h}{mu} = \frac{h}{\sqrt{2emV}}$$

Thus

$$\lambda = \frac{1.17 \times 10^{-15}}{\sqrt{mV}} \text{ nm}$$

after substituting for h and e, while m is given by eq. (8.17) and Fig. 8.5. The electrons from a gun having an anode voltage of 100 kV have a wavelength of 3.5 pm, five orders of magnitude less than the wavelength of visible light, so that a great improvement in resolving power is to be expected. The first transmission electron microscope was built in 1932 using magnetic lenses and has been developed considerably since then, with the scanning electron microscope appearing in 1964.

The general layout of a modern transmission electron microscope is shown in Fig. 8.32; it is analogous to an optical microscope. The electron source is a heated tungsten filament within the electron gun and the beam is focused on to the specimen by means of a condenser lens. As in the optical microscope a transparent specimen is used which normally has a thickness between about 10 and 100 nm. The electron beam may be restricted by an aperture of diameter about 25 μm, and after passing through it the specimen enters the objective lens to form an image which is further magnified by the projector lens or 'eyepiece'. Finally the

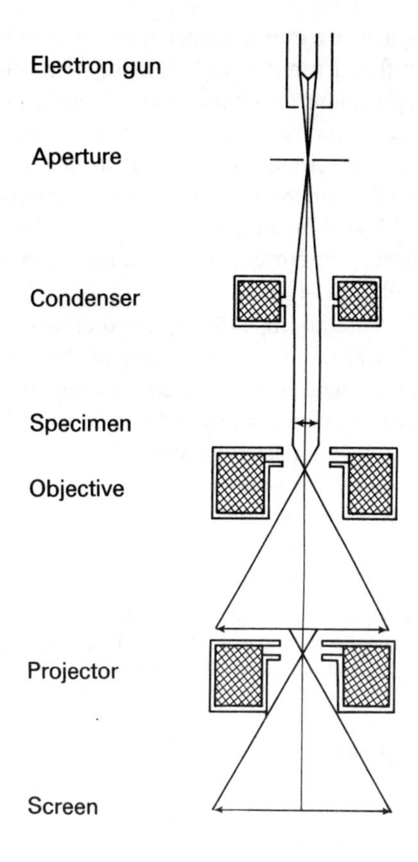

Figure 8.32 Transmission electron microscope.

electrons strike a fluorescent screen to give a visible magnified image of the specimen. The screen may be replaced by a photographic plate which is sensitive to electrons as well as to light.

Each lens consists of a solenoid encased in soft iron and has a soft-iron pole-piece to concentrate the magnetic field. The focal length of the lenses is typically a few millimetres and is adjusted by varying the current through the solenoid. The best resolution is only obtained when both this current and the electron-gun voltage are held constant to within about 1 in 10^6. Then the resolving power is in the range 0.1–1.0 nm, which is a great improvement on the optical microscope but far from the theoretical limit due to imperfections in the lenses (Ref. [3]). Electron microscopes having accelerating voltages up to 1 MV are also available, not only to improve the resolving power but also to allow examination of specimens with thicknesses up to 1000 nm. These have properties more representative of the bulk material than the very thin specimens. The energy lost by the electron beam in passing through even a thin specimen leads to local heating if it is a poor thermal conductor, so that a replica is often made of a refractory material.

Among the many applications of the transmission electron microscope (TEM) an important one is to examine the structure of an integrated circuit (Chapter 7), using a thin section from the circuit. The higher the beam energy the greater is the penetration and the thicker the section can be. For example, the thickness can be 0.5 μm for an 80 keV beam and 1.5 μm for a 200 keV beam. Typical magnifications for electron microscopes with three or more stages are about 500 000 × with a resolution of about 0.2 nm.

In the scanning electron microscope (SEM) a two-dimensional raster is applied to the beam so that it scans across the surface of the specimen (Fig. 8.33). Bombardment by the electron beam causes secondary and primary (backscattered) electrons to be produced from the surface and also X-rays. The secondary or the backscattered current from the detector modulates the intensity of an electron beam in a cathode-ray tube. Two-dimensional movement of the CRT beam is synchronized with the SEM raster so that an image of the specimen surface is formed. Necessary contrast in the image is due to variations in the secondary or the backscattered electron flux from the specimen. The X-rays can provide a chemical analysis of any part of the specimen when peaks in the detector signal are examined. Typical beam energies are 2 to 40 keV and resolutions better than

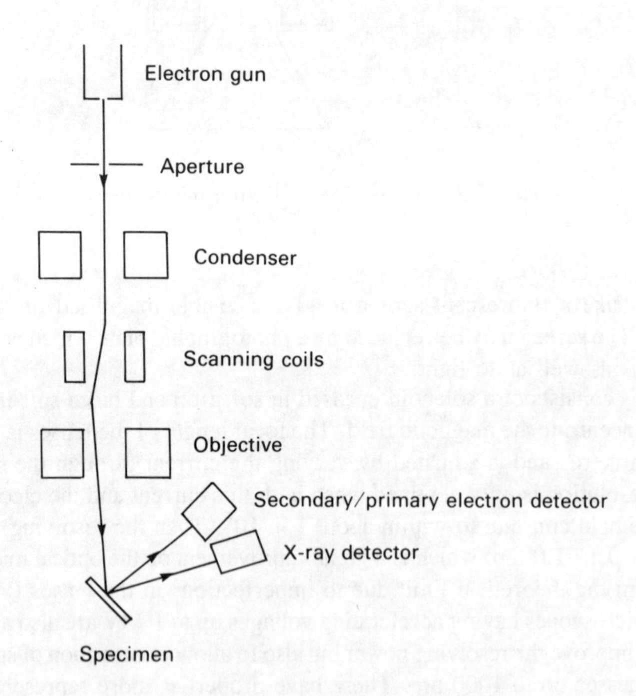

Figure 8.33 Scanning electron microscope.

10 nm can be achieved for an electron beam diameter of 4 nm. Typical depths of field are 2 to 4 μm at 10 000 × magnification and 0.2 to 0.4 nm at 100 × magnification.

The SEM has fewer and simpler lenses than the TEM but has more electronic circuitry, leading to a lower overall cost. With an SEM a specimen can be examined and analysed point by point, which is useful for relatively large specimens like integrated circuits.

Further details on electron microscopy and its applications are given in Ref. [3]. The use of electron microscopy and other diagnostic techniques for integrated circuits are described in Ref. [7] of Chapter 7.

The cold cathode discharge

Let us consider a planar diode with a photosensitive cathode containing gas at a low pressure. If light falls on the cathode photoelectrons are released, so that as the anode voltage is raised from zero the anode current rises with it. When all the available electrons are flowing to the anode, the current becomes independent of the voltage and so is saturated, at a value I_0 dependent on the light intensity (Fig. 8.34). However, when V_A reaches the value V_i, the ionization potential of the gas, the current begins to rise once more with voltage. This is because when $V_A = V_i$ molecules in the layer of gas next to the anode may be ionized, so that one colliding electron can cause another electron to be produced. When $V_A > V_i$ ionization can occur away from the anode at a point where V_A

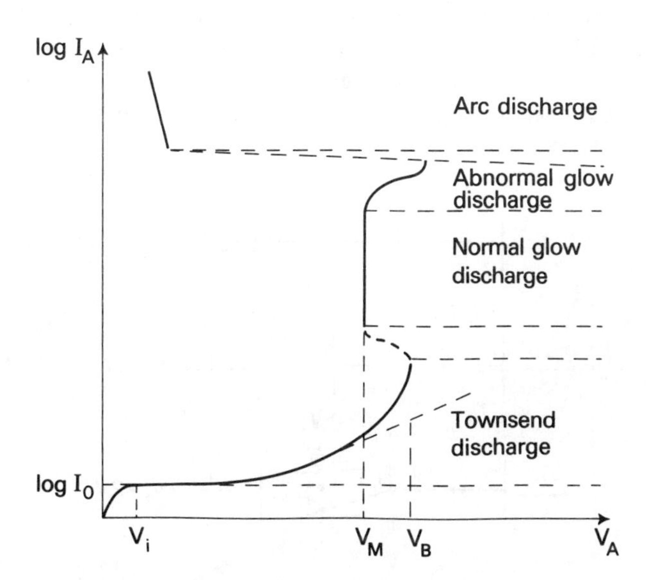

Figure 8.34 Idealized characteristic of a cold-cathode discharge.

$= V_i$, and when $V_A > 2V_i$ both electrons can take part in further ionizing collisions (Fig. 8.35). Thus the presence of the gas has caused the increase of current, which is multiplied in a similar way to the reverse current in a semiconductor diode (Chapter 3). The effect is known as a *Townsend electron avalanche*, after J.S. Townsend, an early investigator. The anode current rises with voltage in the Townsend discharge region of Fig. 8.34 until at $V_A = V_B$, the breakdown voltage becomes *self-sustained*, i.e. I_A is maintained even when the source illuminating the cathode is removed.

It has been found that secondary electrons are released from the cathode due to the bombardment by positive ions from the gas, and at V_B each electron released from the cathode causes the release of one more electron by secondary emission. After breakdown has occurred a plasma is formed, as in the case of the hot-cathode discharge. Consequently, the anode voltage required to maintain the current drops, but to a value V_M well above the ionization potential, and a *normal discharge* is set up. The cathode fall in voltage is relatively high, about 100 V, and the current density is up to about 1 A/cm^2 in the normal glow discharge. A neon-filled tube for example has $V_A = 115$ V and can be used as a voltage reference source, although a semiconductor breakdown diode is normally preferred.

Within the normal glow discharge electrons are excited to high energy levels and light is emitted from the discharge as the electrons fall to lower levels. This characteristic glow is now the main electronic application of the cold cathode discharge. For instance, a neon-filled tube with a built-in series resistor can be directly connected across the 240 V a.c. supply to provide a convenient *mains indicator*. In this case the current changes direction on each half cycle. Again, a *seven-segment gas discharge device* is available, requiring a 200 V d.c. supply but providing a large, bright display. Finally the *plasma display panel* consists of a 512 × 512 matrix of gas discharges providing points of light about 1 mm apart (Ref. [4]). Such a flat panel provides a convenient alphanumeric display

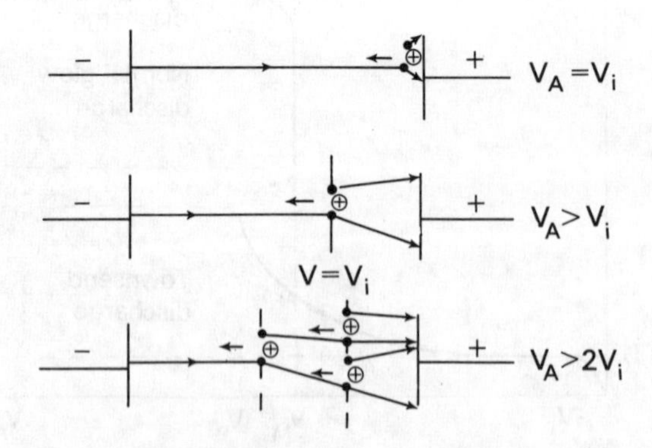

Figure 8.35 Formation of an electron avalanche.

which is more rugged and requires less space than a CRT and so is particularly suitable for military applications.

If the current in a cold cathode discharge is allowed to increase until the glow completely covers the cathode the voltage between the electrodes first rises and then falls to about the ionization potential. The corresponding regions of the characteristic are called the *abnormal glow discharge* and the *arc*. The arc is concentrated into a small area of the cathode, and has a high current density, around 10^7 A/m^2, and a low voltage of about 20 V which falls further as the current is allowed to rise, giving a negative resistance.

Points to remember

* Electron emission from a surface can occur due to photo-, thermionic, field and secondary emission.
* The cathode-ray tube uses thermionic emission and is an essential part of a computer terminal (VDU), an oscilloscope and a television set at present.
* The photomultiplier uses both photo- and secondary emission from a cold cathode to provide low-noise image sensing.
* The electron microscope uses thermionic emission and electron optics and has no solid-state equivalent.
* Cold-cathode gas-discharge devices are used in plasma displays, which provide an alternative to the cathode-ray tube for communicating text and graphics.

References

[1] Oman, M.F. *Introduction to the physics of electronics*, (Prentice Hall), 1974.
[2] Griffiths, D. *Introduction to electrodynamics*, (Prentice Hall), 1981.
[3] Watt, I.M. *The principles and practice of electron microscopy*, (Cambridge University Press), 1985.
[4] Slottow, H.G. 'Plasma displays', *IEEE Transactions on Electron Devices, ED-23*, 1976, p. 766.

Problems

8.1 A photocathode is illuminated with radiation of wavelength 500 nm. The cathode has a work function of 1.2 eV. Calculate the anode voltage required to produce zero anode current.

When the anode voltage is +90 V find the velocity of the electrons at the anode

if the cathode is illuminated with radiation of wavelength 250 nm.
[−1.29 V; 5.74 × 10^6 m/s]

8.2 A photomultiplier has a cathode with a work function of 1.5 eV and ten dynodes each with a secondary emission coefficient of 6. If radiation is incident on the cathode, calculate (i) the maximum wavelength for which collector current will flow; (ii) the maximum initial electron velocity if the wavelength of the radiation is 0.6 μm; (iii) the final collector current if the cathode current is 10^{-10} A.
[0.83 μm; 4.5 × 10^5 m/s; 6.1 mA]

8.3 Two large parallel metal plates are horizontal with separation 5.0 mm and the upper plate is held at +150 V with respect to the lower plate. An electron with initial velocity 10^6 m/s upwards is released at the centre of the lower plate. Calculate (i) the velocity of the electron when it strikes the upper plate; (ii) the time of flight of the electron; (iii) the amount of energy conveyed to the upper electrode.
[7.32 × 10^6 m/s; 1.2 ns; 153 eV]

8.4 In a cathode-ray tube the deflecting plates are 1 cm long, 0.3 cm apart and 20 cm distant from the screen. The final anode voltage is 1 kV and a deflecting potential of 35 V is applied. Calculate the spot deflection on the screen.
[1.167 cm]

8.5 A solenoid of 945 turns is 30 cm long and it is assumed to produce a magnetic field which is uniform over the entire length of the solenoid. An electron travelling in a vacuum is accelerated through a potential difference of 500 V and then enters one end of the solenoid, crossing the axis at an angle of 5°. Calculate (i) the minimum and (ii) the next higher value of solenoid current that will make the electron cross the axis again at the other end of the solenoid.
[400 mA; 800 mA]

9 Microwave devices and electrical noise

Microwave devices

Microwave frequencies extend approximately from 300 MHz to 300 GHz, with the range above 30 GHz being known as millimetre waves since the wavelength is less than 1 cm. In the microwave frequency range the simple wire connections between the various parts of an electronic circuit are no longer adequate, and components such as resistors, capacitors and inductors (*lumped* components) no longer behave as predicted at lower frequencies. This is because the dimensions of an amplifier *circuit*, for instance, are comparable to the wavelength of the signals, which ranges from about 1 m down to about 1 mm. Thus there are appreciable phase differences between the ends of connections, since a wire 10 cm long represents 30% of a wavelength at 1 GHz but only 0.03% of a wavelength at 1 MHz, so that at the lower frequency, phase change along the wire may be ignored. In addition, when lumped components are used the loss of energy by radiation from the circuit becomes excessive and its operation becomes very inefficient.

These problems are overcome by confining the signals within hollow conducting tubes of rectangular or circular cross-section, known as *waveguides*. These prevent energy loss by radiation and can fulfil the circuit functions of resistors, capacitors and inductors by suitable choice of their length and termination. However, up to about 3 GHz coaxial transmission lines have sufficiently low attentuation and provide more flexible interconnections than waveguides, while also preventing radiation loss, and similar considerations apply to optical fibres (Chapter 5). Other microwave components, such as the resonant cavity which behaves like a tuned circuit, are introduced below in connection with electronic devices.

In conventional electronic devices the transit time is about 1 ns from emitter to collector, source to drain or cathode to anode in a bipolar transistor, field-effect transistor or small triode respectively. Thus when the operating frequency is above about 100 MHz the transit time is comparable to the signal period. Also the capacitances between electrodes and the inductances of leads between electrodes and the external circuit all have appreciable reactance. It may be shown that both transit time and stray components (*a*) reduce the input impedance and (*b*) reduce the gain of an amplifier.

One approach to these problems is to reduce the dimensions of conventional transistors, which in turn reduces both the transit time and the stray component values. In this way BJTs operating up to about 2 GHz and FETs operating up to about 20 GHz can be produced, as described in Chapters 4 and 6 respectively, and Schottky diodes are used as mixers up to frequencies over 100 GHz.

Another approach is to apply different operating principles to those used at lower frequencies. At the present time both thermionic valves and semiconductor devices are used in this category. Microwave valves, which in order of importance are travelling wave tubes, klystron amplifiers, magnetrons and klystron oscillators, are described below. They are generally applied where high power at high frequency is required, for example over 500 W at above 10 GHz. The semiconductor devices are described later and can be used for all low power applications below 100 GHz.

Microwave valves

The general principle used in microwave valves, such as the klystron and the travelling-wave tube described below is that the kinetic energy of a beam containing a high density of electrons is converted into energy at microwave frequencies. In order to achieve efficient conversion the beam must be prevented from spreading due to the mutual repulsion of the electrons; beam confinement can be achieved by means of a magnetic field. This will be described qualitatively; an analysis of the method (known as Brillouin focusing) is given in Ref. [1].

The general arrangement is shown in Fig. 9.1(*a*), the permanent magnet providing a non-uniform magnetic field near its poles and a uniform magnetic field B_0 along its axis. The beam is concentrated from a concave cathode by means of the beam-forming electrode and emerges from the electron gun with a radius equal to that of the anode aperture, a. Owing to the shape of the lines of magnetic flux at the ends of the solenoid there will be a radial component of the field B_r (Fig. 9.1(*b*)), so that an electron on the outside of the beam will receive an impulse directing it out of the page as it enters the magnetic field. This will cause the electron to move in a circular path in a plane perpendicular to the page, and this motion combined with its longitudinal motion will cause it to spiral about the axis. The effect is similar to the magnetic focusing described in Chapter 8, but detailed analysis shows that the angular frequency of rotation about the axis is *half* that expected from a magnetic field B_0 in a simple focusing system.

B_0 is given by

$$\frac{6.9 \times 10^{-7}\, I_0}{V_2^{1/2}\, r^2} \text{ tesla}$$

where I_0 is the beam current, V_0 the potential at the axis of the beam, and $r(\leq a)$

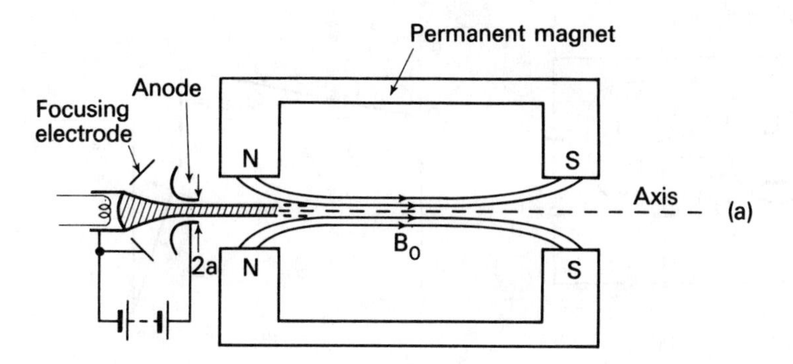

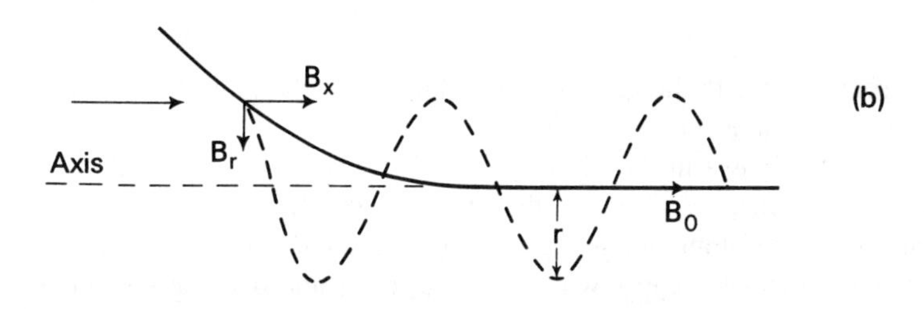

Figure 9.1 Magnetic focusing with a permanent magnet.

the radius of the beam envelope. In practice the optimum field is between 1.5 and 2.0 B_0 since it is difficult to inject the beam correctly into the magnetic field.

There are now two forces on the electron acting away from the axis, owing to the mutual repulsion of the electrons and their circular motion respectively. The magnetic field B_0 provides an opposing force acting towards the axis, so that the radius of the spiral can equal the radius of the aperture of the electron gun. Such an electron beam, having uniform radius along its length, can be used as a source of energy.

Energy exchange between a beam of electrons and an electric field

When an electron moves between electrodes in a vacuum an induced current flows in the external circuit, as discussed in Chapter 8, and the positive charge transferred externally to the electrode equals the negative charge reaching the electrode internally. In a triode two such induced currents flow, i_1 to the grid and i_2 to the anode (Fig. 9.2(*a*)). As the electron moves from cathode to grid, i_1 flows and as it moves from grid to anode, i_2 flows. Provided that the triode has a high amplification factor μ, there is very little charge induced on the anode when the electron is moving in the cathode-grid space and very little induction on the cathode when the electron is moving in the grid-anode space. The variation of i_1 and i_2 with time is shown in Fig. 9.2(*b*), where it may be seen that i_1 reverses as the

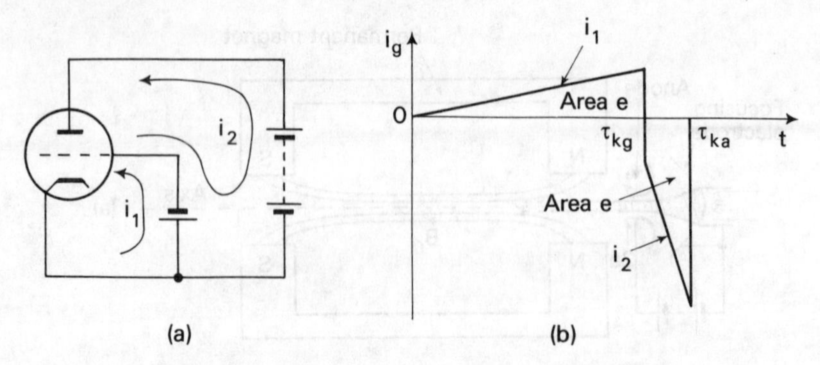

Figure 9.2 (*a*) Induced currents; (*b*) grid current.

electron passes through the grid. The effect of space charge is neglected, so that both currents change linearly with time.

The current i_2 flows in the direction which transfers energy from the anode supply to the electron. i_1 flows in the direction in which energy is transferred to the negative grid supply while the electron is approaching the grid, and energy is drawn from the grid supply while the electron is approaching the anode. For a large number of electrons the total energy will be the sum of the individual contributions, so we can say that when a beam of electrons moves towards a negative electrode energy is transferred *to* the supply. When the beam moves away from a negative electrode towards a positive electrode, energy is removed *from* the supply. This principle is also used in the amplification of a.c. power, especially at microwave frequencies.

Power transfer to an external impedance

Consider a triode with the cathode at a negative potential, the grid earthed directly and the anode earthed through a resistor R (Fig. 9.3(*a*)). This arrangement is particularly relevant to the klystron valve, discussed below. Electrons will pass through the grid to strike the anode, and if the rate of flow is $\mathrm{d}n/\mathrm{d}t$ electrons per second, an induced current $I = e(\mathrm{d}n/\mathrm{d}t)$ will, flow through the resistor R. This will develop a potential drop IR and the power dissipated will be I^2R. However, each electron in the beam meets a retarding field due to the voltage IR between grid and anode, so that it gives up kinetic energy eIR. Thus the total power lost by the electrons is

$$eIR \frac{\mathrm{d}n}{\mathrm{d}t} = I^2R \tag{9.1}$$

which is equal to the power dissipated by the resistor. The remainder of the kinetic energy of the electrons is dissipated as heat at the anode when they strike it.

Since d.c. power is transferred from the power supply to the beam and from the beam to the resistor through the induced current, it would be very useful if a.c. power could be transferred to the resistor in a similar manner. This is achieved by modulating the current with an a.c. signal, which may be applied to the grid with very little power taken from the signal source. If the input signal has an angular frequency ω the instantaneous current may then be written

$$i = I + I_m \sin \omega t \tag{9.2}$$

which shows that the current rises and falls sinusoidally about a mean value I (Fig. 9.3(*b*)). Then the total instantaneous power in the resistor R is

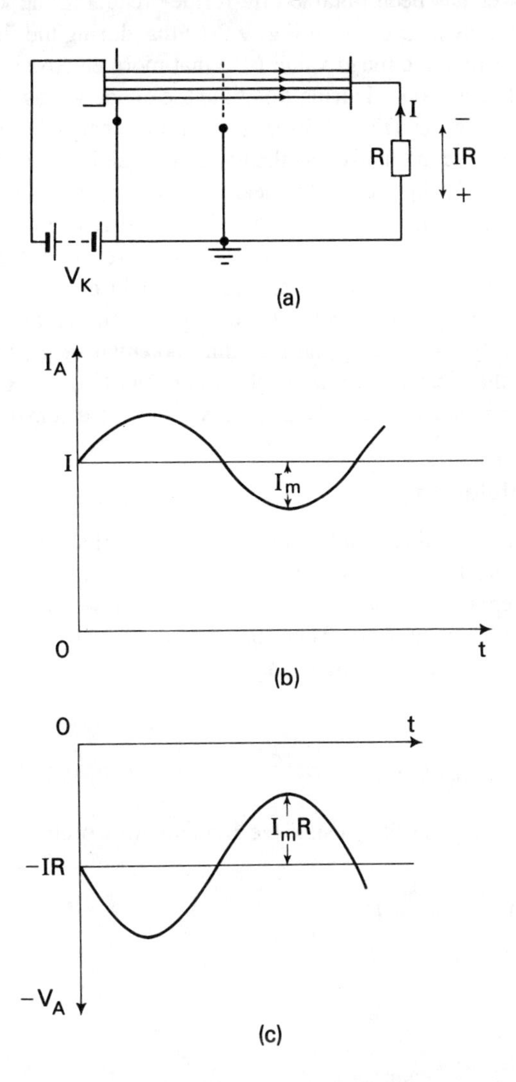

Figure 9.3 Energy exchange between an electron beam and electric fields.

$$P = i^2R = (I + I_m \sin \omega t)^2R \tag{9.3}$$

or

$$P = (I^2 + 2II_m \sin \omega t + I_m^2 \sin^2 \omega t)R \tag{9.4}$$

The mean value of P over a complete cycle is then $I^2R + \frac{1}{2}I_m^2 R$, which is greater than the power in the absence of modulation owing to the term $\frac{1}{2}I_m^2R$. This is the a.c. output power which has been obtained for a very small a.c. input power, so that considerable power amplification can be obtained.

The a.c. power has been obtained by further reducing the kinetic energy of the electrons. It may be seen from Fig. 9.3(*b*) that during the first half-cycle the current is higher than the mean value I, so that more electrons than average are moving towards the anode. During this time the anode voltage is more negative than the mean value $-IR$ (Fig. 9.3(*c*)), so that a.c. energy is removed from the beam. During the second half-cycle the anode voltage is more positive than $-IR$, so that a.c. energy is supplied to the beam. However, in this half-cycle the current is less than I so that fewer electrons than average are moving towards the anode. Thus over a whole cycle the beam gives up more a.c. energy than it receives and so supplies a.c. power to the resistor. It should be noted that the induced current is independent of the magnitude of R provided that the electron velocity is not significantly changed by the retarding potential $-I_mR$, i.e. provided that the voltage on the electron gun is much larger than $(I + I_m)R$. Also R may be either an actual resistor or the dynamic resistance of a tuned circuit.

Velocity modulation

In order to obtain the sinusoidal modulation of current shown in Fig. 9.3(*b*) the velocity of the electrons is modulated in a microwave valve. The single grid of Fig. 9.3(*a*) is replaced by two closely spaced grids between which an alternating voltage, $V_m \sin \omega t$, is applied. Then the electron velocity is due to the sum of the alternating and direct voltages and is given by

$$u = \sqrt{\frac{2e}{m}}\, (V_K + V_m \sin \omega t)^{1/2} \tag{9.5}$$

Normally $V_m \ll V_K$, so that, using the bionomial theorem,

$$u \approx \sqrt{\frac{2eV_K}{m}} \left(1 + \frac{V_m}{2V_K} \sin \omega t\right) \tag{9.6}$$

or

$$u \approx u_0 \left(1 + \frac{V_m}{2V_K} \sin \omega t\right) \tag{9.7}$$

where

$$u_0 = \sqrt{\frac{2eV_K}{m}} \tag{9.8}$$

Thus, from eq. (9.7), the velocity is sinusoidally modulated and the velocity modulation has again been obtained with a small a.c. power input to the grids.

The effect on the beam may be illustrated by means of an Applegate diagram (Fig. 9.4). Since the distance from the modulating grids is plotted against time, the velocities of electrons leaving the grids at different times are shown by straight lines of different slopes. An electron leaving when the input signal is zero will have the normal velocity u_0, but if it leaves during a positive half-cycle it will travel faster, and if it leaves during a negative half-cycle it will travel slower, than u_0. The faster and normal-velocity electrons will catch up the slower ones and they will all arrive at a particular plane in the valve, the *bunching plane*, at the same time. Thus the electron density in the beam can be a maximum here and these maxima arrive at the bunching plane at time intervals equal to the period of the input signal. At times between the electron bunches the electron density is a minimum, and the actual position of the bunching plane will move slightly up and down jwith the amplitude of the input signal. This effect will be very small when $V_m \ll V_{HT}$ since the change in velocity is then also negligible. The two grids are known as a *buncher*, and a geometrical construction will show that there are a series of equally spaced bunching planes along the beam.

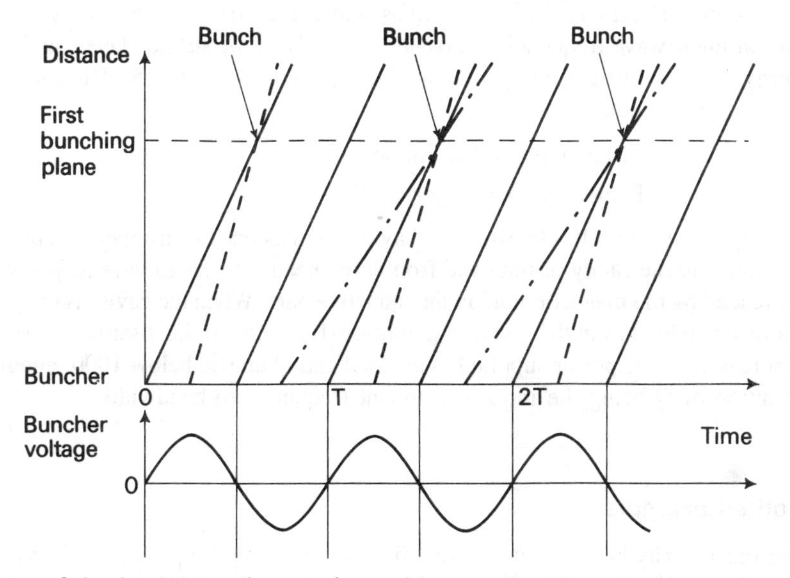

Figure 9.4 Applegate diagram for a klystron amplifier. Buncher voltage = $V_m \sin 2\pi t/T$.

Resonant cavities

In practice the buncher is part of a resonant cavity, which is the microwave equivalent of an *LC* parallel resonant circuit. As the resonant frequency f_0 is raised *L* and *C* must be reduced, since $f_0 = 1/2\pi\sqrt{(LC)}$. Ultimately the inductance becomes a single turn and the capacitance may be in the form of two discs (Fig. 9.5). A further reduction in inductance is obtained by considering more turns in parallel until a solid of revolution is formed. The cavity is completed by replacing the discs with wire grids so that an electron beam can pass through it, while in a practical cavity the solid of revolution is made from continuous copper sheets of low resistivity.

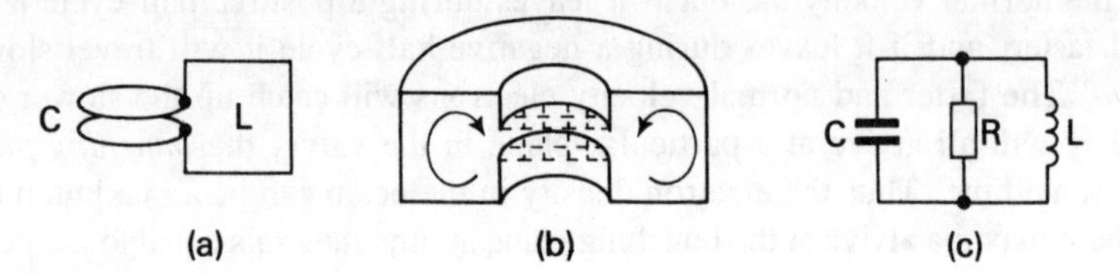

Figure 9.5 Formation of a resonant cavity. (*a*) Circuit; (*b*) section through resonant cavity; (*c*) equivalent circuit.

Inductive and capacitive parts of the cavity can thus be identified which store energy in a magnetic field and an electric field respectively. When the cavity is excited into oscillation at its resonant frequency the electromagnetic energy is exchanged between the electric and magnetic fields, just as in an *LC* circuit at resonance. Energy is also dissipated as heat at the surface of the cavity walls, since at microwave frequencies currents flow only in the surface layer, and this energy is represented as being dissipated in a parallel resistance *R*. The *Q*-factor of the cavity is defined as

$$Q = 2\pi \times \frac{\text{Maximum stored energy}}{\text{Energy dissipated per cycle}} \tag{9.9}$$

and is typically about 10 000 when the cavity is unloaded. Electromagnetic power is supplied to the cavity or removed from it by means of a small wire loop which is threaded by the magnetic field in the inductive part. When the cavity is coupled to a resistive load as in the klystron amplifier (Fig. 9.6(*a*)) the dissipated energy is increased by the power supplied to the load and *Q* falls to below 1000, another definition of *Q* being the ratio of resonant frequency to bandwidth.

Worked example

A resonant cavity in a microwave amplifier has a resonant frequency of 900 MHz. The *Q*-factor of the cavity is 10 000 when it is unloaded and 300 when it is coupled to a resistive load. Obtain the 3 dB bandwidth of the amplifier in each case.

Solution

$$Q = \frac{f_0}{\Delta f}$$

where f_0 is the resonant frequency and Δf is the frequency range over which signals are no more than 3 dB below the peak value at resonance, which defines the bandwidth.

For the unloaded cavity

$$\Delta f = \frac{f_0}{Q} = \frac{9 \times 10^8}{10^4} = 90 \text{ kHz}$$

For the loaded cavity

$$\Delta f = \frac{9 \times 10^8}{300} = 3 \text{ MHz}$$

so the effect of loading is to increase the bandwidth.

The cavity may be tuned mechanically over a small range of resonant frequencies by means of a screwed plunger in the side. Moving the plunger into the cavity effectively increases the length of the inductive part and reduces the resonant frequency, the opposite effects occurring when the plunger is moved outwards.

The klystron amplifier

To form an amplifier two cavities are used having the same resonant frequency, the one nearer to the cathode being the buncher; the other, known as the *catcher* (Fig. 9.6(*a*)), is mounted at a bunching plane. The electron beam is confined by a magnetic field as described above and is velocity modulated by a signal fed into the buncher at its resonant frequency, while the load, represented by a resistance, is coupled into the catcher. The beam will induce currents in the walls of the catcher cavity and excite it into oscillations, which will be maintained if the a.c. field across the catcher grids can extract energy from the beam. This occurs when the bunches of maximum electron density arrive at the catcher to meet a retarding field, and the conditions of minimum electron density coincide with an acclerating field (Fig. 9.6(*b*)). The waveform of the current at the catcher is then of the form shown in Fig. 9.7. The amplitude of the oscillations in the catcher cavity builds up at the expense of the kinetic energy of the electrons. The amplitude becomes constant when the energy lost by the electrons is equal to the sum of the energy supplied to the load and the energy dissipated in the cavity and load.

The input power at the buncher cavity is much less than the output power

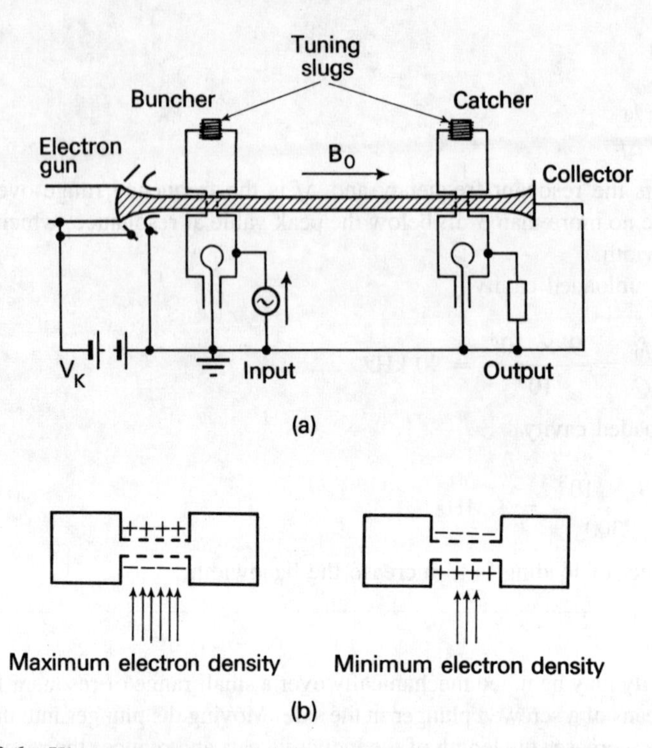

Figure 9.6 Klystron amplifier. (*a*) Two-cavity klystron amplifier; (*b*) oscillation conditions in a resonant cavity.

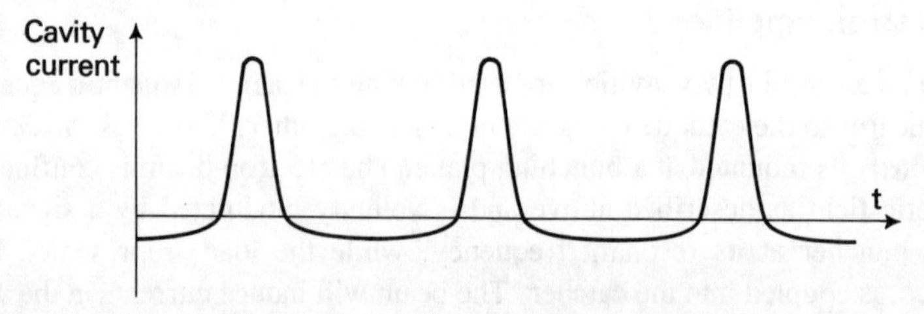

Figure 9.7 Cavity current waveform.

obtained at the catcher cavity, and a power gain of about 10 000, or 40 dB, is obtainable. Since the input voltage is much less than the anode supply voltage, V_{AA}, of the electron gun, the amplifier is insensitive to changes in V_{AA}, but it can operate over only a narrow band of frequencies owing to the high Q of the cavities. The effect of transit time through either cavity is small, since electrons are accelerated to a high velocity *before* passing through the cavity. This contrasts with conventional valves where most of the acceleration takes place *after* they have passed through the control grid, so that the electron transit time through the modulating region is much longer and becomes a limiting factor at much lower frequencies.

Klystron amplifiers are used in transmitters for television broadcasting and are available to cover a frequency range from 0.4 to 1.2 GHz with a power gain of 40 dB and an output power of 25 kW, the electron gun voltage being about 18 kV in this case. For high-power amplifiers of this type two intermediate cavities are mounted at bunching planes between the input and output cavities, making a four-cavity amplifier. The alternating voltages due to oscillations in these extra cavities increase the density of the electron bunches at the output cavity and so increase the output power.

The reflex klystron

The microwave signals used in the amplifier described in the preceding section are generated by another type of valve, known as the *reflex* klystron (Fig. 9.8(*a*)), which does not require a focusing magnetic field since it is much shorter than the klystron amplifier. The reflex klystron has only one cavity to which the load is coupled and electrons passing through it are returned to the cavity when they meet the retarding electric field provided by the *reflector* electrode (Fig. 9.8(*b*)). When the voltage on the electron gun is switched on and the first electrons pass through the cavity they induce currents in its walls and excite it into oscillation. These oscillations will be maintained if the electrons meet a retarding a.c. field when they return to the cavity, as shown in the Applegate diagram of Fig. 9.9. Here the alternating voltage has caused velocity modulation of the electrons on their first passage, and their trajectories have been adjusted by means of the reflector voltage so that bunching occurs three-quarters of a cycle after the normal-velocity electrons have passed through the cavity. At this time there is a retarding alternating potential acting on the electrons which have returned to the cavity, so that they give up their energy to the a.c. field and maintain oscillations in the

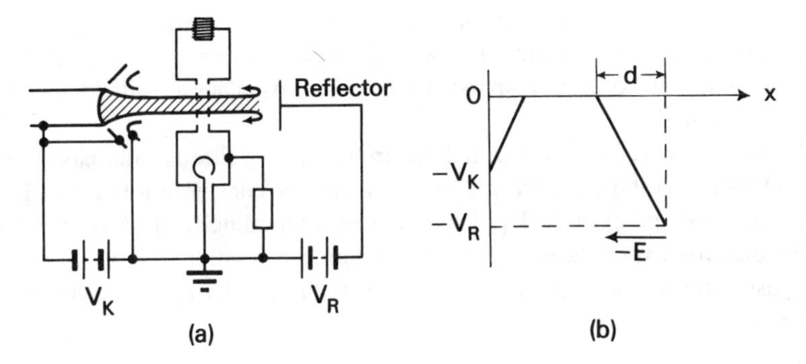

Figure 9.8 Klystron oscillator. (*a*) Reflex klystron oscillator; (*b*) voltage distribution in the reflex klystron.

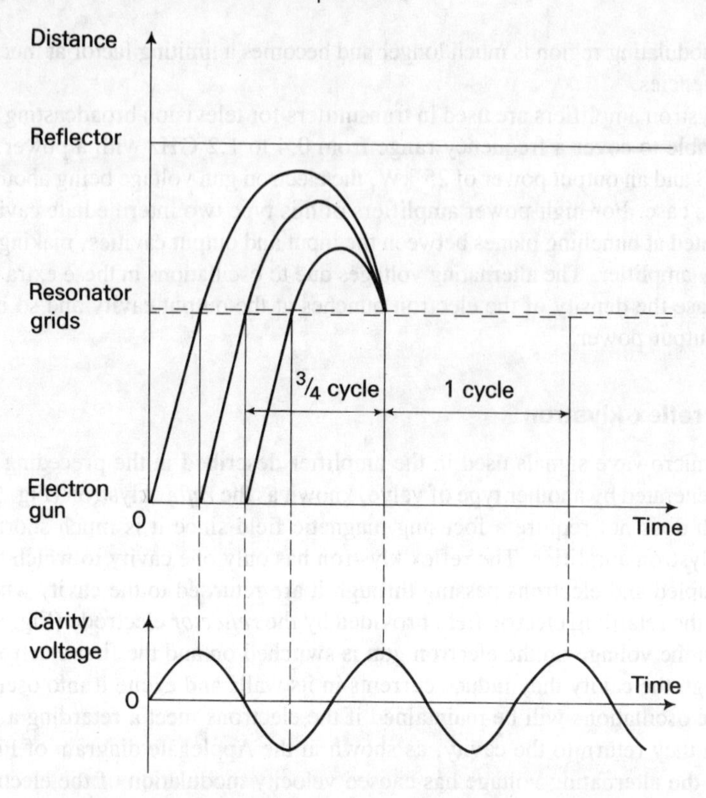

Figure 9.9 Applegate diagram for the reflex klystron.

cavity. The cavity then acts as both a buncher and a catcher. Ideally the returning electrons give up all their energy to the cavity, but some may pass through it and rejoin the oncoming electrons. It is apparent that oscillations will also be maintained if electrons are bunched $1\frac{3}{4}$ cycles after passing through the cavity, which is achieved by reducing the retarding potential on the collector, and in general bunching must occur after $(n + \frac{3}{4})$ cycles, where $n = 0, 1, 2, 3, \ldots$ Each integral value of n corresponds to a *mode* of oscillation, called the $\frac{3}{4}$ mode, the $1\frac{3}{4}$ mode and so on.

The path followed by the normal electrons passing through the cavity when the alternating voltage is zero may be obtained by considering the voltage between reflector and cavity, $-V_R$. This provides a retarding field $-E$ (Fig. 9.8(*b*)), and an electron experiences a force $-Ee$ and an acceleration $-Ee/m$. If the electron passes through the grids at time $t = 0$ with velocity u_0, the velocity after time t is

$$u = u_0 - \frac{Ee}{m}t \tag{9.10}$$

and the distance travelled beyond the grid towards the reflector is

$$s \int_0^t u \, dt = u_0 t - \frac{1}{2} \frac{Ee}{m} t^2 \tag{9.11}$$

This is the equation of a parabola, and the electron returns to the grid when $s = 0$, given by a time

$$t = \frac{2mu_0}{eE} \tag{9.12}$$

from eq. (9.11). Thus, if u_0 is fixed by the electron gun voltage and the distance between reflector and cavity is d, the time spent by an electron between reflector and cavity is given by

$$t_f = \frac{2mu_0 d}{eV_R} = \frac{n + \frac{3}{4}}{f} \tag{9.13}$$

where f is the oscillation frequency to be maintained. This need not necessarily be at the resonant frequency of the cavity, although it must be fairly close to it, since t_f in practice and hence the oscillating frequency, may be adjusted electronically over a small range by changing the reflector voltage V_R without changing the mode of oscillation.

The maximum output power is obtained at the resonant frequency, but where this is 100 mW at 6 GHz, for example, it is possible to obtain an output greater than 40 mW over a tuning range of 60 MHz, with a sensitivity of about 1.5 MHz per volt of change in reflector voltage. The reflex klystron is the most widely used microwave valve, although it is being superseded in many applications by solid-state devices, and typical versions cover the frequency range 6–12 GHz with output powers between 0.1 and 1 W. The anode and cavity voltage is commonly about 300 V, and the reflector voltage, which may be varied for electronic tuning, is between about –100 and –200 V. These voltages are quoted with respect to the cathode, but in practice the cavity is normally earthed.

The cavity magnetron

A device which uses a number of resonant cavities for the generation of power at microwave frequencies is the *cavity magnetron*, illustrated in Fig. 9.10. The anode is a circular block of copper, which commonly contains eight identical cavities separated by eight poles and opening onto an annular space with the cathode at the centre. A strong magnetic field, provided by a permanent magnet, acts at right angles to the d.c. electric field between anode and cathode. Thus electrons leaving the cathode would move in approximately circular paths near the anode and would be prevented from reaching the anode if the magnetic field were strong enough. However, the electrons will induce currents in the cavities and set up oscillations which create a.c. electric fields between the poles. Then

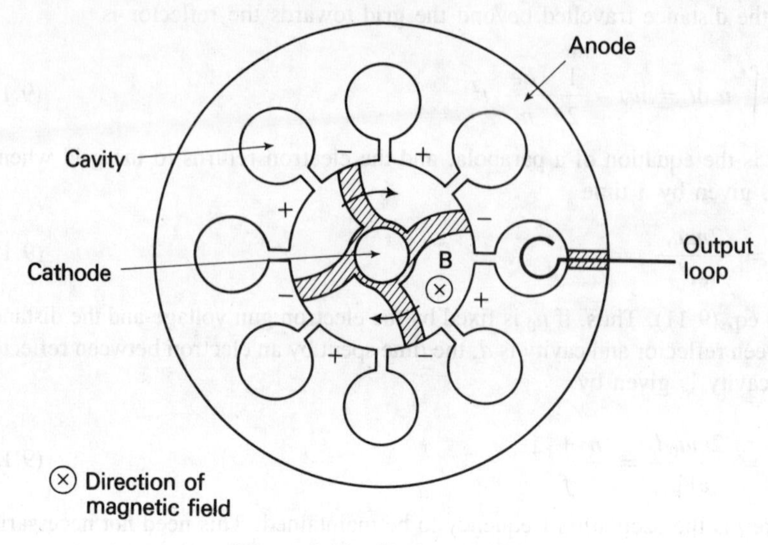

Figure 9.10 Cavity magnetron.

at any instant alternate poles are at potentials which are positive and negative with respect to the direct anode voltage. There is a phase difference of 180° between each pair of poles and so this method of oscillation is called the *π-mode*. As each cavity goes through a complete cycle of oscillation it appears that a wave travels round the anode, such that the wavelength equals the distance between alternate poles round the internal circumference. If the oscillations are to be maintained an integral number of wavelengths must occur round the anode, so that an even number of cavities is required.

The a.c. electric force acting on the electrons and the form of the travelling wave at a particular instant are shown in Fig. 9.11, in which the cavities are drawn in a linear arrangement for convenience. In the actual magnetron the *x*- and *y*-directions correspond to tangential and radial directions respectively. The force acting on the electrons in the *y*-direction due to the d.c. fields is given by

$$F_y = Ee - Beu_x \tag{9.14}$$

where u_x is the component of velocity in the *x*-direction parallel to the anode. The electric field E is the resultant of fields due to the anode voltage and the space charge and acts to increase the velocity of electrons *towards* the anode. The a.c. field at the anode causes the space between the electrodes to be divided into accelerating regions, A, and decelerating regions, D, which move round the anode with the travelling wave. In the A-regions the velocity parallel to the anode, u_x, is increased and in the D-regions u_x is decreased.

Electrons leave the cathode continuously to follow curved paths to the anode. Those entering an A-region experience an increase in u_x, so that F_y becomes

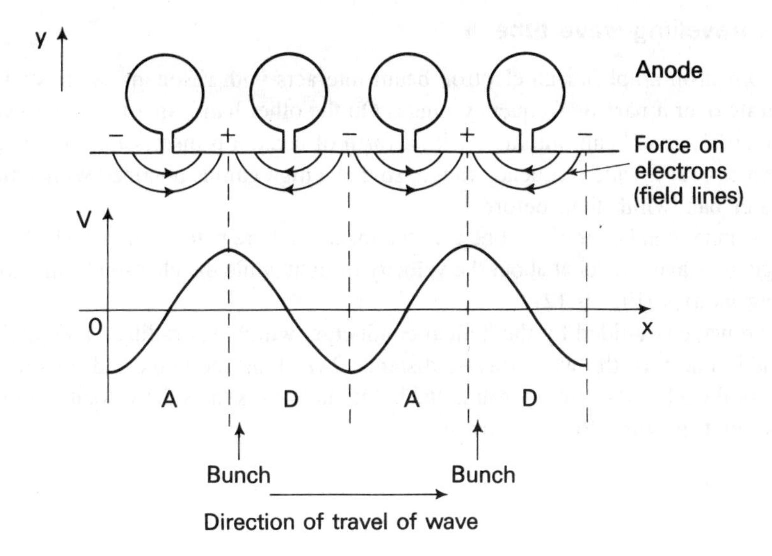

Figure 9.11 Travelling wave in cavity magnetron

negative (eq. (9.14)) and they return to the cathode. Here they cause secondary emission as they give up their energy, which comprises about 90% of the total cathode emission in a practical magnetron. Electrons entering a D-region have u_x reduced which makes F_y positive, so that they continue towards the anode. Thus there is a tendency for electrons to collect in the decelerating regions, where bunches are formed with the wave. Since the bunches are moving in a region of retarding field they give up the energy they have gained from the d.c. field to the a.c. field, so that oscillations are maintained in the cavities. The bunches are in a state of dynamic equilibium, since they lose electrons to the anode and gain electrons from the cathode. The space charge in the bunches then appears like the spokes of a wheel rotating round the cathode (Fig. 9.10).

The density of the space charge in a magnetron is high so that it is normally operated with the anode voltage applied in short pulses, although continuous wave (c.w.) magnetrons are also available. The output with a pulsed voltage consists of bursts of microwave power at the oscillation frequency, which can be varied over a relatively small range by mechanical tuning of the cavities. The anode is normally cooled either by air forced past it or by means of a water jacket. Applications include the transmitters of radar and navigation systems and microwave ovens. A typical communication magnetron oscillates between 9.3 and 9.4 GHz, with a peak anode voltage of 6 kV applied in 1 μs pulses at a rate of 1000 times a second, having a microwave output of 8 kW for each pulse.

Large numbers of magnetrons are used in microwave ovens and a typical device has a continuous output power of 600 W at 2.45 GHz, with a fan providing forced-air cooling. This is achieved with an input power of 1.2 kW at 240 V, 50 Hz giving an overall efficiency of about 50%.

The travelling-wave tube

In a klystron amplifier an electron beam interacts with resonant cavities which operate over a narrow frequency range. On the other hand, in a travelling-wave tube an electron beam interacts with a form of coaxial transmission line, which operates over a wide frequency range, so that a high gain is obtained with a much greater bandwidth than before.

The inner conductor of the line is in the form of a wire helix along which electromagnetic waves travel at about the velocity of light while the electron beam moves along its axis (Fig. 9.12).

The wave is guided by the helical conductor, which has radius r and pitch d. Thus in one turn the wave travels distance $2\pi r$ along the helix and d along the axis of the tube. Its velocity round the helix may be taken as the velocity of light, c, so that the time for one turn is

$$t = \frac{2\pi r}{c} \tag{9.15}$$

The velocity along the axis is

$$u = \frac{d}{t} = \frac{cd}{2\pi r} \tag{9.16}$$

Considering the whole tube,

$$u = c \times \frac{\text{Axial length}}{\text{Wire length}} \tag{9.17}$$

and typically u might be about 0.09 c, which corresponds to the velocity of a 2 kV electron beam.

The axial velocity of propagation is thus effectively reduced by the helix, so that the wave travels slightly more slowly than the electrons in the beam and in the same direction. This causes bunches to be formed which move in the decelerating region of the wave, as in the magnetron. Thus energy is continuously extracted from the beam by the wave, which therefore grows in amplitude as it travels down the tube.

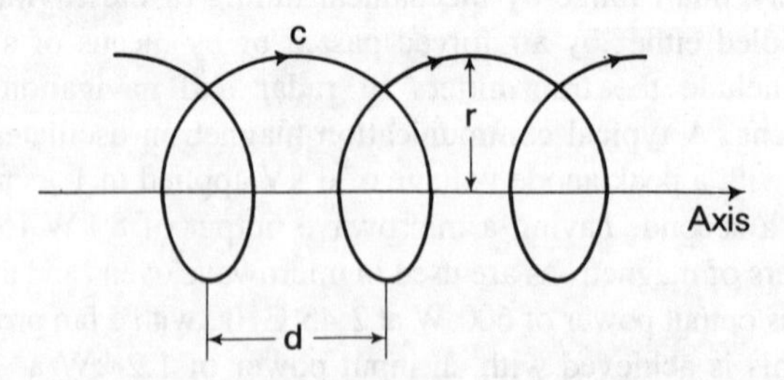

Figure 9.12 Helix of a travelling-wave tube.

The maximum operating frequency is determined by the condition that r is much less than the wavelength of the maximum operating frequency, which ensures that all parts of the wave may be considered in phase round a particular turn. The minimum operating frequency is determined by the length of the tube, which becomes too large to be practical for frequencies below about 500 MHz.

An amplifier using a travelling-wave tube is illustrated in Fig. 9.13. The electron beam is confined by a magnetic field, as described earlier, which is usually provided by a permanent magnet. The beam leaves the electron gun with a velocity determined by the anode voltage V_A and is extracted from the tube by the collector, which is held at a lower voltage V_C. The helix is at a higher voltage V_H, which is adjusted for maximum gain, all voltages being measured with respect to the cathode. The signal is fed into and out of the amplifier by means of waveguides, and the input and output couplers are carefully designed to ensure that the maximum bandwidth is obtained.

Practical travelling-wave tubes typically have a bandwidth of 1 GHz centred in the range 6 to 12 GHz, which makes them very useful in microwave communication amplifiers. Typical values of operating voltage are $V_A = 2$ kV, $V_H = 3$ kV and $V_C = 1$ kV, and a gain of 40 dB with an output power of 10 W is readily obtainable.

Bunching conditions

The a.c. electric force acting on the electrons at a particular instant is shown in Fig. 9.14 together with the form of the travelling wave; V represents the voltage distribution along the helix. Accelerating and decelerating regions occur, as in the magnetron, and electrons tend to bunch where the a.c. field is decelerating

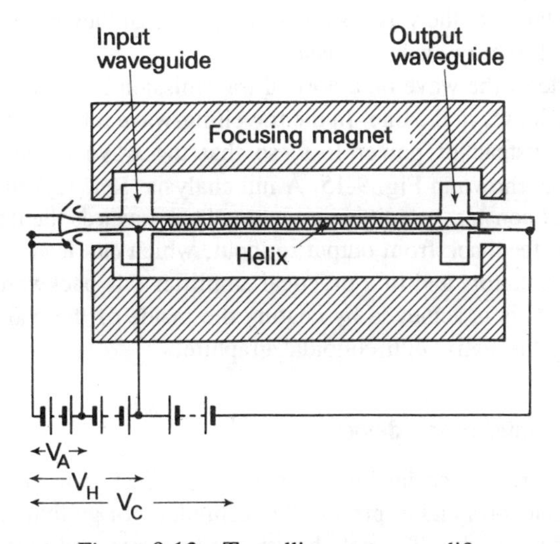

Figure 9.13 Travelling-wave amplifier.

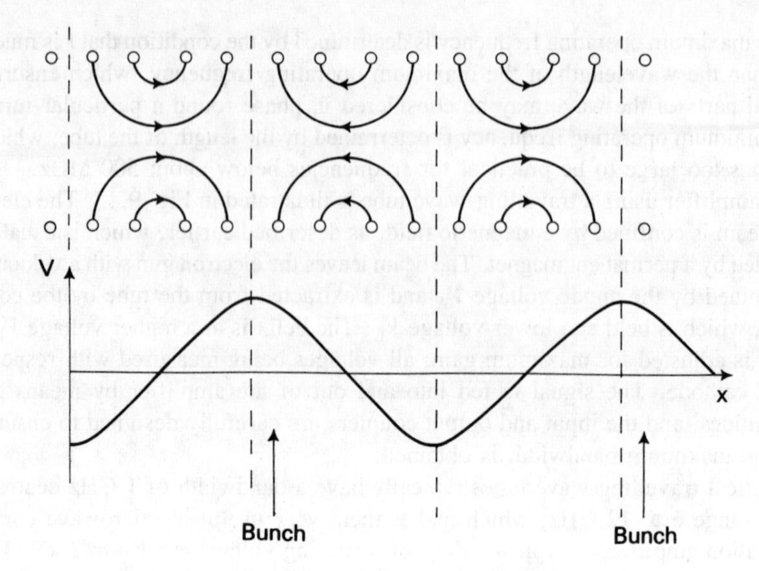

Figure 9.14 Bunching conditions in a travelling-wave tube.

ahead and accelerating behind. If the bunches travelled with the peak of the wave, where $\partial V/\partial x$ is zero, there would be no exchange of energy between the beam and the wave. Thus their velocity is adjusted until they travel slightly faster than the wave, in the decelerating region where $\partial V/\partial x$ is negative. Then, since the a.c. field is retarding, it extracts energy from the electrons, which are therefore slowed down as each bunch occurs along the tube. Eventually the velocity of the electrons relative to the wave becomes zero, so that they move with the peak of the wave and energy exchange ceases.

The amplitude of the wave on a normal transmission line decreases exponentially due to the losses in the line. In the travelling-wave tube there is power gain along the line instead of power loss, so that the wave amplitude *increases* exponentially, as shown in Fig. 9.15. A full analysis (Ref. [1]) shows that there is also a wave of constant amplitude travelling back towards the input. This constitutes positive feedback from output to input, which can lead to the amplifier becoming an oscillator, and in a practical amplifier this backward wave has to be attentuated. This is achieved by coating the inside of the walls of the glass tube containing the helix with colloidal graphite.

Semiconductor microwave devices

As mentioned earlier, special bipolar and field effect transistors are used at microwave frequencies and at present the technology of gallium arsenide FETs is advancing very rapidly. The gate length has been reduced to 0.5 μm and the

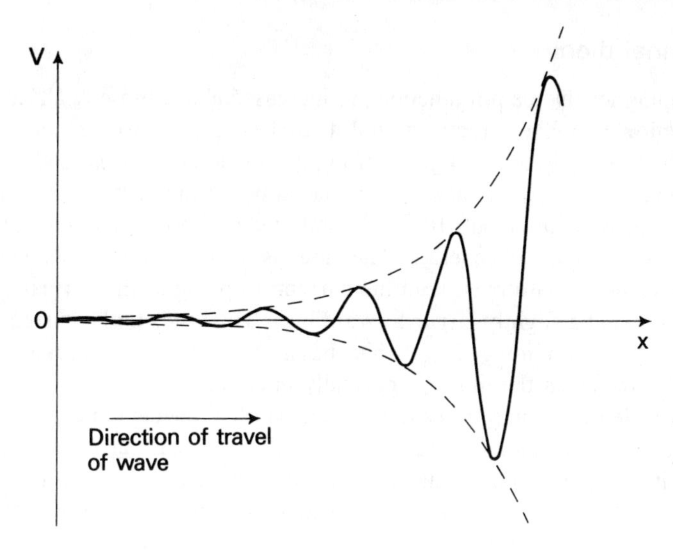

Figure 9.15 Growth of wave along a travelling-wave tube

structures described in Chapter 7 will extend operation into the 100 GHz region. Production of integrated circuit microwave amplifiers for continuous wave applications has now been achieved (Ref. [2]).

The semiconductor microwave devices using different operating principles may all be called diodes, since they have two terminals, although they are *not* all p-n junctions. The principles involved are:

negative resistance;
variable capacitance;
variable impedance;
and non-linear resistance.

Amplifiers, oscillators, attenuators, detectors and mixers can then all be provided in ways different from those used at lower frequencies.

Negative resistance is provided by tunnel, Gunn and IMPATT devices used as amplifiers and oscillators. Variable capacitance is used in the parametric amplifier, based on a varactor diode. Variable impedance occurs in the p-i-n diode which is used as a switch or attenuator. Finally the non-linear *I/V* characteristic of a Schottky diode is used in a microwave receiver. The received signal and a local oscillator signal are both applied to the diode so as to derive a signal at the difference frequency or intermediate frequency. The received signal may be amplitude-modulated or frequency-modulated and further circuitry is required to detect the modulating signal. Schottky diodes and earlier point-contact diodes are used since they have no charge storage effects and the device is known as a *varistor* in this application.

The tunnel diode

If the doping levels in a p-n junction are increased above those in a Zener diode the breakdown voltage is reduced until it reaches zero. However, this is not the limit, which is set by the solubility of impurities in the material and is reached when about 1 atom in 10^4 has been replaced by an impurity atom. This gives a doping density of about $10^{25}/m^3$, and under these conditions the semiconductors are highly degenerate. The diode is then still in a breakdown condition at a small forward bias, with the current dropping to the normal value for a p-n junction at a slightly larger forward bias. This leads to the current voltage characteristic shown in Fig. 9.16, which has a negative-resistance region between V_P and V_V in which the diode is normally operated.

The high density of impurities reduces the width of the depletion layer to about 10 nm so that the potential barrier is sufficiently narrow for an electron to tunnel through it, since it also has a finite height (Appendix 1). This effect is responsible for the shape of the characteristic below the voltage V_V, above which forward current flows by the normal process. The tunnel diode was first described by Esaki in 1958. The energy diagram at zero bias is shown in Fig. 9.17. The Fermi level is continuous through the junction and lies within the conduction band in the n-region and within the valence band in the p-region since the materials are degenerate. Thus even at room temperature the lower levels in the conduction band of the n-region are filled and the upper levels in the valence band of

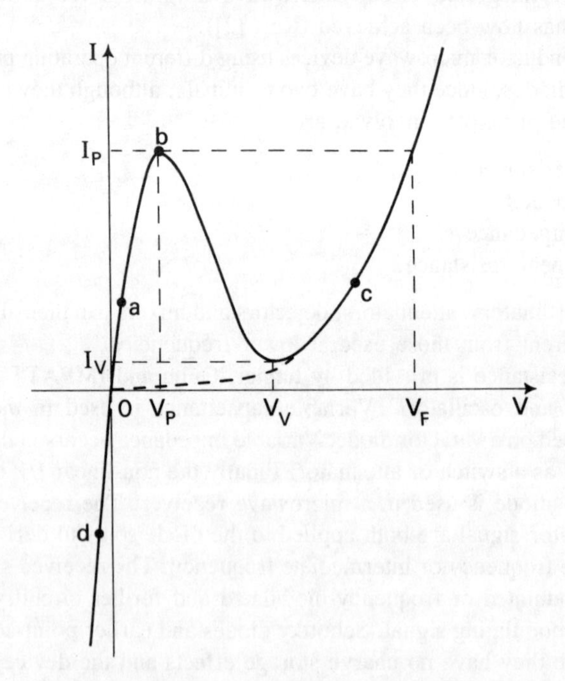

Figure 9.16 ***I–V* characteristic of a tunnel diode.**

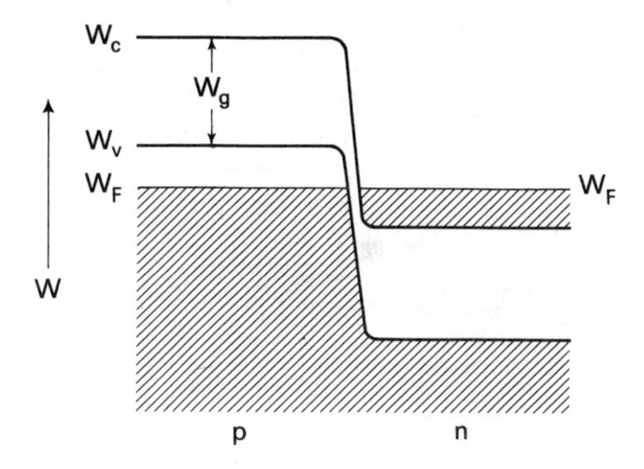

Figure 9.17 Energy diagram for a tunnel diode at zero bias.

the p-region are empty, to about the Fermi level in each case. No tunnelling occurs and the current is zero.

When forward bias is applied the height of the barrier is reduced and energy levels in the n-region move upwards on the energy diagram (Fig. 9.18(*a*) and point *a* on Fig. 9.16). Electrons in the conduction band of the n-region face empty levels in the p-region, and since energy is conserved during tunnelling, electrons move horizontally on the energy diagram. Thus an electron tunnel current flows from the n- to the p-region *through* the potential barrier, and at the same time electrons diffuse from the n-region and holes diffuse from the p-region *over* the potential barrier. Hence the total current is the sum of the tunnel and diffusion currents. As the forward bias is increased a point is reached at which all the conduction-band electrons face empty levels in the p-region and the tunnel current reaches a maximum (Fig. 9.18(*b*) and point b on Fig. 9.16), while for a slightly larger bias the filled and empty levels move away from each other and the current falls. The diffusion current increases slowly with bias until when the filled and empty levels are completely separated the tunnel current has become zero and current flows only by diffusion to give the normal diode characteristic (Fig. 9.18(*c*) and point c on Fig. 9.16). At the forward voltage V_F the diode current due to diffusion has become equal to the peak tunnel current I_P.

Under reverse bias conditions, filled levels in the valence band of the p-region move opposite empty levels in the conduction band of the n-region, so that an electron tunnel current flows from the p- to the n-region. This leads to a high current at a low reverse voltage (Fig. 9.18(*d*) and point d on Fig. 9.16), and it may be said that breakdown due to tunnel current extends from reverse bias to the forward bias V_V, which occurs at the valley point where the tunnel current ceases.

The current scale on Fig. 9.16 depends on the junction area and the doping level, diodes being obtainable with peak currents, I_P, between about 50 μA and

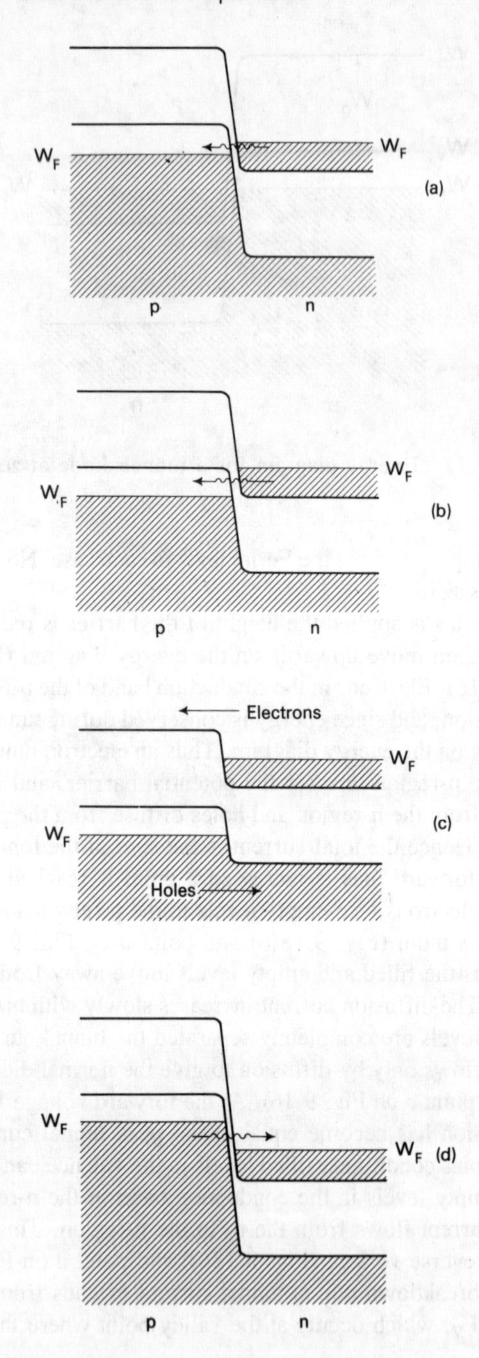

Figure 9.18 **Energy diagrams for a tunnel diode. (*a*), (*b*), (*c*) and (*d*) refer to the corresponding points on Fig. 9.16.**

5 A. Typical values of peak-to-valley current ratio range from 5 to 15. Material used for making tunnel diodes include germanium, silicon and gallium arsenide, with the voltages V_P, V_V and V_F dependent on the material used for the diode, and mean values for germanium, silicon and gallium arsenide are compared in Table 9.1. It may be noted that these voltages increase with the energy gap of the material and that the dynamic ranges, $V_V - V_P$, are about equal for silicon and gallium arsenide diodes and larger than the value for germanium diodes.

Table 9.1 Tunnel diode characteristics

	Ge	Si	GaAs
Peak voltage V_P, mV	55	65	150
Valley voltage V_V, mV	320	420	500
Dynamic range, mV	265	355	350
Forward voltage V_F, mV	480	720	980
Maximum operating temperature, °C	100	200	175
Energy gap, eV	0.72	1.10	1.35

Since the propagation of electrons by tunnelling proceeds at a velocity near that of light, the transit time is extremely short, given approximately by $10^{-8}/3 \times 10^8 \approx 10^{-17}$ s, so that the frequency response is limited only by stray inductance and capacitance and can extend into the gigahertz region. Only majority carriers are involved in the operation of a tunnel diode, so that it is relatively insensitive to changes in temperature. It can operate at temperatures as low as 4 K and up to several hundred degrees Celsius.

An equivalent circuit for a tunnel diode is given in Fig. 9.19. R represents the resistance of the contacts and leads to the p-n junction and has a typical value of about 1 Ω, while L represents the self-inductance of the leads and is typically about 1 nH. The junction itself is represented by the capacitance C, which is a function of the bias voltage and is typically about 1 pF, and by the resistance $-r$, typically about −100 Ω. This is the slope resistance of the region between the voltages V_P and V_V where the current falls for an increase in voltage, so that $\partial I/\partial V$ is a negative quantity, $-1/r$. The self-resonant frequency of the equivalent circuit is given approximately by $1/2\pi\sqrt{(LC)}$ (*see* Problem 9.7), and for the typical values given above is about 5 GHz. For frequencies below this the diode may be represented by R in series with the parallel combination of $-r$ and the effective capacitance. The diode can then be used in parallel with a tuned circuit or mounted in a resonant cavity to form either an amplifier or an oscillator.

Tunnel diode amplifier and oscillator

The equivalent circuit of the amplifier is given in Fig. 9.20, where R_s includes the source resistance and the diode lead resistance. At resonance the impedance

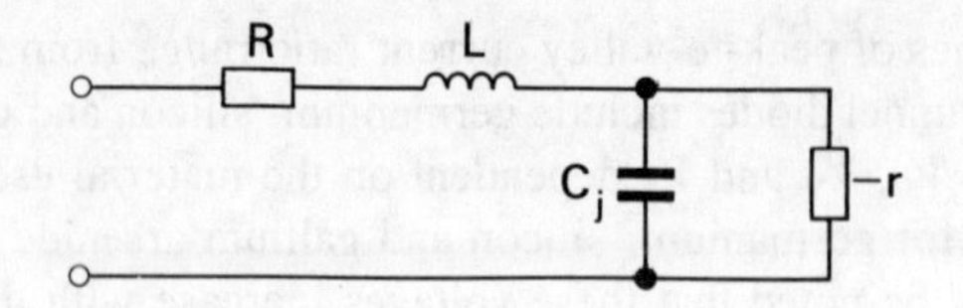

Figure 9.19 Equivalent circuit of a tunnel diode.

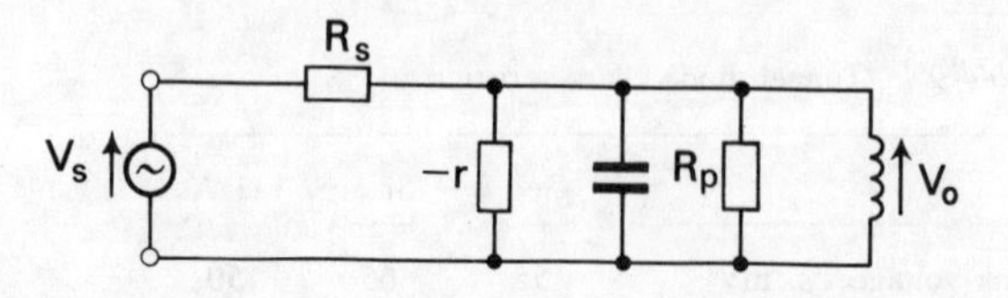

Figure 9.20 Equivalent circuit of a tunnel diode amplifier.

of the tuned circuit or cavity is equal to its parallel resistance R_p, which includes any externally connected resistive load. The output voltage becomes

$$V_o = \frac{V_s(-rR_p/-r + R_p)}{R_s(-rR_p/-r + R_p)}$$

$$V_o = -V_s \frac{rR_p}{R_sR_p - r(R_s + R_p)} \tag{9.18}$$

and the voltage gate is

$$\frac{V_o}{V_s} = -\frac{\dfrac{rR_p}{R_s + R_p}}{\dfrac{R_sR_p}{R_s + R_p} - r} \tag{9.19}$$

dividing through by $R_s + R_p$. For stable amplification there is no phase change introduced by the diode, so that the gain is positive; this occurs when

$$r > \frac{R_sR_p}{R_s + R_p} \tag{9.20}$$

from eq. (9.19). Since it is required that $V_o > V_s$, from eq. (9.18),

$$rR_p > r(R_s + R_p) - R_SR_p \tag{9.21}$$

or

$$R_p > r$$

so that, for stable amplification,

$$R_p > r > \frac{R_s R_p}{R_s + R_p} \tag{9.22}$$

For given values of r and R_p the gain increases with R_s since the denominator of eq. (9.19) becomes smaller as $R_s R_p/(R_s + R_p)$ tends to r.

When

$$\frac{R_s R_p}{R_s + R_p} = r \tag{9.23}$$

the gain becomes infinite, so that an output voltage is obtained for no input voltage. The amplifier has then become an oscillator, and since $V_s = 0$, R_s now represents the resistance of the power supply providing the bias current and voltage to the tunnel diode. The total loss resistance of the circuit is R_s in parallel with R_p, which is exactly compensated by the negative resistance of the diode. The tuned circuit has then become effectively loss free, so that when an oscillatory current has been started in the circuit (by the transient occurring when the bias supply is switched on) it will continue indefinitely.

If we put

$$\frac{R_s R_p}{R_s + R_p} = R_B \tag{9.24}$$

where R_B is the effective bias resistance, the d.c. circuit becomes as shown in Fig. 9.21(*a*), and R_B may be represented by a load line on the diode characteristics (Fig. 9.21(*b*)). R_B includes both the source and load resistances since the tunnel diode amplifier has a direct connection between the input and output circuits. It is clear that the conditions $R_B < r$ and $R_B = r$, representing stable amplification and oscillation respectively, are shown by a load line cutting the characteristics at one point only, Q. If $R_B \sim r$ the load line cuts the characteristics in the three points, Q, Q_1 and Q_2, and the current can adjust itself either to Q_1 or to Q_2, which define the limiting values of current and voltage. This is an unstable biasing condition and is used in switching applications. For instance, the diode can be used to generate pulses by switching between Q_1 and Q_2. A typical tunnel diode amplifier will have a gain of 5–15 dB and an octave bandwidth centred on a resonant frequency between 4 and 15 GHz.

Transferred electron (TE) devices, the Gunn device

In 1963 Gunn reported that a homogeneous sample of n-type gallium arsenide generated microwave oscillations when a steady electric field was set up in it. If the field exceeded a threshold value of about 300 kV/m the current through the sample was found to flow in pulses with a period proportional to its length

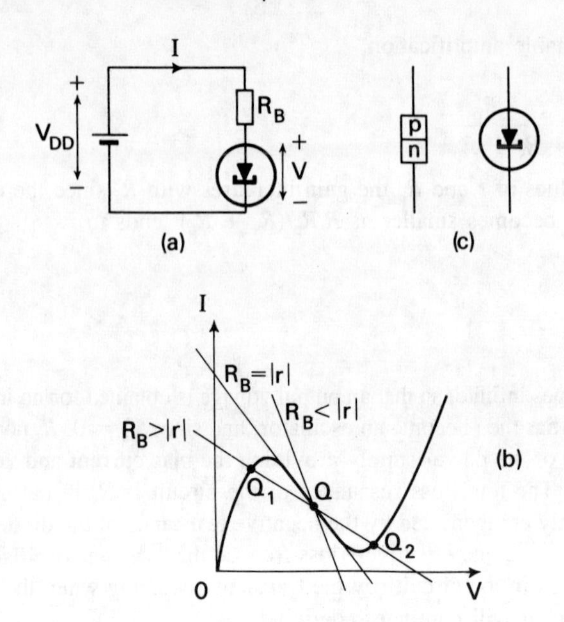

Figure 9.21 Tunnel diode. (*a*) Bias circuit; (*b*) load lines on *I–V* characteristic; (*c*) graphical symbols.

l, which can lie between about 0.01 and 2.5 mm. This is the *Gunn effect*, which takes place in the bulk of the semiconductor instead of at a junction. Since the frequency of the current pulses lies in the gigahertz region it has led to the production of a simple microwave oscillator, requiring a supply of only a few volts for operation. It is known as a *Gunn device* or as a *transferred electron device* due to its operating principles.

The production of pulsed currents from a steady electric field depends on the availability of a material with two conduction bands. These are normally shown on an energy/momentum-vector diagram, which is described in Appendix 1 for one dimension. In a three-dimensional crystal the *k*-axis is divided into sections each of which corresponds to a particular direction in the crystal lattice and is called a *Brillouin zone*. The two conduction bands shown in Fig. 9.22 for gallium arsenide correspond to two different directions in the lattice and are parabolic in form, with their minima coinciding with a zone boundary. This is the same form as for an electron in free space, but in a solid the electron has an effective mass m^* related to the curvature of the W/k graph. In the lower energy band, $m^* = 0.067\ m$, while in the upper energy band $m^* \approx 0.35\ m$ (Ref. [4]), with a separation $\Delta W = 0.36$ eV between the minimum energies of the bands. Since the effective mobility may be considered to fall as mass rises (eq. (2.54)), the corresponding mobilities are found to be about 0.50 m^2/Vs in the lower band and about 0.02 m^2/Vs in the upper band. The density of allowed energy levels is greater in the upper band since it rises with effective mass.

When the field E in the device is low, most of the electrons are in the lower

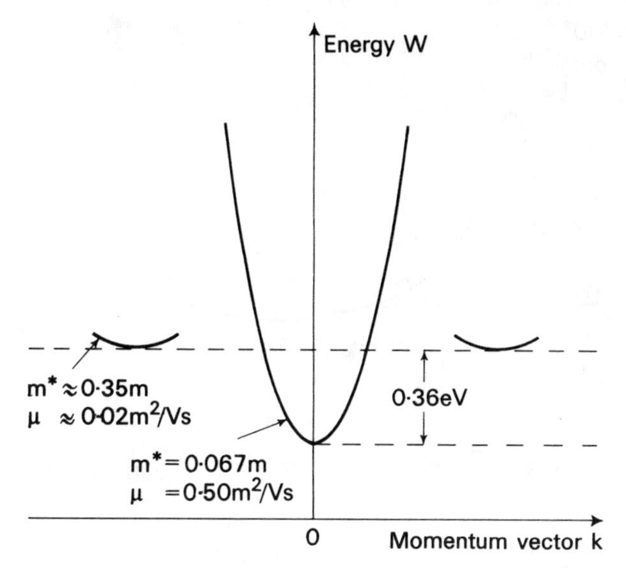

Figure 9.22 Energy/momentum-vector diagram for gallium arsenide.

conduction band since their thermal energy $(3/2)kT$ is less than ΔW at room temperature. As the field is increased the electrons gain energy $E\bar{l}$, where $\bar{l}$ is their mean free path, and more electrons are able to transfer to the upper conduction band, until at a sufficiently high field all the electrons will be in the upper band. Thus at low fields the electron mobility is high and at high fields the mobility falls as all the electrons are transferred to the upper conduction band. This is illustrated in Fig. 9.23, which relates electron velocity and electric field and may be compared with Fig. 2.15 for elemental semiconductors such as germanium and silicon. The slope du/dE at a particular point gives the differential mobility. The regions of high and low positive mobility described above are separated by a third region of *negative* mobility, which is the cause of the Gunn effect.

In fact, Fig. 9.23 also represents a current/voltage characteristic, since for a device of given cross-sectional area and length the current is proportional to the electron velocity (eq. (2.60)) and the voltage is proportional to the electric field. Thus the characteristic has a region of negative resistance, similar to that in the tunnel diode characteristic (Fig. 9.16). However, unlike the tunnel diode it is not possible to bias the device to a stable operating point in this region, since the current flows in pulses (Fig. 9.24(*a*)), and so the curve is a calculated one (Ref. [3]).

The Gunn effect

Gunn showed by means of a capacitively coupled electric probe that the electric field within the semiconductor was not uniform when the threshold field E_T was

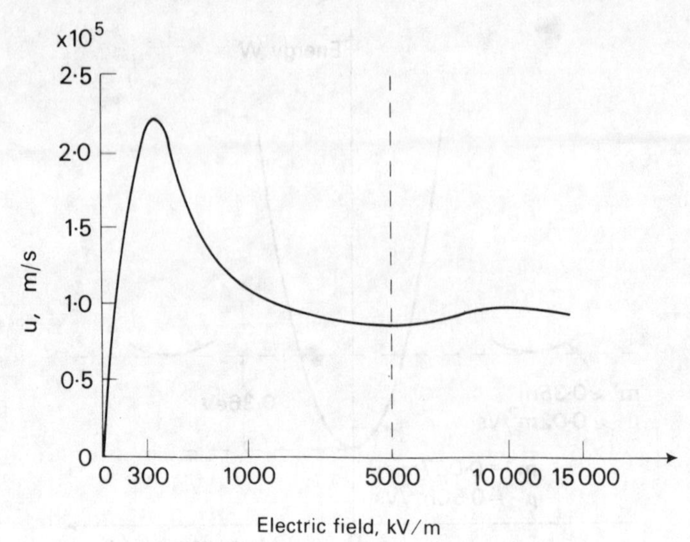

Figure 9.23 Drift-velocity/field characteristic of a Gunn diode.

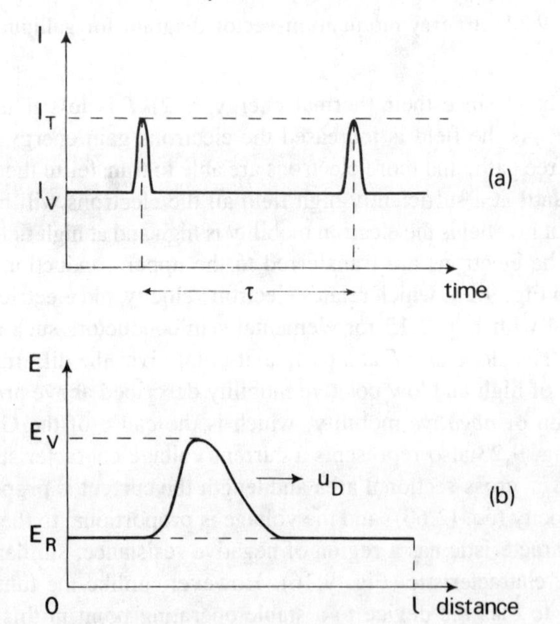

Figure 9.24 Gunn device. (*a*) Current pulses; (*b*) high-field domain.

exceeded but was distributed as in Fig. 9.24(*b*). A narrow region of high electric field known as a *domain*, was found to move along the specimen in the direction of electron flow with a constant drift velocity u_D. While the domain was travelling from cathode to anode the current in the external circuit remained constant at the valley level I_V, but as the domain moved into the anode contact the current rose to the threshold level I_T. As soon as the whole domain had moved into the anode the current fell to I_V again, while a new domain formed at the cathode to repeat the cycle. The velocity u_D was found to remain constant at about 10^5 m/s even when the applied voltage was increased above the threshold value.

It is considered that a domain begins to form at a *nucleating centre*, which may be found where the charge distribution is no longer uniform such as occurs near the cathode contact. Then, if the device is biased at a constant voltage just above the threshold voltage $V_T = E_T l$, the electric field is equal to E_T everywhere except inside a nucleating centre of length x, where it is above E_T. Electrons within the centre are then transferred to the upper conduction band, so that they have a lower mobility than electrons outside the centre, which are still in the lower conduction band. Thus the transferred electrons are retarded relative to the electrons drifting both in front of them and behind them. This leads to the accumulation of space charge in the centre due to both the low-mobility electrons and the high-mobility electrons catching them up from behind. The space charge causes a further increase of the field inside the centre until it reaches a value E_V, while the field in the remainder of the diode falls to a value E_R (Fig. 9.24(*b*)), which correspond to the currents I_V and I_R respectively. The domain has now formed and moves towards the anode with velocity u_D.

Theory of operation

A simple analysis of a Gunn device, based on that suggested by Hilsum (Ref. [5]), brings out some of the important operating features; more comprehensive theories are given in Ref. [4]. The characteristic is idealized into three parts (Fig. 9.25), consisting of two straight lines, having slopes μ_1 at low fields and μ_3 at high fields, which are joined by a third straight line of slope $-\mu_2$. The slope μ_1 corresponds to the mobility of electrons in the lower conduction band, and μ_3 to the mobility in the upper conduction bands. The following approximate values may be used: $\mu_1 = 0.50\ \mathrm{m^2/V\ s}$, $\mu_2 = 0.02\ \mathrm{m^2/V\ s}$, $\mu_3 = 0.02\ \mathrm{m^2/V\ s}$. The field E_V occurs at the valley point on Fig. 9.25 where the mobility changes from a negative to a positive value, and the point of intersection of the domain velocity and low-field lines defines E_R, approximate values being $E_R = 200$ kV/m, $E_T = 300$ kV/m, and $E_V = 5000$ kV/m.

If the increase in field within the centre is ΔE_2 due to the accumulation of space charge, the corresponding increase of potential across it is $\Delta E_2 x$, where x is the width of the domain. Since the bias voltage is constant this must be accompanied by a decrease in potential in the remainder of the device $\Delta E_1(l-x)$, where ΔE_1 is a reduction in the field E (Fig. 9.25). Then

$$\Delta E_2 x = \Delta E_1(l-x)$$

and

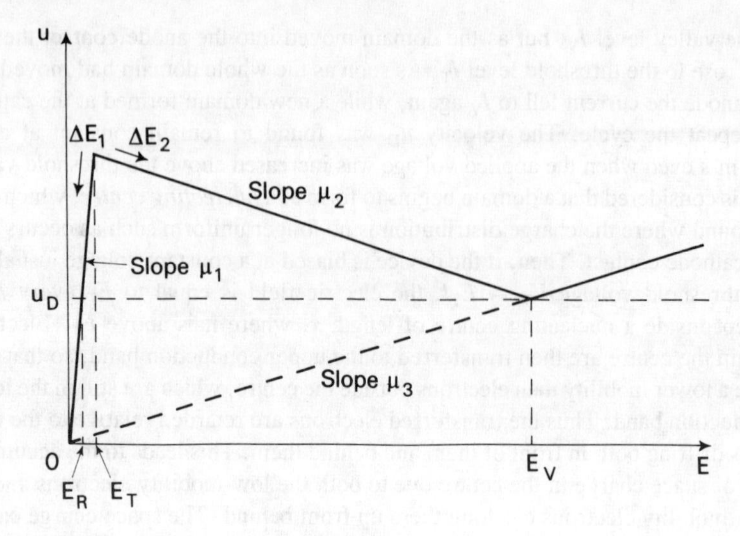

Figure 9.25 Idealized drift-velocity/field characteristic of a Gunn device.

$$\frac{\Delta E_2}{\Delta E_1} = \frac{l - x}{x} \tag{9.25}$$

Since x is considerably less than l, $\Delta E_2 > \Delta E_1$ so that the field within the centre continues to rise rapidly until it has reached E_V and the field within the remainder of the device falls to E_R. At this stage the domain has formed and drifts towards the anode with velocity u_D, the carriers both inside and outside it moving with the same velocity, so that, from Fig. 9.25,

$$u_D = \mu_1 E_R = \mu_3 E_V \tag{9.26}$$

Domain width

The changes in drift velocity from the threshold value are $\mu_1 \Delta E_1$ outside the domain and $\mu_2 \Delta E_2$ inside it, and since both drift velocities start and finish at the same values,

$$\frac{\mu_1 \Delta E_1}{\mu_2 \Delta E_2} = \frac{\mu_1 x}{\mu_2 (l - x)} = 1 \tag{9.27}$$

using eq. (9.25), which leads to

$$\frac{x}{l} = \frac{\mu_2}{\mu_1 + \mu_2} \tag{9.28}$$

for a device biased at the threshold voltage V_T. Inserting typical values for μ_1 and μ_2 gives $x/l = 1/26$. Thus x is much less than l, but it does increase with the bias voltage, since for an applied voltage V greater than V_T the equation

$$V = E_V x + E_R(l - x) \tag{9.29}$$

must be satisfied. Differentiation of this equation yields

$$\frac{dx}{dV} = \frac{l}{E_V - E_R} \approx \frac{l}{E_V} \tag{9.30}$$

which indicates that an increase in voltage does not affect the drift velocity u_D given by eq. (9.26), but instead it causes the width of the domain to increase. Inserting typical values in eq. (9.30) suggests a rate of increase of 0.2 μm/V for voltages above V_T.

Domain formation

Only one domain is formed at a time, since when the field starts to grow in a nucleating centre the field external to it is reduced below E_T and nucleation can occur nowhere else until the domain has reached the anode. Then the field rises to E_T everywhere in the device and a new domain can form. It is assumed that the domain field builds up to its steady-state amplitude in a time much shorter than the transit time, l/u_D. If this were not so it would grow as it drifted along the device and this could prevent its complete formation. Such a situation occurs in short devices of high-resistivity material, which have a low value of free electron density n_0. A minimum value of the product of doping density and length, $n_0 l$, is thus obtained for an oscillator, and has been found to be about $10^{16}/\text{m}^2$ ($10^{12}/\text{cm}^2$).

When $n_0 l \approx 10^{16}/\text{m}^2$ the domain length is about the same as the device length, so that the above analysis applies only to samples where $n_0 l$ is greater than $10^{16}/\text{m}^2$. However, for samples in which $n_0 l$ lies between 10^{15} and $10^{16}/\text{m}^2$ domain formation is prevented and a small signal amplifier is produced.

Modes of operation

The behaviour of a Gunn oscillator depends on the circuit conditions in which it operates. When the circuit is resistive, or the voltage across the device is constant, the period of oscillation is fixed at the transit time of a domain, l/u_D. However, when it is mounted in a resonant cavity the frequency of oscillation can be varied over about an octave by changing the cavity dimensions. In both cases the device is said to be operating in a *transit-time* mode, with the oscillation frequency controlled mainly by the sample length. Another mode of operation in which the frequency is controlled by the circuit rather than the sample length is the LSA (*limited space-charge accumulation*) mode. Here the field acting on the diode consists of the sum of the d.c. and the a.c. fields, and

$$E = E_0 + E_1 \sin 2\pi f t \tag{9.31}$$

The device is biased at a field E_0 well above threshold (Fig. 9.26(a)) and the period of the signal t_0 is less than the domain growth time constant t_g, so the total voltage swings below the threshold before the domain can form. Also t_0 is greater than the domain decay time constant t, so the accumulated space charge is removed during a small fraction of an a.c. cycle. Hence a domain never forms, the current carriers are considered as independent charges and the device behaves as a negative resistance, in a similar manner to the tunnel diode. However, there is a minimum frequency below which the time spent between E_T and E_0 is long enough for space charge to accumulate, and there is a minimum doping level below which the negative resistance is too high. Thus operation in the LSA mode is found to be restricted to values of n_0/f between 2×10^{10} and 2×10^{11} m^{-3} Hz^{-1}, giving a frequency range of about a decade for a device with a given doping density. Other modes of operation are also possible, as discussed more fully in Ref. [4].

An upper limit to the operating frequency is set by the length of the device. Since gallium arsenide can be deposited expitaxially (Chapter 7), an n^+-region is first deposited on an n-substrate (Fig. 9.26(*b*)). An n-epilayer is then deposited

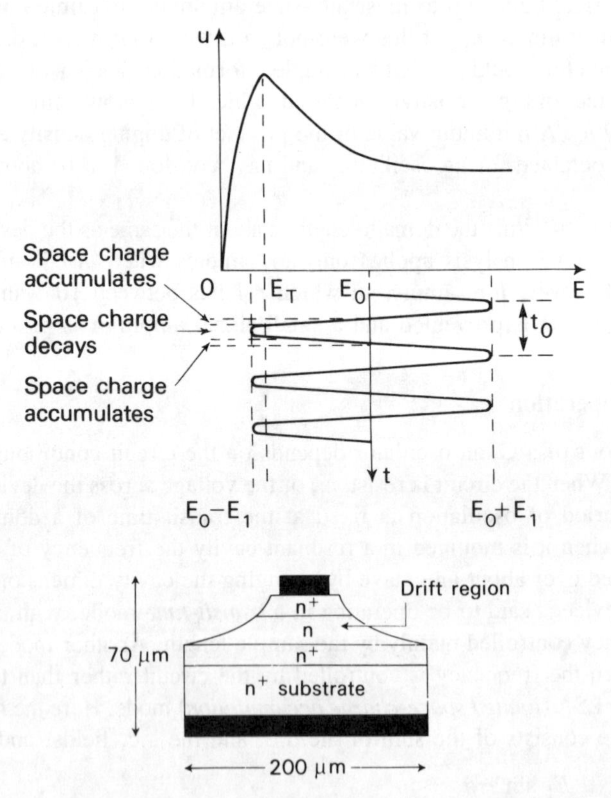

Figure 9.26 (*a*) LSA mode for a Gunn device; (*b*) cross-section through a Gunn device (not to scale).

to form the drift region, in which the Gunn effect takes place, followed by another n^+-region. Ohmic contacts are formed by evaporation of aluminium on to the outside n^+-regions.

Pulsed gallium arsenide oscillators can provide an output of over 1 kW at 2 GHz and about 0.5 W at 50 GHz. Continuous wave oscillators produce 2 W at 8 GHz down to 30 mW at 100 GHz. Continuous wave amplifiers have a gain of 6 to 12 dB over the range 4 to 15 GHz. The LSA mode provides the highest power of oscillations, maintaining a sinewave in the cavity. Pulsed and continuous wave oscillations are also available at outputs up to 1 W and frequencies up to 40 GHz from devices manufactured from indium phosphide.

Parametric amplification

In the most usual type of amplifier, such as that described in Chapter 4, the energy for operation is obtained from a direct voltage supply and an active device, such as a transistor or a valve, provides the essential amplifying element. In a *parametric amplifier*, however, the energy to operate the amplifier is obtained from an alternating voltage supply whose frequency is different from the signal frequency. In addition, the essential amplifying element is a circuit parameter, such as capacitance or inductance, which is varied at the supply or *pump* frequency.

Consider a capacitor C which has a charge q on its plates so that the resulting voltage across it is $v = q/C$. If the plates are pulled apart very quickly, C will fall and q will remain unchanged, so that v must rise proportionately. Thus the energy stored by the capacitor, $\frac{1}{2}Cv^2$, will, rise; this energy is supplied by the agency causing movement of the plates. If v is the instantaneous value of an alternating voltage, the plates may be pushed towards each other when $v = 0$ without any change in voltage.

In principle an inductor L may also be used, which has a current i flowing through it resulting in a flux Φ. Then $i = \Phi/L$, and if L can be reduced very quickly, i will be increased and the magnetic energy stored by the inductance, $\frac{1}{2}Li^2$, will also rise. In practice, it is difficult to obtain a suitable variable inductor, but a variable capacitor is readily available in the form of the depletion-layer capacitance of a reverse-biased junction diode (Chapter 3). The junction capacitance is a function of the reverse voltage, which may be varied at the desired pump frequency.

Suppose such a variable capacitance $C(t)$ forms part of a series circuit with a fixed inductance (Fig. 9.27) and is supplied with a signal V_s at the resonant frequency of the circuit. The output from the circuit is taken as the voltage V across $C(t)$. If the capacitance is constant initially at a value C and the instantaneous value of V is $v = V_m \sin\omega t$ then the mean energy W stored in the capacitance is $\frac{1}{2}CV_m^2$ (Fig. 9.28(*a*)). However, if $C(t)$ is reduced to $C - \delta C$, where $\delta C \ll C$, when v reaches its peak value both v and the stored energy will be increased (Fig. 9.28(*b*) and (*c*)). The capacitance then returns to C when $v = 0$ without affecting the stored energy, and $C(t)$ is changed at *twice* the signal frequency,

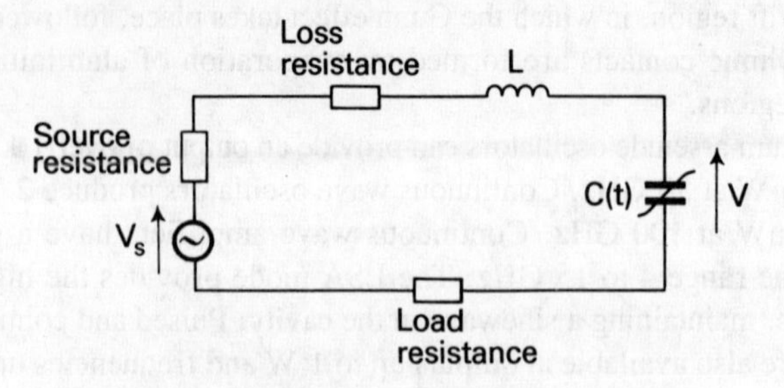

Figure 9.27 Simple parametric amplifier circuit.

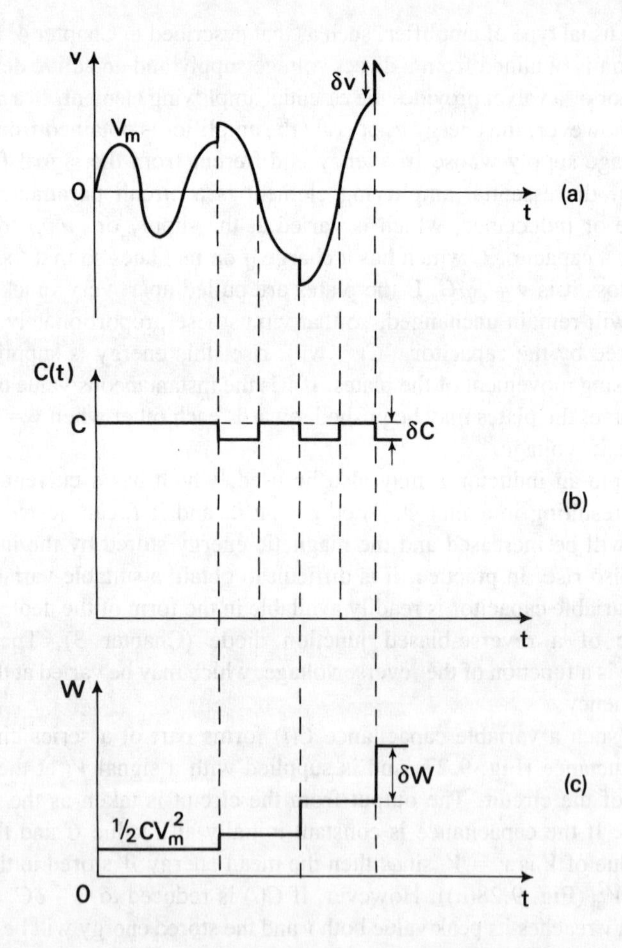

Figure 9.28 Waveforms for a degenerate parametric amplifier. (*a*) Capacitor voltage; (*b*) capacitance variation; (*c*) mean stored energy.

each time $C(t)$ is reduced both the voltage and the stored energy are increased, such an arrangement being called a *degenerate* parametric amplifier.

A practical circuit contains series resistance R_1, which is the sum of the resistances due to the signal source, the losses in the inductor and capacitor and the load resistance. At resonance, with the pump source disconnected, the current in the circuit is then V_s/R_1. When the pump source is connected power is supplied *to* the circuit, which means that a *negative* series resistance, $-R_2$, is introduced.* The total circuit resistance then becomes $R_1 - R_2$, which is still positive, and the current rises to $V_s/(R_1 - R_2)$, causing v to rise also. Care must be taken to control the power supplied by the pump so that the magnitude of R_2 is always less than that of R_1, because if $R_2 = R_1$ the current tends to become infinite and independent of V_s. Thus the circuit would generate its own current and become an oscillator, so that in a practical amplifier the gain is normally not greater than about 30 dB.

When the pump source is connected, v will rise until the energy added per cycle of capacitance change equals the energy dissipated in $R_1 - R_2$ in the same time (Fig. 9.29). This enables the power supplied by the pump to be calculated. Since at any instant the capacitor voltage is $v = q/C$, then for a change in capacitance δC the corresponding change in voltage will be δV (Fig. 9.28(a)), where

$$\delta v = \frac{dv}{dC}\,\delta C \tag{9.32}$$

$$= -\frac{q}{C^2}\,\delta C = -\frac{v}{C}\,\delta C \tag{9.33}$$

and v is the voltage at the instant at which C is changed. At any instant the energy stored in the capacitor is $\frac{1}{2}Cv^2$, so that when both C and v are changed together the corresponding change in energy will be δW, where

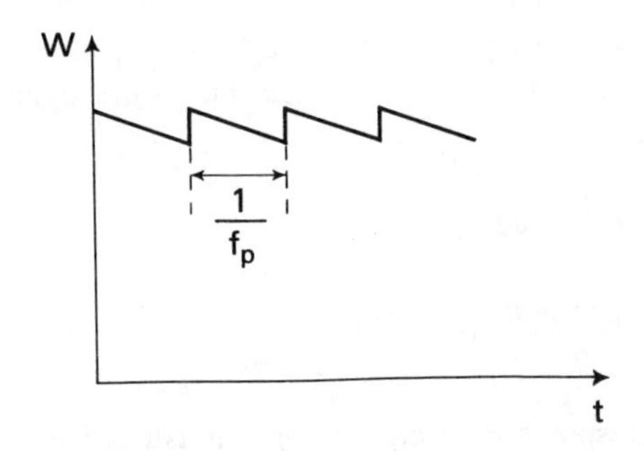

Figure 9.29 Energy stored in a varactor diode at equilibrium.

* Power is absorbed by a positive resistance and so power is supplied by a negative resistance.

$$\delta W = \frac{\partial W}{\partial C}\delta C + \frac{\partial W}{\partial v}\delta v$$

$$= \tfrac{1}{2}v^2\delta C + vC\delta v \tag{9.34}$$

$$= \tfrac{1}{2}v^2\delta C - v^2\delta C$$

or

$$\delta W = -\tfrac{1}{2}v^2\delta C \tag{9.35}$$

It should be noted that δW is positive when δC is negative. Then at equilibrium the energy supplied is dissipated before the next capacitance change in time $1/f_p$, where f_p is the pump frequency. Hence the power supplied, p_p, is given by

$$p_p = -\frac{v^2}{2} f_p \delta C \tag{9.36}$$

and for a pump waveform which is rectangular, as in Fig. 9. 28(*b*), $v = V_m$. thus power is supplied by the pump when δC is negative and this power is proportional to the pump frequency. When δC is positive, power would be absorbed by the pump if v were not zero at the same time.

When the frequency of the pump source is exactly twice that of the signal, as in the above circuit, the amplifier is said to be *degenerate*. It is difficult to maintain the correct phase relationship between the pump and signal waveforms in practice, owing to lack of control over the signal frequency. However, greater flexibility in operation and improved power gain are obtained if a second tuned circuit is connected across the variable capacitance (Fig. 9.30).* This is known as an *idler* circuit, since it contains no source of energy, but it does allow the pump to work at a frequency other than twice the signal frequency by modifying the voltage v across $C(t)$. If the signal source is made zero and the pump source applied on its own the idler and signal circuits are excited into oscillation at their respective resonant frequencies, so that v is the sum of the voltages due to the signal and idler circuits. Suppose these voltages are equal, with $v_L = V_m \sin \omega_i t$ due to the idler circuit and $v_s = V_m \sin \omega_s t$ due to the signal circuit. Then

$$v = v_i + v_s \tag{9.37}$$

$$= V_m(\sin \omega_i t + \sin \omega_s t) \tag{9.38}$$

$$= 2V_m \sin \frac{(\omega_i + \omega_s)t}{2} \cos \frac{(\omega_i - \omega_s)t}{2} \tag{9.39}$$

The waveform corresponding to eq. (9.39) is illustrated in Fig. 9.31(*a*) and it may be seen that the voltage is zero when either

* A pump circuit resonant at f_p can only be used when the pump waveform is sinusoidal, as in the practical form of the amplifier.

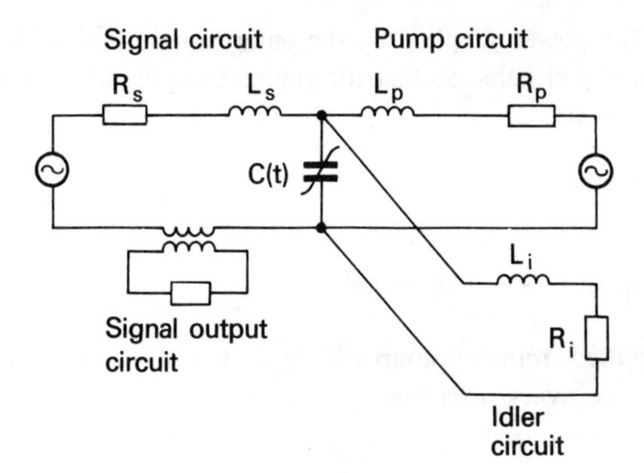

Figure 9.30 Parametric amplifier with idler circuit. At microwave frequencies the tuned circuits are formed by waveguides.

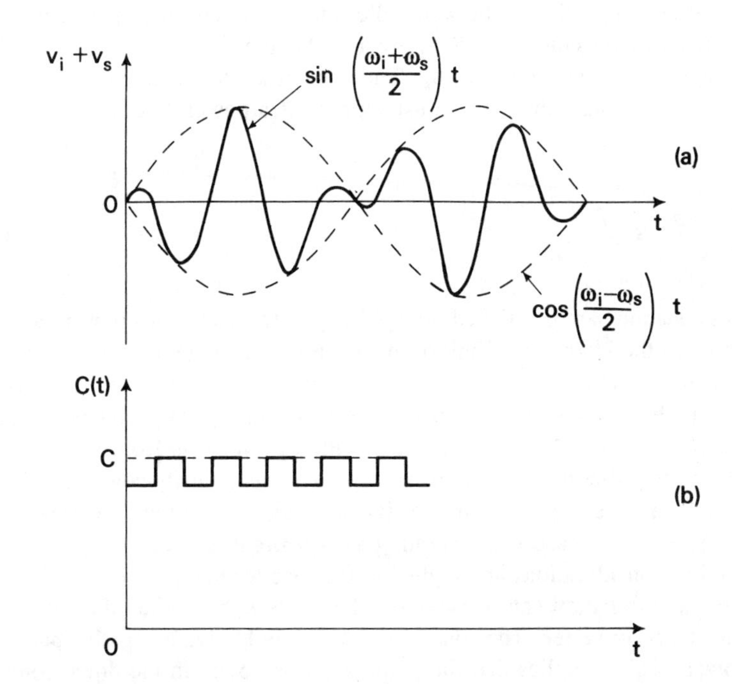

Figure 9.31 Waveforms for a non-degenerate parametric amplifier.

$$\sin\frac{(\omega_i + \omega_s)t}{2} = 0 \quad \text{or} \quad \cos\frac{(\omega_i - \omega_s)t}{2} = 0$$

Thus if $\omega_p = \omega_i + \omega_s$, $C(t)$ can be increased without loss of energy as before, and similarly for $\omega_p = \omega_i - \omega_s$. In either case $C(t)$ may not be decreased exactly at the peak of the waveform, and the amplitudes of consecutive peaks will not be the same. The power supplied by the pump source is then divided between the idler and signal circuits, so that for any given cycle of the pump waveform when $\omega_p = \omega_i + \omega_s$,

$$p_p = p_i + p_s \tag{9.40}$$

$$= -\frac{v^2}{2} f_p \delta C = -\frac{v^2}{2} (f_i + f_s)\delta C \tag{9.41}$$

The power supplied from the pump will depend on the value of v at which $C(t)$ is reduced, but on every occasion

$$\frac{p_i}{p_s} = \frac{f_i}{f_s} \tag{9.42}$$

and pump power is divided between idler and signal circuits in the same ratio as their respective frequencies. Suppose now that the signal source is applied and supplies power P_s at frequency f_s. Then, in order to maintain the condition implied in eq. (9.42), the pump must supply extra power P_i at frequency f_i, so that

$$\frac{P_i + p_i}{P_s + p_s} = \frac{f_i}{f_s} \tag{9.43}$$

and this equation can be satisfied only if $P_i/P_s = f_i/f_s$ also. Thus power gain from the signal to the idler circuit will occur when $f_i > f_s$, together with an increase in frequency, and in this condition the amplifier is known as an *up-converter*. In some applications it may be necessary to reduce the signal frequency, and this will be achieved when $f_i < f_s$, together with a power loss from signal to idler circuit, the amplifier then becoming a *down-converter*. Finally power at the signal frequency can be extracted from the signal circuit, so that the power gain is $(P_s + p_s)/P_s$, in which case it is operating as a *straight amplifier*.

If similar considerations are applied to the case when $\omega_p = \omega_i - \omega_s$, it will be seen that amplification can occur only when $f_i > f_s$, i.e. when the amplifier is used as an up-converter. The condition $f_i > f_s$ would give a negative pump output power, which implies that the pump absorbs power in the down-converter condition. Similarly, when the circuit is operated as a straight amplifier, power is absorbed at the signal frequency, which is again undesirable.

Practical form of the parametric amplifier

An important difference between the amplifier described above and a practical amplifier is that the pump waveform will be sinusoidal rather than rectangular. Also v_s and v_i need not be equal, which leads to the waveform of Fig. 9.32(*a*).

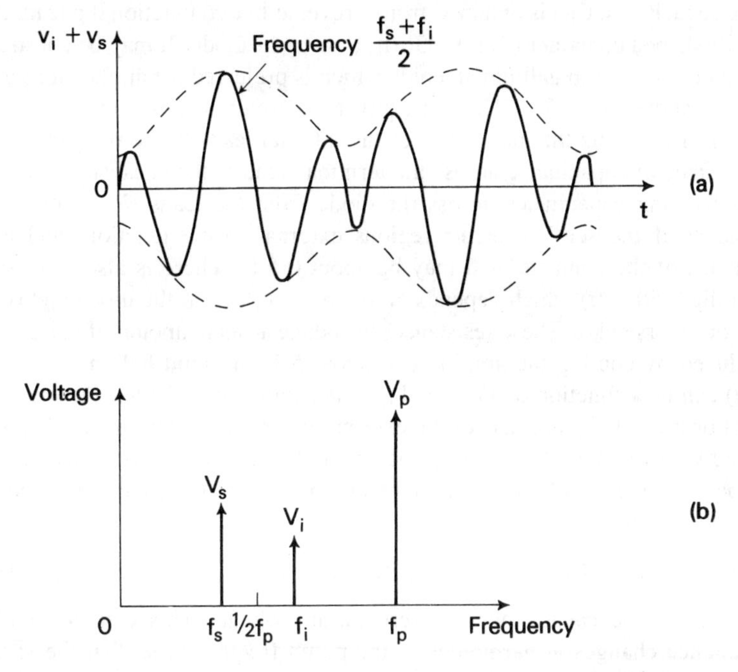

Figure 9.32 Conditions for a practical parametric amplifier with $f_p = f_s + f_i$.

The relationships between the frequencies of the signal, idler and pump voltages are shown in Fig. 9.32(*b*), the voltage relationships being only approximate.

Since the change of $C(t)$ is not instantaneous some variation of charge will occur and the analysis becomes much more complex, but the conclusion is still reached that power is a function of frequency as implied by eq. (9.42). A complete analysis was given by Manley and Rowe in 1956 (Ref. [6]).

The parametric amplifier is very useful in situations where the signal strength is very low, for instance in a satellite communication receiver where the signal power received from the satellite may be only about 10^{-13} W. Thus the noise power introduced by the amplifier must be even lower if the signal is not to be lost in the noise. Since its essential amplifying element is a capacitance, which would introduce no noise at all if it were perfect, a parametric amplifier operated at a low temperature is used as the first stage in the earth receiver. The earth's atmosphere allows electromagnetic waves within certain ranges of frequency to pass through it more easily than others, and so satellite communication systems are operated between 4 and 8 GHz, since this frequency range corresponds to a 'window' of easy transmission. The bandwidth of a typical amplifier is 500 MHz, which is similar to that of a travelling-wave-tube amplifier. This corresponds to a straight amplifier operating with f_s = 4 GHz, f_p = 28 GHz and f_i = 24 GHz, providing a power gain of 400 times, or 26 dB.

The capacitance $C(t)$ is obtained from a reverse-biased junction diode mounted in a pill-shaped container (Fig. 9.33(*a*)), a *varactor* diode. It may be constructed from silicon, or from gallium arsenide which is preferred for applications at frequencies above about 20 GHz. An equivalent circuit is shown in Fig. 9.33(*b*), where L_s represents the inductance of the internal leads to the p-n junction. C_p is the effective capacitance across the terminals due to the encapsulation and C_f is the fringing capacitance across the diode from the leads. R_{sc} is due to the resistance of the semiconductor regions external to the junction, and to the resistance of the contacts, and may be about 0.5 Ω. There is also a resistance in parallel with $C(t)$ which depends on the bias applied to the diode and is very high for reverse bias. These resistances introduce a small amount of noise, which is reduced by cooling the amplifier to about 5 K in liquid helium.

$C(t)$ can be a function of $V_R^{-1/2}$ or $V_R^{-1/3}$, depending on whether the junction is abrupt or graded. V_R consists of the externally applied reverse bias voltage with a pump voltage alternating at frequency f_p, and owing to the non-linear relationship between $C(t)$ and V_R for either type of junction, the capacitance may be written in the form

$$C(t) = C_0 + C_1 \cos \omega_p t + C_2 \cos 2\omega_p t + C_3 \cos 3\omega_p t + \ldots \quad (9.44)$$

where C_0 is the capacitance for zero pump voltage. This equation includes capacitance changes at harmonics of the pump frequency, so that the varactor diode can also be used as a harmonic generator. In the parametric amplifier the varactor diode is effectively operated in a tuned circuit resonant at f_p, so that eq. (9.44) becomes

$$C(t) = C_0(1 + \gamma \cos \omega_p t) \quad (9.45)$$

where $\gamma = C_1/C_0$. In practice γ has values up to about 0.3 and C_0 is typically

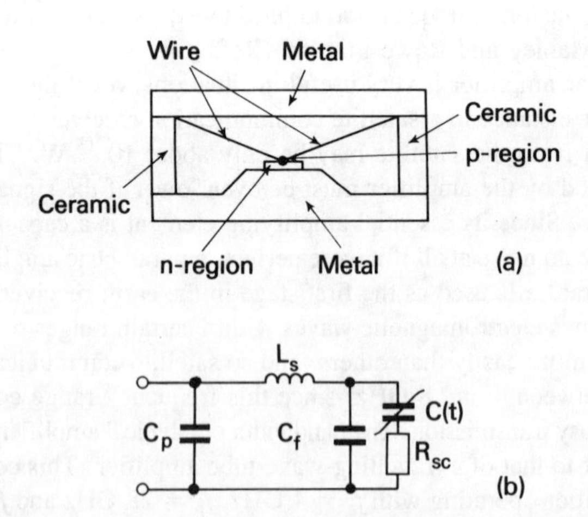

Figure 9.33 Varactor diode. (*a*) Practical form; (*b*) equivalent circuit.

1 pF, both γ and C_0 being functions of V_R. It may be seen from the equivalent circuit (Fig. 9.33(*b*)) that there are two resonant frequencies. The series resonant frequency is

$$f_0 = \frac{1}{2\pi\sqrt{(LC_0)}}$$

which is in the gigahertz region since L is about 1 nH, while the parallel resonant frequency is about $3f_0$. The diode may be operated at f_0 for maximum current, or below f_0 where the overall combination is capacitive. The pump may be a reflex klystron or a solid-state source of microwave power.

Frequency multiplication

At the present time, a common technique for generating low-power microwave signals using solid-state devices is to start with a transistor oscillator working at about 100 MHz. The frequency of the output from the oscillator is then multiplied in stages until the desired frequency is achieved. One method uses the oscillator output as the pump waveform of a varactor diode. This would be mounted in a cavity resonant at 400 MHz, say, from which is extracted the power at $4f_p$ generated by the varactor diode (eq. (9.44)). Two similar frequency multipliers then raise the frequency to 1.6 GHz and 6.4 GHz respectively.

A second method uses a *step recovery diode* as a frequency multiplier. When a p^+-n diode is used as a rectifier the excess charge established within the n-region during the conduction half-cycle needs a short time to disperse when the diode is cut off. This depends on the lifetime of holes in the n-region, and at microwave frequencies conduction continues after the alternating voltage has reversed polarity (Fig. 9.34). Such a waveform has a high harmonic content,

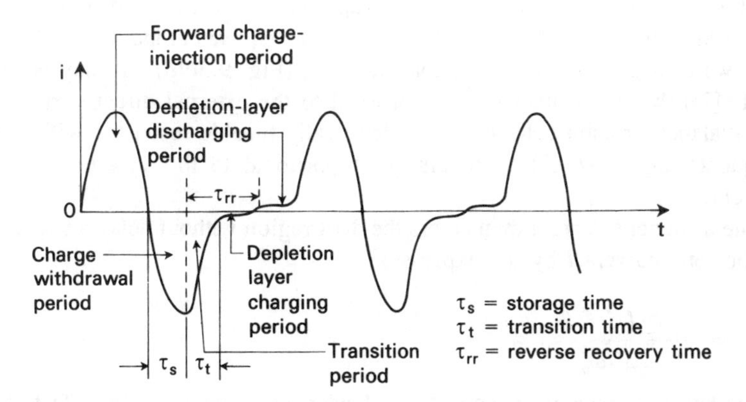

Figure 9.34 Microwave current waveform of a sinusoidally switched step-recovery diode.

and the step recovery diode can give useful frequency multiplication up to ten times the signal frequency. Where frequency multiplication is not required, the storage property is undesirable so that a Schottky diode is used for rectification at microwave frequencies. Here the current is carried by majority carriers and the current can be switched off within about 0.1 ns since there is no recombination.

IMPATT diodes

Continuous wave microwave power can also be generated directly by means of a reverse-biased p-n junction, as first suggested by Read in 1958. He proposed a p^+-n-i-n^+ structure (Fig. 9.35(*a*)), where *i* stands for an intrinsic region which has very few free charges within it. In practice, since it is not possible to maintain instrinsic resistivity in the i-region during processing, it becomes lightly doped p- or n-type and a p^+-n-n^+ or p^+-p-n-n^+ structure is used (Figs. 9.35(*c*) and (*e*)). These are known as single drift and double drift IMPATT (IMPact Avalanche Transit Time) diodes respectively. The Read diode may be considered as an ideal IMPATT diode and these devices produce a negative resistance by combining avalanche breakdown with electron transit time effects. In the Read and single drift IMPATT diodes the p^+-n junction is reverse biased giving a high field at the junction which falls towards the n^+-region (Figs. 9.35(*b*) and (*d*)). The peak field is about 5×10^7 V/m so that breakdown occurs near this peak in the *avalanche region*. Electron-hole pairs are generated at the p^+-n junction and electrons move into the *drift region*. The field remains high enough to cause constant carrier velocity through most of the drift region, giving a fixed transit time τ.

An a.c. field is superimposed on the d.c. field when oscillations commence and the current in the avalanche region has two alternating components, the avalanche current i_a and the displacement current i_d due to the effective capacitance across the avalanche region. The ionization of atoms by the impact of electrons is relatively slow and the current will reach a maximum value only *after* the total field has reached its maximum. Thus i_a lags behind the a.c. field, in practice by 90°, and so is represented as flowing through an equivalent inductance L_a, while i_d flows through an equivalent capacitance C_a (Fig. 9.36(*a*)). It may be shown (Ref. [7]) that L_a is inversely proportional to the external direct current I, so the avalanche region behaves as an electrically tunable resonator with angular frequency $\omega_a = (L_a C_a)^{-1/2}$ inversely proportional to the square root of the current.

The avalanche current then enters the drift region without delay and is related to the total current I by the expression

$$i_a = \frac{I}{1 - \omega^2/\omega_a^2} \tag{9.46}$$

which becomes negative when $\omega > \omega_a$. Under these conditions the effect of transit time is to introduce an extra 90° phase lag between the current I and the a.c.

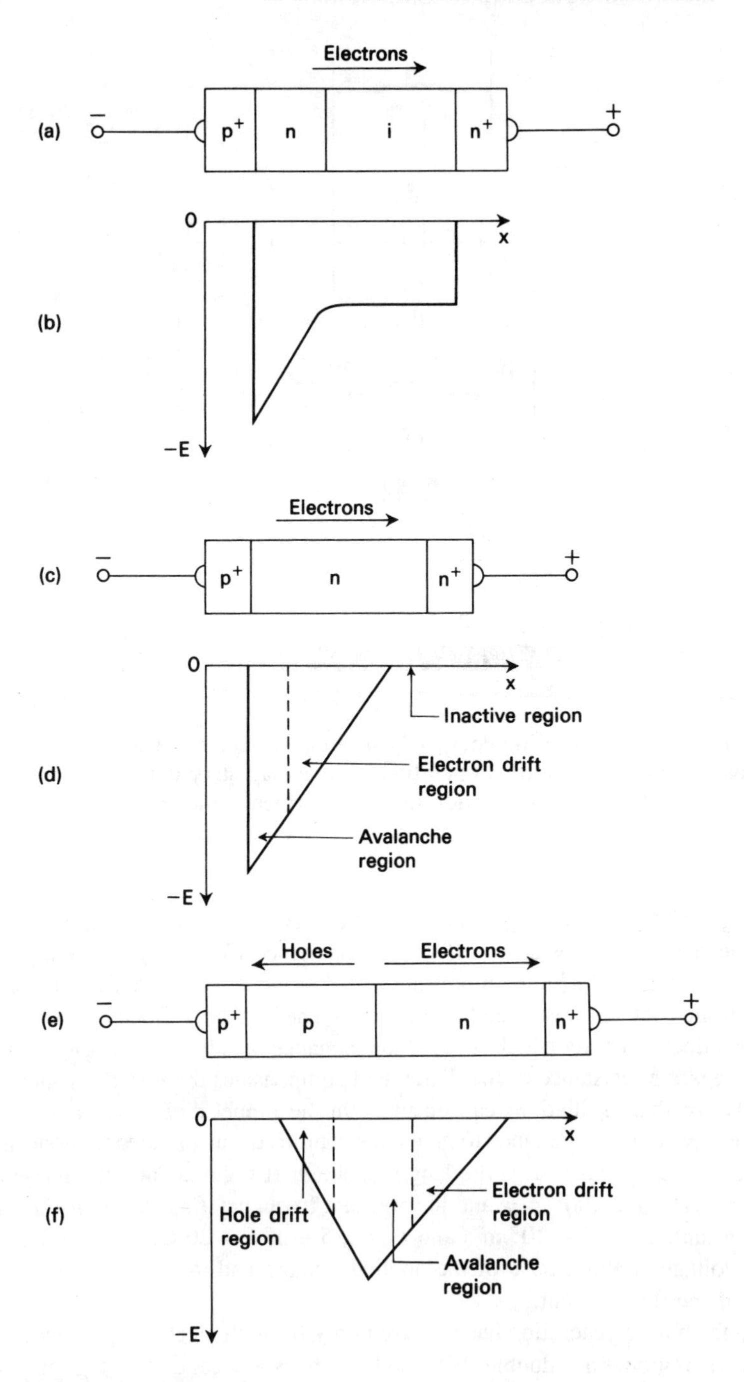

Figure 9.35 IMPATT diodes. (*a*) Read diode structure; (*b*) electric field; (*c*) single drift structure; (*d*) electric field; (*e*) double drift structure; (*f*) electric field.

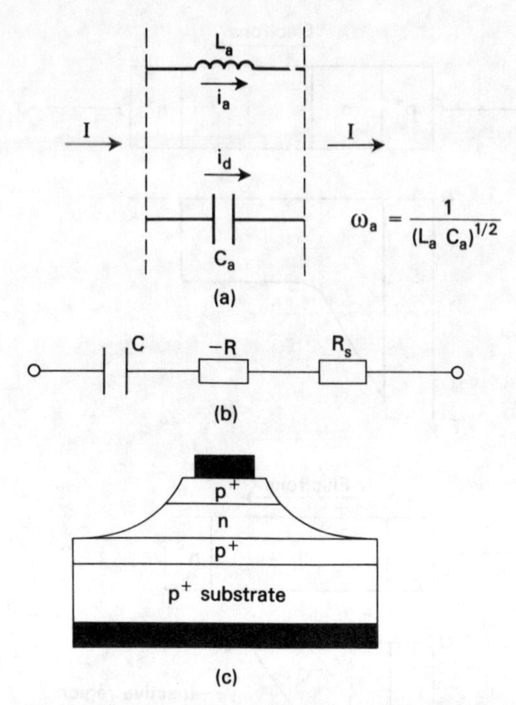

Figure 9.36 IMPATT diode. (*a*) Equivalent circuit of avalanche region; (*b*) equivalent circuit for normal operation with $\omega > \omega_a$, stray inductance and capacitance is omitted; (*c*) cross-section through a typical structure.

voltage. Then the drift region behaves as a negative resistance $-R$, and the avalanche region behaves as an effective capacitance C, leading to the equivalent circuit of Fig. 9.36(*b*). R_s represents the resistance of the inactive region and the ohmic contacts at each end of the diode. The IMPATT diode can be mounted in an inductive cavity which is tuned to resonance by adjusting the diode current. The negative resistance of the diode then compensates the loss resistance of the cavity, so that oscillations can occur as in the tunnel diode oscillator.

The cavity should be tuned to give a resonant frequency related to diode length, since $\tau = L/u_d$, where L is the length of the drift region. The total phase angle is $\pi = w\tau$ so that $2\pi f\tau = \pi$ and the resonant frequency $f = 1/2\tau = u_d/2L$, which is constant. For $u_d = 10^5$ m/s and $L = 2.5$ μm, $f = 20$ GHz. In practice, the a.c. voltage is about 50% of the applied voltage and the a.c. current is about $2/\pi$ of the d.c. current.

So far power generation has occurred only through electrons passing through the drift region. In a double drift diode there is a second drift region through which holes pass, giving increased power generation, and higher output and efficiency than the single drift diode (Fig. 9.35(*f*)). Both types of IMPATT diode are produced in gallium arsenide as well as in silicon, with continuous wave output powers up to 10 W at 10 GHz. For outputs below 1 W the gallium arsenide devices operate up to 50 GHz and the silicon devices up to 100 GHz. Pulsed outputs are obtainable between 100 W near 10 GHz and 5 W at 50 GHz from silicon devices.

TRAPATT diodes

When the current in an IMPATT diode is allowed to increase there is a corresponding increase in the length of the avalanche region until it fills the drift region. The charge density is so high that the conditions approximate to those in the plasma of a gas discharge (Chapter 8). The plasma is then extracted by means of the electric field, after which the voltage across the diode terminals returns to the breakdown value until the plasma has re-formed. This is known as the TRApped Plasma Avalanche Transit Time or TRAPATT mode and the same structure can be operated as an IMPATT diode at lower currents or as a TRAPATT diode at higher currents but lower frequencies (Fig. 9.36(*c*)). Pulsed TRAPATT oscillators have been operated at 10 W output at 100 GHz and 1000 W output at 10 GHz.

The p-i-n diode

An ideal p-i-n diode has an intrinsic (i) region sandwiched between a p^+- and an n^+-region (Fig. 9.37(*a*)). As mentioned under IMPATT diodes the i-region is in practice lightly doped, either p- or n-type. When a reverse bias *below* the breakdown value is applied all the carriers are swept out of the i-region and the diode has capacitance given by

$$C_j = \epsilon A/d$$

where A is the diode area and d the width of the i-region. Consequently C_j remains almost constant as the reverse bias changes, in contrast to the reverse bias capacitance of a p-n junction, and is typically about 1 pF. A true i-region is depleted of carriers even at zero bias but if the region is lightly doped, then carriers are present at zero bias and a small reverse bias has to be applied before they are swept out. Thus R_i and C_i are introduced into the equivalent circuit (Fig. 9.37(*b*)) to represent the resistance and capacitance of the i-region outside the swept-out regions. R_s is the series resistance of the p- and n-regions and is typically less than 1 Ω, while any series inductance is normally negligible at operating frequencies.

The circuit of Fig. 9.37(*b*) may be transformed to that of Fig. 9.37(*c*) for the normal conditioon of $(\omega C_i R_i)^2 \gg 1$. This applies for low reverse bias and reduces to the circuit of Fig. 9.37(*d*) for normal reverse bias where $R_i \rightarrow 0$ and $C_i \rightarrow \infty$. The normalized capacitance and resistance are shown in Fig. 9.38(*a*) as a function of reverse bias (Ref. [4]). Here C_j remains about constant at the valve defined above for a reverse bias beyond about 5 V. R_i tends towards zero as it

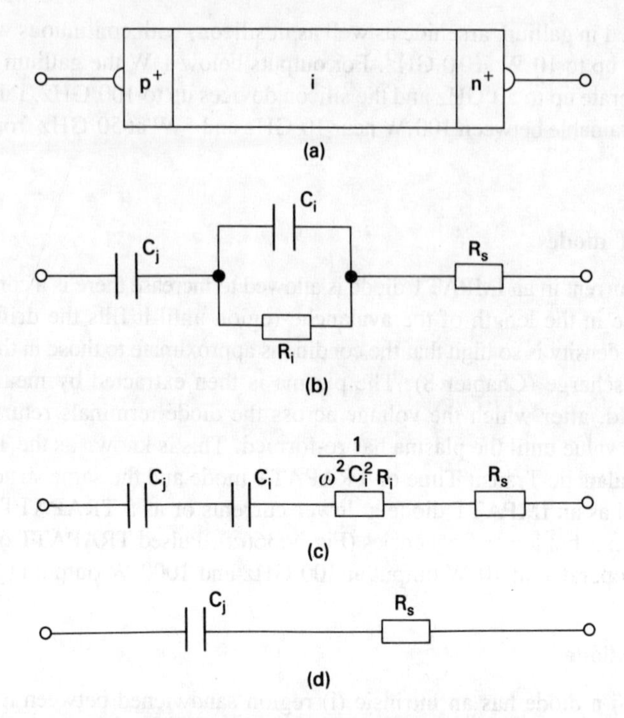

Figure 9.37 (*a*) p-i-n diode; (*b*) equivalent circuit; (*c*) equivalent circuit for low reverse bias; (*d*) equivalent circuit for normal reverse bias. Stray inductance and capacitance is omitted.

includes carriers swept out of the i-region, so that $R_i + R_s$ becomes constant at R_s.

Under forward bias the i-region will receive holes injected from the p-region and electrons injected from the n-region. Where the densities of these current carriers are p and n respectively, which are proportional to the forward current I_F, the conductivity of the i-region is

$$\sigma_i = e(\mu_p p + \mu_n n) \tag{9.47}$$

and the resistance of the *i* region is

$$R_i = \frac{d}{\sigma_i A} \tag{9.48}$$

σ_i is proportional to I_F, so R_i is proportional to I_F^{-1} as indicated in Fig. 9.38(*b*), where I_F is forward current.

The p-i-n diode is applied as a microwave switch, since it has low impedance at moderate forward current and constant capacitance representing high impedance for reverse bias. When the diode is connected across a waveguide the signal is

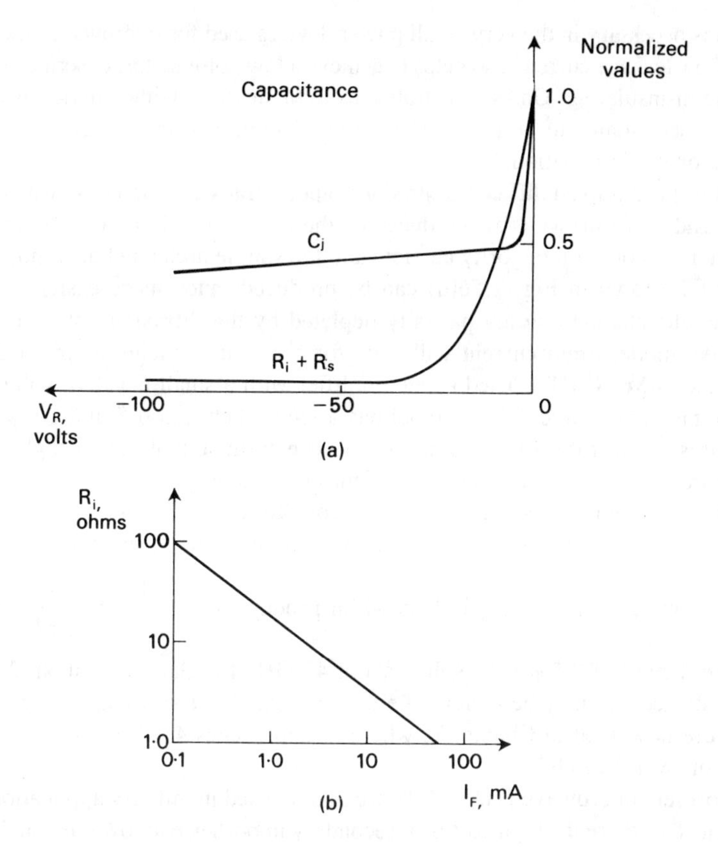

Figure 9.38 p-i-n diode. (*a*) Reverse capacitance and resistance each normalized to the value at $V_R = 0$, typically about 2.0 pF and 100 Ω respectively. (*b*) Variation of intrinsic resistance R_i with forward current I_F.

then either reflected back or allowed to proceed almost unchanged. This allows the device to modulate signals up to the GHz range. The forward resistance can also be varied continuously from small to large values by changing the bias, so that the diode can be used as a variable attenuator. A further application under reverse bias is the p-i-n photodiode, discussed in Chapter 5.

Gallium arsenide FETs

As discussed in Chapter 7, gallium arsenide is preferable to silicon due to its higher electron mobility and energy gap. For microwave applications the high mobility of 0.85 $m^2/V\ s$ leads to high electron velocity at low fields (Fig. 9.23) and hence to short transit times and wide frequency response. The large energy gap of 1.43 eV reduces leakage current and allows a higher working temperature, which is necessary in the very small power devices used for microwave amplification. Low leakage current also helps to achieve a low noise factor, described below. The semi-insulating GaAs substrates lead to devices with smaller parasitic capacitances than can be achieved with conducting substrates having isolating diodes or oxide isolation.

So far it has proved difficult to produce bipolar transistors in GaAs with a good yield, and performance is also reduced by the low hole mobility of 0.04 m^2/V s. A FET relies only on majority carriers and the simple n-channel depletion-mode MESFET shown in Fig. 7.56(*a*) can be produced much more easily.

Since the channel is only partially depleted by the diffusion potential of the Schottky diode, drain current will flow for zero gate voltage, as in the JFET. Thus the D-MESFET is used as an amplifier with a small negative voltage between gate and source. The channel has a length between 0.5 and 1.0 μm and since it is so short it is mostly pinched off. The drain-source field is high enough to ensure electron velocity saturation. This occurs at about 1.4×10^5 m/s, compared with about 6.5×10^4 m/s for silicon (Ref. [15], Chapter 7).

For a saturation velocity u_s and channel length L, the electron transit time $\tau = L/u_s$, which leads to a gain bandwidth product $f_T = \dfrac{1}{2\pi\tau} = \dfrac{u_s}{2\pi L}$. For a channel length of 0.5 μm f_T is then about 45 GHz for GaAs and about 21 GHz for Si devices. f_T may be increased either by reducing L or by using the HEMT structure described in Chapter 7, where u_s approaches 4×10^5 m/s leading to an f_T of over 100 GHz.

At present microwave GaAs FETs are mainly used in military applications and satellite TV where the high cost is of secondary importance to low noise and broad band properties (Ref. [3]). One or two stages on a single chip of area up to 16 mm^2 can provide up to 10 dB gain between 2 and 20 GHz, with noise factors between 2 and 5 dB and a power output of 1 W or less. Higher power amplifiers with up to four stages on a single chip of area up to 45 mm^2 can provide up to 32 dB of gain. The bandwidth is narrower and centred on frequencies between 7 and 28 GHz, and the output power can be up to 3 W.

Electrical noise

The microwave devices described above are often used in situations where the signal power is very small, for instance in the receiver of a satellite communication system or a radio telescope. The minimum signal that can be observed is set by the level of *electrical noise* in the system, which is a random fluctuation of voltage and current or electromagnetic fields. It is present in all electronic devices and components and also in the atmosphere. An indication whether satisfactory amplification can be obtained is given by the ratio of signal power to noise power, because if this ratio is only just above unity it will be difficult to decide whether the signal is present at all. Thus when weak signals are to be amplified it is essential that the noise power introduced by the devices and components should be as small as possible, and this is true at all frequencies and not just at microwave frequencies.

Two of the main sources of noise in electronic devices are *thermal noise* and *shot noise*. Thermal noise is due to the random motion of the current carriers in a metal or semiconductor which increases with temperature (see Chapter 2). If a resistor is short-circuited, the mean current is zero but the instantaneous current varies about zero in a random fashion and with very small amplitude. Measurements show that all frequencies within the bandwidth B of the measuring instrument are equally present in the random current. The phenomenon was studied in detail by Johnson and Nyquist in 1928. They showed that the thermal noise power available from a resistor is given by the expression

$$P_n = kTB \tag{9.49}$$

where k is Boltzmann's constant and T is the absolute temperature.

The resistor can be represented by a noise-free resistor R in series with a generator producing $\overline{v^2}$, the mean-square noise voltage (Fig. 9.39(a)). If the resistor is short-circuited the mean-square noise current flowing is $\overline{v^2}/R$, which represents the sum of the individual noise currents at all the frequencies present within the bandwidth considered. Maximum noise power will be extracted if a second noise-free resistor R is connected across the first (Fig. 9.39(b)). The total power is then $\overline{v^2}/2R$ and the power dissipated in the second resistor is half the total power, given by

$$P_n = \frac{\overline{v^2}}{4R} = kTB \tag{9.50}$$

so that

$$\overline{v^2} = 4kTRB \tag{9.51}$$

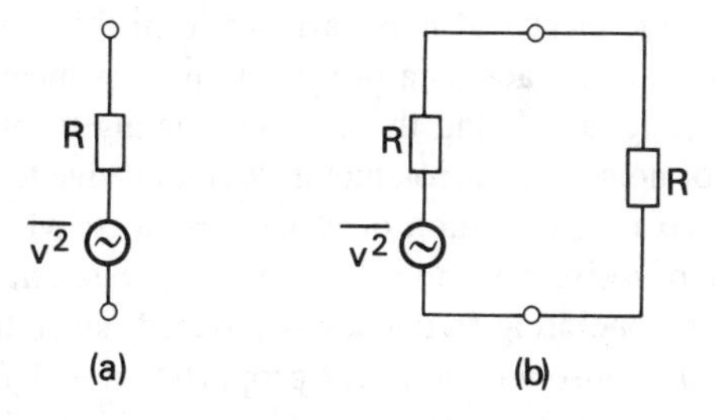

Figure 9.39 Thermal noise. (a) Noise voltage of a resistor; (b) condition for maximum transfer of noise power.

Worked example

A resistor is operated at room temperature over a bandwidth of 5 MHz. Determine the mean-square and r.m.s. noise voltages.

Solution

The mean-square noise voltage is given by

$$\overline{v^2} = 4kTRB$$
$$= 4 \times 1.38 \times 10^{-23} \times 293 \times 10^3 \times 5 \times 10^6$$
$$= 8.09 \times 10^{-11}\ \mathrm{V}^2$$

The r.m.s. value is the square root of the mean-square value, or 9.0 μV.

Equation (9.51) applies to the effective resistance of any circuit however complicated and represents its noise contribution in terms of voltage fluctuation. Pure reactances, having zero resistance, contribute no electrical noise.

Shot noise is due to the random flow of electrons in an electric current and is due to the particle nature of electric charge. Thus, for example, in a saturated diode at a given cathode temperature, the current has a fixed average value with very small random fluctuations about this value. These constitute a mean-square noise current, which was shown by Schottky to be

$$\overline{i_n^2} = 2eI_S B \tag{9.52}$$

where I_S is the average saturated current. Under space-charge-limited conditions i_n^2 is reduced, since the electrons no longer move independently, the current being controlled by the electrons in the space charge. Then the noise current is

$$\overline{i_n^2} = \Gamma^2 2eI_S B \tag{9.53}$$

where $\Gamma \leq 1.0$ and is called the *space-charge reduction factor*. It has a typical value of about 0.3 for the oxide cathode of a receiving valve. Shot noise is also extended to include the electrical noise arising from the random nature of the diffusion of current carriers across a p-n junction. The mean-square current is given by eq. (9.52), I_S now being the current flowing through the junction.

Another source of noise in semiconductor devices is due to the random nature of the generation and recombination of charge carriers, which in effect causes a small, fluctuation of resistance and is known as *recombination noise*. A further source appears as the operating frequency is reduced, since the total noise rises above the level of shot nose with a power proportional to $1/f$. This is known as *flicker noise* and is strongly dependent on surface conditions. It becomes apparent below about 1 kHz for bipolar transistors and JFETs, and below 100 kHz for MOSTs, which rely upon surface phenomena for their operation. *Partition noise* is due to the random manner in which a stream of electrons divides amongst a number of electrodes, for instance between the screen grid and anode of a pentode.

Noise factor

The noise introduced by an amplifier will be due to a combination of some or all of the sources described above. They will give rise to an output noise voltage $\overline{v_0^2}$, which may be conveniently represented by an input noise voltage given by

$$\overline{v_0^2} = \frac{\overline{v_0^2}}{A^2} \tag{9.54}$$

where A is the voltage gain of the amplifier. If the noise voltage introduced with the signal is $\overline{v_g^2}$, and the mean-square signal voltage is $\overline{v_s^2}$, the signal/noise ratio at the input is

$$\frac{S}{N} = \frac{\overline{v_s^2}}{\overline{v_r^2} + \overline{v_g^2}} \tag{9.55}$$

This ratio depends on the signal level, which is not necessarily constant, and the bandwidth of the amplifier. A useful parameter which is independent of signal level and expresses the degradation of the signal/noise ratio when the amplifier is introduced is the *noise factor* (or *noise figure*) given by

$$F = \frac{\overline{v_r^2} + \overline{v_g^2}}{\overline{v_g^2}} = 1 + \frac{\overline{v_r^2}}{\overline{v_g^2}} \tag{9.56}$$

Then in a noise-free amplfier $\overline{v_r^2} = 0$ and $F = 1$. The noise factor in decibels is $10 \log_{10}F$, which becomes 0 dB for a noise-free amplifier.

Noise input temperature, T_i

When the noise factor is close to unity, as in the cooled parametric amplifier, it is useful to express F in terms of absolute temperature. $\overline{v_g^2}$ is standardized as the noise from a source at a temperature of 290 K and the amplifier noise voltage $\overline{v_r^2}$ can be obtained in terms of the noise input temperature of the amplifier, T_i. Then, from eq. (9.56),

$$\frac{\overline{v_r^2}}{\overline{v_g^2}} = F - 1 = \frac{R_r}{P_g} \tag{9.57}$$

But $P_g = 290\ kB$ and $P_r = T_i kB$, so that

$$F - 1 = \frac{T_i}{290} \tag{9.58}$$

and

$$T_i = 290(F - 1) \text{ kelvin} \tag{9.59}$$

so that a noise-free amplifier would have $T_i = 0$ K. The electrical noise present in the atmosphere may also be described in terms of noise temperature, which reaches a minimum value of a few kelvins for signals of frequency between about 4 and 8 GHz. Thus this frequency range is very suitable for satellite communications, since the noise received with the signal is smaller than at frequencies outside the range.

Microwave amplifier characteristics

The general variation of noise figure with log frequency is shown in Fig. 9.40. At low frequencies $1/f$ noise predominates and this falls as frequency rises to f_1. Above f_1 the noise figure is independent of frequency, so this is the 'white' noise region which is maintained until f_2. Above f_2 the noise figure rises again and an upper limit may be set a few dB above the white noise level. An amplifier would be operated in the white noise region as far as possible, since this defines a minimum noise level. Some characteristics of semiconductor microwave amplifiers are compared in Table 9.2.

In general the maximum output power is available only at the low end of the frequency range. All types except the IMPATT amplifier are capable of operating over an octave bandwidth, which is half the centre frequency of the band. It is anticipated that GaAs FETs will replace tunnel diode amplifiers, except in specialized applications. The noise figures refer to the white noise region of Fig. 9.40, and for comparison the noise figure of a travelling wave-tube amplifier is about 25 dB. The noise figure of a cooled parametric amplifier is about 0.3 dB, representing a noise input temperature of 20 K, so the latter type is unlikely to be surpassed in low signal applications such as satellite communications.

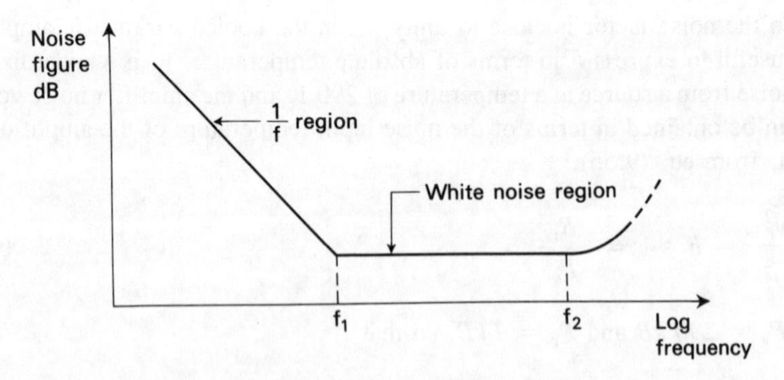

Figure 9.40 Variation of noise figure with frequency.

Table 9.2 Characteristics of single-stage semiconductor microwave amplifiers

	Gain	Centre frequency	Noise figure
	dB	GHz	dB
Bipolar transistor	4–10	1–10	2–6
Field-effect transistor	3–30	10–28	2–5
Tunnel diode	5–15	4–15	5
Transferred electron device	7–10	5–37	15
IMPATT diode	5–10	6–10	20

Points to remember

* Electron beams are used in specially designed thermionic valves to provide high power at microwave frequencies.
* Electron bunching occurs in klystrons, magnetrons and travelling-wave tubes to ensure that energy is passed from an electron beam to an electromagnetic wave.
* Negative resistance in various forms leads to power transfer from the d.c. supply to a.c. power in the load in the tunnel diode, the Gunn diode and the IMPATT diode.
* Variable capacitance is used to 'pump' microwave power from an a.c. supply at one frequency to the load at another frequency in the parametric amplifier.
* The p-i-n diode is an electronically controlled microwave switch or attenuator.
* In all systems where the electrical signal is very small, including microwave amplifiers, the minimum signal is limited by various types of electrical noise which can be described through noise factor or noise temperature.
* Gallium arsenide FETs are being used increasingly as microwave amplifiers, both as discrete devices and in integrated circuits, due to the better properties of gallium arsenide compared to silicon.

References

[1] Liao, S.Y. *Microwave devices and circuits*, (Prentice Hall International), 1980.
[2] Bierman, H. 'Move over silicon! Here comes GaAs', *Electronics*, vol. 58, 1985, pp. 39–46.

[3] Butcher, P.N. 'The Gunn Effect', *Reports on Progress in Physics*, 1967, vol. 30, p. 97.
[4] Sze, S.M. *Physics of Semiconductor Devices*, 2nd ed. (Wiley), 1981.
[5] Hilsum, C. 'A simple analysis of transferred electron oscillators', *Brit. J. Appl. Phys.*, 1965, vol. 16, p. 1401.
[6] Manley, J.M. and Rowe, H.E. 'Some general properties of non-linear elements', *Proc. IRE*, 1956, vol. 44, pp. 904–13.
[7] Gilden, M. and Hines, M.E. 'Electronic tuning effects in the Read microwave avalanche diode', *IEEE Trans. Electron. Devices*, ED13, 1966, No. 1, p. 169.

Problems

9.1 An electron beam is formed by accelerating electrons to a uniform velocity of 10^7 m/s. The beam then passes through two grids spaced 0.5 mm apart, between which is applied a 6 V r.m.s. sinusoidal voltage of frequency 1 GHz. (i) Stating any assumptions made, show that the beam leaving the output grid is velocity modulated, and calculate the depth of modulation. (ii) What would be the effect of increasing the signal frequency to 20 GHz?
[(i) 1.5%; (ii) transit time = 1 cycle, so there is no velocity modulation]

9.2 The electron beam in a klystron has a mean velocity of 2×10^7 m/s. Determine the peak percentage velocity modulation of the beam produced by a sinusoidal signal having an r.m.s. value of 15 V. Derive the expression used.
[0.93%]

9.3 A reflex klystron oscillator has a reflector mounted 2 mm away from the resonator grids and is mechanically tuned to a frequency of 5 GHz. The resonator is maintained at 300 V and the reflector voltage can be varied between 0 and –500 V with respect to the cathode. Determine the number of possible modes of oscillation that can be obtained and the corresponding values of the reflector voltage.
[Three modes, with V_R = 367, 125 and 11 V]

9.4 Draw a diagram to show the essential details of a travelling-wave-tube amplifier, and describe the function of each part. Explain briefly how the energy is transferred from the electron beam to the r.f. circuit.

The helix in such an amplifier has a pitch of 0.75 mm, an angle of 6°7′ and has 192 turns. Estimate the minimum electron-beam accelerating voltage, and determine the operating frequency when the number of r.f. wavelengths on the helix is 30.

What happens when the beam voltage is less than the value estimated?
[2.91 kV; 6.67 GHz]

9.5 Show how the losses in a parallel tuned circuit at resonance are reduced when a negative resistance is connected in parallel and explain briefly how this principle is applied in microwave amplifiers and oscillators.

Describe the construction and operating principles of (i) a tunnel diode and (ii) a Gunn diode. Indicate how a negative resistance is produced in each case.

Explain which device would be preferred for a 10 W pulsed oscillator working at a frequency of 10 GHz.

9.6 The helix of a travelling-wave tube has 200 turns and an axial length of 15 cm. It operates with 37 wavelengths along the axis and a final anode voltage of 3 kV. Calculate the helix radius and the operating frequency.
[1.1 mm; 8.02 GHz]

9.7 Using the equivalent circuit of Fig. 9.19, show that the input impedance of a tunnel diode is given by the expression

$$Z_{in} = R - \frac{r}{1 + (\omega Cr)^2} + j\left[\omega L - \frac{\omega Cr^2}{1 + (\omega Cr)^2}\right]$$

Determine the frequencies for which the real and imaginary components become zero if $R = 0.5\ \Omega$, $L = 1$ nH, $C_i = 10$ pF and $r = -25\ \Omega$, and explain the significance of the results.
[4.45 GHz; self-oscillation occurs; 1.59 GHz; diode is a pure negative resistance]

9.8 Discuss the factors in a tuned tunnel diode amplifier which limit (i) the output voltage, and (ii) the load resistance.

A tunnel diode with slope resistance $-100\ \Omega$ and junction capacitance 2 pF is operated at 1 GHz in a resonant cavity with capacitance of 1.9 pF and loss resistance 10 kΩ. If the source resistance is 150 Ω and the voltage gain is 30, determine the load resistance and the bandwidth. (*Hint*: express all quantities in parallel, fed from an input current V_s/R_s. Bandwidth is given by $1/2\pi CR$, where C is total capacitance and R is total parallel resistance.)
[289 Ω; 9.1 MHz]

9.9 Explain briefly the operating principles of a Gunn diode used as (i) a microwave oscillator; (ii) a microwave amplifier.

Such a diode, 10 μm long, requires a threshold bias of 3 V and produces current pulses at a repetition rate of 10 GHz. If the maximum field in a domain is 5×10^6 V/m and the negative mobility is -0.01 m^2/Vs, sketch the velocity-field characteristic and estimate the minimum domain field.
[2.04×10^5 V/m]

9.10 A Gunn diode of length 2 μm has a drift velocity–field characteristic which may be approximated by three linear regions, with slopes as follows:

0 to 300 kV/m, +0.50 m^2/V s
300 to 5000 kV/m, -0.01 m^2/V s
above 5000 kV/m, +0.02 m^2/V s

Sketch the characteristic and estimate:

(i) the domain drift velocity;
(ii) the frequency of current pulses;
(iii) the maximum and minimum fields in a domain;
(iv) the width of a nucleating centre.

[10^5 m/s, 50 GHz, 5000 and 200 kV/m, 0.04 μm]

9.11 A coil and a variable-capacitance diode are connected in series with a sinusoidal source of e.m.f. The capacitance of the diode is modulated by a 'pump' source having an e.m.f. of square waveform.

If the signal frequency is 50 MHz, the diode sensitivity is 2 pF/V, the peak-to-peak amplitude of the square wave is 0.5 V and the power dissipated in the load and losses at equilibrium is 0.1 mW, estimate (i) the pump frequency, and (ii) the peak voltage across the variable capacitance.
[100 MHz; 1.41 V]

9.12 An amplifier with a gain of 20 dB has a bandwidth of 2 MHz and at the output the mean-square noise voltage is found to be 10^{-10} V^2. It is supplied from a source of resistance 50 Ω at a temperature of 20 °C. Calculate the noise temperature of the amplifier and explain briefly how it might be achieved at an operating frequency near 1 GHz.
[180 K; parametric amplifier required]

10 Dielectric materials and components

Dielectric materials are insulators, with resistivities between about 10^{12} and 10^{20} Ω m. Their properties determine the performance of capacitors, cables and general insulation for which a wide range of materials is used, in the form of a solid, liquid or gas.

The *relative permittivity* ϵ_r is an important property of these materials which affects the capacitance of any circuit element. The capacitance of the parallel plate geometry illustrated in Fig. 10.1(*a*) is given by

$$C = \frac{\epsilon_r \epsilon_0 A}{d} = \epsilon_r C_0 \tag{10.1}$$

where $C_0 = \epsilon_0 A/d$, A being the plate surface area and d the spacing between the plates. Thus for given dimensions C is proportional to ϵ_r, which can be affected by the frequency of the applied field, temperature and pressure.

Polarization

The electrons in an insulator are mainly bound to atoms and so are not free to wander through the material under the action of an electric field. However, the field can cause a small separation of the effective positive and negative charges within each atom (Fig. 10.1(*a*)) to form an *induced electric dipole*. For two charges $+Q$ and $-Q$ separated by distance x the *electric dipole moment* μ is a vector quantity, where $\mu = Qx$. Thus when the field is applied charges appear on the dielectric surfaces as shown in Fig. 10.1(*b*). They are known as *bound charges*, since electrons remain bound to individual atoms, and they disappear when the field is removed. A similar effect occurs for materials with permanently separated charges, as in ionic crystals or materials with asymmetric molecules. In these cases the material contains *permanent electric dipoles* which are randomly oriented until aligned by the field to produce surface charges as before.

Whether the dipoles are induced or permanent, the *polarization* P is the bound charge per unit area of dielectric surface and so has dimensions C/m^2, the same

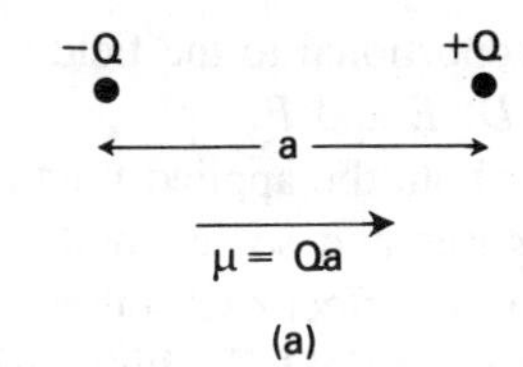

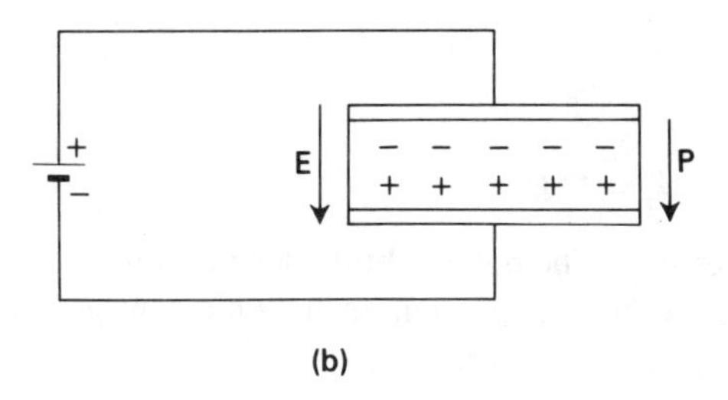

Figure 10.1 (*a*) Electric dipole; (*b*) polarization of a dielectric showing bound charges.

as electric flux density D. For two plane parallel plates forming a capacitor in vacuum the flux density is

$$D_0 = \epsilon_0 E \tag{10.2}$$

where E is the field applied between the plates. When a dielectric is introduced into the gap, the polarization induces extra charge on the plates of the capacitor so that D is increased and becomes

$$D_i = \epsilon_0 E + P$$

$$= \epsilon_r \epsilon_0 E$$

Hence

$$P = \epsilon_0 (\epsilon_r - 1) E \tag{10.3}$$

and the fact that ϵ_r is a constant shows that polarization is proportional to the average field between the plates, being a temporary electrical strain caused by the field.

Polarizability

The polarization can also be expressed as the dipole moment per unit volume, which has dimensions C/m^2, the same as P. If there are N elementary dipoles per m^3, each of moment μ, then

$$P = N\mu \tag{10.4}$$

μ itself may also be assumed proportional to the field within the dielectric, E_i, and is a vector quantity as are D, E and P.

In general E_i is determined by both the applied field and the induced dipoles and so $E_i > E$. However, in gases and some liquids, where the interatomic distance is large, the dipoles do not affect each other and the internal field E_i may be taken as equal to the applied field E with negligible error.

The constant of proportionality for μ is the *polarizability* α so that

$$\mu = \alpha E_i \tag{10.5}$$

Then

$$P = N\alpha E_i = \epsilon_0(\epsilon_r - 1)E \tag{10.6}$$

the *Clausius* relationship. The polarizability depends on the atomic structure, for instance on the electronic arrangement or the ionic configuration, and so relates relative permittivity to these properties since

$$\epsilon_r = 1 + \frac{N\alpha E_i}{\epsilon_0 E} \tag{10.7}$$

from eq. (10.6).

In general it may be shown that

$$E_i = E + \frac{\gamma P}{\epsilon_0} \tag{10.8}$$

where γ is a constant which depends on the precise arrangement of the atoms in the material. For a solid crystal having a cubic structure, Lorentz showed in 1878 that $\gamma = 1/3$ so that

$$E_{i\ \text{Lorentz}} = E + \frac{P}{3\epsilon_0} \tag{10.9}$$

An expression can now be derived for the relative permittivity in a static electric field. Combining eqs (10.3) and (10.8) leads to

$$E_i = E(1 + \gamma(\epsilon_r - 1)) \tag{10.10}$$

Then from eq. (10.7) and re-arranging

$$\epsilon_r = 1 + \frac{N\alpha}{\epsilon_0 - \gamma N\alpha} \tag{10.11}$$

For solids N, α and γ are only slightly affected by temperature so that ϵ_r is virtually independent of temperature in this case. For solids where $\gamma = 1/3$ it follows from eq. (10.11) that

$$\frac{\epsilon_r - 1}{\epsilon_r + 2} = \frac{N\alpha}{3\epsilon_0} \tag{10.12}$$

the *Clausius-Mosotti* relationship, which can be used to determine α.

For most solids $(\epsilon_r - 1) > 1$ while for gases $(\epsilon_r - 1) \ll 1$. This is due to the wide differences in the typical values of N, about $10^{29}/m^3$ for solids and about $10^{25}/m^3$ for gases at atmospheric pressure. Using eq. (10.11) with a typical value for α of about 10^{-40} F m^2 in each case leads to a value for $(\epsilon_r - 1)$ of 1.8 for a solid and 1.1×10^{-4} for a gas, so that ϵ_r is virtually 1 for gases.

Types of polarization

In practice polarization may be caused by one or more mechanisms, so that α may have one or more components. There are four main types of polarization called electronic, ionic, orientational and interfacial, respectively, and each is characterized by a frequency above which it becomes ineffective. This in turn controls the frequency response of a particular dielectric material and so affects the choice of a capacitor for a given application.

Electronic (optical) polarization

In a simple model the atom may be considered to consist of a positive nucleus and a negative electron cloud of equal charge magnitude. The charge centres coincide in the absence of an electric field but when a field is applied the electrons move relative to the nucleus and a dipole is created. If the field is alternating the electron motion can be oscillatory up to very high frequencies due to the low mass of electrons. In fact the upper frequency is around 10^{15} Hz, in the ultraviolet region. Polystyrene, dry air and the rare gases such as helium, argon and neon are materials where this is the essential polarization mechanism, and such materials are classified as *non-polar*.

Ionic (molecular) polarization

Ionic crystals contain permanent dipole moments, due to positive and negative ions held in the lattice, even in the absence of an applied field. The field then has two effects (i) to shift the electron cloud relative to the nucleus in each ion and (ii) to shift the positive and negative ions relative to each other. Thus both electronic and ionic polarization are present together. Due to the larger mass of ions compared with electrons, the upper frequency of ionic motion is in the infrared region at about 10^{14} Hz. Polytetrafluoroethylene (PTFE) is one material where this limitation occurs, such materials being classified as *polar*.

Orientational polarization

Other materials, such as liquids, contain permanent dipoles whose orientation is random when no field is applied and so are known as dipolar. The dipoles tend to align themselves with the field, although this tendency is opposed by the random motion of the dipoles due to the temperature. This effect occurs in transformer oils, certain ceramics and electrolytic dielectrics, described in a later section. The effect could only occur in solids where a charge can change its position, as in some glasses containing impurity ions with higher than average charge which form a dipole with a vacant site. The upper frequency of dipole motion is within the range 10 kHz to 1 GHz, which covers the communications band and so is very important in practice.

Interfacial polarization

Defects such as missing atoms, impurity centres and dislocations occur in a real crystal and one or more of the few free electrons moving in the field may be trapped near an electrode by such a defect. This results in localized accumulation of charge, inducing an image charge on the electrode which leads to an effective dipole. The mechanism operates from d.c. to about 100 Hz and occurs in aluminium electrolytics, mica and paper dielectrics and sometimes in polystyrene. It can cause the appearance of a charge on a capacitor immediately after it has been discharged. This is because free charges within the dielectric may take several seconds to escape from traps after the applied field has been removed.

Complex relative permittivity and dielectric loss

In an alternating field ϵ_r will have real and imaginary parts ϵ_r' and ϵ_r'' corresponding to α' and α'', the real and imaginary parts of the polarizability. Then

$$\epsilon_r = \epsilon_r' - j\epsilon_r'' = 1 + \frac{N}{\epsilon_0}(\alpha' - j\alpha'')$$

so that

$$\epsilon_r' = 1 + \frac{N\alpha'}{\epsilon_0} \tag{10.13}$$

and

$$\epsilon_r'' = \frac{N\alpha''}{\epsilon_0} \tag{10.14}$$

The frequency variations of ϵ_r' and ϵ_r'' will correspond to variations of α' and α'' respectively, as shown in Fig. 10.2. When $\alpha' = 0$, $\epsilon_r' = 1$ at frequencies well above ω_0, and an ideal capacitor would behave as though its dielectric were a vacuum. Similarly $\epsilon_r'' = 0$ at all frequencies except near and at ω_0, so that at frequencies well below ω_0, $\epsilon_r = \epsilon_r'$.

The complex relative permittivity is directly related to the power loss in a dielectric which occurs as a result of the polarization mechanism. This is additional to any loss in ohmic conductance G_p due to electron-hole pairs. In Fig. 10.3(*a*)

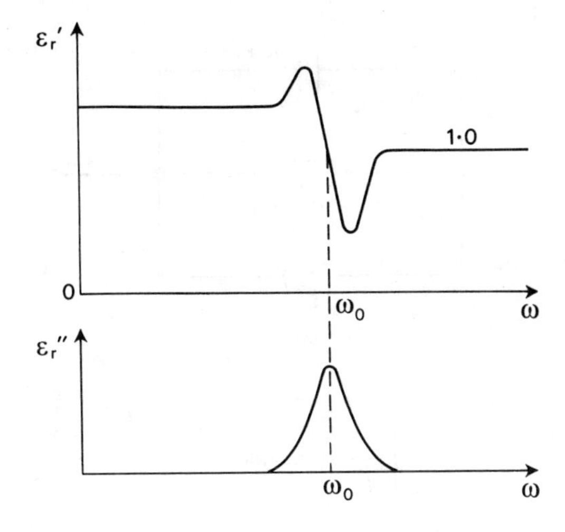

Figure 10.2 Variation of relative permittivity with angular frequency for electronic polarization.

C_0 represents the capacitance between the plates before the dielectric is inserted. Then the admittance becomes

$$\begin{aligned} Y &= j\omega C_0(\epsilon_r' - j\epsilon_r'') \\ &= \omega\epsilon_r'' C_0 + j\omega\epsilon_r' C_0 \end{aligned} \tag{10.15}$$

This expression represents a capacitance $\epsilon_r' C_0$ in parallel with a conductance $\omega\epsilon_r'' C_0$ due to the polarization and leads to the equivalent circuit of Fig. 10.7(*a*). Then the total conductance is

$$G_T = G_p + \omega\epsilon_r'' C_0 \tag{10.16}$$

which remains virtually independent of frequency for $\omega\epsilon_r'' C_0 \ll G_p$. However, at high frequencies and especially near the resonant frequency ω_0 the polarization term predominates and the losses are high.

When a voltage V is applied across the capacitor, the current I is divided between the losses and the capacitance itself so that

$$I = G_T V + j\omega\epsilon_r' C_0 V$$

Putting $\omega\epsilon_r' C_0 = B$ leads to

$$I/V = Y = G_T + jB$$

and the corresponding admittance diagram for Y is shown in Fig. 10.3(*b*). Here the small angle δ is used as a measure of the perfection of the capacitor, for if $\delta = 0$ the loss conductance $G_T = 0$ also. tan δ is known as the *loss factor* of the capacitor and is given by

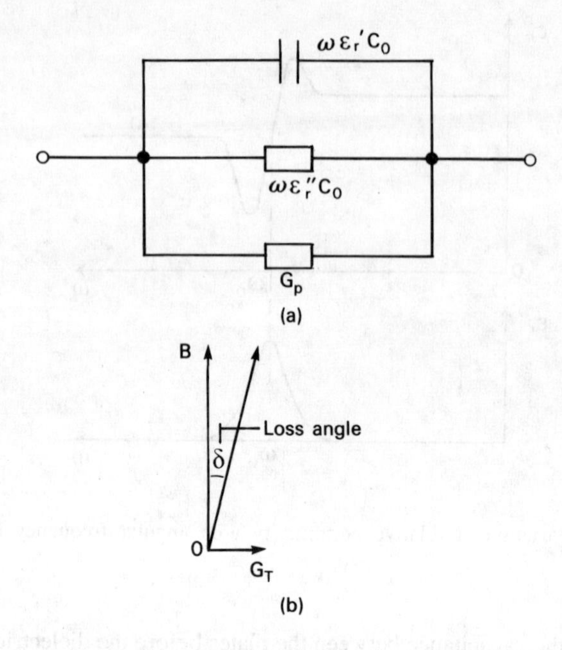

Figure 10.3 (*a*) Equivalent circuit of a capacitor with dielectric and conductive losses; (*b*) vector diagram of admittance of capacitor in (*a*).

$$\tan \delta = \frac{G_T}{B} \tag{10.17}$$

At a resonant frequency where $\omega\epsilon_r'' C_0 \gg G_p$ eq. (10.15) reduces to

$$\tan \delta = \epsilon_r''/\epsilon_r' \tag{10.18}$$

Typical values of tan δ, remote from resonant frequencies, are given in Table 10.1.

Variation of ϵ_r with frequency

The frequency variation of ϵ_r' for a material in which all the polarization effects discussed above are present is shown in Fig. 10.4(*a*). The drop in ϵ_r' at each characteristic frequency leads to a fall in capacitance and a peak in dielectric loss due to ϵ_r''. Thus a dielectric is chosen so that ϵ_r' is constant over the required frequency range, which gives $\epsilon_r'' = 0$.

The upper frequency response can extend up to about 10^{10} Hz for polystyrene and about 10^9 Hz for mica and low ϵ_r ceramics such as magnesium silicate, although the actual response obtainable with a capacitor will be limited by its construction and encapsulation which introduce stray capacitance and inductance. At low frequencies polystyrene and tantalum capacitors can be used down to about

Table 10.1 Properties of common capacitor materials

Capacitor Type	Relative permittivity	Loss factor at 1kHz	Maximum working voltage
	ϵ_r	tan δ	V
Ceramic, low ϵ_r	6–20	0.001	50–500
Ceramic, high ϵ_r	1000–3000	0.02	50–500
Mica	3–7	0.001	400
Paper	5	0.01	200–1000
Polycarbonate	2.8	0.003	63–400
Polypropylene	2.3	0.0005	400
Polystyrene	2.5	0.0005	30–630
Polyethylene terephthalate	2–5	0.01	250–750
Aluminium (electrolytic)	—	0.05	6.3–500
Wet tantalum (electrolytic)	—	0.05	30–125
Dry tantalum (electrolytic)	—	0.05	6–150
Solid tantalum (electrolytic)	—	0.05	3–50

1 Hz and aluminium electrolytics down to about 10 Hz, with other materials having a higher limit. The fundamental high frequency limitation would be set by ionic polarization and the low frequency limitation by interfacial polarization. Figure 10.4(*b*) shows approximate frequency ranges for common types of capacitors.

Table 10.1 gives typical values of characteristics for some common capacitor materials. In the high ϵ_r ceramics, such as barium titanate, electric charges are very loosely bound so that large dipole moments can be induced. The losses are relatively high and the materials are also ferro-electric, that is charge and voltage are non-linearly related through a hysteresis loop, which also leads to a large variation of capacitance with temperature. These materials are further discussed under piezoelectric devices in a later section. The values of loss factor at 1 kHz in Table 10.1 are mainly due to the insulation resistance, which is normally around 10 GΩ or above at room temperature for all capacitors except electrolytics.

Dielectric strength

If the capacitor voltage is increased from zero the leakage current remains low until the breakdown voltage is reached. Here the current starts to rise rapidly, as in a reverse-biased p-n diode (Chapter 3), and the corresponding electric field is the *dielectric strength*, typically between 10^6 and 10^8 V/m.

The time for which the breakdown voltage is applied is also important since most dielectrics will withstand a higher voltage for short pulse durations. However, the breakdown voltage is reduced as temperature is increased due to the release of electrons and holes, so application of long high-voltage pulses can lead to breakdown. Heating is also caused by polarization losses, so breakdown tends

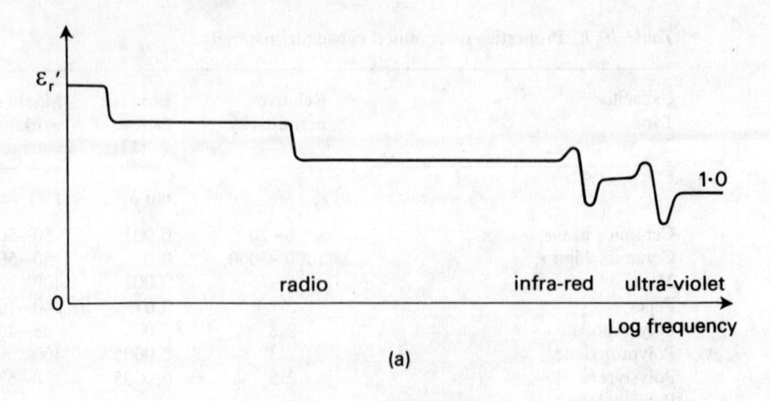

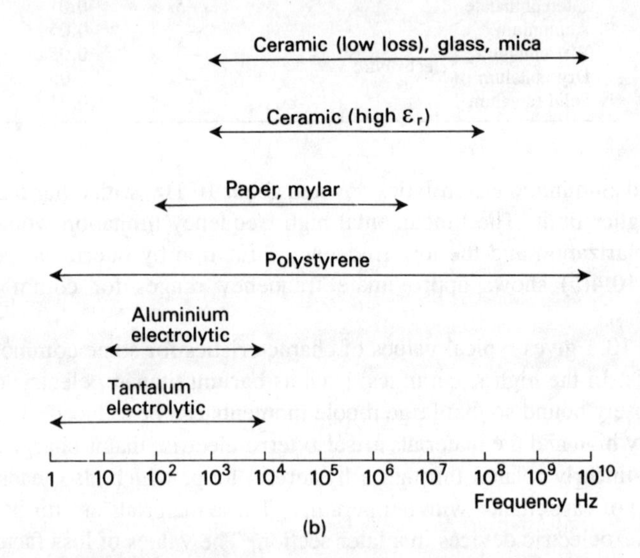

Figure 10.4 (*a*) Variation of relative permittivity with frequency when all polarization mechanisms are present; (*b*) frequency ranges for common types of capacitor.

to occur at a lower voltage if the frequency is increased towards the maximum value. Again moisture has the effect of reducing breakdown voltage by providing extra conducting paths between the electrodes.

Some typical maximum working voltages are given in Table 10.1 for fixed capacitors, which are somewhat lower than the breakdown voltage V_B and apply at normal ambient temperature. Where E_B is the dielectric strength and d is the dielectric thickness then $V_B = E_B d$. The voltage rating can be related to other properties of the capacitor by considering the stored energy. From eq. (10.1) this is given by

$$\frac{CV^2}{2} = \frac{\epsilon_r\epsilon_0 V^2 A}{2d}$$

which leads to

$$\frac{CV^2}{2} = \frac{\epsilon_0}{2} \times \epsilon_r \left(\frac{V}{d}\right)^2 \times Ad \qquad (10.19)$$

At the breakdown voltage the stored energy per unit volume, CV_B^2/Ad, is proportional to the product of the relative permittivity and the square of the dielectric strength. Thus where size must be kept to a minimum the dielectric is chosen to give as large a value as possible to this product. For a given dielectric however, an increase in voltage rating or an increase in capacitance is accompanied by an increase in volume.

Capacitor construction

The construction of a capacitor depends on the type of dielectric, the size of capacitance and the maximum operating frequency (Ref. [1]). For flexible dielectrics such as paper and plastic films, the component can be formed from a thin strip of dielectric foil sandwiched between two metal foils to which contacts are made (Fig. 10.5(*a*)). The capacitance then depends on the length and width of the strip and the total size is reduced by rolling the sandwich tightly. For rigid dielectrics such as ceramics the capacitor can be tubular (Fig. 10.5(*b*)) or disc-shaped. Mica capacitors are normally stacks of thin mica sheets interleaved with metal electrodes.

Capacitors produced using these and other methods are available in the range 1 pF to 10 μF. For larger capacitances the volume of the component becomes unacceptably large and *electrolytic capacitors* are used. In one type the dielectric is a film of aluminium oxide Al_2O_3, about 0.1 μm thick deposited electrolytically on an aluminium foil anode (Fig. 10.6(*a*)). Contact is made to the other side of the film through paper soaked in an electrolyte of glycol and ammonium tetraborate and an aluminium foil cathode is used as a contact electrode to the electrolyte.

The metal-oxide contact is rectifying, with the oxide behaving as a p-type semiconductor, so reverse bias must be applied with the anode positive. The capacitance between electrolyte and anode can then have values up to about 1000 μF due to the thinness of the oxide. Since the dielectric is polarized, the upper operating frequency is about 10 kHz. Non-polarized forms are available with a film on each electrode, but these give only half the capacitance for a given volume. The forward-biased film has a very low resistance, so that each film in ineffective during the half cycle when the other is effective.

The *etched-foil* type has a similar construction to the plain-foil type, but an increased anode area due to surface etching of the anode before depositing the film. The capacitance is then increased about eightfold and the loss factor approximately doubled.

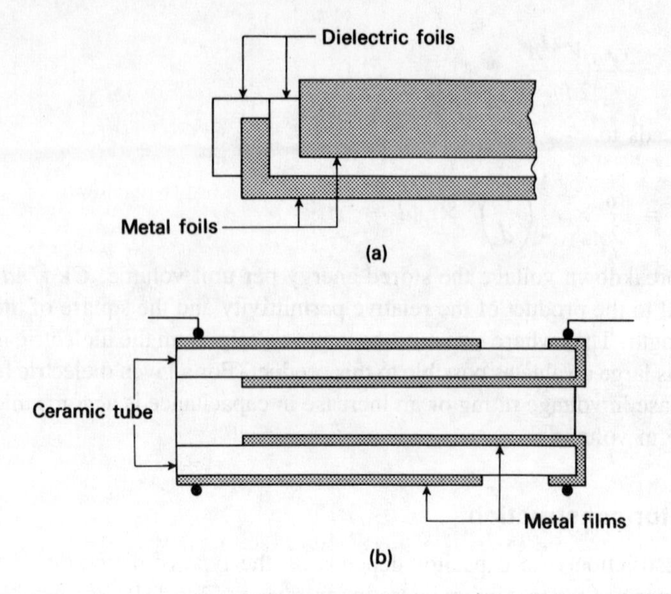

Figure 10.5 Capacitor construction. (*a*) Flexible dielectric; (*b*) rigid dielectric.

Tantalum may be used instead of aluminium to form both plain- and etched-foil electrolytic capacitors (dry tantalum capacitor). Another version uses a solid electrolyte with an oxide film dielectric, Ta_2O_3, as shown in Fig. 10.6(*b*), giving a small capacitor due to the very large surface area of the sintered tantalum (solid tantalum capacitor). Finally the wet tantalum capacitor has the oxide dielectric with pure tantalum disc anode and tantalum or silver cathode. The performance is similar to that of a paper capacitor but with much higher capacitance values.

Piezoelectric transducers

Piezoelectricity refers to the appearance of electric charges on the surfaces of a crystal under pressure (from the Greek 'piezo', to press). It occurs only in those dielectrics without a centre of symmetry in the crystal, that is, where the distributions of positive and negative ions are asymmetrical about a point. An applied stress produces displacement of the ions whether there is a centre of symmetry or not, but only in the asymmetrical case would there be a resultant dipole moment. The effect occurs in 20 out of the 21 classes of crystal without a centre of symmetry. Quartz, Rochelle salt and barium titanate are common piezoelectric crystals (Ref. [2]).

The bulk properties of such materials can be described by two equations:

$$T = yS - pE \tag{10.20}$$

$$D = \epsilon E + pS \tag{10.21}$$

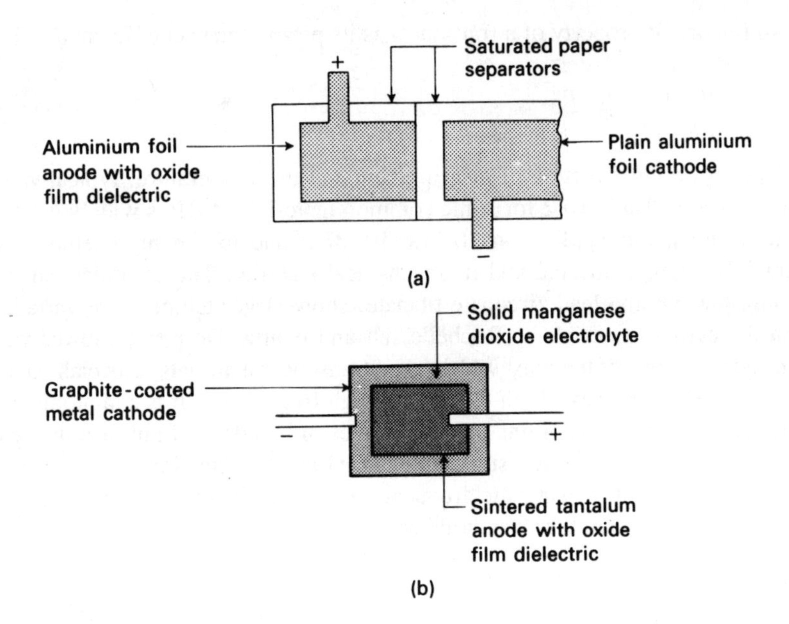

Figure 10.6 Electrolytic capacitor construction. (*a*) Aluminium; (*b*) solid tantalum.

where T is stress, S strain, y Young's moduls, p the piezoelectric constant and $\epsilon = \epsilon_r\epsilon_0$. When $p = 0$ the equations reduce to $T = yS$, Hooke's Law, and $D = \epsilon E$, Gauss' Law. However, for a piezoelectric crystal where $p \neq 0$, if the electric field $E = 0$ we have

$$D = pS = \frac{p}{y}T \tag{10.22}$$

so that a flux density D is produced by applying stress T to the crystal. On the other hand if an electric field is applied when $T = 0$ we have

$$S = \frac{p}{y}E \tag{10.23}$$

so that strain S is produced by electrical means.

Piezoelectric materials are used as electromechanical transducers, which allow an exchange between electrical and mechanical energy. For instance the pickup of a record player converts vibrations from the groove of a record into electrical signals. A microphone performs a similar function for the longitudinal air movements produced by sound waves. An accelerometer is attached to a vibrating structure to produce an electrical measure of its acceleration and displacement. Ignition systems, whether for a car engine or to produce a gas flame, can depend on striking a piezoelectric crystal.

An important property of a transducer is its piezoelectric coefficient d, where

$$d = \left.\frac{\partial S}{\partial E}\right|_{T=0} = \frac{p}{y} \tag{10.24}$$

It depends on the direction of measurement and the temperature, typical values being given in Table 10.2 for some common materials. ADP is widely used for underwater sound applications below 10 MHz due to the high temperature (100 °C) it will withstand and its mechanical stability. The ceramics, such as barium titanate and lead zirconate titanate, show larger temperature variations than the earlier materials — Rochelle salt and quartz. Despite its lower value of d, quartz is one of the most widely used transducer materials, especially above about 10 MHz, and has a high Q-factor at high frequencies. It is commonly used in transducers for transmitting acoustic waves in liquids, as in ultrasonic cleaning and sonar for underwater surveying and submarine detection. Quartz crystals are also used in stabilizing the frequency of an oscillator, as in broadcasting transmitters and laboratory standards where accuracies of up to 1 part in 10^9 are required.

Table 10.2 Piezoelectric coefficients of common materials at 20 °C

Material	d m/V	Application
Rochelle salt	3.5×10^{-10}	Microphones, pickups
Quartz	2×10^{-12}	Crystal oscillators, underwater sound transducers
Ammonium dihydrogen phosphate ADP	1.5×10^{-6}	Underwater sound transducers
Barium titanate	2.5×10^{-10}	Accelerometers
Lead zirconate titanate PZT	6×10^{-10}	Car ignition, gas lighters

Equivalent circuit of a quartz crystal

A plate is cut from a single crystal of quartz and opposite faces are coated with a thin gold film to allow electrical contact, as exemplified in the circuit symbol (Fig. 10.7(a)). The angle of cut and size of plate determine the resonant frequencies when an a.c. field is applied and crystals have been produced in the frequency range 1.2 kHz to about 100 MHz using harmonics. If the dimension parallel to the direction of vibration is L a mechanical resonance is set up whenever L is an integral number of half wavelengths or

$$L = \frac{n\lambda}{2} \tag{10.25}$$

λ refers to the sound waves in the material which have a velocity v, so that the frequency of the lowest mode of vibration is $f_1 = v/\lambda = v/2L$ and depends on the crystal dimensions. Resonance can occur only when the applied frequency is very close to f_1, so that the mechanical vibrations have a very high Q-factor typically between 10 000 and 500 000.

The equivalent circuit of a quartz crystal is shown in Fig. 10.7(b). The series circuit formed from L, C and r represents the mechanical resonance. C_0 represents the capacitance between the electrodes and R the loss resistance of the quartz. There is a parallel resonant frequency f_2 in addition to the series resonance at f_1 and the difference between f_1 and f_2 is very small, due to the high Q-factor. In practice $1/(\omega C_0)$ is much greater than r at series resonance, so the impedance is resistive at f_1 but remains capacitive at frequencies below f_1.

Then, ignoring R except at parallel resonance, the impedance is given by

$$Z = \frac{-\dfrac{\mathrm{j}}{\omega C_0}\left(r + \mathrm{j}\omega L - \dfrac{\mathrm{j}}{\omega C}\right)}{r + \mathrm{j}\omega L - \dfrac{\mathrm{j}}{\omega C} - \dfrac{\mathrm{j}}{\omega C_0}} \tag{10.26}$$

At series resonance $\omega_1 L = 1/\omega_1 C$ so the impedance

$$Z_s = r \quad \text{at} \quad f_1 = \frac{1}{2\pi(LC)^{1/2}}$$

At parallel resonance

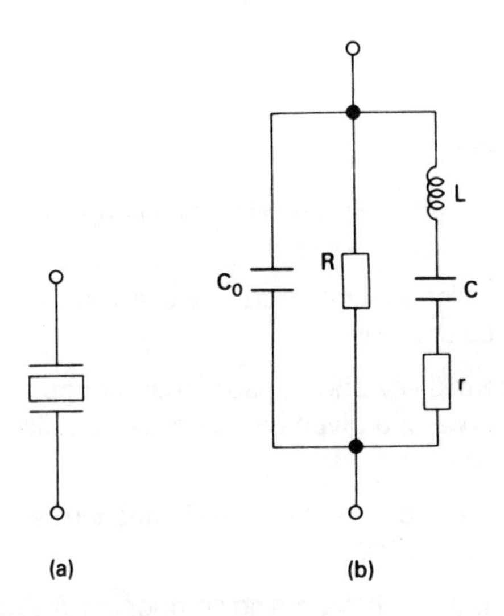

Figure 10.7 Quartz crystal. (a) Circuit symbol; (b) equivalent circuit.

$$\omega_2 L = \frac{1}{\omega_2}\left(\frac{1}{C_0} + \frac{1}{C}\right)$$

at

$$f_2 = \frac{1}{2\pi\left(\dfrac{LC_0C}{C_0 + C}\right)^{1/2}}$$

Here the impedance

$$Z_p = \frac{-\dfrac{j}{\omega_2 C_0}\left(r + \dfrac{j}{\omega_2 C_0}\right)}{r}$$

$$\simeq \frac{1}{\omega_2^2 C_0^2 r} \text{ in parallel with } R. \tag{10.27}$$

The frequency responses of the crystal are shown in Fig. 10.8. The actual value of f_1 is a function of temperature, so that for the highest accuracy the crystal is normally held at constant temperature in a small oven. In controlling the frequency of an oscillator the crystal is normally used either at f_1, where it behaves as a resistance r, or at just above f_1, where it behaves as an inductor. Should the oscillator frequency change due to variations in other components in the circuit the crystal impedance will change to maintain oscillations, but the corresponding change in frequency near f_1 will be very low.

Points to remember

* Capacitance is proportional to relative permittivity which depends on electric dipoles.
* Dipole characteristics are controlled by electronic, molecular, orientational and interfacial polarizations.
* Both relative permittivity and capacitance change with frequency. The frequency response of a given type of capacitor ultimately depends on the polarization mechanism.
* An exchange between electrical and mechanical energy occurs in piezoelectric transducers.
* A quartz crystal is an essential timing component in all computer systems.

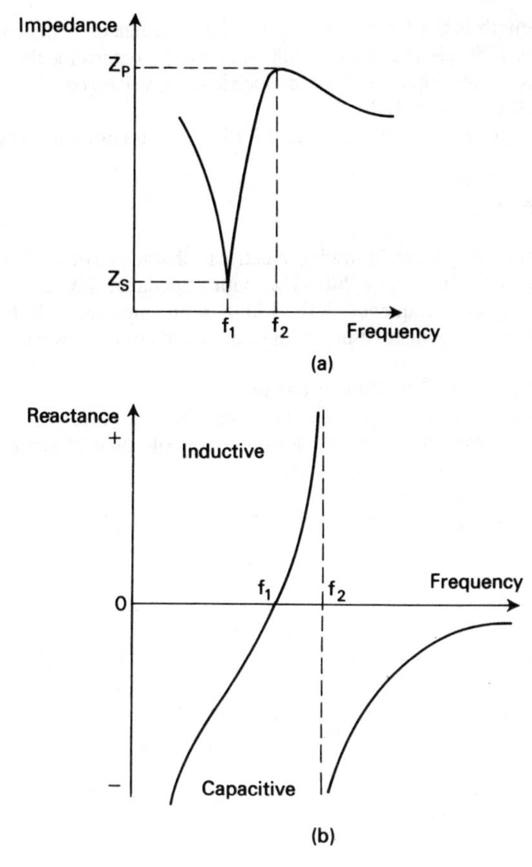

Figure 10.8 Frequency responses of a quartz crystal. (*a*) Impedance; (*b*) reactance.

References

[1] *Developments in small capacitors*, (ERA Technology), 1980.
[2] Taylor, G.W. (Ed.), *Piezoelectricity*, (Gordon and Breach), 1985.

Problems

10.1 The capacitance of an evacuated test cell is increased by 0.005% when it is filled with a gas. If the gas contains 2×10^{25} atoms/m^3, estimate its electronic polarizability.
[2.2×10^{-41} F m^2]

10.2 A solid contains 5×10^{28} atoms/m^3, each with polarizability 2×10^{-40} F m^2. Assuming a Lorentz internal field, show that $E_i/E = 1.6$.

10.3 A test cell for liquid dielectrics consists of two concentric cylindrical electrodes separated by a 2.0 mm gap. The diameter of the inner electrode is 0.15 m, its

effective length is 0.10 m and the cell is fully screened. If the capacitance of the filled cell is 1750 pF and tan $\delta = 0.0074$ at 50 Hz, determine the complex relative permittivity of the liquid and the dielectric loss resistance.
[$8.4 - j\,0.063$; 2.4×10^8]

10.4 Using the equivalent circuit of Fig. 10.15, show for a quartz crystal that

$$f_2 - f_1 \simeq f_1 \frac{C}{2C_0}.$$

10.5 A quartz crystal has the following electrical characteristics:
series resonant frequency 200 kHz, with impedance 200 Ω;
parallel resonant frequency 200.25 kHz, with impedance 40 MΩ.
Determine the component values of the equivalent circuit, assuming an infinite loss resistance.
[28.5 H; 2.2×10^{-14} F; 200 Ω; 8.9 pF]

10.6 Compare the frequency response of capacitors having polystyrene, electrolytic and ceramic dielectrics. State, with reasons, a suitable type of capacitor for:
(i) the tuned circuit in a 100 MHz oscillator;
(ii) an accurate integrator;
(iii) a smoothing fitter for a rectifying power supply.

11 Magnetic components

Magnetic materials used in the cores of inductors and transformers in communications circuits are required to have low losses, that is a high quality factor Q, to avoid degrading circuit performance. In these circuits hysteresis loss is negligible due to the small signals which lead to small flux changes in the core. Eddy current losses are reduced in a core of high resistivity, as shown below, which leads to the widespread use of ferrite cores.

Ferrites are, in general, insulators made from mixed oxides of divalent metals and iron, with a general formula (M) Fe_2O_4. Ferrites used as cores for high-frequency inductors and transformers normally contain manganese-zinc or nickel-zinc mixtures. Magnetic bubble memories are based on gadolinium garnet with gallium substituted for iron, $Gd_3Ga_5O_{12}$. In microwave applications, other spinel ferrites and garnet ferrites such as yttrium-iron-garnet are used (not considered here), while $BaFe_{12}O_{19}$ is used as a material for permanent magnets.

Losses in a ferrite-cored inductor

Ferrites are used between about 1 kHz and 1 MHz for the magnetic cores of transformers and high Q-factor inductors, since at these frequencies the eddy current loss is too high in iron cores. The eddy current loss depends on the size and shape of the material as well as the frequency of the magnetic flux and it may be shown (Ref. [1]) that the loss is given by

$$P_{ed} = \frac{(\pi \hat{B} f d)^2}{\rho \alpha} \text{ W/m}^3 \tag{11.1}$$

This expression assumes constant relative permeability μ_r and a sinusoidal flux density of frequency f and peak value $\hat{B}$ perpendicular to a plane containing dimension d (Fig. 11.1). The current flows in the direction which sets up a magnetic field opposing the applied field. α depends on the shape and is 6 for laminations of thickness d metres and 16 for a cylinder of diameter d metres (Ref. [2]). ρ is the bulk resistivity.

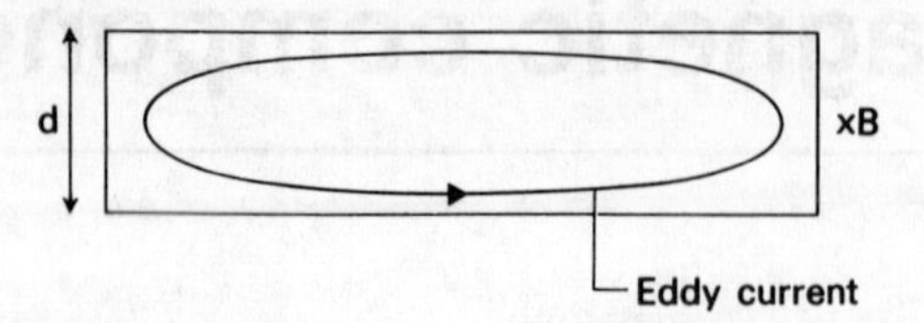

Figure 11.1 Eddy current in a conductor, alternating quantities.

The self inductance of a coil of N turns wound on an ideal toroid, carrying current I and embracing flux ϕ is given by

$$L = \frac{N\phi}{I} \tag{11.2}$$

If the toroid has length l and cross section A then

$$L = \frac{NBA}{I} = \frac{NA}{I}\,\mu_0\mu_r\,\frac{NI}{l}$$

So

$$L = \frac{\mu_0\mu_r N^2 A}{l} \text{ henries} \tag{11.3}$$

In practice the coil is wound on a plastic former which is enclosed within a ferrite pot core (Fig. 11.2(*a*)). The air gap in the centre of the pot is partially shunted by a cylindrical magnetic core (Fig. 11.2(*b*)). The reluctance of the air gap varies with the position of the shunting core so that the inductance of the winding can be adjusted. Equation (11.3) is then modified to

$$L = \frac{\mu_0\mu_e N^2 A_e}{l_e} \tag{11.4}$$

where μ_e, A_e and l_e are effective quantities.

Design charts have been devised relating L, Q, f, number of turns and conductor diameter for specified core materials (Ref. [2]). These charts are based on the summation of loss tangents and represent the performance of ferrite-cored inductors in parallel resonant circuits. Such components are used at flux densities which are only about 1% of saturation, so the curvature of the $B-H$ characteristic is neglected and the losses expressed in terms of series resistance and loss angle. This leads to an equivalent circuit shown in Fig. 11.3 and where losses are low they can be expressed in terms of the tangents of the loss angles for the series and parallel branches, in a similar manner to dielectrics. For the series arm the loss tangent is $R_T/\omega L$ and for the parallel arm it becomes $1/\omega C_s R_s$. Then, the total loss tangent is given by

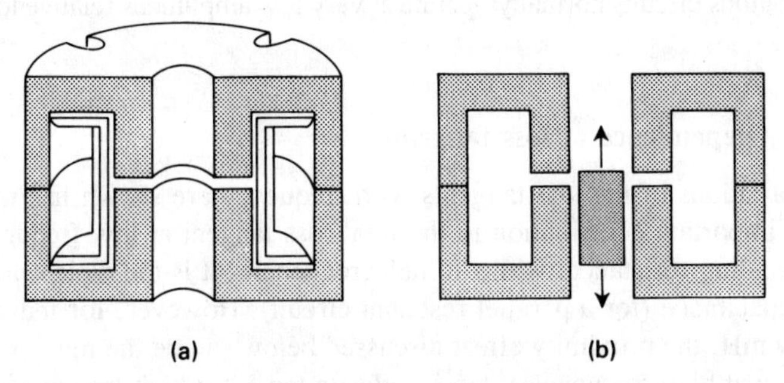

Figure 11.2 (*a*) Half section of ferrite pot core with plastic former (Ref. [2]); (*b*) section through pot core showing shunting core (Ref. [2]).

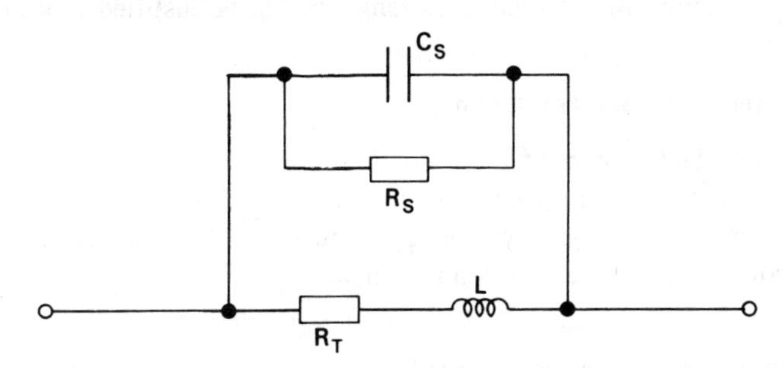

Figure 11.3 Equivalent circuit of an inductor.

$$\tan \delta_T = \Sigma \tan \delta = \frac{R_T}{\omega L} + \frac{1}{\omega C_s R_s} = \frac{1}{Q} \tag{11.5}$$

The total series loss resistance R_T is obtained from the sum of the individual loss resistances, which are related to loss mechanisms as follows

R_h	hysteresis	core losses
R_{ed}	eddy currents	
R_r	residual loss	
R_{dc}	d.c. winding resistance	winding losses
R_{se}	skin effect	
R_{pe}	proximity effect	
R_s	dielectric loss in stray capacitance C_s	

Then

$$R_T = R_h + R_{ed} + R_r + R_{dc} + R_{se} + R_{pe} \tag{11.6}$$

R_s appears separately in parallel as shown in Fig. 11.3. The hysteresis loss resistance R_h is usually negligible because inductors and transformers used in communications circuits normally operate at very low amplitudes relative to saturation values.

Frequency dependence of loss tangents

Typical variations of the loss tangents with frequency are shown in Fig. 11.6. The most important contribution to the total loss tangent at low frequencies is the d.c. winding resistance, while at high frequencies it is the dielectric loss in the self capacitance (for a parallel resonant circuit). However, for inductors of only a few mH, the proximity effect discussed below can be the most important contribution at high frequencies. $\tan \delta_{se}$ adds to $\tan \delta_{dc}$ at high frequencies only.

It is difficult to analyse theoretically the contribution of residual loss but the frequency variation of the other loss tangents can be justified more easily.

Eddy current loss tangent, $\tan \delta_{ed}$

From eq. (11.1) $P_{ed} \propto B^2f^2$ and also $P_{ed}/I^2 = R_{ed}$. But for a given inductor $B \propto I$ so that R_{ed} is proportional to f^2. Then $R_{ed}/\omega L = \tan \delta_{ed} = \text{const.}\, f$ and so $\tan \delta_{ed}$ rises with frequency. $\tan \delta_{ed}$ is much less for a ferrite core than for an iron core, since the ferrite is an insulator.

d.c. winding resistance loss tangent, $\tan \delta_{dc}$

$\tan \delta_{dc} = R_{dc}/\omega L$ so $\tan \delta_{dc}$ depends on $1/f$ since R_{dc} is independent of frequency.

Skin effect loss tangent, $\tan \delta_{se}$

An individual conductor carrying alternating current will be surrounded by an alternating magnetic field which will induce opposing eddy currents within the conductor itself (Fig. 11.4). Near the axis of the conductor these currents will tend to reduce the main current but near the surface they will increease it. The

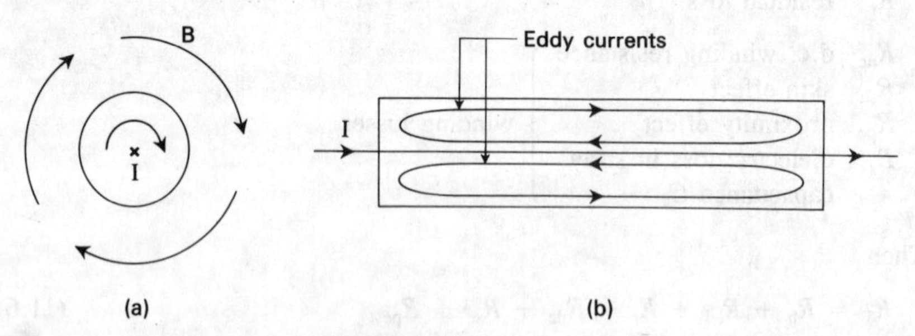

Figure 11.4 Skin effect. (*a*) Section through conductor; (*b*) plan view of conductor showing eddy currents. All quantities are alternating.

result is a non-uniform distribution of current across the conductor with more current at the surface than in the middle. For a.c. the induced e.m.f.s. increase with frequency until the current is flowing only in a thin layer, or skin, at the surface. Thus the effective cross-section is reduced and the a.c. resistance increased by an amount $R_{se} = FR_{dc}$ (Ref. [3]). F is the *skin effect factor* which is proportional to (frequency)^2. Then $R_{se}/\omega L = \tan \delta_{se} = \text{const.}\ f$.

In practice the effect is reduced by using stranded wire, giving bunched conductors with a more uniform current distribution over the bunch than over a single equivalent conductor.

Proximity effect loss tangent, $\tan \delta_{pe}$

There will also be an alternating magnetic field due to the inductor winding as a whole, which will usually be normal to the axis of an individual conductor (Fig. 11.5(*a*)). Eddy currents will flow in the conductor to oppose this field, which again lead to a non-uniform current distribution larger on one side than the other (Fig. 11.5(*b*)). The effect depends on the geometry of the magnetic core, conductor diameter, number of strands and frequency. $\tan \delta_{pe} = \text{const.}\ f$, as for the skin effect, but $\tan \delta_{pe} > \tan \delta_{se}$ at a particular frequency.

The proximity effect may be reduced by decreasing the conductor diameter, but this leads to an increase in $\tan \delta_{dc}$. Instead the strands are insulated from each other and twisted so that the eddy current e.m.f.s. e cancel each other in adjacent loops (Fig. 11.5(*c*)), with the wires soldered together at each end.

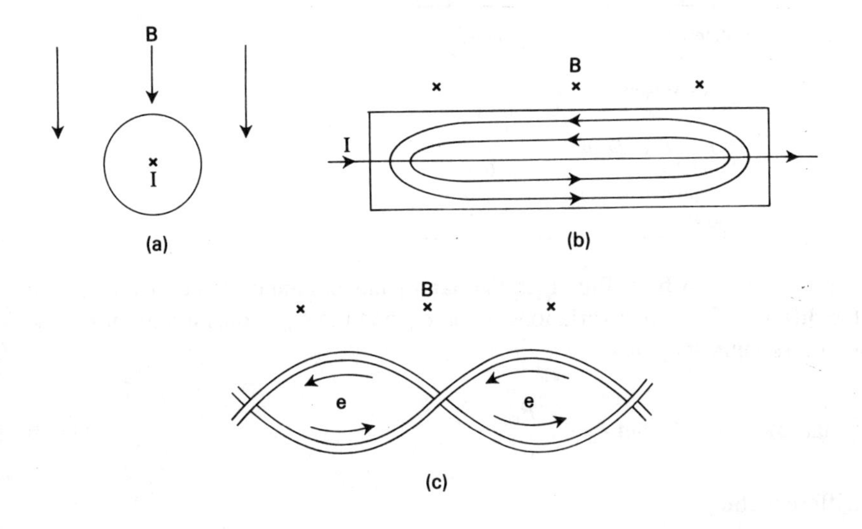

Figure 11.5 Proximity effect. (*a*) Section through conductor; (*b*) plan view of conductor showing eddy currents; (*c*) twisting of conductors to produce cancellation of e.m.f. All quantities are alternating.

Dielectric loss in the stray capacitance

In a practical inductor, distributed stray capacitance will exist between turns and between layers of winding, as well as between the windings and the core. The stray capacitance C_s will therefore have its own dielectric loss resistance R_s associated with it (Fig. 11.3). This gives a loss tangent for C_s of $\tan \delta_d = 1/(\omega C_s R_s)$ and the loss conductance is given by

$$\frac{1}{R_s} = \omega C_s \tan \delta_d \tag{11.7}$$

The contribution to the total loss tangent of the inductor is then

$$\begin{aligned}\tan \delta_{cp} &= \omega L/R_s \\ &= \omega^2 L C_s \tan \delta_d\end{aligned} \tag{11.8}$$

so that $\tan \delta_{cp} = \text{const.}\ f^2$.

Normally an inductor is designed to work well below its self-resonant frequency so that $\omega^2 L C_s \ll 1$. $\tan \delta_d$ depends mainly on the conductor insulation and some typical values are given in Table 11.1. A possible method of reducing C_s, and hence $\tan \delta_{cp}$, is to divide the winding into sections connected in series. This reduces the capacitance between adjacent layers and also within each section where there will be fewer layers, ideally only one.

Table 11.1 Loss tangents for various winding insulation materials (Ref. [2])

Material	$\tan \delta_d$
Single conductor:	
Dry enamel	0.01–0.02
Bunched conductors:	
Natural silk	0.015
Polypropylene fibre	0.003
Rayon	0.035

It may be seen from Fig. 11.6 that $\tan \delta_T$ has minimum value mainly due to the different frequency variations of $\tan \delta_{dc}$ and $\tan \delta_{cp}$. Considering only these two loss tangents gives

$$\tan \delta_T = \omega^2 L C_s \tan \delta_d + \frac{R_{dc}}{\omega L} \tag{11.9}$$

Differentiating

$$\frac{d \tan \delta_T}{d\omega} = 2\omega L C_s \tan \delta_d - \frac{R_{dc}}{\omega^2 L} \tag{11.10}$$

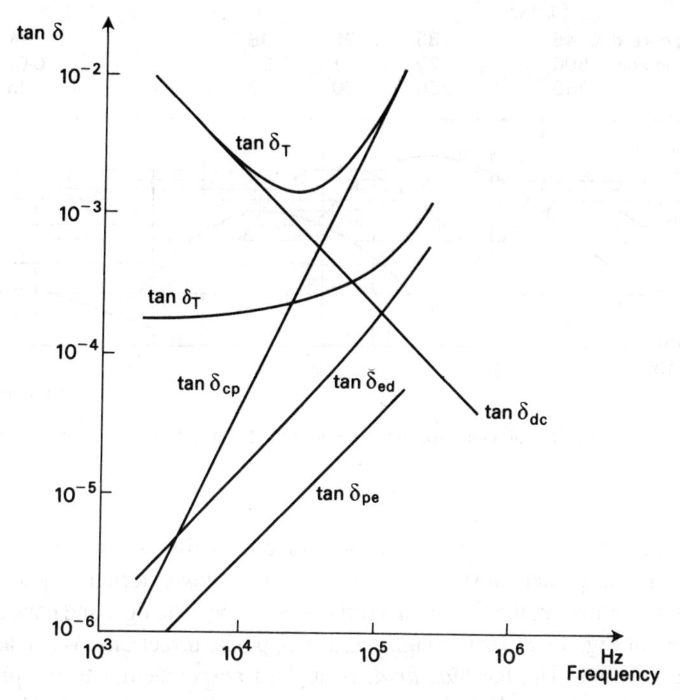

Figure 11.6 Typical variation of loss tangents with frequency for an inductor.

which becomes zero where

$$\omega = \left(\frac{R_{dc}}{2L^2 C_s \tan \delta_d} \right)^{1/3} \tag{11.11}$$

Equation (11.11) thus provides the frequency at which tan δ_T is a minimum and the actual value of tan δ_T at this frequency is obtained by substitution in eq. (11.9).

The value of Q-factor at any frequency is given by 1/tan δ_T, so that Q will reach a maximum at the frequency obtained from eq. (11.11). The variation of Q with frequency depends on the core diameter, inductance and effective permeability as illustrated in Fig. 11.7. Not only do individual Q-factors reach a maximum with frequency but also the maximum obtainable Q-factor is greatest between about 40 kHz and 200 kHz. The lower peaks at low and high frequencies are mainly due to d.c. winding resistance and residual loss respectively.

Magnetic bubble memory

An application of ferrites to computer memories uses a material such as gadolinium gallium garnet, $Gd_3Ga_5O_{12}$, in which the natural direction of magnetization is

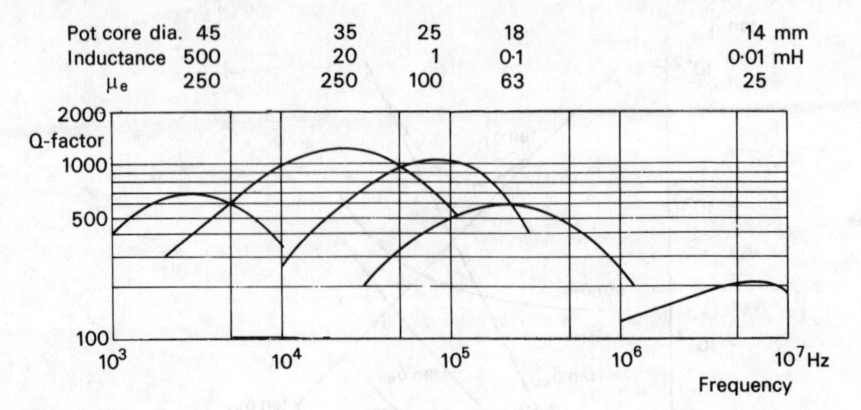

Figure 11.7 Variation of *Q*-factor with frequency for various coil designs (Ref. [2]).

perpendicular to the surface. This is deposited as a film a few μm thick on to a non-magnetic garnet substrate. In the absence of any external magnetic field, domains are formed in the film with a serpentine shape. The light and dark domains illustrated in Fig. 11.8(*a*) are magnetized in opposite directions. When an external magnetic field H_b, the *bias field*, is applied perpendicular to the plate surface, those domains with their magnetization direction opposite to H_b contract in area. As H_b is increased these domains, *or magnetic bubbles*, decrease to a diameter *d* of 1 to 10 μm (Fig. 11.8(*b*)). The bubbles interact strongly within a distance of at least four bubble diameters. Thus reducing *d* allows the number of bubbles in a given area of film to be increased.

If a second external magnetic field H_d, the *drive field*, is applied parallel to the plate surface, bubbles can be moved through the plate. Linear tracks of a soft magnetic material, such as permalloy, are deposited on the plate surface using integrated circuit technology, so that the bubbles can be constrained to move along well-defined tracks. These tracks are designed so that bubbles remain underneath them, and they are magnetized by the drive field so that bubbles move continuously in a fixed direction. This requires that H_d is rotated, which is achieved through generation by an electromagnet rather than through any mechanical motion. Typical values of H_b and H_d are 8000 A/m and 4000 A/m peak respectively. H_b is supplied by a permanent magnet and its value is chosen to ensure stability of the bubbles.

One practical arrangement is the *T-bar* track, shown in Fig. 11.9. When H_d is parallel to a permalloy arm a magnetic field is induced in it as shown for four directions in Fig. 11.9(*a*) to (*d*). Each magnetic bubble behaves like a small magnet with one of its poles at the plate surface. As shown this is the *S*-pole, so it is attracted to an *N*-pole in the track (Fig. 11.9(*a*)). A conducting loop of aluminium-copper is deposited under the first bar and bubbles are launched from the key-shaped track on the left when a current is passed through the loop (Fig. 11.9(*b*)).

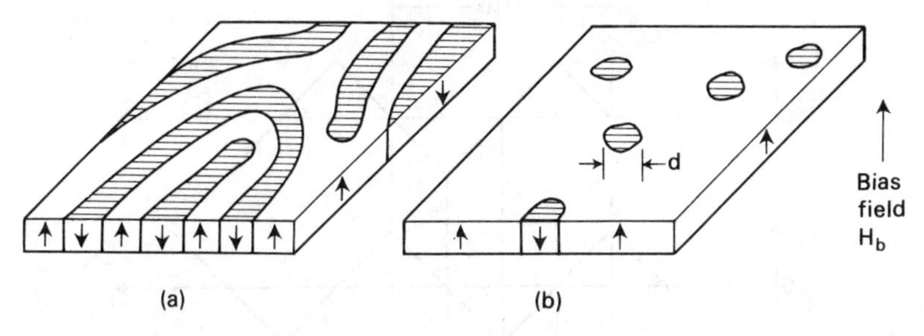

Figure 11.8 Domains in (*a*) garnet plate with no external magnetic field; (*b*) formation of magnetic bubble domains when a bias field is applied.

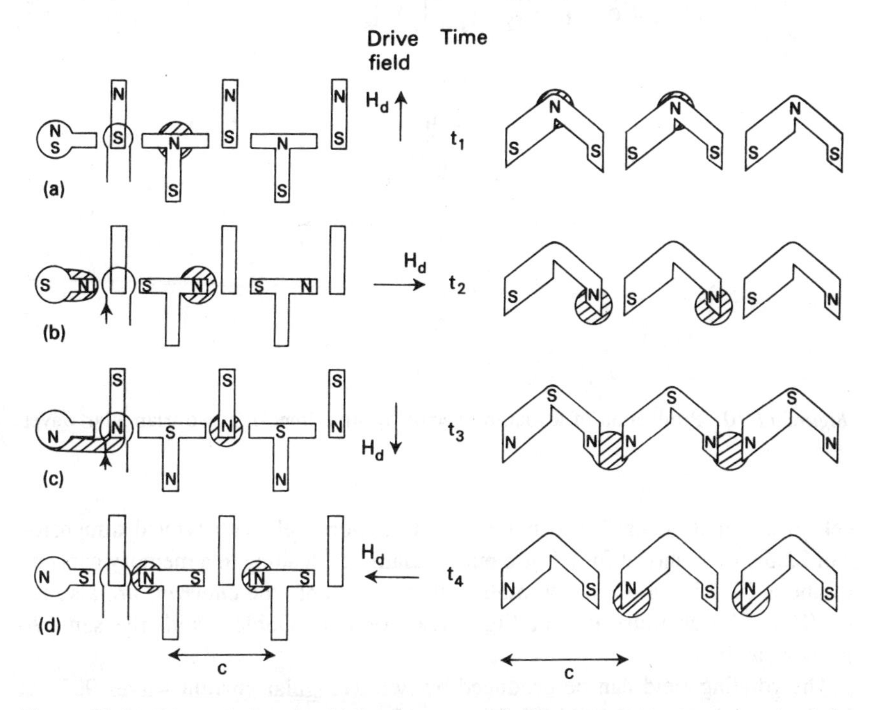

Figure 11.9 T-bar track and bubble generator, showing the effect of rotating the drive field. The later chevron pattern is also shown.

This occurs when H_d has rotated through 90°, causing the first bubble to move along the arm of the T. Current is maintained in the loop for a further 90° rotation of H_d (Fig. 11.9(*c*)), when the new bubble is drawn under the first bar and the first bubble is under the second bar. After 270° rotation of H_d the current in the loop is turned off and both bubbles are held under the arms of adjacent T's (Fig. 11.9(*d*)). Thus one complete rotation of H_d corresponds to both the formation of a bubble, if required, and the transfer of an existing bubble from one

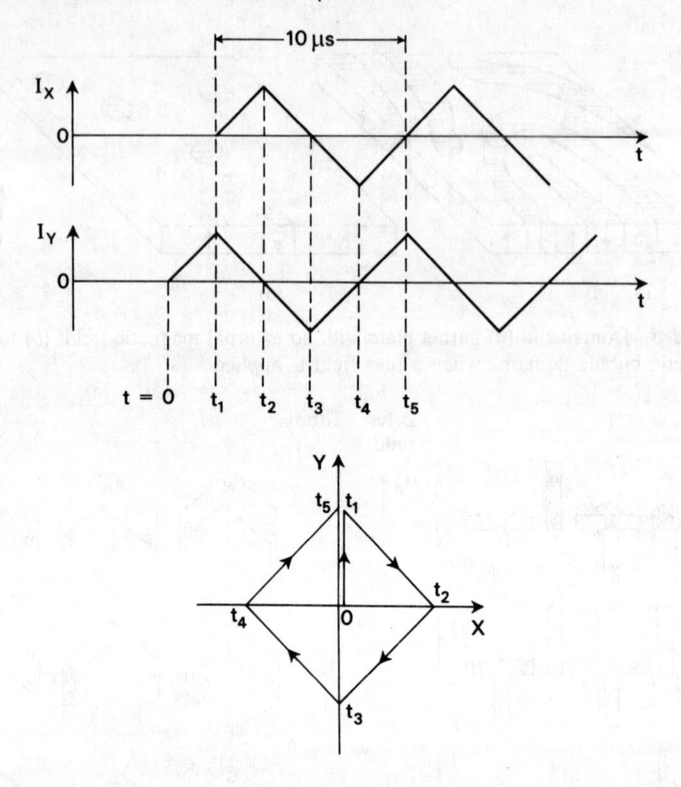

Figure 11.10 **Production of a rotating vector by superimposing two triangular waves.**

cell to the next. Each T and bar forms a memory cell with typical dimension c of 22.5 μm as shown for a 5 μm bubble diameter, leading to a memory capacity of about 10^6 bits/in. Other permalloy shapes, notably the *chevron*, are also used (Ref. [4]) as illustrated in Fig. 11.9, for two bubbles, with the same h_d phasors as before.

The rotating field can be produced by two triangular current waves 90° out of phase as shown in Fig. 11.10. The waveform is suitable for digital generation by an external circuit and a current up to 0.5 A peak is required. Times t_1 to t_4 correspond to the four positions of H_d shown in Fig. 11.9 and the frequency is 100 kHz, so a bubble is formed within 10 μs.

Memory organization

The track of Fig. 11.9 is arranged into an endless loop to form a continuous shift register. The access time for a single loop would be unacceptably long, so the memory is organized into a number l of *minor* loops all connected to a single *major* loop (Fig. 11.11). In any bit position the presence of a bubble indicates a

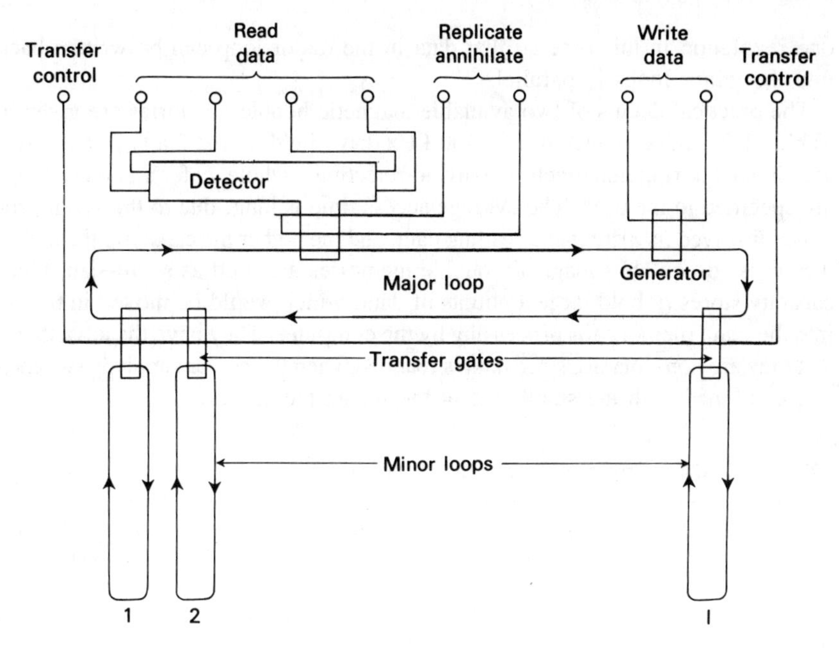

Figure 11.11 Organization of a typical bubble memory.

logical 1 and its absence a logical 0. The minor loops are spaced such that one unused bit position exists between all data bits in the major loop after being transferred from a minor loop. Thus data moves round the major loop at a rate of 50 kilobits/second and round each minor loop at 100 kilobits/second.

Writing is achieved by introducing data into the major loop through the *generator*, which is similar to the current loop of Fig. 11.9. l bits can be transferred simultaneously to the top bit position of one of the minor loops. Thus a block of data is entered serially until the first bit is aligned with the first minor loop when a current of about 40 mA is passed through the transfer control to move all the bits in the block out of the major loop in parallel.

Reading requires that bits in the minor loops must first be rotated to place the desired block of data at the top bit position in the loop. The transfer gates are again activated to move these bits into the major loop. The bubbles are then shifted to the *replicate* section of the major loop. Here data is duplicated so that it can be passed to the detector and also returned to the major loop. Alternatively it may be just passed to the detector, before being *annihilated* by running into a permalloy guard rail. The detector uses permalloy magneto-resistive elements connected in two arms of a bridge to produce differential electrical signals as a bubble passes through. This improves the rejection of unwanted noise signals and gives an output of a few mV per bubble at a rate of 50 kbit/s. The duplicate data remaining in the major loop are rotated until the first of the bits reaches the position above minor loop 1. The data in the minor loops have also made one revolution in this time so that data in the major loop can be written back into the minor loops in parallel.

The practical details of two available magnetic bubble memories are given in Table 11.2, both of which have a 100 kHz drive field. Some faulty minor loops are allowed during manufacture to avoid rejecting a whole wafer and these loops are specified to the user. The average access time is long, due to the serial processes involved in writing and reading data, and the higher the capacity, the longer the access time. Thus magnetic bubble memories are used as small-size, high-capacity stores to hold large amounts of data, which would be moved in blocks into the main memory for processing by the computer. They have the advantages that they are non-volatile, since data is retained when the power supply is switched off, combined with the small size of the integrated circuit.

Table 11.2 Characteristics of magnetic bubble memories

	Number of minor loops	Number of bit positions		Maximum capacity bits	Average random access time, ms
		Major loop	Minor loop		
Medium capacity	144–157	640	641	100 637	4
Large capacity	2048–2096	4096	512	1 073 152	40

Points to remember

* A ferrite-cored inductor has four types of loss due to the winding and another three due to the core, each being represented by a resistance.
* The total inductor loss varies with frequency due to the variations of individual losses.
* A magnetic bubble is an elementary magnet whose movement in a shift register is controlled by a rotating field to form a high-capacity memory.

References

[1] Hammond, P. *Electromagnetism for Engineers*, (Pergamon Press), 2nd ed., 1978.
[2] Snelling, E.C. *Soft Ferrites: Properties and Applications*, (Iliffe), 2nd ed., 1988.
[3] Casimir, H.B.G. and Ubbink, J. 'The skin effect', *Philips Tech. Rev.*, 1967, vol. 28, pp. 271, 300.
[4] Chang, H. *Magnetic-bubble memory technology*, (Dekker), 1978.

Problems

11.1 In a 50 mH inductor with a ferrite core the only significant losses are due to (i) a d.c. resistance of 7.0 Ω and (ii) a self-capacitance of 10 pF which itself has a loss tangent of 0.02. Determine the maximum value of the *Q*-factor and the frequency at which this occurs. You may assume that the self-resonant frequency of the coil is much higher than this frequency.
[910; 30.5 kHz]

11.2 A 25 mH ferrite-cored inductor has a self-resonant frequency of 200 kHz and it is operated in a parallel tuned circuit which is resonant at 20 kHz. The inductor windings have a d.c. resistance of 6.0 Ω and the loss tangent of their insulation is 0.02. Assuming that other losses are negligible determine (i) the *Q*-factor at 20 kHz; (ii) the frequency at which maximum *Q*-factor occurs; and (iii) the maximum *Q*-factor.
[472; 34 kHz; 590]

11.3 A magnetic bubble memory has one major loop and 150 minor loops each with a capacity of 640 bits. One unused bit position exists between all data bits after transfer to the major loop and the detector is two bit positions from the first minor loop.

A block of data 150 bits long is held at address 300. If the frequency of the drive field is 100 kHz, estimate the total time for the block to be read out of the memory.
[6.02 ms]

Appendix 1
Wave mechanics: an introduction

De Broglie's concept of a guiding wave for a moving particle was developed by Schrödinger in 1926 into a comprehensive theory called *wave mechanics*. The concepts of travelling waves are derived first and then Schrödinger's equation is developed.

Travelling waves

The expression for a wave motion travelling with velocity u in the positive x-direction is given in general by

$$\psi = \mathrm{f}(x - ut) \tag{A.1.1}$$

This equation is a function both of distance and time, ut being the distance travelled by the wavefront in time t. For a sinusoidal waveform the wave repeats itself after a distance $x = \lambda$, the wavelength, so that eq. (A.1.1) becomes

$$\psi = \psi_m \sin\left(\frac{2\pi}{\lambda}\, x - ut\right) \tag{A.1.2}$$

where ψ_m is the peak value (Fig. A.1.1). The travelling wave can be represented by a phasor of length ψ_m rotating clockwise. The locus of the tip of the phasor projected on to the vertical axis gives the value of the displacement ψ at a particular time, and the waveform may be considered to move to the right with velocity u as the phasor rotates with angular frequency ω in radians per second. If f is the frequency of the wave in cycles per second, or hertz (Hz),

$$\omega = 2\pi f \tag{A.1.3}$$

a complete cycle corresponding to rotation of the phasor through 2π radians, or 360°. As the phasor rotates and the distance x travelled by a point on the wave increases, the angle kx increases *negatively* and when $x = \lambda$, $kx = 2\pi$, or

$$k = \frac{2\pi}{\lambda} \tag{A.1.4}$$

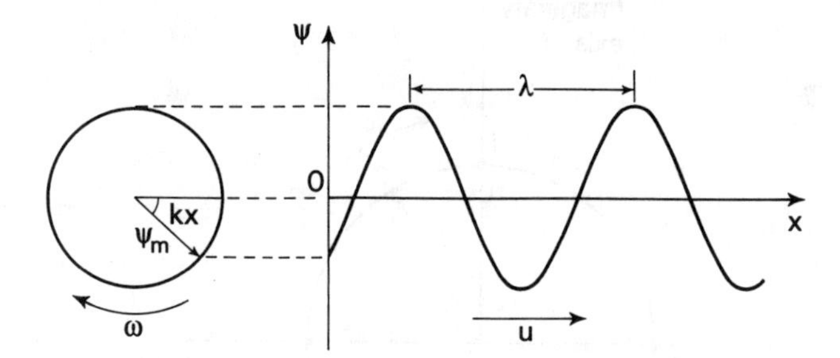

Figure A.1.1 Phasor representation of a travelling wave.

where k is the *phase-change coefficient*. Similarly, movement of the wave to the *left* occurs for anticlockwise rotation, with the angle kx increasing *positively*. In either case the time for a complete cycle is $1/f$, so that

$$\lambda = \frac{u}{f} \tag{A.1.5}$$

We need to express eq. (A.1.2) in a standard form in terms of distance only, and differentiating twice leads to

$$\frac{\partial^2 \psi}{\partial x^2} + \left(\frac{2\pi}{\lambda}\right)^2 \psi = 0 \tag{A.1.6}$$

or

$$\frac{\partial^2 \psi}{\partial x^2} + k^2 \psi = 0 \tag{A.1.7}$$

These are two forms of the *wave equation*, which is applicable to all types of wave motion with suitable interpretation of the displacement ψ.

A more convenient expression than eq. (A.1.2) for a travelling wave may be obtained by using the operator $j = \sqrt{-1}$. Multiplication of a phasor by j does not change its length, but rotates it *anticlockwise* through 90°, while multiplication by –j rotates it *clockwise* through 90°. In Fig. A.1.2 the phasor OP has unit length and it is shown on an *Argand* diagram, where real numbers are plotted on the horizontal axis and imaginary numbers, which incorporate j, are plotted on the vertical axis. OP then is fully described in amplitude and phase angle θ by the expression

$$a + jb = 1 \tag{A.1.8}$$

where $a + jb$ is a *complex number*, with a real part, a, and an imaginary part, b. Then, by Pythagoras, the length OP, known as the *modulus* of the phasor, is given by

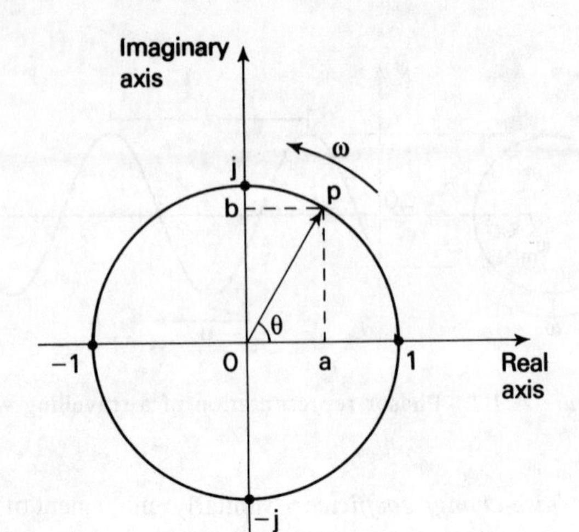

Figure A.1.2 Rotation of a phasor with operator j.

$$1 = \sqrt{(a^2 + b^2)} \quad \text{(A.1.9)}$$

and the phase angle is

$$\theta = \tan^{-1}(b/a) \quad \text{(A.1.10)}$$

Using trigonometry and the exponential series so that the complex number may be expressed in exponential form,

$$a + \mathrm{j}b = \cos\theta + \mathrm{j}\sin\theta = \exp(\mathrm{j}\theta) \quad \text{(A.1.11)}$$

Then at a given time the expression for a wave travelling to the left is $\psi_m \exp(\mathrm{j}kx)$, and for a wave travelling to the right it is $\psi_m \exp(-\mathrm{j}kx)$. The complete solution of the wave equation (A.1.7) is given by

$$\psi = A\exp(-\mathrm{j}kx) + B\exp(\mathrm{j}kx) \quad \text{(A.1.12)}$$

which may be confirmed by differentiating this equation twice and substituting back into eq. (A.1.7). A and B are constants depending on the boundary conditions of a particular problem and are determined later in selected cases; either A or B may be zero.

Schrödinger's equation

Before deriving Schrödinger's equation, we must relate the properties of the particle to its wavelength. Einstein's theory of relativity led to the result that energy and mass were related by

$$W = mc^2 \quad \text{(A.1.13)}$$

where c is the velocity of electromagnetic waves. Hence for a quantum of radiation,

$$hf = mc^2 \tag{A.1.14}$$

so that the 'mass' of a quantum is

$$m = \frac{hf}{c^2} \tag{A.1.15}$$

But the momentum of a quantum, p, is

$$p = mc = \frac{hf}{c} = \frac{h}{\lambda} \tag{A.1.16}$$

so that

$$\lambda = \frac{h}{p} \tag{A.1.17}$$

— the expression applied by de Broglie to any moving particle. The kinetic energy of a particle moving with velocity u is

$$K = \tfrac{1}{2} mu^2 = \frac{p^2}{2m} = \frac{h^2}{2m\lambda^2} \tag{A.1.18}$$

Since the total energy of a system is the sum of its kinetic and potential energies

$$W = K + V \tag{A.1.19}$$

and from eq. (A.1.18),

$$\frac{1}{\lambda^2} = \frac{2mK}{h^2} = \frac{2m}{h^2}(W - V) \tag{A.1.20}$$

Substituting into the wave equation (A.1.6),

$$\frac{\partial^2\psi}{\partial x^2} + \frac{8\pi^2 m}{h^2}(W - V)\psi = 0 \tag{A.1.21}$$

This is Schrödinger's wave equation in one dimension, and the question arises of the interpretation of the displacement of the wave, ψ.

With other types of wave motion the energy density of the wave is proportional to the square of the displacement, so that, if eq. (A.1.9) represents the guiding wave of a beam of electrons, say, then the intensity of the beam in a volume element dr must be proportional to ψ^2 dr, where dr = dx dy dz. The more intense the beam the greater is the probability of finding an electron in the element dr, so that ψ^2 may be interpreted as being proportional to the probability of finding an electron at the point represented by allowing dr to become very small. Since multiplication of ψ by a constant does not alter eq. (A.1.9), ψ^2 may be chosen so that it is *equal* to the probability of finding the electron, and ψ is then the displacement of a wave whose intensity gives the probability of finding the electron at a particular point.

Electron in free space

The solution of Schrödinger's equation is very difficult except in a few cases. The simplest of these is that of an electron in free space, which has zero potential energy whatever its position, or $V = 0$. Schrödinger's equation then reduces to

$$\frac{\partial^2 \psi}{\partial x^2} + \frac{8\pi^2 m}{h^2} W\psi = 0 \qquad \text{(A.1.22)}$$

which becomes the same as eq. (A.1.7) when

$$k = \frac{(8\pi^2 mW)^{1/2}}{h} \qquad \text{(A.1.23)}$$

and a solution of eq. (A.1.22) for a wave travelling to the right is

$$\psi = A \exp(-jkx) \qquad \text{(A.1.24)}$$

where A is a constant. Since k is a real number, W can only have positive values, all of which are allowed. Thus in free space an electron may have any value of kinetic energy and is free of quantum restrictions, which agrees with experiment.

Another way of describing this continuous variation of energy is to express W in terms of k, using eq. (A.1.23) so that

$$W = \frac{h^2 k^2}{8\pi^2 m} \qquad \text{(A.1.25)}$$

k is a function of momentum p since

$$k = \frac{2\pi}{\lambda} = \frac{2\pi}{h} p \qquad \text{(A.1.26)}$$

using eq. (A.1.17), so that for matter waves k is termed the *momentum vector* or *k-vector*. Since in one dimension p can be positive or negative, k considered as a vector can also be positive or negative, so that W varies parabolically with k, as shown in Fig. A.1.3.

Electron in a potential well

On the other hand, an electron in an atom will be affected by a variation of potential energy such as that shown in Fig. 1.3 for a hydrogen atom. This Coulomb potential well may be very crudely approximated by the rectangular potential well of Fig. A.1.4. Let us suppose for simplicity that the electron moves between the

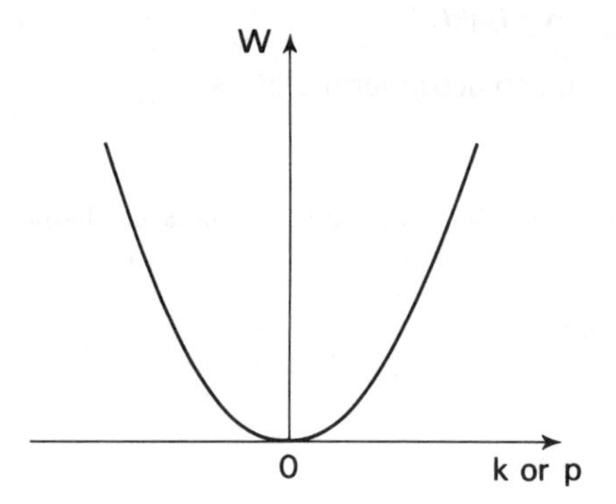

Figure A.1.3 Energy/momentum-vector diagram for an electron in free space.

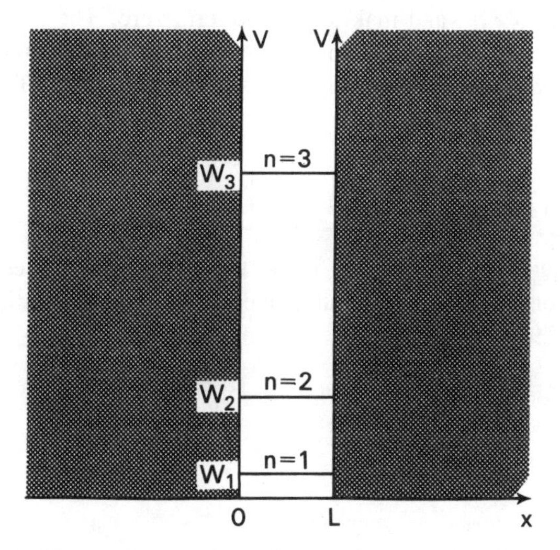

Figure A.1.4 Energy levels of an electron in a rectangular potential well.

walls of the well in straight lines parallel to the x-axis and that the walls are perfectly reflecting; also that the potential energy is given by

$$V = 0 \quad \text{for} \quad 0<x<L$$

$$V \to \infty \quad \text{for} \quad x\leq 0 \quad \text{or} \quad x \geq L$$

This implies that the electron is to be found only within the well, or that

$$\psi = 0 \quad \text{for} \quad x\leq 0 \quad \text{or} \quad x\geq L$$

Again Schrödinger's equation reduces to eq. (A.1.22) within the potential well, with a general solution given by eq. (A.1.12), from which at $x = 0$, $\psi = A+B$, and since $\psi = 0$, $A = -B$. Then at $x = L$,

$$\psi = B[\exp(jkL) - \exp(-jkL)] \tag{A.1.27}$$

which, expressed in trigonometric terms, gives

$$\psi = j2B \sin kL \tag{A.1.28}$$

For $\psi = 0$ at $x = L$ either $B = 0$, which implies no displacement, or

$$kL = \frac{2\pi L}{\lambda} = n\pi$$

This makes

$$L = n\frac{\lambda}{2} \tag{A.1.29}$$

where $n = 1, 2, 3 \ldots$, and makes the sine term zero. The j in eq. (A.1.28) indicates a phase angle of 90° which is not apparent if the magnitude (modulus) of ψ is considered, since this is always a positive quantity, $|\psi|$. Then, using eq. (A.1.29),

$$|\psi| = 2B \sin\left(\frac{n\pi x}{L}\right) \tag{A.1.30}$$

which is the equation of a *standing wave*, the sum of two waves travelling in opposite directions, since the sine function contains two exponential terms. But, from eq. (A.1.29)

$$\frac{n\pi}{L} = \frac{(8\pi^2 m W_n)^{1/2}}{h} = k \tag{A.1.31}$$

or

$$W_n = \frac{n^2h^2}{8mL^2} \tag{A.1.32}$$

This means that the energy of the electron is restricted to discrete levels such as W_1, W_2 ... shown in Fig. A.1.4, which are similar to the discrete energy levels of the Bohr hydrogen atom (Fig. 1.3). The value of B depends on the probability that a particular level will be occupied, which in turn depends on the statistics used such as Fermi-Dirac or Maxwell-Boltzmann. For a level where it is certain that an electron will be found $\int |\psi|^2 \, dx = 1$, since $dr \rightarrow dx$ in the one-dimensional case.

Electron tunnelling

Let us now suppose that an electron is incident upon an energy barrier of finite height V_0 and finite width d (Fig. A.1.5(*a*)). If the energy of the electron is less than V_0 and d is large it cannot penetrate the barrier, any more than it can escape from a potential well. However, if d is made sufficiently small then there is a correspondingly small probability that the electron will penetrate the barrier, a phenomenon known as *tunnelling*. In regions I and III, outside the barrier $V = 0$ and Schrödinger's equation then becomes

$$\frac{\partial^2 \psi}{\partial x^2} + \frac{8\pi^2 m}{h} W\psi = 0 \tag{A.1.33}$$

while in region II, inside the barrier,

$$\frac{\partial^2 \psi}{\partial x^2} + \frac{8\pi^2 m}{h} (W - V_0)\psi = 0 \tag{A.1.34}$$

where $W - V_0$ is negative since $V_0 > W$.

The solution of eq. (A.1.33) for region I is then

$$\psi_1 = A_1 \exp(-jk_1 x) + B_1 \exp(jk_1 x) \tag{A.1.35}$$

where

$$k_1 = \left(\frac{8\pi^2 mW}{h^2}\right)^{1/2} \tag{A.1.36}$$

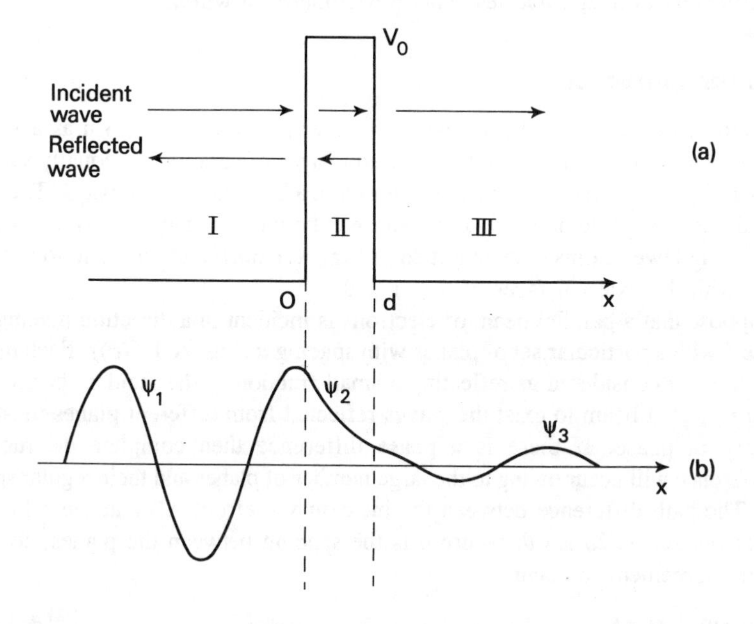

Figure A.1.5 Electron tunnelling. (*a*) Potential barrier and de Broglie waves; (*b*) wave functions.

The first term represents a wave of amplitude A_1 travelling towards the barrier, while the second term represents a wave of amplitude B_1 reflected from the barrier (Fig. A.1.5(a)). For region II the solution of eq. (A.1.34) is

$$\psi_2 = A_2 \exp(-k_2 x) + B_2 \exp(k_2 x) \tag{A.1.37}$$

where

$$k_2 = \left[\frac{8\pi^2 m(V_0 - W)}{h^2}\right]^{1/2} \tag{A.1.38}$$

with the js absent in eq. (A.1.37), and again represents two waves travelling in opposite directions. Finally for region III we have

$$\psi_3 = A_3 \exp(-jk_1 x) \tag{A.1.39}$$

since there is no reflected wave.

Considering only the waves travelling to the right represented by the A terms, it may be seen that the amplitudes of ψ_1 and ψ_3 remain constant with distance. However, the amplitude of ψ_2 falls exponentially with distance and will tend to zero when the barrier width d is sufficiently large. Since both the wave functions and their derivatives must be continuous they will appear as in Fig. A.1.5(b), so that, for a narrow barrier, ψ_3 will possess a finite amplitude representing a probability that an electron will penetrate the barrier and appear on the other side. In practice electron tunnelling is only significant for barriers less than a few electronvolts in height and less than a nanometre in width.

Electron diffraction

The atoms in a crystal form a regular 3-dimensional array, known as a *space lattice*. All the atoms may then be included in a set of parallel and equally spaced planes (Fig. A.1.6(a)), which can be chosen in a large number of ways. The sets containing a large number of atoms will then be more widely spaced than those containing fewer atoms (shown dotted in Fig. A.1.6(a)) and will tend to correspond with the external faces of the crystal.

Suppose that a parallel beam of electrons is incident in a direction making an angle θ with a particular set of planes with spacing a (Fig. A.1.6(b)). Each plane may then be considered as reflecting a small fraction of the incident beam, but for a reflected beam to exist the waves reflected from different planes must be exactly in phase. If there is a phase difference then complete destructive interference will occur owing to the large number of planes and their regular spacing. The path difference between the incident wavefront AB and the reflected wavefront AC is $2a \sin\theta$, where a is the spacing between the planes, so that for reinforcement to occur

$$2a \sin\theta = n\lambda \tag{A.1.40}$$

where $n = 1, 2, 3, \ldots$ This equation was first derived by Bragg in 1912 to explain

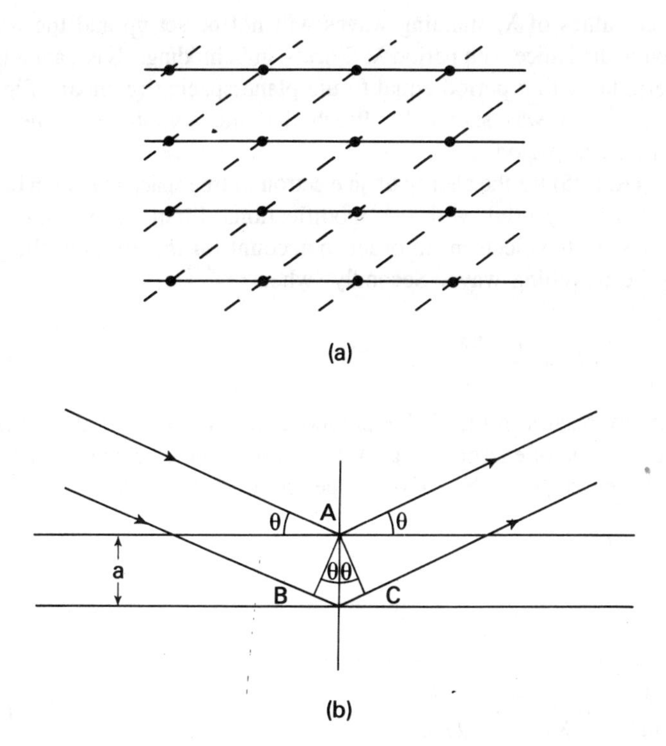

Figure A.1.6 Electron diffraction. (*a*) Crystal planes; (*b*) reflection of electron beam.

the diffraction of X-rays by a crystal. The incident beam is said to be *diffracted* at a number of different angles, each of which corresponds to one value of n. Bragg's law was used to interpret the results of experiments in which a beam of electrons was diffracted by a crystal, either from its surface or by a film about 10 nm thick, and in this way the wave nature of electrons was confirmed.

Electron in a crystal

An electron moving through a crystal will be influenced by the periodic potential due to the atomic cores, which are regularly spaced at distance r_0 (Fig. A.1.6 and Fig. 2.1). The electron may then be considered as a wave moving at right angles to a set of planes with spacing a (not necessarily equal to r_0), so that Bragg's law becomes

$$n\lambda = 2a \tag{A.1.41}$$

At values of λ where this equation is satisfied, reinforcement of the waves will occur, so that a standing wave will be set up between the planes. Thus the electron will be unable to penetrate the lattice and its particle velocity u will be zero.

At all other values of λ, standing waves will not be set up and the wave will travel through the lattice with period λ. Since V in Schrödinger's equation (A.1.21) will be periodic with a period equal to the plane spacing (compare Fig. 2.4(*b*) where $a = r_0$), it was shown by Bloch that the amplitude of the wave is modulated also at period a.

Equation (A.1.25) for the energy of an electron in free space may then be applied to an electron in a crystal, with two modifications. Firstly m becomes m^*, the effective mass of the electron, in order to account for the effect of the periodic lattice on the travelling wave. Secondly, when

$$\lambda = \frac{2a}{n} \quad \text{or} \quad k = \frac{n\pi}{a} \tag{A.1.42}$$

a discontinuity occurs in the $W - k$ parabola, which corresponds to the centre of a forbidden band of energies (Fig. A.1.7). The particle velocity of the electron is related to the slope of the curve, since from eq. (A.1.25),

$$\frac{\mathrm{d}W}{\mathrm{d}k} = \frac{h^2 k}{4\pi^2 m^*} \tag{A.1.43}$$

and

$$u = \frac{p}{m^*} = \frac{h}{\lambda m^*} = \frac{hk}{2\pi m^*} \tag{A.1.44}$$

so that

$$\frac{\mathrm{d}W}{\mathrm{d}k} = u\frac{h}{2\pi} \tag{A.1.45}$$

But in the forbidden band $u = 0$, so that $\mathrm{d}W/\mathrm{d}k = 0$ also, as shown at the band edges in Fig. A.1.7. The effective mass of the electron is related to the rate of change of slope with k, since from eq. (A.1.43),

$$\frac{\mathrm{d}^2W}{\mathrm{d}k^2} = \frac{h^2}{4\pi^2 m^*}$$

or

$$m^* = \frac{h^2}{4\pi^2} \bigg/ \frac{\mathrm{d}^2W}{\mathrm{d}k^2} \tag{A.1.46}$$

m^* is constant wherever energy increases with k^2 (eq. (A.1.25)), which applies towards the top of the energy/k-vector graph. This corresponds to the conduction band of a material where the electron will behave as though it were in free space, but with a modified mass. However, as the energy is reduced to the top

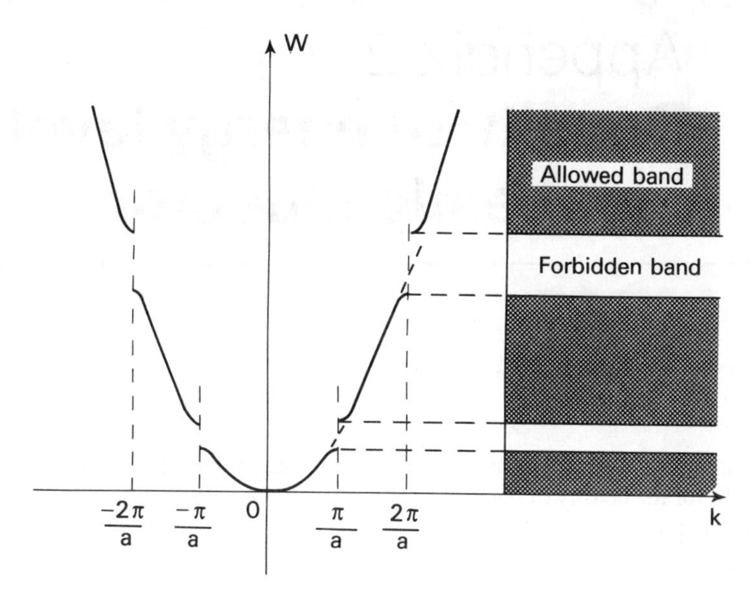

Figure A.1.7 Energy/momentum-vector diagram for an electron in a crystal.

of a forbidden band $d^2W/dk^2 \rightarrow 0$, which suggests that $m^* \rightarrow \infty$ and corresponds to $u \rightarrow 0$, since it is not possible to move an infintely large mass. The concept of effective mass in the allowed band *below* a forbidden band is difficult to grasp, since d^2W/dk^2 is negative at the top of the allowed band, zero in the middle and positive again at the bottom, and in these regions the interaction between the periodic lattice and the electron becomes very complicated.

The foregoing theory suggests that in an insulator the wavelength of electrons is related to the atomic spacing by eq. (A.1.41) at energies in the forbidden band below the conduction band. Thus standing waves are set up and no electron flow can occur. However, in a metal the wavelength of electrons at the top of the valence band is not related to the atomic spacing by this equation, so that the electron waves can travel freely through the crystal.

Appendix 2
Density of energy levels in a semiconductor

In this appendix a derivation is given for a relationship which was expressed in eq. (2.28) for the density of the energy levels in the conduction and valence bands of a conductor:

$$(N_c N_v)^{1/2} = GT^{3/2}$$

where G is a constant. We need first an expression for the *number* of energy levels in a crystal, which is a function of energy and is directly proportional to the volume of the crystal. Thus the number of energy levels per unit volume, $\delta(W)$, is independent of volume and is known as the *density of states* function. If there are δS energy levels in a small energy range δW, then

$$S(W) = \frac{\delta S}{\delta W} \rightarrow \frac{dS}{dW} \qquad (A.2.1)$$

as $\delta W \rightarrow 0$.

It is convenient to express the kinetic energy of an electron in terms of its momentum, p, which has direction as well as magnitude, and in a 3-dimensional crystal it may be expressed in terms of components p_x, p_y and p_z. These are measured along the x-, y- and z-axes respectively to form a 3-dimensional *momentum space* (Fig. A.2.1). Each point in this space represents the momentum of an electron, which can be equally well represented by a vector of length p drawn from the origin to the point. Then all the electrons with energies between 0 and W are represented by points lying within a sphere of radius p, given by

$$p = \sqrt{(2m_e W)} \qquad (A.2.2)$$

which is obtained from eq. (A.1.18), m_e being the effective mass of the electron. We are assuming here that each electron sees a uniform potential field set up by the atomic cores and the other electrons. This corresponds to uniform potential energy, so that $V = 0$ within the crystal and the energy of an electron is only kinetic.

The surface area of the sphere is $4\pi p^2$, so that for a small increment of

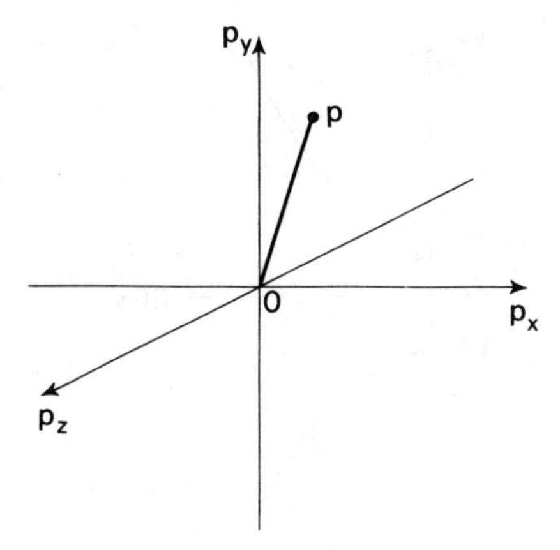

Figure A.2.1 Momentum space.

momentum δp the number of energy levels within a momentum shell of volume $4\pi p^2\ \delta p$ may be considered in order to obtain the density of states. From quantum theory the minimum momentum-distance product is h, so that the minimum momentum-volume product is h^3. Then the number of separate energy levels in the momentum shell is, in the limit,

$$\mathrm{d}S = 2 \times \frac{4\pi p^2}{h^3}\ \mathrm{d}p \tag{A.2.3}$$

since each value of momentum may be associated with *two* electrons having opposite spins. From eq. (A.2.2),

$$\mathrm{d}p = \frac{1}{2}\sqrt{\frac{2m_e}{W}}\ \mathrm{d}W \tag{A.2.4}$$

so that, on substitution in eq. (A.2.3),

$$\frac{\mathrm{d}S}{\mathrm{d}W} = \frac{8\pi\sqrt{2}m_e^{3/2}W^{1/2}}{h^3} \tag{A.2.5}$$

This equation may be applied to electrons in the conduction band by measuring their energy with respect to W_c at the bottom of the band (Fig. A.2.2), so that, in the conduction band,

$$\frac{\mathrm{d}S}{\mathrm{d}W} = S(W) = \frac{8\pi\sqrt{2}m_e^{3/2}(W - W_c)^{1/2}}{h^3} \tag{A.2.6}$$

Similarly the energy of the holes in the valence band can be measured *down* from the energy W_v at the top of the band, so that, in the valence band,

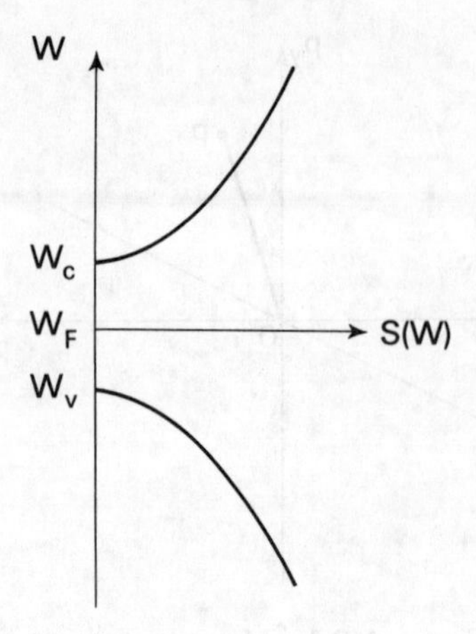

Figure A.2.2 Density of state functions for a semiconductor.

$$S(W) = \frac{8\pi\sqrt{2}m_h^{3/2}(W_v - W)^{1/2}}{h^3} \tag{A.2.7}$$

where m_h is the effective mass of a hole. These expressions are illstrated in Fig. A.2.2.

The number density of electrons dn within an energy range dW will depend on both the density of available states dS and the probability that each state is occupied, $p_F(W)$. Thus

$$\mathrm{d}n = p_F(W)\mathrm{d}S \tag{A.2.8}$$

and for the whole range of energy being considered the number density of electrons is

$$n = \int p_F(W) \frac{\mathrm{d}S}{\mathrm{d}W} \mathrm{d}W \tag{A.2.9}$$

For a crystal $p_F(W)$ is given by the Fermi function (eq. (2.17)), which for electrons in the conduction band of a semiconductor reduces to

$$p_F(W) = \exp\left(-\frac{W - W_F}{kT}\right) \tag{A.2.10}$$

Then, substituting in eq. (A.2.9) and integrating from the bottom of the conduction band,

$$n = \frac{8\pi\sqrt{2}m_e^{3/2}}{h^3} \int_{W_c}^{\infty} (W - W_c)^{1/2} \exp\left(-\frac{W - W_F}{kT}\right)\mathrm{d}W$$

$$= \frac{8\pi\sqrt{2}m_e^{3/2}}{h^3} \exp\left(-\frac{W_c - W_F}{kT}\right)$$

$$\times \int_{W_c}^{\infty} (W - W_c)^{1/2} \exp\left(-\frac{W - W_c}{kT}\right) dW \qquad \text{(A.2.11)}$$

The integral in this equation can be expressed in a standard form if we make $y^2 = (W - W_c)/kT$. Then the integration becomes of the type

$$\int_0^{\infty} y^2 \exp(-y^2) dy = \frac{\sqrt{\pi}}{2} (kT)^{3/2} \qquad \text{(A.2.12)}$$

in this case so that

$$n = \frac{2}{h^3} (2\pi m_e kT)^{3/2} \exp\left(-\frac{W_c - W_F}{kT}\right) \qquad \text{(A.2.13)}$$

which, by comparison with eq. (2.21), can be written

$$n = N_c \exp\left(-\frac{W_c - W_F}{kT}\right) \qquad \text{(A.2.14)}$$

where

$$N_c = \frac{2}{h^3} (2\pi m_e kT)^{3/2} \qquad \text{(A.2.15)}$$

Thus N_c can be conveniently regarded as the density of states lying at the energy W_c which gives the same answer as integration over the whole conduction band of energies. By a similar analysis the density of states lying at the energy W_v in the valence band is

$$N_v = \frac{2}{h^3} (2\pi m_h kT)^{3/2} \qquad \text{(A.2.16)}$$

using eq. (2.34)

$$p = N_v \exp\left(-\frac{W_F - W_v}{kT}\right) \qquad \text{(A.2.17)}$$

The product of N_c and N_v is then obtained from eqs (A.2.15) and (A.2.16), so that

$$(N_c/N_v)^{1/2} = \frac{2}{h^3} (2\pi k)^{3/2} (m_e m_h)^{3/4} T^{3/2} = GT^{3/2} \qquad \text{(A.2.18)}$$

Thus the value of G depends on the value of the effective masses of electrons and holes. If it is assumed that $m_e = m_h = m$, the mass of an electron in free space, then $G = 4.83 \times 10^{21}$ m^{-3} $K^{-3/2}$ electrons or holes. However, in practice the numerical value of G is 8.8×10^{22} for germanium and 8.9×10^{22} for silicon and it is assumed that the discrepancy is due to the effective masses being unequal to m.

In an intrinsic semiconductor n = p, so that the position of the Fermi level may be found by equating eq. (A.2.13) to eq. (A.2.16) combined with eq. (A.2.17), which gives

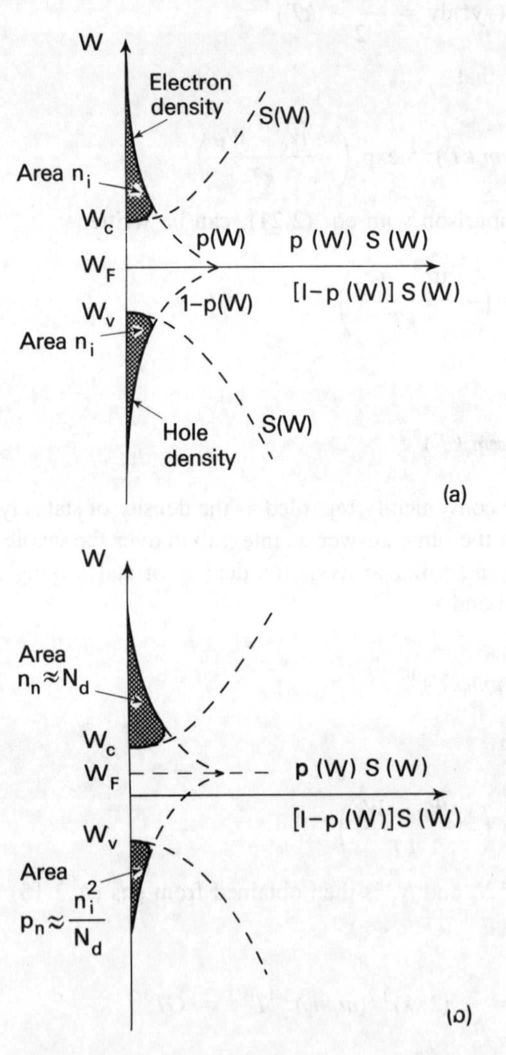

Figure A.2.3 Semiconductor carrier densities. (*a*) Intrinsic semiconductor; (*b*) n-type semiconductor.

$$m_{\mathrm{e}}^{3/2} \exp\left(-\frac{W_{\mathrm{c}} - W_{\mathrm{F}}}{kT}\right) = m_{\mathrm{h}}^{3/2} \exp\left(-\frac{W_{\mathrm{F}} - W_{\mathrm{v}}}{kT}\right) \tag{A.2.19}$$

and

$$W_{\mathrm{F}} = \frac{W_{\mathrm{c}} + W_{\mathrm{v}}}{2} + \tfrac{3}{4}kT \ln \frac{m_{\mathrm{h}}}{m_{\mathrm{e}}} \tag{A.2.20}$$

This shows that if $m_{\mathrm{e}} = m_{\mathrm{h}}$ the Fermi level lies in the middle of the energy gap for all temperatures, but that if $m_{\mathrm{e}} \neq m_{\mathrm{h}}$ it is only exactly in the middle of the energy gap for $T = 0$ K.

For intrinsic semiconductors it has already been shown in Chapter 2 that the position of the Fermi level is a function of both the doping density and the temperature, the effective masses being involved through N_{c} and N_{v}. The carrier densities for intrinsic and n-type semiconductors as a function of energy are illustrated in Fig. A.2.3. The values of $p(W)$ and $1 - \mathrm{p}(W)$ are both one-half at the Fermi level, but the exponential variation of these probabilities is much more rapid than shown in these diagrams. However, they illustrate that the variation of the electron density close to W_{c} and the variation of the hole density close to W_{v} are each controlled mainly by the corresponding $S(W)$ function. The variations of the carrier density at energies above W_{c} and below W_{v} are controlled by the respective probability functions. Thus there is an energy in the conduction band at which the electron density is a maximum and an energy in the valence band at which the hole density is a maximum. The area under the electron density curve must correspond to the total number of electrons, and the area under the hole density curve must correspond to the total number of holes. In an intrinsic semiconductor these areas are equal since the Fermi level is in the centre of the conduction band. In an n-type semiconductor the Fermi level is nearer the bottom of the conduction band, so that the total number of electrons is much greater than the total number of holes. In a p-type semiconductor the Fermi level is nearer the top of the valence band so that the situation is reversed and the number of holes greatly exceeds the number of electrons.

Appendix 3
Computer simulation using the SPICE program

The following is based on the *SPICE2G.O User's Guide* by A. Vladimirescu, A. R. Newton and D.O. Pederson of the Department of Electrical Engineering and Computer Sciences, University of California, Berkeley, 94720, USA.

SPICE is a general-purpose circuit simulation program for nonlinear d.c., nonlinear transient and linear a.c. analyses. It has built-in models for the semiconductor devices and the user need specify only the pertinent model parameter values. Full details about SPICE facilities and for running SPICE programs are given in the *User's Guide*.

Device parameters

Models are available for the diode which can be a junction diode or a Schottky barrier diode, the bipolar transistor (BJT) which can be n-p-n or p-n-p, and the junction FET which can be n- or p-channel. There are three MOSFET models; MOS1 (level 1) is described by a square-law $I-V$ characteristic and is the only one given in this Appendix. MOS2 is an analytical model while MOS3 is a semi-empirical model, both of which include second-order effects. All the MOSFET models are available for n- or p-channel, depletion or enhancement device simulation.

Model parameters that are not specified in the program are assigned the default values given below for each model type. This generally means that internal resistances, capacitors and transit times are set to zero, so that a.c. and transient effects are not simulated. Any number of devices in parallel may be specified by the area factor and the * indicates which parameters are affected. If the area factor is omitted a value of 1.0 is assumed.

Most of the parameters are discussed in Chapters 3, 4 and 6, except for the flicker noise parameters KF and AF which appear in each model. These two parameters are used in the small-signal a.c. analysis to determine the equivalent noise current generator connected across the device. The noise current has the form

$$\bar{i}_n^2 = \frac{KF \times I^{AF}}{fC_1} + C_2$$

where C_1 and C_2 are constants for each device and I is the diode, collector or drain current. $\bar{i}_n^2$ thus falls as frequency rises as discussed under flicker noise in Chapter 9.

Diode model parameters

	Name	Parameter	Units	Default	Example	Area
1	IS	Saturation current	A	1.0E-14	1.0E-14	*
2	RS	Ohmic resistance	Ω	0	10	*
3	N	Emission coefficient	—	1	1.0	
4	TT	Transit-time	s	0	0.1 ns	
5	CJO	Zero-bias junction capacitance	F	0	2 pF	*
6	VJ	Junction potential	V	1	0.6	
7	M	Grading coefficient	—	0.5	0.5	
8	EG	Activation energy	eV	1.11	1.11 Si 0.69 Sbd 0.67 Ge	
9	XTI	Saturation-current temperature exponent	—	3.0	3.0 jn 2.0 Sbd	
10	KF	Flicker noise coefficient	—	0		
11	AF	Flicker noise exponent	—	1		
12	FC	Coefficient for forward-bias depletion capacitance formula	—	0.5		
13	BV	Reverse breakdown voltage	V	infinite	40.0	
14	IBV	Current at breakdown voltage	A	1.0E-3		

BJT model parameters

	Name	Parameter	Units	Default	Example	Area
1	IS	Transport saturation current	A	1.0E-16	1.0E-15	*
2	BF	Ideal maximum forward beta	—	100	100	
3	NF	Forward current emission coefficient	—	1.0	1	
4	VAF	Forward Early voltage	V	infinite	200	
5	IKF	Corner for forward beta high-current roll-off	A	infinite	0.01	*
6	ISE	B-E leakage saturation current	A	0	1.0E-13	*
7	NE	B-E leakage emission coefficient	—	1.5	2	
8	BR	Ideal maximum reverse beta	—	1	0.1	
9	NR	Reverse current emission coefficient	—	1	1	
10	VAR	Reverse Early voltage	V	infinite	200	
11	IKR	Corner for reverse beta high-current roll-off	A	infinite	0.1	*
12	ISC	B-C leakage saturation current	A	0	1.0E-13	*

BJT model parameters (cont.)

	Name	Parameter	Units	Default	Example	Area
13	NC	B-C leakage emission coefficient	—	2	1.5	
14	RB	Zero bias base resistance	Ω	0	100	*
15	IRB	Current where base resistance falls halfway to its minimum value	A	infinite	0.1	*
16	RBM	Minimum base resistance at high currents	Ω	RB	10	*
17	RE	Emitter resistance	Ω	0	1	*
18	RC	Collector resistance	Ω	0	10	*
19	CJE	B-E zero-bias depletion capacitance	F	0	2 pF	*
20	VJE	B-E built-in potential	V	0.75	0.6	
21	MJE	B-E junction exponential factor	—	0.33	0.33	
22	TF	Ideal forward transit time	s	0	0.1 ns	
23	XTF	Coefficient for bias dependence of TF	—	0		
24	VTF	Voltage describing VBC dependence of TF	V	infinite		
25	ITF	High-current parameter for effect on TF	A	0		*
26	PTF	Excess phase at frequency = 1.0/(TF*2PI) Hz	degree	0		
27	CJC	B-C zero-bias depletion capacitance	F	0	2 pF	*
28	VJC	B-C built-in potential	V	0.75	0.5	
29	MJC	B-C junction exponential factor	—	0.33	0.5	
30	XCJC	Fraction of B-C depletion capacitance connected to internal base node	—	1		
31	TR	Ideal reverse transit time	s	0	10 ns	
32	CJS	Zero-bias collector-substrate capacitance	F	0	2 pF	*
33	VJS	Substrate junction built-in potential	V	0.75		
34	MJS	Substrate junction exponential factor	—	0	0.5	
35	XTB	Forward and reverse beta temperature exponent	—	0		
36	EG	Energy gap for temperature effect on IS	eV	1.11		
37	XTI	Temperature exponent for effect on IS	—	3		
38	KF	Flicker-noise coefficient	—	0		
39	AF	Flicker-noise exponent	—	1		
40	FC	Coefficient for forward-bias depletion capacitance formula	—	0.5		

JFET model parameters

	Name	Parameter	Units	Default	Example	Area
1	VTO	Threshold voltage	V	–2.0	–2.0	
2	BETA	Transconductance parameter	A/V**2	1.0E-4	1.0E-3	*
3	LAMBDA	Channel length modulation parameter	1/V	0	1.0E-4	
4	RD	Drain ohmic resistance	Ω	0	100	*

JFET model parameters (cont.)

	Name	Parameter	Units	Default	Example	Area
5	RS	Source ohmic resistance	Ω	0	100	*
6	CGS	Zero-bias G-S junction capacitance	F	0	5 pF	*
7	CGD	Zero-bias G-D junction capacitance	F	0	1 pF	*
8	PB	Gate junction potential	V	1	0.6	
9	IS	Gate junction saturation current	A	1.0E-14	1.0E-14	*
10	KF	Flicker noise coefficient	—	0		
11	AF	Flicker noise exponent	—	1		
12	FC	Coefficient for forward-bias depletion capacitance formula	—	0.5		

MOSFET model parameters

A total of 42 parameters is given for MOSFET simulation. Those given below are a selection for the MOS1 (level 1) model as used in the simulations of Chapters 6 and 7. The last 8 parameters are taken out of the ordering in the *User's Guide*. Device dimensions and areas are specified within the program and details of all 42 parameters are given in *The Simulation of MOS Integrated Circuits using SPICE 2* by A. Vladimirescu and S. Liu, Memorandum No. UCB/ERL M80/7 February 1980 from the Electronics Research Laboratory, College of Engineering, University of California, Berkeley 94720, USA.

	Name	Parameter	Units	Default	Example
1	LEVEL	Model index	—	1	
2	VTO	Zero-bias threshold voltage	V	0.0	1.0
3	KP	Transconductance parameter	A/V**2	2.0E-5	3.1E-5
4	GAMMA	Bulk threshold parameter	V**0.5	0.0	0.37
5	PHI	Surface potential	V	0.6	0.65
6	LAMBDA	Channel-length modulation (MOS1 and MOS2 only)	1/V	0.0	0.2
7	RD	Drain ohmic resistance	Ω	0.0	1.0
8	RS	Source ohmic resistance	Ω	0.0	1.0
9	CBD	Zero-bias B-S junction capacitance	F	0.0	2 pF
10	CBS	Zero-bias B-S junction capacitance	F	0.0	2 pF
11	IS	Bulk junction saturation current	A	1.0E-14	1.0E-15
12	PB	Bulk junction potential	V	0.8	0.87
13	CGSO	Gate-source overlap capacitance per meter channel width	F/m	0.0	4.0E-11
14	CGDO	Gate-drain overlap capacitance per meter channel width	F/m	0.0	4.0E-11
15	CGBO	Gate-bulk overlap capacitance per meter channel length	F/m	0.0	2.0E-10

MOSFET model parameters (cont.)

	Name	Parameter	Units	Default	Example
16	RSH	Drain and source diffusion sheet resistance	Ω/sq	0.0	10.0
17	CJ	Zero-bias bulk junction bottom capacitance per square-metre of junction area	F/m**2	0.0	2.0E-4
18	MJ	Bulk junction bottom grading coef.	—	0.5	0.5
19	TOX	Oxide thickness	m	1.0E-7	1.0E-7
20	NSUB	Substrate doping	1/cm**3	0.0	4.0E-15
21	TPG	Type of gate material: +1 opposite to substrate −1 same as substrate 0 Al gate	—	1.0	
22	XJ	Metallurgical junction depth	m	0.0	1 μm
23	UO	Surface mobility	cm**2/V s	600	700
24	KF	Flicker noise coefficient	—	0.0	1.0E-26
25	AF	Flicker noise exponent	—	1.0	1.2
26	FC	Coefficient for forward-bias depletion capacitance formula	—	0.5	

Circuit simulation

Circuits and SPICE simulation programs are given below for:

(i) nMOS and CMOS inverters with inverter load;
(ii) standard TTL NAND gate with load;
(iii) nMOS and CMOS NAND gates with inverter loads and track capacitance.

In each case the ground node is labelled 0, as required by SPICE, and other nodes are labelled in sequence. Diode nodes are in the order

anode cathode

followed by the device name. BJT nodes are in the order

collector base emitter

followed by the device name. MOST nodes are in the order

drain gate source substrate

followed by the device name and dimensions in m or m^2.

In each case the device parameters are defined on the MODEL line, where the + sign indicates a continuation, with dimensions given in the appropriate list. Resistors and capacitors are specified in the order

name node 1 node 2 value

Input voltages VDD and VIN are specified with respect to ground. For VIN the pulse input is specified as

initial voltage final voltage delay time rise time fall time
pulse width period

Transfer characteristics are specified by

.DC VIN initial voltage final voltage voltage increment
.PLOT DC voltage node(s)

Transient responses are specified by

.TRAN step time final time
.PLOT TRAN voltage node(s)

nMOS inverter with inverter load

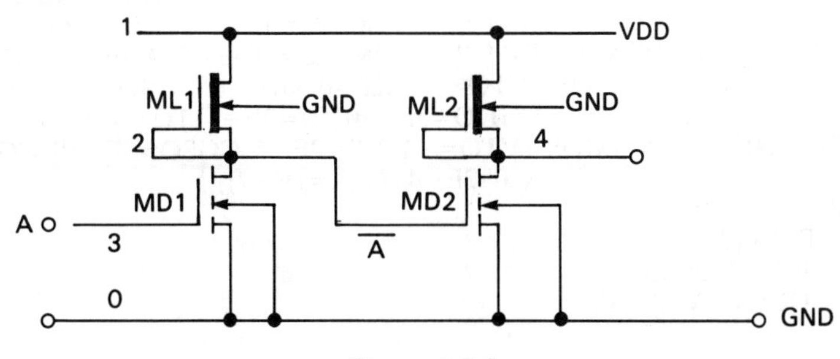

Figure A.3.1

```
VDD 1 0 5V
VIN  3 0 PULSE (0  5  1NS  1NS  1NS  10NS  100NS)
ML1  1 2 2  0  DEPLTN   L=8UM W=4UM AS=60E-12 AD=60E-12
ML2  1 4 4  0  DEPLTN   L=8UM W=4UM AS=60E-12 AD=60E-12
MD1  2 3 0  0  ENHANC  L=4UM W=8UM AS=60E-12 AD=60E-12
MD2  4 2 0  0  ENHANC  L=4UM W=8UM AS=60E-12 AD=60E-12
.MODEL DEPLTN NMOS (VTO=-3 KP=2E-5 CGSO=3E-10 CGDO=3E-10
+                   CJ=2E-4 TOX=1E-7)
.MODEL ENHANC NMOS (VTO=+1 KP=2E-5 CGSO=3E-10 CGDO=3E-10
+                   CJ=2E-4 TOX=1E-7)
.DC  VIN  0.0  5.0 0.1
.PLOT  DC V(2)
.TRAN  0.2NS  20NS
.PLOT  TRAN  V(3)  V(2)
.END
```

The transfer characteristic is shown in Fig. 6.18 and the transient response in Fig. 6.20.

CMOS inverter with inverter load

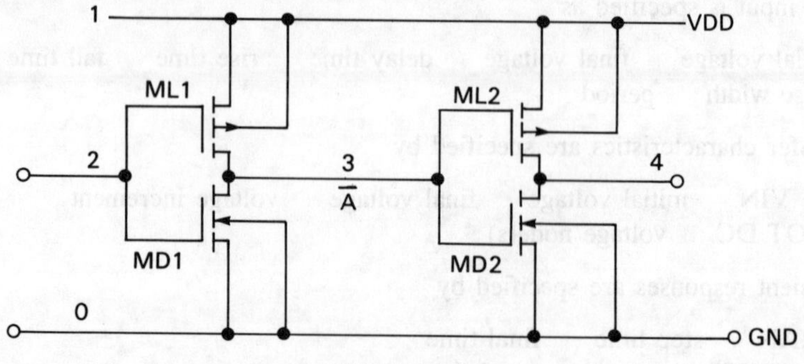

Figure A.3.2

```
VDD 1 0 5V
VIN  2 0 PULSE (0 5 1NS 1NS 1NS 10NS 100NS)
ML1 3 2 1 1 ENHANCP  L=4UM  W=24UM  AS=100E-12  AD=100E-12
MD1 3 2 0 0 ENHANCN  L=4UM  W=24UM  AS=100E-12  AD=100E-12
ML2 4 3 1 1 ENHANCP  L=4UM  W=24UM  AS=100E-12  AD=100E-12
MD2 4 3 0 0 ENHANCN  L=4UM  W=24UM  AS=100E-12  AD=100E-12
.MODEL ENHANCP PMOS (VTO=-1 KP=6.667E-6 CGSO=3E-10
+                    CGDO=3E-10 CJ=2E-4 TOX=1E-7)
.MODEL ENHANCN NMOS (VTO=+1 KP=2E-5 CGSO=3E-10 CGDO=3E-
+                    CJ=2E-4 TOX=1E-7)
.DC   VIN   0.0   5.0   0.1
.PLOT   DC V(3)
.TRAN   0.2NS   20NS
.PLOT   TRAN   V(2)   V(3)
.END
```

The transfer characteristic is shown in Fig. 6.24 and the transient response in Fig. 6.26.

Standard TTL NAND gate with load

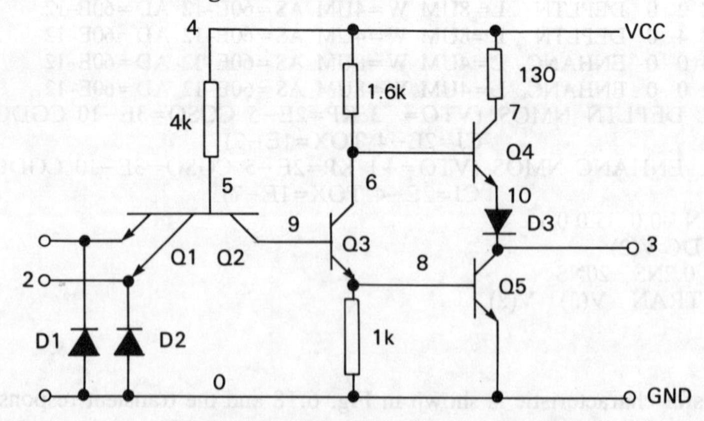

Figure A.3.3 Standard TTL NAND gate.

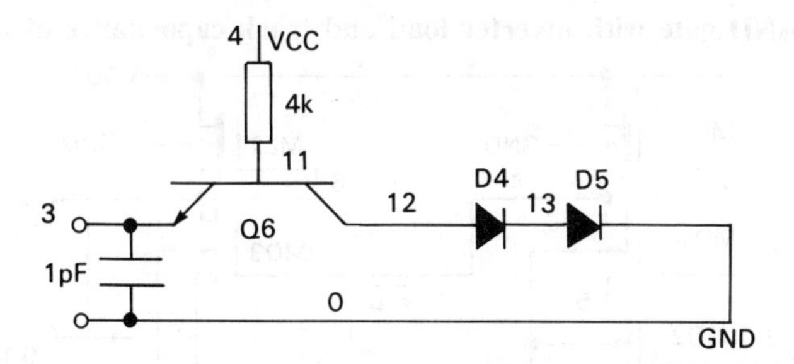

Figure A.3.4 TTL load.

```
VCC  4   0   5V
VIN  1   0   PULSE (0  3.5  5NS  1NS   1NS  20NS  100NS)
RB   4   5   4K
R1   4   6   1.6K
R2   8   0   1K
R3   4   7   130
R4   4   11  4K
C1   3   0   1PF
Q1   9   5   1  QMOD
Q2   9   5   4  QMOD
Q3   6   9   8  QMOD
Q4   7   6   10  QMOD
Q5   3   8   0  QMOD
Q6   12  11  3  QMOD
D1   0   1   DMOD
D2   0   4   DMOD
D3   10  3   DMOD
D4   12  13  DMOD
D5   13  0   DMOD
.MODEL  QMOD  NPN(BF=50  RB=70  RC=40  CCS=2PF  TF=0.1NS
+TR=10NS  CJE=0.9PF  CJC=1.5PF  PC=0.85  VA=50)
.MODEL  DMOD  D(TT=0.3NS  CJO=3PF)
.DC  VIN  0.0  3.5  0.1
.PLOT  DC  V(3)
.TRAN  0.5NS  60NS
.PLOT  TRAN  V(1)  V(3)
.END
```

In the simulation, node 2 is connected to node 4. The transfer characteristic is shown in Fig. 7.28 and the transient response in Fig. 7.29.

nMOS NAND gate with inverter load and track capacitance of 0.1 pF

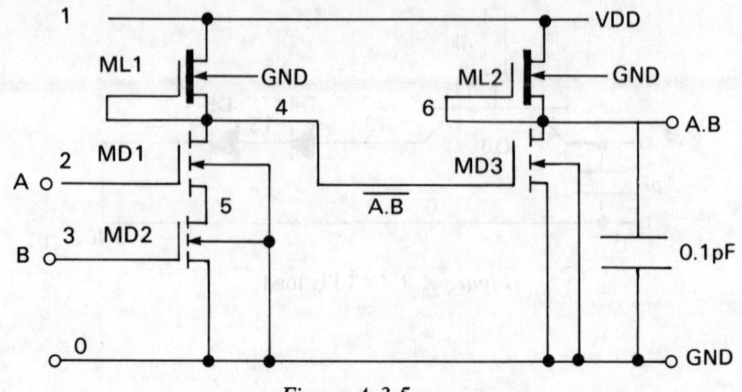

Figure A.3.5

```
VDD 1 0 5V
VIN  3 0 PULSE (0  5  1NS  1NS  1NS  10NS  100NS)
ML1  1 4 4  0  DEPLTN   L=8UM W=4UM AS=60E-12 AD=60E-12
ML2  1 6 6  0  DEPLTN   L=8UM W=4UM AS=60E-12 AD=60E-12
MD1 4 1 5  0  ENHANC  L=4UM W=8UM AS=60E-12 AD=60E-12
MD2 5 3 0  0  ENHANC  L=4UM W=8UM AS=60E-12 AD=60E-12
MD3 6 4 0  0  ENHANC  L=4UM W=8UM AS=60E-12 AD=60E-12
.MODEL DEPLTN NMOS (VTO=-3 KP=2E-5 CGSO=3E-10 CGDO=3E-1(
+                              CJ=2E-4 TOX=1E-7)
.MODEL ENHANC NMOS (VTO=+1 KP=2E-5 CGSO=3E-10 CGDO=3E-1
+                              CJ=2E-4 TOX=1E-7)
.DC   VIN  0.0   5.0 0.1
.PLOT  DC V(4)
.TRAN  0.2NS  20NS
.PLOT  TRAN  V(3)  V(4)
.PLOT  TRAN  V(3)  V(6)
.END
```

In the simulation, node 2 is connected to node 1.

CMOS NAND gate with inverter load and track capacitance of 0.1 pF

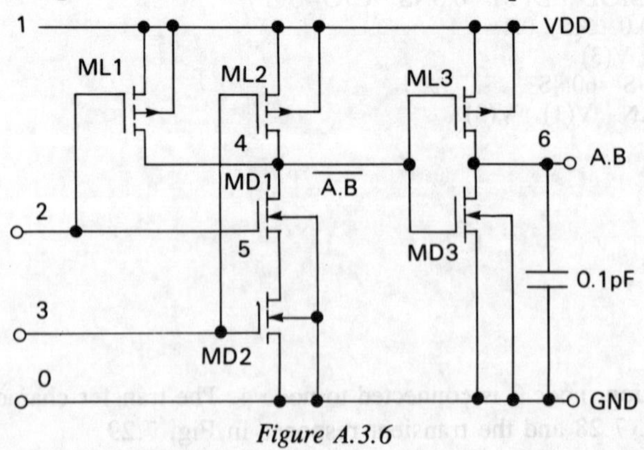

Figure A.3.6

```
VDD 1 0 5V
VIN  3 0 PULSE (0 5 1NS 1NS 1NS 10NS 100NS)
ML1 4 1 1 1 ENHANCP  L=4UM W=24UM AS=100E-12 AD=100E-12
MD1 4 1 5 0 ENHANCN  L=4UM W=24UM AS=100E-12 AD=100E-12
ML2 4 3 1 1 ENHANCP  L=4UM W=24UM AS=100E-12 AD=100E-12
MD2 5 3 0 0 ENHANCN  L=4UM W=24UM AS=100E-12 AD=100E-12
ML3 6 4 1 1 ENHANCP  L=4UM W=24UM AS=100E-12 AD=100E-12
MD3 6 4 0 0 ENHANCN  L=4UM W=24UM AS=100E-12 AD=100E-12
.MODEL ENHANCP PMOS (VTO=-1 KP=6.667E-6 CGSO=3E-10
+                    CGDO=3E-10 CJ=2E-4 TOX=1E-7)
.MODEL ENHANCN NMOS (VTO=+1 KP=2E-5 CGSO=3E-10 CGDO=3E-10
+                    CJ=2E-4 TOX=1E-7)
.DC  VIN  0.0  5.0  0.1
.PLOT  DC  V(4)
.TRAN  0.2NS  20NS
.PLOT  TRAN  V(3)  V(4)
.PLOT  TRAN  V(3)  V(6)
.END
```

In the simulation, node 2 is connected to node 1.

Physical constants

Charge of electron, $e = 1.602 \times 10^{-19}$ C
Mass of electron, $m = 9.109 \times 10^{-31}$ kg
Specific charge of electron, $e/m = 1.759 \times 10^{11}$ C/kg
Speed of light, $c = 2.998 \times 10^{8}$ m/s
Permittivity of vacuum, $\epsilon_0 = 8.851 \times 10^{-12}$ F/m
Permeability of vacuum, $\mu_0 = 4\pi \times 10^{-7}$ H/m
Planck constant, $h = 6.626 \times 10^{-34}$ J s
Boltzmann constant, $k = 1.380 \times 10^{-23}$ J/K
Avogadro constant, $N_A = 6.023 \times 10^{23}$/mol

Symbols

Symbol	Meaning
a	grade constant
A	amplification
	area
B	admittance
	bandwidth
	magnetic flux density
C	capacitance
	capacitance per unit area
	Curie constant
d	distance
	quantum number for $l = 2$
D	charge per unit area
	diffusion coefficient
E	electric field
f	frequency
	quantum number for $l = 3$
g, G	conductance
	generation rate
h	hybrid parameter
H	magnetic field
i	$\sqrt{-1}$
J	current density
k	momentum vector
K	kinetic energy
	thermal conductance
l	angular momentum quantum number
	free path
	length
$\bar{l}$	mean free path
L	diffusion length
	length
y	Young's modulus
w, W	width
W	energy
m	magnetic quantum number
m_0	rest mass of electron
m^*	effective mass of electron
M	current multiplication factor
	magnetization
n	number density of free particles
	principal quantum number
N	density of energy levels
	number density of atoms
p	momentum
	number density of holes
	piezoelectric constant
	probability
	quantum number for $l = 1$
P	power
	pressure
	polarization
q, Q	charge
r	distance
	radius
R	resistance
s	distance
	quantum number for $l = 0$
	spin quantum number
S	surface area
	strain
t	time
T	temperature
	stress
u	speed
	velocity
$\bar{u}$	mean speed
ζ	quantum yield
κ	thermal conductivity
λ	wavelength

Z	atomic number
	impedance
	thermoelectric figure of merit
α	common-base current gain
	ionizing power
	polarizability
α_P	Peltier coefficient
α_S	Seebeck coefficient
β	common-emitter current gain
γ	emitter efficiency
	crystal constant
	space-charge reduction factor
δ	base transport factor
	loss angle
	secondary emission coefficient
	small change
ϵ	permittivity
	transfer inefficiency
μ	amplification factor
	feedback factor
	mobility
	permeability
	dipole moment
μ_B	Bohr magneton
ρ	charge density
	resistivity
σ	conductivity
τ	time constant
ϕ	work function
Φ	magnetic flux
χ	depth of conduction band
	magnetic susceptibility
ψ	diffusion potential
	probability function
ω	angular frequency

Subscripts

a, A	anode
a	acceptor
b, B	base
B	breakdown
BO	breakover
c	conduction band
c, C	collector
d	delay
	donor
	drift
d, D	drain
D	domain
e	effective
	electron
e, E	emitter
f	fall
f, F	forward
F	Fermi
g	gap
g, G	gate
h	high
	hole
	horizontal
	hysteresis
H	holding
i	idler
	input
	ion
j	junction
m	mutual
M	Maxwell
	maintaining
n	electron
	noise
	n-type
o	output
oc	open-circuit
p	hole
	p-type
	pump
r, R	retarding
	reverse
r	rise
s, S	saturated
	signal
	source
	substrate
	storage
sc	short-circuit
T	threshold
V	valley
v	valence band
	vertical
0	initial
	leakage (current)
	mid-band

Voltage and current symbols

I_A	direct anode current	V_H	direct helix voltage
I_B	direct base current	V_K	direct cathode voltage
I_C	direct collector current	V_M	maintaining voltage
I_D	direct drain current	V_R	direct reflector voltage
I_E	direct emitter current		retarding voltage
I_G	direct gate current	V_{AA}	anode supply voltage
V_A	direct anode voltage	V_{BB}	base supply voltage
V_C	direct collector voltage (micro-wave valves)	V_{CC}	collector supply voltage
		V_{DD}	drain-supply voltage
		V_{SS}	source-supply voltage

The symbols for alternating quantities are illustrated below in terms of collector current. Similar types of symbol and subscript are used for voltages and for other electrodes.

i_c	incremental (instantaneous) value of varying component	I_{cm}	peak value of varying component
i_C	total (direct plus incremental) value	I_c	r.m.s. value of varying component

SI symbols and prefixes

		Derivation
A	ampere	
C	coulomb	A s
cd	candela	
F	farad	C/V
H	henry	V s/A
Hz	hertz	cycle per second
J	joule	kg m^2/s^2
K	kelvin	
kg	kilogram	
lm	lumen	cd sr
lx	lux	lm/m^2
m	metre	
N	newton	J/m
rad	radian (plane angle)	
s	second	
sr	steradian (solid angle)	
S	siemens	(mho, or reciprocal ohm)
T	tesla	Wb/m^2
V	volt	W/A
W	watt	J/s
Wb	weber	V s
Ω	ohm	V/A
eV	electronvolt	1.602×10^{-19} J
°C	degree Celsius	K + 273

Prefixes

T	tera-	10^{12}	m	milli-	10^{-3}
G	giga-	10^{9}	μ	micro-	10^{-6}
M	mega-	10^{6}	n	nano-	10^{-9}
k	kilo-	10^{3}	p	pico-	10^{-12}
d	deci-	10^{-1}	f	femto-	10^{-15}
c	centi-	10^{-2}	a	atto-	10^{-18}

Properties of common semiconductors

Groups	Semiconductor	Energy gap eV 300 K	Energy gap eV 0 K	Mobility $m^2/V\ s$ electrons	Mobility holes	Relative permittivity
IV	Carbon	5.47	5.51	0.18	0.16	5.5
	Germanium	0.66	0.75	0.39	0.19	16
	Silicon	1.12	1.16	0.15	0.06	12
III–V	Gallium arsenide	1.43	1.52	0.85	0.04	11
	Gallium phosphide	2.24	2.40	0.01	0.008	10
	Indium arsenide	0.33	0.46	3.30	0.046	15
	Indium phosphide	1.29	1.34	0.46	0.015	14
II–VI	Cadmium sulphide	2.42	2.56	0.03	0.005	10
	Cadmium selenide	1.70	1.85	0.08		10
IV–VI	Lead sulphide	0.41	0.34	0.06	0.07	17

Index